Modeling, Simulation and Control in Energy Systems

Modeling, Simulation and Control in Energy Systems

Guest Editors

Dimitris Ipsakis
Andreas Yiotis

MDPI

Basel • Beijing • Wuhan • Barcelona • Belgrade • Novi Sad • Cluj • Manchester

Guest Editors
Dimitris Ipsakis
School of Production Engineering and Management
Technical University of Crete
Chania
Greece

Andreas Yiotis
School of Rural, Surveying and Geoinformatics Engineering
National Technical University of Athens
Athens
Greece

Editorial Office
MDPI AG
Grosspeteranlage 5
4052 Basel, Switzerland

This is a reprint of the Special Issue, published open access by the journal *Processes* (ISSN 2227-9717), freely accessible at: https://www.mdpi.com/journal/processes/special_issues/HI863SV90Y.

For citation purposes, cite each article independently as indicated on the article page online and as indicated below:

Lastname, A.A.; Lastname, B.B. Article Title. *Journal Name* **Year**, *Volume Number*, Page Range.

ISBN 978-3-7258-7617-4 (Hbk)
ISBN 978-3-7258-7618-1 (PDF)
https://doi.org/10.3390/books978-3-7258-7618-1

Contents

About the Editors

Dimitris Ipsakis

Dimitris Ipsakis is an Associate Professor at the School of Production Engineering and Management at the Technical University of Crete. He was born in 1982 in Thessaloniki, and originates from Istanbul and Ano Rodonia, Pella by descent. He received his diploma in Chemical Engineering and his PhD from the Department of Chemical Engineering at the Aristotle University of Thessaloniki in 2005 and 2011, respectively. Additionally, he holds an MSc degree from the Department of Mathematics at the Aristotle University of Thessaloniki. He teaches courses related to control systems, energy economics, and energy system design and modeling. During his academic career, he has published several articles in prestigious journals and international conferences. His work focuses on modeling, control, and economic evaluation of energy and industrial process systems, including renewable energy systems, hydrogen (H_2) technologies, and CO_2 valorization. Furthermore, his publications have received more than 2,000 citations, with his h-index being 21 (March 2026). In the past, he has worked as a researcher at the Centre for Research & Technology Hellas (2006–2013, 2016–2019), at the Department of Chemical Engineering of Aristotle University of Thessaloniki (2014–2016), and at Texas A&M University at Qatar, Department of Chemical Engineering (September–November 2014).

Andreas Yiotis

Andreas Yiotis holds a diploma (1997) and a PhD in Chemical Engineering (2003) both from the National Technical University of Athens. From 2010–12, he was a Marie Curie fellow at the Centre National de la Recherche Scientifique in Orsay, France, and from 2013–19, he was a post-doctoral researcher at the Centre for Scientific Research 'Demokritos' in Athens, Greece. He is currently serving as an Assistant Professor at the School of Rural, Surveying and Geoinformatics Engineering of the National Technical University of Athens. Prior to this, he served as an Assistant Professor (2019–2023) and Associate Professor (2024–2025) at the School of Mineral Resources Engineering of the Technical University, teaching courses relevant to fluid mechanics, petroleum engineering and energy transition technologies. His research focuses on transport phenomena in porous media, including immiscible flow in geologic formations, flow and diffusion through membranes and fuel cell components, and the design of thermally integrated fuel cells and metal hydride systems. He has authored two books in fluid mechanics (2024) and hydrocarbon reservoir engineering (2023) and has over 40 publications in peer-reviewed journals, attracting more than 1500 heterocitations (h-index=20).

processes

Editorial

Special Issue on "Modeling, Simulation and Control of Energy Systems"

Dimitris Ipsakis [1,*] and Andreas Yiotis [2]

[1] School of Production Engineering and Management, Technical University of Crete, 73100 Chania, Greece
[2] School of Mineral Resources Engineering, Technical University of Crete, 73100 Chania, Greece; agiotis@tuc.gr
* Correspondence: dipsakis@tuc.gr; Tel.: +30-28210-37362

1. Introduction

Clean energy systems are at the forefront of current discussions on the protection of our environment against global warming. Decarbonization strategies along with novel energy systems have diverted research groups to the fields of renewables, alternative fuels, energy efficiency improvement, autonomous systems design, waste management and energy policy regulations. To this end, frameworks for the modeling, simulation and control of energy systems have become essential tools for identifying the complex dynamics of these strategies and predicting system performance under stochastic disturbances and system variations (either short-term or long-term). Such energy systems may range from TRL (technology readiness level) 1 to 9, and crucial efforts have been devoted to their efficient scale-up and commercialization. Additional challenges are the advanced modeling and process control techniques that have advanced from pilot- to industrial-scale applications. This Special Issue entitled "Modeling, Simulation and Control in Energy Systems" collected high-quality studies that address challenges facing the areas of energy production, management, utilization and storage. Three core topics were identified: (i) power system optimization, grid management and renewable integration; (ii) intelligent control, prediction and data-driven modeling in energy systems; and (iii) mechanical, thermal and fluid system modeling and engineering design. Overall, nineteen (19) studies were published and are available online at: https://www.mdpi.com/journal/processes/special_issues/HI863SV90Y. The list of contributions refers to articles numbered 1–19.

2. An Overview of the Published Papers

2.1. Power System Optimization, Grid Management and Renewable Integration

Articles related to this topic focused on the improvement in performance, economics, reliability, and sustainability of modern power systems through optimization, forecasting, coordination and intelligent control. The methods that were analyzed involved (among others) intelligent optimization algorithms (PSO, ALO, GA), economic modeling, risk assessment, multi-energy coupling, cloud–edge computing and renewable integration.

Serrano et al. (List of Contributions 1) proposed a hybrid optimization method using the Greedy Randomized Adaptive Search Procedure (GRASP) and Tabu Search to improve capacity management in smart grids through the optimal siting of distributed generators and coordinated load shedding. Applied on a real 1806-bus network, this approach enhances voltage levels, reduces losses and strengthens reliability by enabling islanded operation and more efficient load restoration. The results showed that combining DG allocation with demand-side management outperformed strategies using only one technique, providing a promising tool for modern grid planning and performance improvement. Wang et al. (List of Contributions 2) discussed a pricing strategy tailored for electricity prosumers who both generate and consume energy. Users were clustered by demand preference, price sensitivity and risk tolerance, enabling customized packages. Two pricing schemes were proposed: fluctuating time-of-use pricing and discount-based purchase–sale packages. A hybrid Self-Adaptive Weight and Reverse Learning Particle Swarm Optimization (SAW and RL-PSO) algorithm was used to determine optimal pricing while accounting for market uncertainty via CvaR (Conditional Value at Risk) risk modeling. The results showed reduced prosumer costs, improved load balancing and increased retailer revenue, improving renewable energy consumption and market stability. In a study by Li et al. (List of Contributions 3), a robust economic dispatch strategy that integrates virtual power plants (VPPs) to manage renewable-energy uncertainty in modern grids was developed. Renewable output variability was modeled using interval-based prediction errors, and a mixed-integer nonlinear VPP optimization problem was solved using an improved Ant Lion Optimizer. Testing on an IEEE-30 bus system showed considerably reduced computation time. Chen et al. (List of Contributions 4) developed a linear two-port model of pipeline heat transfer (derived from an implicit upwind difference scheme) to compress partial differential equation (PDE) dynamics into offline coefficient matrices. Embedded in an electro-thermal coupling system (EDETCS), the approach preserves accuracy while drastically speeding computation. Case studies showed lower operating costs and reduced wind curtailment, improving system flexibility and economic performance. In List of Contributions 5, an adaptive cloud–edge scheduling method that mines edge-side computing resources to relieve cloud master station load for power distribution security tasks is proposed. The authors modeled multi-stage relay security protection, formulated a Karush–Kuhn–Tucker (KKT)-solved linear program to minimize the maximum expected cloud pressure and validated the approach with software simulations. The results showed up to ~35% reductions in peak cloud master pressure and improved resource utilization across varied network scenarios. Zhang et al. (List of Contributions 6) proposed a dual-sequence Monte Carlo method that integrates electrical–thermal cable modeling, emergency-rating behavior and insulation aging to assess underground cable reliability. By introducing design-risk and aging-risk coefficients, the framework quantifies how long- and short-term emergency loading affects failure probability and network performance. Applied to an enhanced IEEE 14-bus network system, the method showed that emergency ratings cut expected unserved energy by ~59% but also accelerated aging on critical cables, providing a balanced tool for risk-aware planning and asset management. Zhang et al. (List of Contributions 7) developed a multi-objective dispatch strategy for park-level integrated energy systems by tightly coupling P2G, CCS and CHP units to enhance flexibility and reduce carbon emissions under uncertain wind and solar output. A reward–penalty ladder carbon-trading mechanism was introduced, and renewable scenarios were generated via kernel density estimation and Frank Copula modeling. Using an NSDBO solver, the simulations showed significant gains in renewable utilization, carbon emissions and total costs decreases. Finally, in (List of Contributions 8), the authors developed an optimization model for selecting grid loss-reduction strategies under China's transmission and distribution pricing reforms. The model minimizes life-cycle costs by combining investment expenses, direct power-loss costs and carbon-emission penalties, while enforcing voltage, branch-flow, and budget constraints. Using an adaptive genetic algorithm, the method efficiently found optimal strategy combinations among many possibilities.

https://doi.org/10.3390/pr14050827

2.2. Intelligent Control, Prediction and Data-Driven Modeling in Energy Systems

In this field, relevant articles were identified using advanced control techniques, machine learning, hybrid ML–numerical models, or deep learning to predict system behavior, support energy management, or improve decision-making. All articles shared a focus on modeling accuracy, feature selection, AI-based forecasting and real-time control decision-making.

Kaiser et al. (List of Contributions 9) conducted a review that examined the limitations of conventional CFD modeling for PEM fuel cells and highlighted how machine learning can enhance parameter estimation, mass-transfer prediction and electrochemical modeling. The authors analyzed the current challenges in water and thermal management, durability and material characterization, showing how hybrid machine learning (ML) CFD approaches can improve reliability and reduce computational costs. In another interesting study (List of Contributions 10), an indirect primary-side control strategy for wireless power transfer electrical vehicle (WPT-EV) chargers was constructed. This system holds battery charging current constant despite wireless-link latency and coil misalignment. A PI controller was paired with an adaptive hill-climbing algorithm that updated an efficiency-based mapping only when new battery measurements arrive. The simulation and laboratory tests showed tight current tracking, smooth phase-shift actuation and robust performance efficiency. Moreover, in (List of Contributions 11), the authors developed a robust hybrid method combining improved Complementary Ensemble Empirical Mode Decomposition for denoising/decomposition, Hilbert Transform for instantaneous features and a PSO algorithm for classification. Validated on real smart-meter and fault-recorder data, the approach achieved >95% identification accuracy under heavy noise. The method is computationally efficient and well-suited for real-time grid monitoring and multi-disturbance scenarios. Tong et al. (List of Contributions 12) proposed a hybrid framework for accurate transformer top-oil temperature prediction by integrating Variational Mode Decomposition (VMD) decomposition, Kernel principal component analysis, and a time-aware Shapley Additive Explanations–Multilayer Perceptron (SHAP–MLP) method for dynamic feature selection. A Time-Series Forecasting System (SOFTS) deep-learning model, optimized with hierarchical Bayesian search, captures multiscale dependencies and improves robustness under fluctuating operating conditions. Finally, Huang et al. (List of Contributions 13) developed a convolutional neural network–bidirectional gated recurrent unit hybrid to predict NOx concentration and boiler thermal efficiency from 3000 operational samples of a 300 MW boiler. It couples IFDA (an improved flow-direction algorithm) for single-objective tuning and NSGA-II for multi-objective optimization of airflows and O_2 to reduce NOx while raising efficiency. The results show the CNN-IFDA-BiGRU model substantially lowers prediction error, and the non-dominated sorting genetic algorithm (NSGA-II) yields ~5.01% NOx reduction and a 0.32% efficiency gain on average.

2.3. Mechanical, Thermal and Fluid System Modeling and Engineering Design

Ending this editorial review, the studies included this last topic involved the simulation, analysis and optimization of mechanical, thermal, combustion and multiphase fluid systems. Other areas studied included hardware implementation, control dynamics, thermal/fluid behavior and component-level performance analysis.

Provatas and Ipsakis (List of Contributions 14) present an educational yet practical framework for designing and simulating feedback controllers for an active vehicle suspension system. Using a quarter-car model, the authors compared P, PI, PID, genetic-algorithm-optimized PID and internal model controllers (IMCs) across disturbance and tracking scenarios. The results show that the optimal PID and IMC deliver superior comfort and stability, with the optimized PID offering best set-point tracking while IMCs demand

https://doi.org/10.3390/pr14050827

less actuator effort. The work stands out for its step-by-step methodology, making advanced control concepts accessible for engineering students. In (List of Contributions 15), the authors analyzed how different blade tip clearances affect internal flow behavior, gas distribution and efficiency in gas–liquid multiphase pumps. Using Eulerian two-fluid CFD modeling with four clearance values (0–0.9 mm), the authors found that larger clearances increased tip leakage flow almost linearly, causing increased energy loss and a noticeable drop in head coefficient. Whereas small clearances led to uneven flow and localized gas accumulation, wider gaps improved uniformity but also reduced efficiency due to leakage. The findings guide optimal clearance selection in high-performance multiphase pump design. Korukcu (List of Contributions 16) introduced the Ozan Combustion Calculator (OCC), a MATLAB-based GUI designed to compute combustion parameters for 16 fuels among common alkanes and alcohols. The tool calculates adiabatic flame temperature, energy release and exergy destruction for both dry and moist air while visually displaying combustion equations—an improvement over earlier software. Validation against methods in the literature showed high accuracy. The study also reveals and offers valuable insights for combustion modeling and education. In (List of Contributions 17), a ferrofluid-based sealing system was used to increase the reliability of pressure-vessel safety valves. A composite structure was developed combining mechanical end-face sealing with a magnetic-fluid seal driven by an electromagnet. Using numerical modeling and through validation with experiments, key geometric parameters—seal gap, groove width, tooth height, tooth width and number of sealing stages—were optimized. The results showed that the sealing pressure first increased then declined with tooth height, groove width and stage number. In another interesting and applied research study (List of Contributions 18), PID, H∞, model predictive control (MPC) and multiple input–multiple output Model Reference Adaptive Control (MRAC) control strategies were compared in offshore-produced water treatment using pilot-scale plant data. Focusing on the multivariable control of separator water level and hydrocyclone pressure drop ratio, the study shows that MRAC provides superior tracking accuracy and adaptability, whereas MPC achieves the highest de-oiling efficiency with reduced control effort. The results highlight clear trade-offs among robustness, efficiency, complexity and industrial feasibility, offering practical guidance for selecting control architectures in offshore water treatment systems. Finally, the authors of (List of Contributions 19) developed a Space Vector Pulse-Width Modulation (SVPWM)-based variable-frequency inverter (31–300 Hz) implemented on a low-cost 8-bit microcontroller. By varying the number of switching periods per electrical cycle (930–1838) and linearizing computations, the design achieves precise timing with minimal computational load. Experimental and MATLAB/Simulink evaluations show the performance comparable to that of other implementations, demonstrating industrial-level capability using inexpensive hardware.

3. Conclusions

In conclusion, the papers collected in this Special Issue highlight the growing role of advanced modeling, simulation and intelligent control techniques in addressing the technical, economic and environmental challenges facing modern energy and engineering systems. From power system optimization and renewable integration to data-driven prediction, adaptive control and detailed multi-physics modeling, these contributions demonstrate how interdisciplinary approaches can enhance system efficiency, reliability and sustainability. Together, they provide valuable insights for researchers and practitioners and point toward future directions where digitalization, artificial intelligence and integrated energy solutions will play an increasingly central role.

Intelligent Control, Prediction and Data-Driven Modeling in Energy Systems

https://doi.org/10.3390/pr14050827

Other notable articles on this topic were not included in this Special Issue [1–10].

Power System Optimization, Grid Management and Renewable Integration

Other notable articles on this topic were not included in this Special Issue [11–17].

Mechanical, Thermal and Fluid System Modeling and Engineering Design

Other notable articles on this topic and not included in this Special Issue include [18–22].

Funding: This research received no external funding.

Acknowledgments: I would like to thank all the contributors, the Editor-in-Chief, and the editorial staff of *Processes* for their enthusiastic support of this Special Issue.

Conflicts of Interest: The authors declare no conflict of interest.

List of Contributions:

1. Serrano, H.d.O.M.; Reiz, C.; Leite, J.B. Capacity Management in Smart Grids Using Greedy Randomized Adaptive Search Procedure and Tabu Search. *Processes* **2023**, *11*, 2464. https://doi.org/10.3390/pr11082464.
2. Wang, X.; Liu, C.; Wu, B.; Wang, W.; Sun, Y.; Peng, J.; Liu, X.; Zhang, K. Design Strategy of Electricity Purchase and Sale Combination Package Based on the Characteristics of Electricity Prosumers in Power System. *Processes* **2024**, *12*, 2836. https://doi.org/10.3390/pr12122836.
3. Li, X.; Zhang, C.; Yi, Q.; Xu, J. Robust Economic Management Strategy for Power Systems Considering the Participation of Virtual Power Plants. *Processes* **2025**, *13*, 25. https://doi.org/10.3390/pr13010025.
4. Chen, J.; Lin, Q.; Yang, Z.; Liu, Q.; Zou, H. A Fast Calculation Method for Economic Dispatch of Electro-Thermal Coupling System Considering the Dynamic Process of Heat Transfer. *Processes* **2025**, *13*, 175. https://doi.org/10.3390/pr13010175.
5. Li, L.; Lu, S.; Sun, H.; Wu, R. Adaptive Scheduling Method of Heterogeneous Resources on Edge Side of Power System Collaboration Based on Cloud–Edge Security Dynamic Collaboration. *Processes* **2025**, *13*, 366. https://doi.org/10.3390/pr13020366.
6. Zhang, J.; Wang, B.; Ma, H.; He, Y.; Wang, H.; Zhang, H. Reliability Evaluation Method for Underground Cables Based on Double Sequence Monte Carlo Simulation. *Processes* **2025**, *13*, 505. https://doi.org/10.3390/pr13020505.
7. Zhang, Z.; Li, X.; Zhang, L.; Zhao, H.; Wang, Z.; Li, W.; Wang, B. Optimized Dispatch of Integrated Energy Systems in Parks Considering P2G-CCS-CHP Synergy Under Renewable Energy Uncertainty. *Processes* **2025**, *13*, 680. https://doi.org/10.3390/pr13030680.
8. Li, W.; Xu, Q.; Wang, X.; Liu, Z.; Li, T.; Zhang, D. Optimal Strategy for Grid Loss Reduction Under Electricity Transmission and Distribution Reform Considering Low-Carbon Benefits. *Processes* **2025**, *13*, 1406. https://doi.org/10.3390/pr13051406.
9. Kaiser, R.; Ahn, C.Y.; Kim, Y.H.; Park, J.C. Towards Reliable Prediction of Performance for Polymer Electrolyte Membrane Fuel Cells via Machine Learning-Integrated Hybrid Numerical Simulations. *Processes* **2024**, *12*, 1140. https://doi.org/10.3390/pr12061140.
10. Lassioui, A.; El Ancary, M.; El Idrissi, Z.; El Fadil, H.; Rachid, K.; Rachid, A. Primary-Side Indirect Control of the Battery Charging Current in a Wireless Power Transfer Charger Using Adaptive Hill-Climbing Control Technique. *Processes* **2024**, *12*, 1264. https://doi.org/10.3390/pr12061264.
11. Liu, K.; Han, J.; Chen, S.; Ruan, L.; Liu, Y.; Wang, Y. Power Quality Disturbance Identification Method Based on Improved CEEMDAN-HT-ELM Model. *Processes* **2025**, *13*, 137. https://doi.org/10.3390/pr13010137.
12. Tong, Z.; Xu, Y.; Meng, X.; Zheng, Y.; Peng, T.; Zhang, C. An Advanced Power System Modeling Approach for Transformer Oil Temperature Prediction Integrating SOFTS and Enhanced Bayesian Optimization. *Processes* **2025**, *13*, 2888. https://doi.org/10.3390/pr13092888.
13. Huang, C.; Zheng, Y.; Zhao, H.; Zhu, J.; Fu, Y.; Tang, Z.; Zhang, C.; Peng, T. Intelligent Deep Learning Modeling and Multi-Objective Optimization of Boiler Combustion System in Power Plants. *Processes* **2025**, *13*, 2340. https://doi.org/10.3390/pr13082340.

https://doi.org/10.3390/pr14050827

14. Provatas, V.; Ipsakis, D. Design and Simulation of a Feedback Controller for an Active Suspension System: A Simplified Approach. *Processes* **2023**, *11*, 2715. https://doi.org/10.3390/pr11092715.
15. Deng, Y.; Li, Y.; Xu, J.; Kuang, C.; Zhang, Y. The Influence of Blade Tip Clearance on the Flow Field Characteristics of the Gas–Liquid Multiphase Pump. *Processes* **2023**, *11*, 3170. https://doi.org/10.3390/pr11113170.
16. Korukçu, M.Ö. A Graphical User Interface for Calculating Exergy Destruction for Combustion Reactions. *Processes* **2024**, *12*, 294. https://doi.org/10.3390/pr12020294.
17. Li, Z.; Wang, Z.; Shen, C.; Li, W.; Jiao, Y.; Cheng, C.; Min, J.; Li, Y. Simulation and Experimental Design of Magnetic Fluid Seal Safety Valve for Pressure Vessel. *Processes* **2024**, *12*, 2040. https://doi.org/10.3390/pr12092040.
18. Kashani, M.; Jespersen, S.; Yang, Z. Evaluation of Advanced Control Strategies for Offshore Produced Water Treatment Systems: Insights from Pilot Plant Data. *Processes* **2025**, *13*, 2738. https://doi.org/10.3390/pr13092738.
19. Cerda-Villafana, G.; Birchfield, A.; Moreno-Vazquez, F.J. A 31–300 Hz Frequency Variator Inverter Using Space Vector Pulse Width Modulation Implemented in an 8-Bit Microcontroller. *Processes* **2025**, *13*, 1912. https://doi.org/10.3390/pr13061912.

References

1. Ipsakis, D.; Damartzis, T.; Papadopoulou, S.; Voutetakis, S. Dynamic Modeling and Control of a Coupled Reforming/Combustor System for the Production of H_2 via Hydrocarbon-Based Fuels. *Processes* **2020**, *8*, 1243. [CrossRef]
2. Savvopoulos, S.V.; Voutetakis, S.S.; Kuhn, S.; Ipsakis, D. Theoretical Feedback Control Scheme for the Ultrasound-Assisted Continuous Antisolvent Crystallization of Aspirin in a Tubular Crystallizer. *Ind. Eng. Chem. Res.* **2021**, *60*, 6221–6234. [CrossRef]
3. Ai, Y.; Peng, M.; Zhang, K. Edge computing technologies for Internet of Things: A primer. *Digit. Commun. Netw.* **2018**, *4*, 77–86. [CrossRef]
4. Sapari, N.M.; Mokhlis, H.; Laghari, J.A.; Bakar, A.H.A.; Dahalan, M.R.M. Application of load shedding schemes for distribution network connected with distributed generation: A review. *Renew. Sustain. Energy Rev.* **2018**, *82*, 858–867. [CrossRef]
5. Etukudor, C.; Couraud, B.; Robu, V.; Früh, W.-G.; Flynn, D.; Okereke, C. Automated Negotiation for Peer-to-Peer Electricity Trading in Local Energy Markets. *Energies* **2020**, *13*, 920. [CrossRef]
6. Meintanis, I.; Halikias, G.; Giovenco, R.; Yiotis, A.; Chrysagis, K. Identification and Model Predictive Control Design of a Polymer Extrusion Process. In *Computer Aided Chemical Engineering*; Elsevier: Amsterdam, The Netherlands, 2017; pp. 1609–1614. [CrossRef]
7. Stamatakis, E.; Yiotis, A.; Giannissi, S.; Tolias, I.; Stubos, A. Modeling and simulation supporting the application of fuel cell & hydrogen technologies. *J. Comput. Sci.* **2018**, *27*, 10–20. [CrossRef]
8. Antal, M.; Toderean, L.; Cioara, T.; Anghel, I. Hybrid Deep Neural Network Model for Multi-Step Energy Prediction of Prosumers. *Appl. Sci.* **2022**, *12*, 5346. [CrossRef]
9. Bai, Y.; Ren, X.; Li, H. State Accurate Representation and Performance Prediction Algorithm Optimization for Industrial Equipment Based on Digital Twin. *Intell. Autom. Soft Comput.* **2023**, *37*, 2999–3018. [CrossRef]
10. Ma, S.; Ding, W.; Liu, Y.; Zhang, Y.; Ren, S.; Kong, X.; Leng, J. Industry 4.0 and cleaner production: A comprehensive review of sustainable and intelligent manufacturing for energy-intensive manufacturing industries. *J. Clean. Prod.* **2024**, *467*, 142879. [CrossRef]
11. Liu, J.; Hou, Z.; Liu, B.; Zhou, X. Mathematical and Machine Learning Innovations for Power Systems: Predicting Transformer Oil Temperature with Beluga Whale Optimization-Based Hybrid Neural Networks. *Mathematics* **2025**, *13*, 1785. [CrossRef]
12. Futter, G.A.; Gazdzicki, P.; Friedrich, K.A.; Latz, A.; Jahnke, T. Physical modeling of polymer-electrolyte membrane fuel cells: Understanding water management and impedance spectra. *J. Power Sources* **2018**, *391*, 148–161. [CrossRef]
13. Dickinson, E.J.F.; Smith, G. Modelling the Proton-Conductive Membrane in Practical Polymer Electrolyte Membrane Fuel Cell (PEMFC) Simulation: A Review. *Membranes* **2020**, *10*, 310. [CrossRef] [PubMed]
14. Ipsakis, D.; Voutetakis, S.; Seferlis, P.; Papadopoulou, S.; Stoukides, M. Modeling and analysis of an integrated power system based on methanol autothermal reforming. In *2009 17th Mediterranean Conference on Control and Automation*; IEEE: New York, NY, USA, 2009; pp. 1421–1426. [CrossRef]
15. Bermeo-Ayerbe, M.A.; Ocampo-Martinez, C.; Diaz-Rozo, J. Data-driven energy prediction modeling for both energy efficiency and maintenance in smart manufacturing systems. *Energy* **2022**, *238*, 121691. [CrossRef]
16. Bong, H.L. Hybrid Adaptive Peak Load Threshold Controller for Battery Energy Storage System: An Industrial Case Study. *Int. J. Electr. Electron. Eng. Telecommun.* **2025**, *14*, 188–198. [CrossRef]

https://doi.org/10.3390/pr14050827

17. Hsu, C.-C.; Jiang, B.-H.; Lin, C.-C. A Survey on Recent Applications of Artificial Intelligence and Optimization for Smart Grids in Smart Manufacturing. *Energies* **2023**, *16*, 7660. [CrossRef]
18. Li, Z.; Li, W.; Wang, Q.; Xiang, R.; Cheng, J.; Han, W.; Yan, Z. Effects of medium fluid cavitation on fluctuation characteristics of magnetic fluid seal interface in agricultural centrifugal pump. *Int. J. Agric. Biol. Eng.* **2021**, *14*, 85–92. [CrossRef]
19. Zhang, J.; Li, Y.; Vafai, K.; Zhang, Y. An investigation of the flow characteristics of multistage multiphase pumps. *Int. J. Numer. Methods Heat Fluid Flow* **2018**, *28*, 763–784. [CrossRef]
20. Ntousakis, E.; Loukakis, K.; Petrou, E.; Ipsakis, D.; Papaefthimiou, S. Optimizing an Urban Water Infrastructure Through a Smart Water Network Management System. *Electronics* **2025**, *14*, 2455. [CrossRef]
21. Yiotis, A.G.; Kainourgiakis, M.E.; Charalambopoulou, G.C.; Stubos, A.K. A generic physical model for a thermally integrated high-temperature PEM fuel cell and sodium alanate tank system. *Int. J. Hydrog. Energy* **2015**, *40*, 14551–14561. [CrossRef]
22. Anastasiou, A.; Zarikos, I.; Yiotis, A.; Talon, L.; Salin, D. Steady-State Dynamics of Ganglia Populations During Immiscible Two-Phase Flows in Porous Micromodels: Effects of the Capillary Number and Flow Ratio on Effective Rheology and Size Distributions. *Transp. Porous Media* **2024**, *151*, 469–493. [CrossRef]

https://doi.org/10.3390/pr14050827

processes

MDPI

Article

Capacity Management in Smart Grids Using Greedy Randomized Adaptive Search Procedure and Tabu Search

Hugo de Oliveira Motta Serrano, Cleberton Reiz * and Jonatas Boas Leite

Department of Electrical Engineering, Faculty of Electrical Engineering, Sao Paulo State University, Campus of Ilha Solteira, São Paulo 15385-000, Brazil; hugo.motta@unesp.br (H.d.O.M.S.); jb.leite@unesp.br (J.B.L.)
* Correspondence: cleberton.reiz@unesp.br

Abstract: Over time, distribution systems have progressed from small-scale systems to complex networks, requiring modernization to adapt to these increasing levels of active loads and devices. It is essential to manage the capacity of distribution networks to support all these new technologies. This work, therefore, presents a method for evaluating the impact of optimal allocation and sizing of DGs and load shedding for response demand programs on distribution networks to improve the reliability and financial performance of electric power systems. The proposed optimization tool uses the Greedy Randomized Adaptive Search Procedure and Tabu Search algorithms. The combined optimization of DG allocation simultaneously with load shedding, reliability indices, load transference, and the possibility of islanded operation significantly improves the quality of the planning proposals obtained by the developed method. The results demonstrate the efficiency and robustness of the proposed method, improving the voltage profile by up to 2.02%, relieving the network capacity, and increasing the load restoration capability and reliability. Statistical analysis is also carried out to highlight the performance of the proposed methodology.

Keywords: distribution network; capacity management; GRASP; Tabu Search; distributed generator; load shedding

1. Introduction

The distribution networks must supply the load demand that varies daily and seasonally. Since the demand is largely unpredictable, and making programmable interruptions is not a good option because of its very high cost, the generation system capacity must be capable of supplying the maximum demand or peak demand [1]. The management of these resources becomes increasingly important due to distributed energy resources (DERs) integration into the grid. The active and reactive power injected by DERs into the grid not only affects the local voltage but can also influence charging conditions and power flow on the network [1]. Distribution networks can experience a significant increase in peak loads, which, in the absence of smart functionality to reduce them, may result in a substantial investment increase in network infrastructure [2]. An adequate management mechanism is thus necessary to optimize the use of the network's capacity, ensuring sustainable and more efficient distribution systems [2].

The installation of a distributed energy resource management system (DERMS) can aid network operations performed using the advanced distribution management system (ADMS) [3]. This strategy can significantly reduce the uncertainties caused by dispersed DERs that may not follow a unified or coordinated pattern of operation. The integration of DERMS and ADMS provides, additionally, the operability of programs for demand-side management (DSM).

The concept of demand-side management emerged in the 1980s as a response to the unpredictable growth in demand for electricity in that period. The development of a broad market based on real-time demand and supply data was a result of this situation. Future consumer demand used to be handled traditionally. However, with the introduction of this concept, fluctuations and variations in energy demand began to be better accommodated to deal with future uncertainties [4].

The implementation of demand-side management increases the complexity of power systems since the adequate performance of demand-side management requires monitoring the loads and generators of the system [5]. With this, additional expenses are generated with the incentive to participate in demand-side management programs. The benefits of demand-side management far outweigh its disadvantages [6]. The DSM has become essential in implementing smart grids, especially in the residential sector, as it significantly contributes to reducing the peak of electricity demand.

Some studies have been carried out on this subject in recent years, such as [7], which applies the Optimal Power Flow (OPF) technique in distribution networks to maximize system benefits by considering the capacity of loads, by assessing the optimal allocation of demand-side resources, and by providing ancillary services like an operational reserve. The authors of [8] developed a procedure to determine the best buses for demand response based on distribution factors, power transfer, available transfer capacity, and optimal power flow. The results help reduce line congestion, increase benefits for customers and the independent system operator, and avoid line interruptions and blackouts, thus increasing system reliability.

The DSM has been strategically used by aggregators and power utilities to manage energy consumption through incentives. Such incentives include reducing electricity tariffs when aggregated demand is below average. The goal is to encourage the final consumer to move flexible loads for such periods [9].

Demand-side management takes three different approaches: strategic load growth, energy efficiency, and demand response [4,6]. In this work, the demand response is applied, which, according to [10], refers to changes in the electricity consumption patterns of end-users in response to changes in electricity prices over time or incentives offered to reduce electricity use during periods of high wholesale market prices or compromised system reliability.

Several recent kinds of research have been carried out regarding the demand response [11–13] to minimize the load curve, considering the presence of DGs. The authors of [11] considered both demand-side management and network reconfiguration simultaneously. The mathematical model's objective function seeks to minimize the total cost over the planning horizon, encompassing DG's investment, operation, fuel, and demand-side management costs. The model is converted into a three-tier programming model, and optimal DG planning results are achieved by using a hybrid solution strategy.

To deal with the problem of photovoltaic and wind energy consumption and minimize the total cost of energy purchase, an optimal real-time dispatch model is established in [12], considering the grid security constraints and supply and demand constraints to ensure safe network operation. The authors of [13] addressed a priority-based load shifting to bridge the gap between demand and supply. The power consumer can then cooperate with the utility by managing its demand, that is, by scheduling flexible appliances based on consumption priority.

Network reconfiguration can include the possibility of islanded operation to increase network reliability. During a fault event, the closest protection device to the fault trips, isolating the downstream sections to minimize the number of customers de-energized [14]. DG integration allows the restoration of downstream feeder sections, with part of distribution network loads, by forming microgrids. Such a strategy reduces the number of customers de-energized and increases restoration capacity and reliability indexes and the profit of distribution companies. The authors of [14] performed demand-side management through the hosting capacity of a meshed distribution network using the Repeating Particle Swarm metaheuristic under a 123-bus network with the integration of multiple and large-scale photovoltaic systems.

Recent works in the literature have focused on the demand response strategy to reduce the load flows through the distribution lines by avoiding network congestion and, thus, improving the system's reliability [15–18]. In [15], the frequency-based load shedding using the polynomial regression method and mixed integer linear programming (MILP) is applied as a demand response. The MILP is used for the ideal location of non-vital loads, discarding them to ensure the functionality of critical loads.

The authors of [16] reviewed load-shedding schemes in distribution networks considering DGs. This paper reports that the conventional technique is a trial-and-error procedure used to develop the number and size of load reduction steps. An adaptive technique is proposed to improve the power imbalance estimation and determine the optimal location of the load to be eliminated. The computational intelligence techniques are based on Artificial Neural Networks, metaheuristics, and Fuzzy Logic using the under-frequency load shedding scheme (UFLS) and the voltage shedding scheme. In [17], the UFLS technique based on the combination of Particle Swarm Optimization (PSO) and Bacterial Foraging (BF) metaheuristics was proposed to reduce the amount of disconnected load and increase the lowest possible oscillation frequency.

The authors of [18] identify DG's optimal size and location with a minimum cost as demand response through the simulated annealing (SA) metaheuristic to a microgrid to achieve the stipulated reliability criteria. In [19], a mixed integer nonlinear programming (MINLP) is proposed, minimizing the system's annual energy losses, while [20] proposes a mixed-integer quadratically constrained programming problem, solving using a genetic algorithm (GA).

However, none of these works considered the optimal allocation of DGs simultaneously with load shedding for power loss minimization and, consequently, the total cost of the distribution network, improving reliability through capacity management. In contrast to works that manage the demand side through the hosting capacity or just cut the load on the network without considering the optimal allocation of a distributed generator, this work presents the following contributions:

- The study and problem definitions of the operation and management of DGs, simultaneously with load shedding, to manage the capacity of distribution networks;
- Use of metaheuristics Greedy Randomized Adaptive Search Procedure (GRASP) with Tabu Search (TS) to solve the proposed problem of distribution network capacity management;
- Statistical analysis to validate the proposed formulation and solution technique, promoting the proper integration of DERs into the electrical grids and, thus, maximizing their benefits and maintaining the quality of supply and the reliability of the electrical distribution system.

This paper is set out as follows. Section 2 describes the mathematical modeling of the problem. Section 3 presents the solution techniques using GRASP and TS metaheuristics. Section 4 presents the numerical results, and Section 5 presents the conclusions.

2. Materials and Methods

Power losses are intrinsic conditions of power energy systems and directly impact their performance. Nevertheless, there are several techniques to reduce them, increasing the voltage profile and energy quality. Optimal DG allocation and sizing have the potential to improve the power system performance, providing decentralization of energy sources, power loss reduction, and postponement in reinforcement investments. The integration of DGs in the distribution system also enables the islanded operation. Thus, during a fault event, the network reconfiguration allows DGs to remain in operation as a microgrid, improving the energy quality by reducing the power outage. Figure 1a highlights such a strategy, where DG1 and DG2 can supply loads nearby.

The DSM also improves power system performance by using load shedding. This scheme gives customers the option of a low-cost service, where the load shedding is a condition of their contract type. The DSM relieves capacity and increases load restoration.

In this way, during a fault event, the load shedding can also be advantageous by increasing the chance of islanded operation and load transference to neighboring feeders, as shown in Figure 1b.

The management of the network capacity, sizing and locating of DGs, and load shedding were carried out using the solution of an optimization model. Figure 2 shows an abstract of the proposed methodology, highlighting some advantages of optimal DG planning and demand-side management.

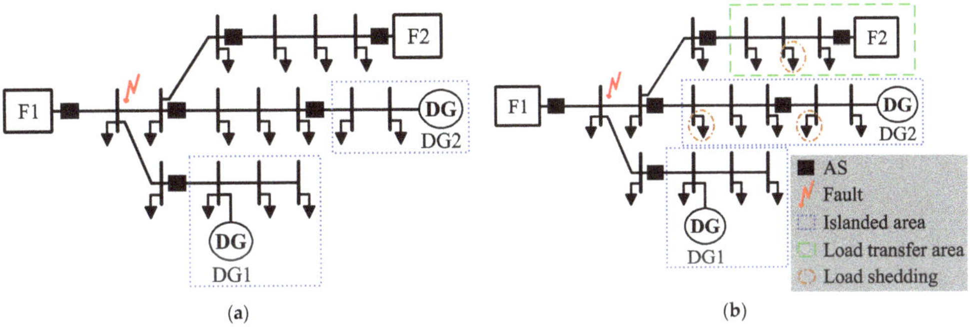

Figure 1. Benefits of applying several strategies for improving energy quality and hosting capacity: (**a**) using only the islaned operation and (**b**) combining islanded operation with load shedding and load transference to neighbor feeders.

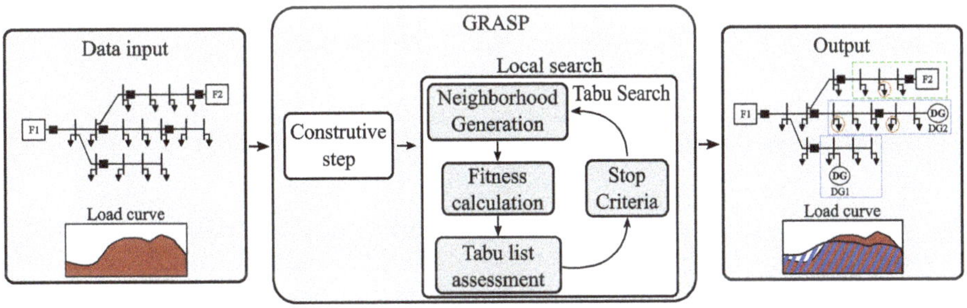

Figure 2. Graphical abstract highlighting the whole process using the proposed method.

2.1. Objective Function

Comprising three elements, the objective function in (1) is composed as follows: cost of losses C^{LOSS}, cost of load shedding C^{SHED}, and cost of energy not supplied C^{ENS}.

$$OF = C^{LOSS} + C^{SHED} + C^{ENS} \tag{1}$$

The distribution network losses are determined as given by (2), assuming the cost of losses for a horizon of one year [21], where k_e is the cost of losses, $I_{j,p,y}$ is the current flow, and $R_{j,p}$ is the resistance of the j-th distribution line at the p-th phase to a distribution network with nl lines. Each year of the planning horizon is represented by y from the set $\boldsymbol{\varphi}$. The cost of load shedding is represented by (3), where $X_{i,p,k,y}$ is the percentage of load shedding; $P^{MAX}_{i,p,k,y}$ is the maximum cuttable power in i-th bus and p-th phase to a distribution network with nb buses; and C^{SHED}_k is the cost of load shedding to the k-th type of customer, which can be residential, R, commercial, C, or industrial, I.

$$C^{LOSS} = k_e \sum_{y \in \boldsymbol{\varphi}} \sum_{j=1}^{nl} \sum_{p \in \{a,b,c\}} \left| I_{j,p,y} \right|^2 R_{j,p} \tag{2}$$

$$C^{SHED} = \sum_{y\in\varphi}\sum_{i=1}^{nb}\sum_{p\in\{a,b,c\}}\sum_{k\in\{R,C,I\}}\left(X_{i,p,k,y}P_{i,p,k,y}^{MAX}C_k^{SHED}\right) \quad (3)$$

The cost of energy not supplied considering the indices of permanent faults is given by (4). We assume that a section is defined as a set of branches and buses belonging to a device's protective zone. More details can be found in [22].

$$C^{ENS} = \sum_{k\in\{R,C,I\}} C_k^{ENERGY}\sum_{y\in\varphi}\frac{P_y^{ENS}}{(1+IRR)^y} \quad (4)$$

In (4), C_k^{ENERGY} is the energy price according to the type of customer, P_y^{ENS} is the energy not supplied during permanent faults for each year y of the planning horizon $\boldsymbol{\varphi}$, and IRR is the internal rate of return. In (5), the estimation of interrupted energy P_y^{ENS} depends on the failure rate λ_h^p for permanent faults and the length of each branch h belonging to the set of branches from section i, $\boldsymbol{\pi}_i$. The total loads in section i and downstream are represented by $P_{i,y}^p$, during year y. The set $\boldsymbol{\beta}$ includes all sections defined by the protection and control devices.

$$P_y^{ENS} = \sum_{i\in\beta}\sum_{h\in\pi_i}\left(\lambda_h^p L_h\right)\left[\left(P_{i,y}^p - P_{i,y}^{MG} - P_{i,y}^{TS}\right)t_R^p + P_{i,y}^{TS}t_{SW}\right] \quad (5)$$

$$P_{i,y}^{MG} = \sum_{m\in\boldsymbol{M}} P_{i,y,m}^{MG} \quad (6)$$

$$P_{i,y,m}^{MG} = 0,\ \ f = h\,\forall\, h\in\sigma_m \quad (7)$$

$$P_{i,y}^{TS} = \sum_{w\in\boldsymbol{W}} P_{i,y,w}^{TS} \quad (8)$$

$$P_{i,y,w}^{TS} = 0,\ \ f = h\,\forall\, h\in\tau_w \quad (9)$$

During a fault event, the islanded operation can occur, depending on the power balance available between loads and DGs in each section. $P_{i,y}^{MG}$ is the microgrids' loading that can be restored, as given in (6). This expression is the sum of loads $P_{i,y,m}^{MG}$ within the microgrid m from the set $\boldsymbol{M}$. For a fault inside the microgrid, that is, a fault in a branch $h\in\sigma_m$, or when the microgrids' loading is bigger than DG capacity, the islanded operation cannot be performed, and DGs must be disconnected, as given in (7).

The same occurs for load transference to neighbor feeders. $P_{i,y}^{TS}$ represents the sum of transferred loads downstream automatic tie switches in section i. Such estimation is based on the sum of all loads that can be restored by using each automatic tie switch from the set $\boldsymbol{W}$ (8). For a fault within τ_w or when the sum of transferred loads is bigger than the neighbor feeder capacity, the load transference cannot be performed, as given in (9). The duration of power outages caused by permanent faults is denoted as t_R^p, while t_{SW} is the time of power outage necessary for the system operator to restore the loads using automatic switches.

2.2. Constraints Set

The active and reactive power balance equations for each bus of radial networks are, respectively, represented by equality constraints (10) and (11). The voltage magnitude limits of the i-th bus are represented by constraint (12), while the current flow limits for each phase, $I_{p,i,y}$, are represented by (13), with an upper limit of $\overline{I_p}$. Active and reactive power of DGs must not exceed upper and lower boundaries as, respectively, given by constraints (14) and (15).

$$P_{p,i,y}(V,\theta) - P^{DG}_{p,i,y} + P^{D}_{p,i,y} = 0, \ \forall i \in nb, \ y \in \boldsymbol{\varphi}, \ p = a,b,c \quad (10)$$

$$Q_{p,i,y}(V,\theta) - Q^{DG}_{p,i,y} + Q^{D}_{p,i,y} = 0, \quad \forall i \in nb, \ y \in \boldsymbol{\varphi}, \ p = a,b,c \quad (11)$$

$$\underline{V_p} \leq V_{p,i,y} \leq \overline{V_p}, \ \forall i \in nb, \ y \in \boldsymbol{\varphi}, \ p = a,b,c \quad (12)$$

$$0 \leq I_{p,i,y} \leq \overline{I_p}, \ \forall i \in nb, \ y \in \boldsymbol{\varphi}, \ p = a,b,c \quad (13)$$

$$\underline{P^{DG}} \leq P^{DG}_{p,i,y} \leq \overline{P^{DG}}, \ \forall i \in nb, \ y \in \boldsymbol{\varphi}, \ p = a,b,c \quad (14)$$

$$\underline{Q^{DG}} \leq Q^{DG}_{p,i,y} \leq \overline{Q^{DG}}, \ \forall i \in nb, \ y \in \boldsymbol{\varphi}, \ p = a,b,c \quad (15)$$

3. Solution Technique

The proposed optimization model, a mixed integer nonlinear programming model, is solved using the GRASP metaheuristic with TS in the local search phase. It combines the advantages of two widely known techniques, speeding up the search process with the greedy stage of GRASP and reducing the chances of becoming stuck in local minima due to the flexibility provided by the tabu list of prohibited solutions.

The GRASP metaheuristic consists of two phases: constructive and local search, performed sequentially and iteratively [23]. In the first phase, different initial greedy and random solutions are generated. This step directs the method to regions more likely to obtain local optima, accelerating the exploration process. Then, the found solutions are refined in the local search phase using the TS algorithm [24], improving them even more.

The tabu list restricts the choice of solutions that have been visited recently. Then, the optimization process allows the TS to select worse-quality solutions for local exploration. Such an aspect allows for extensive exploration, obtaining local optimal solutions from different regions of the search space. From left to right, Figure 3 shows the basic flowchart of GRASP and TS.

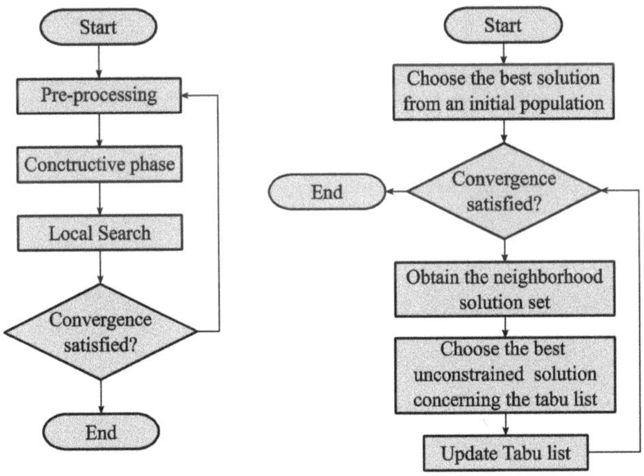

Figure 3. Basic flowcharts of GRASP and TS.

In this work, the convergence criterion is based on the iteration number. The algorithm repeats up to a limit of iterations, or if the sequence of iterations without improvement reaches a certain limit, both previously defined by the user.

3.1. Codification

In coding the individuals for the generators, a set of real numbers is used, in which the integer part characterizes the active power $P^{DG}_{j,y}$ of the DG, and the decimal part represents the power factor $pf_{j,y}$ of the generator, which must be equal to or higher than $\underline{pf}$. If an i-th bus has allocated the j-th DG, $P^{DG}_{i,y} = P^{DG}_{j,y}$ and $Q^{DG}_{i,y} = P^{DG}_{j,y}\left(1 - pf^2_{j,y}\right)^{1/2}/pf_{j,y}$. In (16) and (17), $\alpha_{j,y}$ and $\beta_{j,y}$ are integer random numbers that vary from 0 to $\overline{\alpha}$ and $\overline{\beta}$, respectively. Increments in active power and power factor are ΔP^{DG} and Δpf. As for the load shedding, in (18), integer values represented by the variable $\gamma_{j,y}$ from 0 to $\overline{\gamma}$ are used to characterize the percentage of load shedding $X_{j,y}$.

$$P^{DG}_{j,y} = \alpha_j \Delta P^{DG}, \ \ \alpha_{j,y} = 0, 1, \ldots, \overline{\alpha} \tag{16}$$

$$pf_{j,y} = \underline{pf} + \beta_{j,y}\Delta pf, \ \ \beta_{j,y} = 0, 1, \ldots, \overline{\beta} \tag{17}$$

$$X_{j,y} = \gamma_j \Delta X, \ \ \gamma_{j,y} = 0, 1, \ldots, \overline{\gamma} \tag{18}$$

Figure 4 presents the coding of a candidate solution as a vector where each element is associated with the position of the buses DG_n and LS_n for generator allocation and load shedding, respectively, where n represents the number limit of each one. Figures 5 and 6 display detailed flowcharts of the whole proposed method.

DG_0	...	DG_n	LS_0	...	LS_n
$P^{DG}_{0,0} \cdot pf_{0,0}$	...	$P^{DG}_{n,y} \cdot pf_{n,y}$	$X_{0,0}$	...	$X_{n,y}$
n DGs $\forall y \in \varphi$			n LOAD SHEDDING $\forall y \in \varphi$		

Figure 4. Coding vector of a candidate solution.

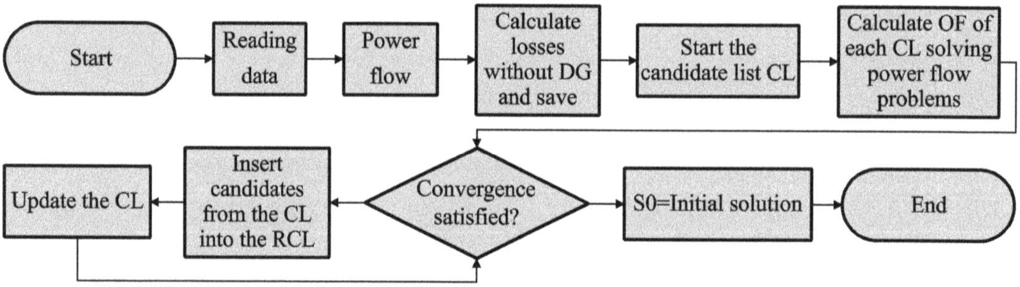

Figure 5. Constructive step flowchart.

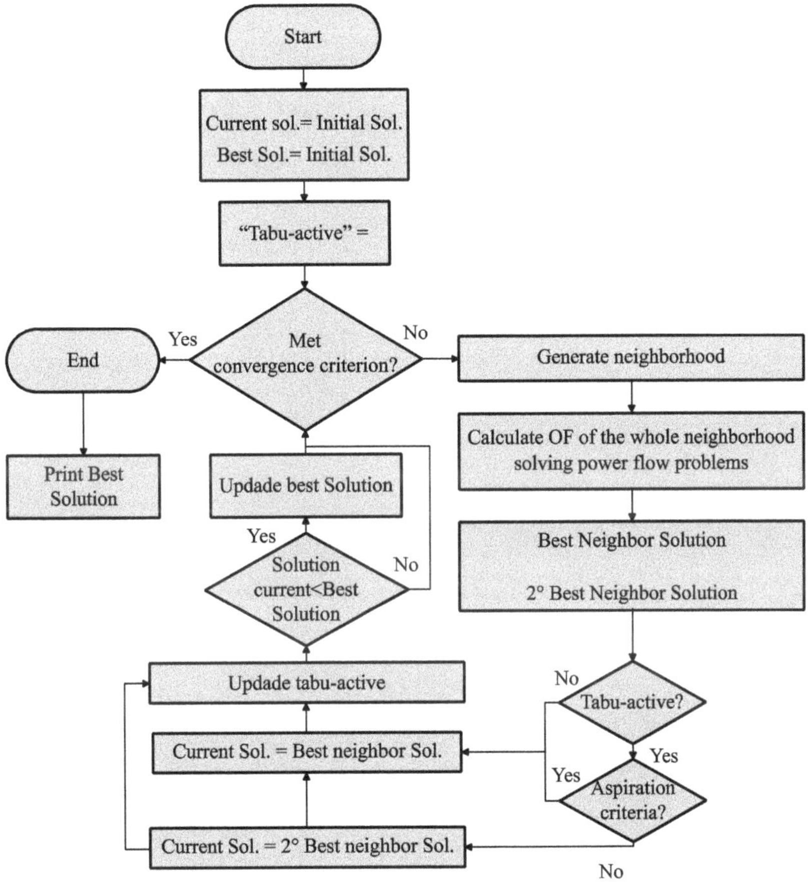

Figure 6. Local search phase flowchart.

3.2. *Constructive*

During the constructive phase, as depicted in Figure 5, the pre-processing stage is initiated, which involves running a three-phase power flow analysis. This analysis provides the voltage profile for every phase at all buses, current flows through branches, and the sum of losses concerning the feeder under study. The power flow analysis ensures the active and reactive power balance, as specified by constraints (10) and (11). A set of candidate buses with the lowest voltages and highest demand is chosen after power flow estimation for DG allocation and load shedding.

The initial candidate construction occurs by inserting elements one by one from the solution of the search space (CL). The search space is narrowed down afterward based on Equation (19), resulting in a list called the restricted candidate list (RCL), which includes the top-performing candidates. An element is randomly selected in each iteration to be included in the candidate solution under the building process.

$$\underline{g(c)} \leq g(c) \leq \underline{g(c)} + a\left[\overline{g(c)} - \underline{g(c)}\right] \tag{19}$$

GRASP has a greedy function $g(c)$ to assess candidate solutions c, providing the cost concerning the addition of an element to the initial solution. Assuming a problem of minimization, the solutions with the lowest and greatest cost among candidates in the CL are, respectively, represented as $\underline{g(c)}$ and $\overline{g(c)}$. A candidate c in the CL is placed in the RCL if and only if (19).

3.3. Local Search Phase

The TS metaheuristic is employed in this stage to enhance the solution obtained from the constructive process. As depicted in Figure 6, the search traverses the neighborhood of the current configuration, locating the setup that enhances the objective function's value. The algorithm's performance is dependent on the quality of these configurations.

After computing the objective function's values throughout the neighborhood, the configuration with the lowest objective function value is chosen. Solutions from the Tabu list are compared with the chosen one. If it is, the algorithm assesses the aspiration criterion, verifying if the solution meets such criteria. The main objective is to remove the prohibition on the attribute of a high-quality neighboring solution that shares the same configuration as a previously visited solution. If the objective function value of the discovered neighboring solution is superior to that of the current solution, the restriction is revoked, and the transition is executed. In contrast, if the aspiration criterion is not satisfied, the configuration with the second-lowest objective function value is chosen.

4. Numerical Results and Discussion

The tests were performed on a real-world distribution feeder with 1806 buses [25], providing power to 834 loads, 781 in the low-voltage network, 53 in the medium-voltage network, and 47 distribution transformers. This test system has 11 automatic switches used for service restoration.

Smart grid customers are classified into residential (R), commercial (C), and industrial (I), where each type represents 50%, 30%, and 20% of distribution network loads. The cost of losses is defined as $k_e = 0.04$ USD/kWh [19]. Other optimization settings and parameters are shown in Table 1. The maximum capacity for each DG installed is 3.6 MW. The objective function proposed in (1) is penalized whenever constraints are violated during the solutions assessment. In the constructive phase, there are limiting parameters ΔP^{DG}, $\overline{\alpha}$, $\underline{pf}$, Δpf, and $\overline{\beta}$, ΔX and $\overline{\gamma}$, also shown in Table 1. ENS and load-shedding costs are based on [22] and [11], respectively. This work considers a planning horizon of one year.

Table 1. Optimization settings and parameters.

Parameters	Value	Parameters	Value	Parameters	Value
C_R^{SHED}	0.04 USD/kWh	C_R^{ENS}	1.5 USD/kWh	$\overline{\alpha}$	5
C_C^{SHED}	0.03 USD/kWh	C_C^{ENS}	3.0 USD/kWh	$\overline{\beta}$	9
C_I^{SHED}	0.02 USD/kWh	C_L^{ENS}	4.64 USD/kWh	$\overline{\gamma}$	5
$\underline{V}$	0.95 pu	$\overline{V}$	1.05 pu	$\underline{pf}$	0.90
ΔX	4%	ΔP^{DG}	160 kW	Δpf	0.01

To our best knowledge, there is no similar work in the literature that includes all aspects addressed in this paper. The proposed methodology's performance is evaluated using statistical analysis. Table 2 shows results from the statistical analysis, while Tables 3 and 4 highlight DG configuration and load shedding found by the worst solution, in that order.

Table 2. Statistical analysis.

a	CVP	$\overline{OF}$ (MUSD)	TA (MUSD)	SD (MUSD)
0.0	10.2	5.545	1.832	0.566
0.2	14	5.619	2.510	0.788
0.8	6	5.268	1.461	0.321
1.0	17	5.715	2.692	0.973

Table 3. Worst solution: DG configuration.

DG_n (bus)	13388	13446	13418	13392	13428	14120
α_n	4	0	0	0	0	0
P_n^G (kW)	640	0	0	0	0	0

Table 4. Worst solution: load shedding setup.

LS_n (bus)	200800	201128	201131	200914	201134	202004
γ_n	4	0	0	0	0	0
X_n	16%	0	0	0	0	0

In the simulations performed in the constructive phase, the twelve candidate buses are selected based on their position in a sequential manner. First, adjustments were made to find the solution with the most adequate value for the initial solution. Thus, we obtained Pearson's coefficient of variation (CVP), the average of objective functions (OF_μ), standard deviation (SD), and total amplitude (TA).

When analyzing Table 2, we can see that the most homogeneous solution was the one in which the algorithm is near random ($a = 0.8$), that is, with a variation of 6% in its solutions. The worst solution was obtained when the constructive phase was purely greedy ($a = 0.0$). In addition, if the solution is completely random ($a = 1.0$), the initial solution may not be very good, causing the algorithm to stop at a local optimum in the neighborhood search, which can generate more heterogeneous solutions. The algorithm stops when one of two conditions is met: either the incumbent solution does not improve for 300 generations, or it reaches the maximum number of 800 generations. In either case, the incumbent solution is considered the best-found solution.

Fifty simulations were performed to evaluate the performance of the algorithm with the results obtained during the tests, i.e., $a = 0.8$. Figures 7–9 are related to the statistical analysis of results provided by the proposed method, where each one shows a probability graph of objective functions, a box diagram with the scattering of objective functions, and the objective function histogram, in that order.

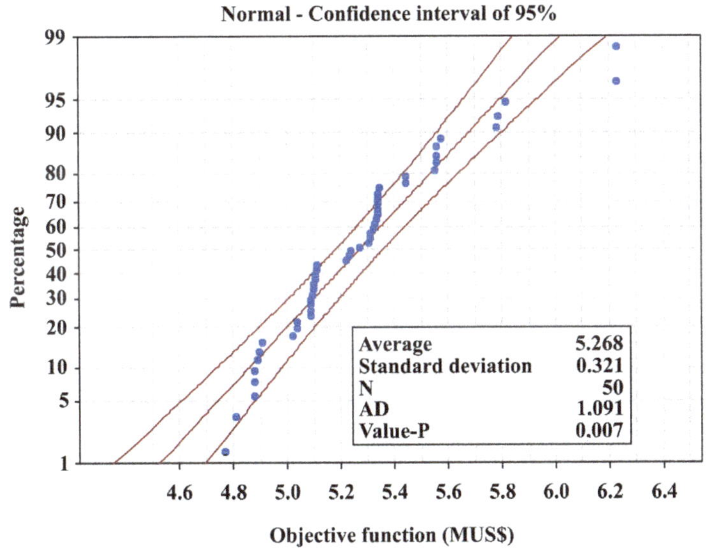

Figure 7. Probability graph of objective functions.

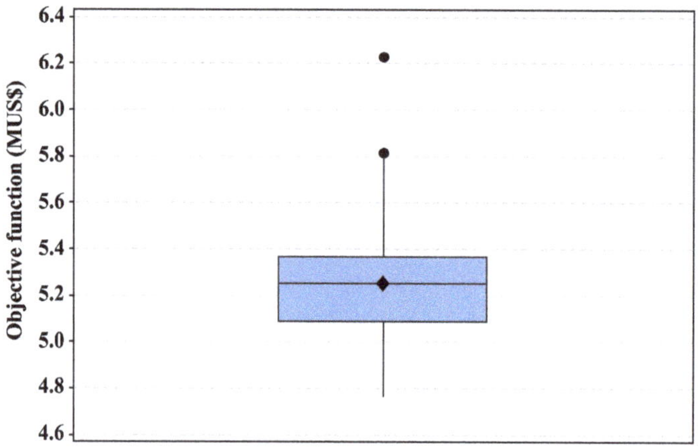

Figure 8. Box diagram: scattering of objective functions.

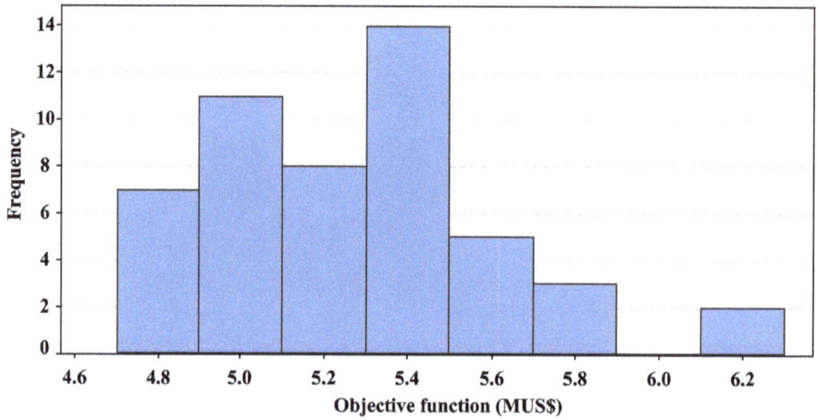

Figure 9. Objective function histogram.

The graph in Figure 8 shows that most of the data are on the right side. The symbol in the center represents the average value of MUSD 5.268, and the line represents the middle value of MUSD 5.251. Finally, 75% of all values are below the third quartile of MUSD 5.369. The graph shows two outliers (atypical value), which is because these solutions do not have any load shedding and have only installed a 640 kV generator, as shown in Tables 3 and 4.

The data do not follow a normal distribution because two solutions presented the value of the objective function as very high, with a much greater amplitude than the others, and the p-value is less than the significance level of 0.05, as shown in Figure 7.

From the histogram of the objective functions in Figure 9, we can see that the data are multimodal and have outliers since two solutions have only one DG installed, and the load cut has not been performed.

A diagram of the test system is shown in Figure 10, including DGs and the load-shedding locations. Low-voltage loads have been omitted from the figure. The dashed grey area shows an example of the low-voltage loads downstream of a distribution transformer. The positions of the worst DG and load-shedding setup are highlighted in red, while the best DG and load-shedding setup are emphasized in green.

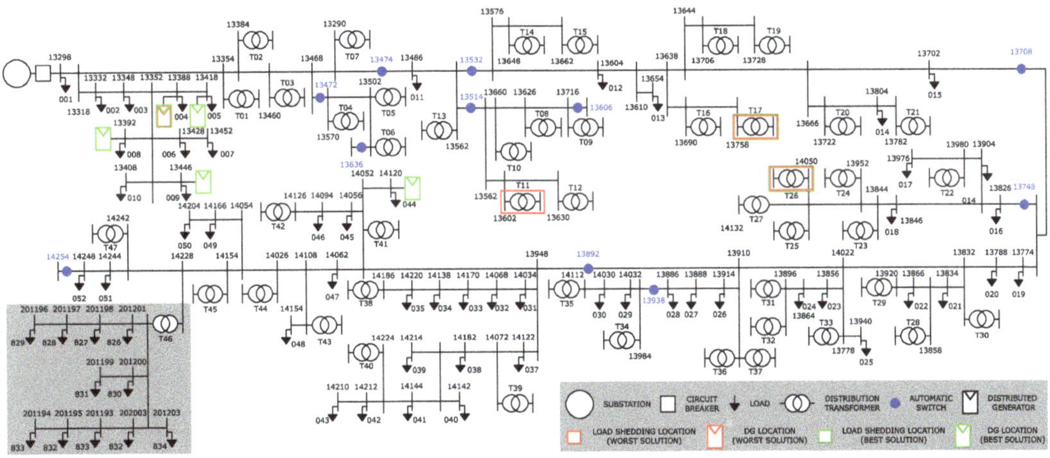

Figure 10. Test system, including DGs and load-shedding location from the best and worst solution.

The best solution presented the value of the objective function of MUS 4.764, with an injection of 2.24 MW in the network and cuts made in the buses, as shown in Tables 5 and 6. In the worst solution, from the six candidate buses found in the constructive phase, only one has load shedding within distribution transformer T11. On the other hand, the best solution includes all six customers for load shedding, where three of them have 20% shedding. Similarly, the worst solution for DGs has only one installed, while the best result includes four DGs.

Table 5. Best solution: DG configuration.

DG_n (bus)	13388	13446	13418	13392	13428	14120
α_n	2	4	4	1	0	3
P_n^G (kW)	320	640	640	160	0	480

Table 6. Best solution: load-shedding setup.

LS_n (bus)	201127	201128	200913	200914	201146	201131
γ_n	5	5	3	3	2	5
X_n	20%	20%	12%	12%	8%	20%

The biggest difference between the worst and best solutions is the number of DG allocations and load shedding, highlighting the importance of implementing such strategies. Nevertheless, the worst and best solutions present very similar locations for load shedding and identical locations for DG installation. Even the worst solution presents attractive positions to install DGs or perform load shedding. Therefore, the difference is related to the number of α_n and γ_n.

As shown in Figure 10, all DG allocations provided by the solution technique are in the medium-voltage network. Likewise, we assume that load shedding must occur in low-voltage buses, which could have high-priority loads. Allowing load shedding in medium-voltage loads could lead the metaheuristic to promote only load shedding in such areas.

Most found solutions have allocated a DG close to the substation, except for a DG at bus 14120. A DG installed far from the substation promotes a relevant increase in the voltage profile, according to the DG's capacity. On the other hand, the allocation of DGs close to the substation provides a low increase in the voltage profile. The solution technique seeks to install more DGs close to the substation to minimize energy losses and, simultaneously, to avoid overvoltage problems.

As indicated in Table 5, none of the generators has $\alpha_n = \overline{\alpha}$ due to the impact on the system's voltage profile. The solution technique identified a more suitable setup when using smaller-capacity generators to avoid overvoltage problems.

The loads downstream of transformers T17 and T26 are shown in Figure 11, highlighting the main areas with load shedding. Three load sheddings were considered in the same buses for both worst and best solutions. The worst case has $\gamma_n = 0$ for loads 200914, 201128, and 201131, while the best case has $\gamma_n = 3$, 5, and 5, respectively. The constructive process finds initial solutions, and the local search refines their quality, maintaining good quality of load-shedding locations, even for the worst solution.

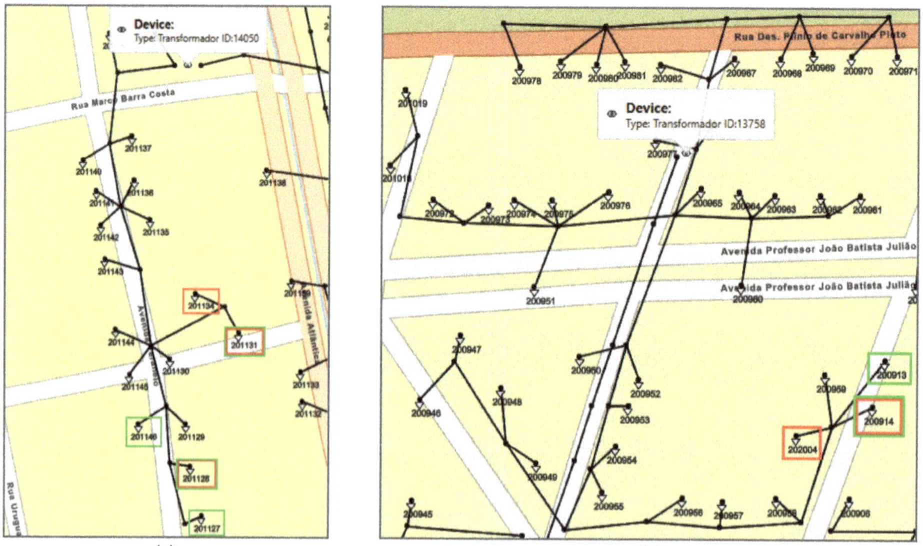

Figure 11. Details from loads in (**a**) TR17 and (**b**) TR26.

Figure 12 shows the voltage profile of the initial condition or base case (orange marks) and the best solution (blue marks). The voltage difference between phases is small. However, solving the problem of managing the capacity considering unbalanced smart grids may provide a more appropriate solution than works without such an important aspect.

As shown in Figure 12, allocating DGs in the distribution network increases the voltage levels up to 2.02% at the bus 200800 within distribution transformer T11 due to the combined optimization of DG allocation simultaneously with load shedding, reliability indices, load transference, and the possibility of islanded operation. Such voltage improvement provides power loss reduction. Phase C in the base case presents the lower voltage levels. Nevertheless, DG allocation and load shedding in the best solution provide the best voltage improvement for the same phase.

The proposed solution allocates the generators in regions where the generators can supply the loads and perform islanded operations during fault events. So, including reliability optimization aspects in the objective function increases the overall benefits from optimal allocation and sizing of DGs and load shedding.

The results show that all parts of the objective function work cooperatively to increase response demand and reliability. For instance, the islanded operation is improved when load allocation and load shedding are considered simultaneously. The implementation of generators allows islanding, while load shedding increases the chances of success of the islanded operation, reducing the demand in the region that will be islanded and increasing reliability and customer satisfaction.

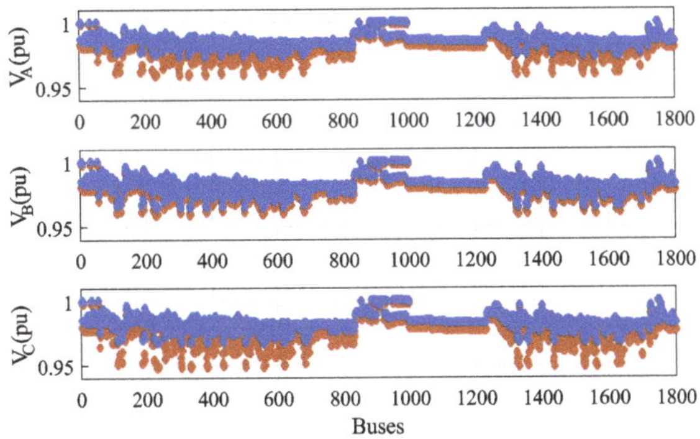

Figure 12. Voltage profile for different cases.

5. Conclusions

This work applied a methodology based on the hybrid GRASP and TS algorithm to evaluate the effects of the optimal allocation of DGs and load shedding in the smart grid while also considering reliability aspects. The results show that the objective function produces a higher value if load shedding is not carried out together with the optimal DG allocation. Optimal benefits from DG and load shedding can be achieved by conducting adequate planning to determine the most suitable bus for minimum shedding, as well as the ideal size and location of DGs, which may vary depending on the system.

The results found highlight the advantages of performing demand-side management, ensuring a better distribution of the load curve. The benefits include a better voltage profile and network capacity relief. Load restoration is also increased by the islanded operation and load transference to neighbor feeders. Load shedding benefits load restoration as well. The combination of these techniques allows for enhancing reliability indices and, consequently, energy quality to customers.

Most DGs installed by the proposed method are closer to the substation. Such an aspect highlights the challenges of the massive integration of DGs into the distribution network. Massive DG placement increases voltage levels and may surpass the voltage upper limit imposed by regulatory agencies. Thus, the proposed method allocates the DGs near the substation to ensure compliance with these limits. Furthermore, although there is a slight voltage difference between phases, the previous conclusions highlight that solving the problem of managing the capacity considering unbalanced smart grids may provide a more appropriate solution than works without such an important aspect.

Distribution networks with degraded voltage profiles can benefit more from the proposed methodology. In this scenario, the DG installation will be larger and better distributed along the distribution network. Thus, the number of restored loads can increase, benefiting reliability indices as well.

Further studies should consider the existence of critical loads into feeder sections, enforcing different rewards to demand response programs and, consequently, improving DSM. Tools for improving distribution network performance should be integrated into the model, such as smart inverters of electronically coupled DGs that can already be installed in the distribution network and belong to a private owner. Furthermore, integrating some technologies, such as energy storage systems, may require future modifications in the proposed method since such a device requires evaluating the energy balance to support the islanded operation of microgrids.

Author Contributions: Conceptualization, H.d.O.M.S. and J.B.L.; Investigation, H.d.O.M.S. and C.R.; Methodology, H.d.O.M.S., C.R. and J.B.L.; Software, H.d.O.M.S., C.R. and J.B.L.; Supervision, J.B.L.; Validation, H.d.O.M.S. and C.R.; Writing—original draft, H.d.O.M.S. and C.R.; Writing—review and editing, C.R. and J.B.L. All authors have read and agreed to the published version of the manuscript.

Funding: This paper was possible thanks to the scholarship granted from the Brazilian Federal Agency for Support and Evaluation of Graduate Education (CAPES)—Finance code 001, and the São Paulo Research Foundation (FAPESP), grants 2015/21972-6 and 2019/07436-5.

Institutional Review Board Statement: This study did not require ethical approval.

Data Availability Statement: There is no additional data beyond that provided throughout the article.

Conflicts of Interest: The authors declare no conflict of interest.

Nomenclature

C^{ENS}	Cost of energy not supplied in monetary unit.
C^{LOSS}	Cost of power losses in monetary unit.
C^{SHED}	Cost of load shedding in monetary unit.
C_k^{ENERGY}	Energy price according to the type of customer, USD/kWh.
C_k^{SHED}	Cost of load shedding to k-th type of customer, USD/kW, which can be residential, R, commercial, C, or industrial, I.
$I_{j,p,y}$	Current flow through the j-th distribution line at p-th phase, A.
$I_{p,i,y}$	Current injected in the i-th bus, phase p, A.
L_h	Line length of the branch h, km.
$P_{i,p,k,y}^{MAX}$	Maximum cuttable power in i-th bus and p-th phase, kW.
$P_{i,y,m}^{MG}$	Loads belonging to the microgrid m in section i, year y that cna be restored, kW.
$P_{i,y}^{MG}$	Microgrids' loads in section i, year y that can be restored, kW.
$P_{i,y}^{TS}$	The sum of transferred loads downstream automatic tie switches in section i, year y, kW.
$P_{i,y}^{P}$	Total loads in the section i and down-stream, year y, kW.
$P_{p,i,y}$	Active power injected in the i-th bus, phase p, kW.
$P_{p,i,y}^{D}$	Active power demand in the i-th bus, phase p, kW.
$P_{p,i,y}^{DG}$	Active power injected by the DG in the i-th bus, phase p, kW.
P_y^{ENS}	Energy not supplied during permanent faults for each year y, kW.
$Q_{p,i,y}$	Reactive power injected in the i-th bus, phase p, kVAr.
$Q_{p,i,y}^{D}$	Reactive power demand in the i-th bus, phase p, kVAr.
$Q_{p,i,y}^{DG}$	Reactive power injected by the DG in the i-th bus, phase p, kVAr.
$R_{j,p}$	Line resistance of the j-th distribution line, phased p, Ω.
T_{SW}	The time of power outage necessary to the system operator restores the loads by using automatic switches, s.
$V_{p,i,y}$	Voltage magnitude of the i-th bus, phase p, pu.
$X_{i,p,k,y}$	Percentage of load shedding in the bus i, phase p, costumer type k.
$X_{j,y}$	Percentage of load shedding in the candidate bus n.
k_e	Cost of losses, USD/kWh.
t_R^p	The duration of power outages caused by permanent faults, s.
$\alpha_{j,y}$	Integer multiplying factor of the base value for the DG power capacity.
$\beta_{j,y}$	Integer multiplying factor of the power factor base value.
$\gamma_{j,y}$	Integer multiplying factor of the load shedding base value.
λ_h^p	Failure rate of line h per kilometer year.
π_i	Set of branches belonging to section i.
σ_m	Set of branches belonging to the microgrid m.
τ_w	Set of branches belonging to the microgrid m.
ΔX	Base value of the load shedding.
Δpf	Base value of the power factor.
IRR	Internal rate of return.
$g(c)$	Greedy function to assess the candidate solution c.
ΔP^{DG}	Base value of the DG power capacity, kW.
$\boldsymbol{\beta}$	Set of all sections defined by the protection and control devices.
$\boldsymbol{\varphi}$	Set of year of the planning horizon.

References

1. Carr, J.A.; Balda, J.C.; Mantooth, H.A. A Survey of Systems to Integrate Distributed Energy Resources and Energy Storage on the Utility Grid. In Proceedings of the 2008 IEEE Energy 2030 Conference, Atlanta, GA, USA, 17–18 November 2008; pp. 1–7.
2. Blokhuis, E.; Brouwers, B.; van der Putten, E.; Schaefer, W. Peak Loads and Network Investments in Sustainable Energy Transitions. *Energy Policy* **2011**, *39*, 6220–6233. [CrossRef]
3. Wang, J.; Lu, X.; Chen, C. *Guidelines for Implementing Advanced Distribution Management Systems-Requirements for DMS Integration with DERMS and Microgrids*; Argonne National Lab (ANL): Argonne, IL, USA, 2015.
4. Jabir, H.; Teh, J.; Ishak, D.; Abunima, H. Impacts of Demand-Side Management on Electrical Power Systems: A Review. *Energies* **2018**, *11*, 1050. [CrossRef]
5. Limaye, D.R. Implementation of Demand-Side Management Programs. *Proc. IEEE* **1985**, *73*, 1503–1512. [CrossRef]
6. Gellings, C.W. The Concept of Demand-Side Management for Electric Utilities. *Proc. IEEE* **1985**, *73*, 1468–1470. [CrossRef]
7. Hayes, B.; Hernando-Gil, I.; Collin, A.; Harrison, G.; Djokic, S. Optimal Power Flow for Maximizing Network Benefits From Demand-Side Management. *IEEE Trans. Power Syst.* **2014**, *29*, 1739–1747. [CrossRef]
8. Dehnavi, E.; Abdi, H. Determining Optimal Buses for Implementing Demand Response as an Effective Congestion Management Method. *IEEE Trans. Power Syst.* **2016**, *32*, 1537–1544. [CrossRef]
9. Saffre, F.; Gedge, R. Demand-Side Management for the Smart Grid. In Proceedings of the 2010 IEEE/IFIP Network Operations and Management Symposium Workshops, Osaka, Japan, 19–23 April 2010; pp. 300–303.
10. U.S. Department of Energy. *Benefits of Demand Response in Electricity Markets and Recommendations for Achieving Them*; U.S. Department of Energy: Washington, DC, USA, 2006.
11. Zhang, S.; Cheng, H.; Wang, D.; Zhang, L.; Li, F.; Yao, L. Distributed Generation Planning in Active Distribution Network Considering Demand Side Management and Network Reconfiguration. *Appl. Energy* **2018**, *228*, 1921–1936. [CrossRef]
12. Zhang, H.; Gong, C.; Ju, W.; Pan, G.; Wang, W. Optimization Dispatch Modeling for Demand Response Considering Supply and Demand Balance and Security Constraints. In Proceedings of the 2021 Power System and Green Energy Conference (PSGEC), Shanghai, China, 20–22 August 2021; pp. 166–170.
13. Arun, C.; Aswinraj, R.; Bijoy, M.T.; Nidheesh, M.; Rohikaa Micky, R. Day Ahead Demand Response Using Load Shifting Technique in Presence of Increased Renewable Penetration. In Proceedings of the 2022 IEEE 7th International conference for Convergence in Technology (I2CT), Mumbai, India, 7–9 April 2022; pp. 1–6.
14. Abideen, M.Z.U.; Ellabban, O.; Ahmad, F.; Al-Fagih, L. An Enhanced Approach for Solar PV Hosting Capacity Analysis in Distribution Networks. *IEEE Access* **2022**, *10*, 120563–120577. [CrossRef]
15. Sarwar, S.; Mokhlis, H.; Othman, M.; Shareef, H.; Wang, L.; Mansor, N.N.; Mohd Khairuddin, A.S.; Mohamad, H. Application of Polynomial Regression and MILP for Under-Frequency Load Shedding Scheme in Islanded Distribution System. *Alex. Eng. J.* **2022**, *61*, 659–674. [CrossRef]
16. Sapari, N.M.; Mokhlis, H.; Laghari, J.A.; Bakar, A.H.A.; Dahalan, M.R.M. Application of Load Shedding Schemes for Distribution Network Connected with Distributed Generation: A Review. *Renew. Sustain. Energy Rev.* **2018**, *82*, 858–867. [CrossRef]
17. Awad, H.; Hafez, A. Optimal Operation of Under-Frequency Load Shedding Relays by Hybrid Optimization of Particle Swarm and Bacterial Foraging Algorithms. *Alex. Eng. J.* **2022**, *61*, 763–774. [CrossRef]
18. Mitra, J.; Vallem, M.R.; Singh, C. Optimal Deployment of Distributed Generation Using a Reliability Criterion. *IEEE Trans Ind. Appl.* **2016**, *52*, 1989–1997. [CrossRef]
19. Atwa, Y.M.; El-Saadany, E.F.; Salama, M.M.A.; Seethapathy, R. Optimal Renewable Resources Mix for Distribution System Energy Loss Minimization. *IEEE Trans. Power Syst.* **2010**, *25*, 360–370. [CrossRef]
20. Dang, C.; Wang, X.; Wang, X.; Li, F.; Zhou, B. DG Planning Incorporating Demand Flexibility to Promote Renewable Integration. *IET Gener. Transm. Distrib.* **2018**, *12*, 4419–4425. [CrossRef]
21. Shekari, T.; Gholami, A.; Aminifar, F.; Sanaye-Pasand, M. An Adaptive Wide-Area Load Shedding Scheme Incorporating Power System Real-Time Limitations. *IEEE Syst. J.* **2018**, *12*, 759–767. [CrossRef]
22. Reiz, C.; De Lima, T.D.; Leite, J.B.; Javadi, M.S.; Gouveia, C.S. A Multiobjective Approach for the Optimal Placement of Protection and Control Devices in Distribution Networks with Microgrids. *IEEE Access* **2022**, *10*, 41776–41788. [CrossRef]
23. Resende, M.G.C.; Ribeiro, C.C. Greedy randomized adaptative search procedures: Advances and extensions. In *Handbook of Metaheuristics*; Springer: New York City, NY, USA, 2019; Chapter 6; pp. 169–220.
24. Glover, F.L.M. *Handbook of Combinatorial Optimization*; Springer: Boston, MA, USA, 1998; Volume 3.
25. Laboratory of Electrical Power System Planning—LAPSEE. Distribution Testing System of 1806 Buses. Available online: https://www.feis.unesp.br/Home/departamentos/engenhariaeletrica/lapsee807/home/distribution_network_1806_lines.rar (accessed on 30 November 2022).

processes

MDPI

Article

Design Strategy of Electricity Purchase and Sale Combination Package Based on the Characteristics of Electricity Prosumers in Power System

Xiaotian Wang [1,*], Chuang Liu [1], Binbin Wu [1], Wei Wang [1], Yi Sun [2,*], Jie Peng [2], Xinya Liu [2] and Kai Zhang [2]

1 State Grid Hebei Electric Power Co., Ltd., Marketing Service Center, Shijiazhuang 050035, China; liuchuang_2024@163.com (C.L.); yunchaosu0250834@163.com (B.W.); wangwei_03112024@163.com (W.W.)
2 School of Electrical and Electronic Engineering, North China Electric Power University, Beijing 102206, China; jtjiang0720@163.com (J.P.); sdtx2016014605@163.com (X.L.); kaizhang0537@163.com (K.Z.)
* Correspondence: 13102002597@163.com (X.W.); sy@ncepu.edu.cn (Y.S.)

Abstract: With the progress in renewable energy and smart grid technologies, electricity users are evolving into prosumers, capable of both consuming and generating electricity through distributed photovoltaic (DPV) systems. Concurrently, the liberalization of the electricity retail market has prompted retailers to design customized electricity packages based on users' needs and preferences, aiming to enhance service quality, efficiency, and user retention. However, previous studies have not fully addressed the multidimensional characteristics and electricity consumption behaviors that influence package selection. This paper initially dissects user characteristics across three key dimensions: electricity demand preferences, price sensitivity, and risk tolerance. Therefore, leveraging utility functions and autonomous choice behavior models, we propose two innovative electricity purchase and sale combination packages: a fluctuating pricing package and a discount-based pricing package. Furthermore, we introduce the Self-Adaptive Weight and Reverse Learning Particle Swarm Optimization (SAW&RL-PSO) algorithm to address the complexities of these choices. Simulation results indicate that the methodologies presented significantly enhance user benefits and retailer revenues while also effectively managing electricity usage fluctuations and the challenges of integrating large-scale DPV systems into the electrical grid.

Keywords: electricity purchase–sale combination package; characteristics of electricity consumption analysis; prosumers; self-selection; improved particle swarm optimization

1. Introduction

The swift development of new electricity systems has spurred reforms on the electricity sales side, and electricity pricing has gradually evolved from uniform formulation to market competition mode [1]. In this context, many electricity retailers have been rapidly established and developed, and the market transaction mode has gradually shown a diversified and distributed pattern [2]. Compared with the traditional single electricity price model, differentiated electricity price packages can provide more targeted prosumer pricing, attract users to actively participate in the electricity market, improve market competitiveness, and achieve the expected improvement of the benefits of electricity retailers and users [3].

1.1. Related Work and Motivation

With the ongoing rise in the adoption of distributed renewable energy on the demand side, the study of prosumers has emerged as a focal point of research in recent years [4]. Prosumers are no longer traditional electricity consumers but independent individuals with the dual identity of electricity prosumers. Therefore, prosumers have the capability to engage in the retail electricity market as consumers, which can lead to a reduction in the cost of electricity acquisition, and as producers in the retail electricity market, which can enhance the revenue from electricity sales [5]. However, demand-side prosumers have the characteristics of being massive and dispersed, their individual resources are limited, and the grid market environment is complex, so they are not well-suited for direct competition in the retail electricity market and, thus, often rely on intermediaries like electricity retailers to participate in market activities.

In the traditional differentiated time-of-use electricity purchase pricing or incentive electricity purchase pricing model, the agent needs to carry out dozens of iterative interactions with each user individually to solve the refined and differentiated optimal electricity sales price for each user [6]. With the gradual increase in the scale of users, the separate iterative interactions with each user will significantly increase the communication and computational pressure of the central platform, and the operation and maintenance costs will become too high. Therefore, it is crucial to analyze and cluster the characteristics of prosumers' electricity consumption and propose differentiated classified electricity packages for the response behaviors of various users. The authors in [7,8] categorized consumers into categories such as peak-day, peak-night, and stable types based on their electricity usage patterns. In [9], the authors analyze the personalized portrait of EV users using four labels, namely, users' essential attributes, charging behavioral attributes, attitudinal preference attributes, and other attributes of the society, and form three types of typical users. In [10], the authors summarize three user classification methods: users' industry attribute classification, typical electricity consumption characteristics classification, and risk preference classification, and sort out the types of electricity packages suitable for different types of users. However, the above literature focuses on the electricity consumption curve and does not quantify the attribute values of users' risk preferences, which lacks a more comprehensive description of users' characteristics and will cause errors in the simulation of users' behavior by the electricity retailers, which will lead to a lack of attractiveness of the electricity packages to users and will not be able to achieve the expected results of the electricity retailers.

At the same time, most existing electricity retailers focus on studying the users' electricity purchase price development strategy and rarely take the prosumers as the main body of the study of their strategy for the development of electricity prices for purchase and sale. In [11], the authors developed a demand-side response model that incorporates both the peak avoidance response coefficient and incentives for electricity sales packages to optimize energy distribution and then solved the price mechanism of electricity sales packages to maximize the interests of electricity retailers. In [12], the authors take into account the electricity retailers' electricity purchasing costs and risks on multiple time scales and spatial scales under the environment of a two-tiered electricity market and constructed a prosumer response model with four packages: fixed electricity price, time-of-use electricity price, sharing ratio packages, and back cover packages. In [13], the authors take the average monthly household electricity consumption as an indicator and design a residential electricity package that considers the differentiation of user utility in off-peak seasons. All of the above literature improved users' utility and electricity retailers' expected revenue. Still, because they only optimize the electricity purchase packages for electricity consumers, it is challenging to meet the needs of prosumers to enhance the appeal of electricity packages to boost their market competitiveness. Therefore, there is an urgent need to develop a well-structured and rational pricing strategy for the procurement and sale of electricity that caters to the needs of prosumers.

Currently, the electricity sale price for prosumers is generally determined by a fixed grid connection fee, which follows a uniform pricing scheme that might not adequately encourage prosumers to be responsive [14]. Meanwhile, in terms of product combination sales, there are existing forms of bundling, discount, and purchase and gift sales [15]. Based on this, the authors in [16] propose two models of full-time and time-of-use electricity purchase and sale bundled packages, both of which use electricity discounting for electricity billing. However, the strategy outlined in [16] maintains a constant conversion ratio between electricity buying and selling across all periods without considering the varying dynamics of supply and demand over different times, thus making it challenging for electricity retailers to design a targeted electricity package that aligns with the local supply and demand conditions, thereby maximizing benefits.

In summary, there are some obvious deficiencies in the existing studies on the design of electricity retail packages and the analysis of electricity prosumers' behavior, including (1) insufficiently comprehensive descriptions of prosumer characteristics and a lack of studies to quantify the values of attributes such as prosumer risk preferences; (2) concentration on the development of a single electricity purchasing price package, while neglecting the design of electricity purchasing and selling packages that are tailored to the characteristics of the prosumer; (3) lack of an effective risk assessment model when dealing with uncertainty factors in the electricity market; (4) the traditional particle swarm optimization (PSO) algorithm easily falls into local extremes when solving the electricity package pricing problem, and there are problems of premature convergence and dimensional catastrophe; (5) existing studies seldom take into account the impact of regional supply and demand fluctuations on the design of electricity sales packages and fail to maximize the benefits.

1.2. Contributions

In summary, this paper's contributions can be encapsulated in the following manner:

(1) Aiming at the existing literature only for the users' preference for electricity to design electricity package classification leads to the problem of insufficient package relevance and applicability, taking into account the various aspects of the users' behavior in choosing electricity plans. From the users' preference for electricity demand, price sensitivity, and risk tolerance ability of the three characteristics of the index to quantify and analyze the users' characteristics, an enhanced k-means clustering technique for categorizing user types is employed, thereby enhancing the alignment between user traits and the characteristics of electricity packages.

(2) The existing literature often designs electricity purchase packages for electricity users but seldom designs electricity purchase and sale packages for prosumers. Considering the fluctuation in regional supply and demand, we have formulated short-term electricity purchase and sale combinations of fluctuation pricing and discount pricing to optimize the economic benefits of electricity retailers and the utility of prosumers and, at the same time, promote local renewable energy consumption to a certain extent. Based on the above-differentiated index values, we developed a utility function and an autonomous decision-making model that incorporates economic factors and electricity demand. Additionally, we established a pricing model for electricity retailers' purchase and sale packages, utilizing CVaR (Conditional Value at Risk) to account for the volatility in electricity load and market prices, among other uncertainties.

(3) Addressing the propensity of particle swarm optimization algorithms to become trapped in local optima, we propose a SAW&RL-PSO algorithm, which solves the above model in the form of self-simulation and improves the particle activity to accelerate the convergence of the algorithm and jump out of the local optimal solution. The simulation results confirm that the strategies discussed in this paper are capable of significantly improving the economic performance of electricity retailers targeting prosumers, as well as the utility of electricity consumption for prosumers.

The main variables and symbols used in this paper are defined as shown in Table 1, and the remaining symbols will be defined when they first appear.

Table 1. Summary of Key Notations.

Notation	Description [Units]
d	Electricity demand preference indicators
s	Electricity price sensitivity indicators
$risk$	Electricity price volatility risk tolerance indicator

Table 1. *Cont.*

Notation	Description [Units]
$P_{c,i,t}/P_{v,i,t}/P_{d,i,t}$	Electricity consumption/electricity generation and discounted electricity consumption of type i prosumers at the time t [kWh]
c_{1t}/f_{1t}	Fluctuating pricing package electricity purchase price and residual electricity on-grid price [USD]
$\kappa_{2t,c}/\kappa_{2t,f}$	Discount pricing package electricity consumption discount ratio and power generation incentive ratio
c_{2t}/f_{2t}	Discount pricing package exceeds the specified value of electricity price and exceeds the specified value of surplus electricity on-grid price [USD]
$P_{j,i,t}$	The responsive electricity consumption of type i prosumers when selecting type j package [kWh]
$D'_{j,i}$	Electricity satisfaction of type i prosumers when selecting type j package
$C'_{j,i}$	Electricity cost of type i prosumers when selecting type j package [dollar]
$V_{j,i}$	The utility considering the users' independent choice
$\Pr_{j,i}$	The probability of type i prosumer chooses the type j packages
F_{CVaR}	Risk assessment index of electricity retailer

2. User Characteristic Analysis

In this section, electricity demand preference, electricity price sensitivity, and transaction risk tolerance are selected to provide a relatively comprehensive description of the influencing factors on users' selection of packages and to provide a reference for the design of packages for different types of users in the subsequent section.

2.1. Users' Behavioral Characteristics

2.1.1. Electricity Demand Preference

Considering the differentiation of consumers' preference for electricity demand satisfaction, this paper quantifies the electricity demand preference based on the changes in consumers' electricity consumption behavior before and after participating in the demand response, where the Euclidean distance d serves as a metric to gauge the resemblance between the electricity consumption curves before and after the response, as shown in Equation (1). The smaller the deviation value, the less inclined consumers are to alter their electricity consumption behavioral patterns.

$$d(P_a, P_b) = \sqrt{\sum_{t=1}^{T} (P_{a,t} - P_{b,t})^2} \tag{1}$$

where $P_{a,t}$ and $P_{b,t}$ are the electricity used at t moments before and after the prosumers participate in the response, respectively.

2.1.2. Estimation of Electricity Price Sensitivity

In this paper, the ratio of the rate of change in electricity price and electricity consumption before and after prosumers participate in demand response is used to characterize their sensitivity to electricity price fluctuations [17], as shown in Equation (2), where the more significant the s, the higher the sensitivity of prosumers to electricity price fluctuations.

$$s = \frac{Q_f - Q_0}{C_f - C_0} \cdot \frac{C_0}{Q_0} \tag{2}$$

where C_0 and Q_0 denote the electricity cost and consumption before demand response; C_f and Q_f denote the electricity cost and consumption after demand response, respectively.

2.1.3. Estimation of Risk Tolerance

As a participant in the electricity market, fluctuations in the affordability of electricity price fluctuations mainly come from the users' income level and fluctuations. Usually, consumers with lower income or growth rates have lower risk tolerance and are more inclined toward economic electricity consumption. Therefore, this section introduces the user risk tolerance coefficient *risk*, which represents the ratio of the fluctuation in annual electricity expenditure to the fluctuation in yearly income of prosumers [18], as shown in Equation (3).

$$risk = \frac{\Delta E_{st}/E_{st(t-1)}}{\Delta I_{st}/I_{st(t-1)}} \tag{3}$$

where $E_{st(t-1)}$, $I_{st(t-1)}$ are the actual annual electricity expenditure and actual annual revenue for the previous year, respectively; ΔE_{st}, ΔI_{st} are the fluctuations in yearly electricity expenditure and annual revenue for the current year, respectively.

As can be seen from the above formula, when $risk > 1$ the consumer's electricity expenditure increase is higher than the increase in annual income, indicating that the consumer's ability to withstand changes in electricity costs is weaker and is more inclined to the economic-based approach to electricity use, and, vice versa, is more inclined to meet the demand for an electricity-based approach to electricity use.

2.2. Improved K-Means Clustering

In the k-means clustering technique, which operates in an unsupervised and iterative fashion, k-initial cluster centers are arbitrarily chosen at the outset. Subsequently, the algorithm measures the proximity of each remaining data point to these k centers, categorizes each data point into the cluster associated with the nearest center, and re-determines the clustering centers for each assigned object. The process repeats until the clustering centers no longer change. This method is efficient, has few errors, and is widely used in user energy behavior analysis. However, the inherent randomness in the selection of initial cluster centers in k-means clustering can readily result in a lack of stability in the outcomes. Therefore, this paper uses improved k-means clustering based on local density to analyze the above user behavior indicators [19]; the detailed clustering process will not be repeated here.

Given the shifts in users' behavioral patterns due to fluctuations in family dynamics, the introduction of new electrical appliances, and personal mobility within the trading scope of the electricity retail market, the makeup of user clusters will evolve as new participants join and others exit. Consequently, alterations in user behavior and composition lead to changes in clustering results, which in turn impact the revenue from bundled packages and the overall competitiveness. Therefore, the package designed in this paper is a monthly package that can be both stable and flexible and will be updated after the clustering similarity ψ is lower than the limiting value to meet the users' personalized electricity price needs and to ensure the electricity retailers and the users' self-simulation of the accuracy of the situation.

$$\begin{cases} \psi_a = L_e(A_z, A_0) \\ \psi_b = L_e(B_z, B_0) \\ \vdots \\ \psi_k = L_e(K_z, K_0) \end{cases} \tag{4}$$

where A_0, $B_0 \ldots K_0$ are the cluster center per-unit values of the existing packages of clusters A, $B \ldots K$, respectively; A_z, $B_z \ldots K_z$ are the cluster center per-unit values of the updated clusters of clusters A, $B \ldots K$, respectively; L_e is the distance function.

3. Electricity Package Selection Model

3.1. Electricity Purchase and Sale Combination Packages

3.1.1. Fluctuating Power Purchase and Sale Packages

According to the principle of "self-generation and self-consumption, surplus electricity on-grid", the electricity package with fluctuating pricing utilizes a time-of-use electricity pricing model for the procurement of electricity. The DPV generation offsets part of the electricity consumption at that moment in time by a certain fluctuation discount factor $\kappa_{1t} \in [0,1]$. The surplus electricity is purchased and settled using the surplus electricity on-grid electricity price f_{1t}, and the final settlement cost C_i, as shown in Equations (5) and (6), and consists of the expense incurred from purchasing electricity at rates determined by the time-of-use pricing mechanism (at this time, the electricity generation falls short of consumption and it becomes imperative to buy supplementary electricity from electricity retailers) and the revenue from selling electricity based on the on-grid electricity price (at this time, the amount of electricity generated is more than the amount of electricity consumed, and it is necessary to sell electricity to the electricity retailers).

$$P_{d,i,t} = \kappa_{1t} P_{v,i,t} \tag{5}$$

$$C_i = \sum_{t=1}^{T} \Delta\left[c_{1t}\max\left\{0, P_{c,i,t} - P_{d,i,t}\right\} - f_{1t}\max\left\{0, P_{d,i,t} - P_{c,i,t}\right\}\right] \tag{6}$$

where $P_{c,i,t}$, $P_{v,i,t}$, $P_{d,i,t}$ are the electricity consumption, electricity generation, and discounted electricity consumption of type i prosumers at the time t, respectively. c_{1t} is the electricity purchase price at the time t, and f_{1t} is the residual electricity on-grid electricity price at the time t.

In the above formula, when $\kappa_{1t} = f_{1t}/c_{1t}$, the package settlement cost is the same as the settlement cost of purchasing and selling electricity alone; when $\kappa_{1t} > f_{1t}/c_{1t}$, it indicates that the supply of electricity in the region is less than the demand at this time, and the electricity retailers encourage the producers and consumers to use less electricity and sell more electricity; when $\kappa_{1t} < f_{1t}/c_{1t}$, it indicates that the supply of electricity in the region is more than the demand at this time, and the electricity retailers encourage the producers and consumers to locally consume electricity.

Fluctuating power purchase and sale packages are associated with fluctuations in the region's electricity supply and demand and are subject to a high degree of uncertainty. Therefore, they are more suitable for prosumers with a higher risk tolerance and irregular electricity consumption behavior.

3.1.2. Discounted Power Purchase and Sale Combination Packages

The discount pricing electricity package employs a pricing structure that combines a "discount fee + kWh fee" for the resolution of electricity transactions. The discount fee is the amount of electricity consumed $P_{c,fix.t}$ and generated $P_{v,fix,t}$ within the specified value of the package in the period t. The prosumers are settled according to the agreed electricity price for the discount $\kappa_{2t,c}$ and incentive percentage $\kappa_{2t,f}$ s. In contrast, if the amount of electricity consumed and generated is more than the package, it will be resolved in accordance with the agreed electricity price c_{2t} and f_{2t}, respectively. The cost computation for prosumers under this pricing scheme is depicted in Equation (7), which includes the prosumers' expenditure of purchasing electricity and revenue of selling electricity within the predetermined values, and the extra expenses for purchasing electricity beyond the usage limit and the surplus income from selling electricity beyond the generation limit. This package type is suitable for economic prosumers with relatively stable overall electricity purchases and sales or who are less risk-tolerant.

$$C_i = \sum_{t=1}^{T} \Delta t\left[c_{2t}\kappa_{2t,c}\min\left\{P_{c,fx,t}, P_{c,i,t}\right\} - f_{2t}\kappa_{2t,f}\min\left\{P_{v,fx,t}, P_{V,i,t}\right\} + c_{2t}\max\left\{0, P_{c,i,t} - P_{c,fx,t}\right\} - f_{2t}\max\left\{P_{v,i,t} - P_{V,fx,t}\right\}\right] \tag{7}$$

3.2. Autonomous Choice Behavior of Consumers and Producers Based on Utility Functions

Based on the above-clustered eigenvalues of electricity demand preference, peak/valley electricity price sensitivity, and risk tolerance of prosumers, electricity retailers conduct self-simulation of prosumers' response behavior and autonomous choice behavior, construct an autonomous choice model of users under different electricity packages, then obtain the probability of each prosumer's choice of each electricity package and take the highest probability package of each prosumer as their choice to carry out electricity package formulation.

Based on the analysis of user behavioral characteristics described in Section 1.1, the load response models of multiple types of consumers under different electricity prices are constructed in this section [20,21], as shown in Equation (8).

$$P_{j,i,t} = P_{0,i,t}[1 - S_i \frac{c_{j,t} - c_{0,t}}{c_{0,t}}] \tag{8}$$

where $P_{j,i,t}$ is the responsive electricity consumption of type i prosumers when selecting type j package, $P_{0,i,t}$ is the original electricity consumption of type i prosumers, S_i is the price-sensitive clustering center value of type i prosumers, $c_{j,t}$ is the price of type j package, and $c_{0,t}$ is the original price of electricity.

Considering that different types of prosumers are differentiated in terms of electricity demand and risk preference, this section constructs the utility function of prosumers from the two indicators of electricity demand satisfaction and electricity cost economy. The centroid values of electricity demand and risk preference obtained above are normalized to serve as the weights of the two indicators to portray the utility of prosumers.

Electricity demand satisfaction is characterized using the Euclidean distance before and after the response, and the Euclidean distance $D_{j,i}$ obtained by type i prosumers choosing type j packages is shown in Equation (9).

$$D_{j,i} = \sqrt{\sum_{t=1}^{T} [P_{j,i,t} - P_{0,i,t}]^2} \tag{9}$$

Since the price of electricity is a cost-based indicator, the Sigmoid function is used to standardize it so that the indicator is mapped to the (0,1) interval, as shown in Equation (10). Where $risk_i$ is the clustering center value of type i prosumers on the risk preference indicator, which is used to enhance the importance of economic indicators in the independent choice. When $risk_i > 1$, prosumers are more sensitive to the economy, an increase in cost will lead to a concomitant increase $C'_{j,i}$, thus enhancing the calculated value of the cost indicator term in Equation (13).

$$C'_{j,i} = 1/(1 + e^{-\frac{C_{j,i} - C_{0,i}}{C_{0,i}} \cdot risk_i}) \tag{10}$$

At the same time, considering the impact caused by the difference in the numerical base of electricity demand satisfaction and electricity cost, the following formula is used to standardize (Equation (11)) and normalize (Equation (12)), the electricity demand satisfaction. Where d_i is the clustering center value of type i prosumers in the electricity demand index and $|D_{j,i} - d_i|$ indicates the absolute value of the difference between electricity demand satisfaction and the clustering center, and the more significant the value, the greater the difference in willingness to respond and the lower the utility of the producer–consumer, therefore, redefine $D'_{j,i}$ as the degree of loss of electricity demand.

$$D^{(1)}_{j,i} = \frac{|D_{j,i} - d_i| - \min|D_{j,i} - d_i|}{\max|D_{j,i} - d_i| - \min|D_{j,i} - d_i|} \tag{11}$$

$$D'_{j,i} = \frac{D^{(1)}_{j,i}}{\sum\limits_{j=1}^{J} D^{(1)}_{j,i}} \tag{12}$$

The entropy weight method is used to calculate the weights $w_{j,C}$ and $w_{j,D}$ of the above two indicators, and then we obtain the utility function $U_{j,i}$ of type i prosumers choosing type j packages, as shown in Equation (14). The larger the cost $C'_{j,i}$ and the degree of loss of electricity demand $D'_{j,i}$, the lower the utility of electricity consumption.

$$\begin{cases} w_{j,C} = \dfrac{1 - [-\ln^{-1} J C'_{j,i} \ln C'_{j,i}]}{n - \sum\limits_{j=1}^{J} [-\ln^{-1} J C'_{j,i} \ln C'_{j,i}]} \\ w_{j,D} = \dfrac{1 - [-\ln^{-1} J D'_{j,i} \ln D'_{j,i}]}{n - \sum\limits_{j=1}^{J} [-\ln^{-1} J D'_{j,i} \ln D'_{j,i}]} \end{cases} \tag{13}$$

$$U_{j,i} = 1 - (w_{j,C} C'_{j,i} + w_{j,D} D'_{j,i}) \tag{14}$$

In the electricity retail market, users can make their own choices. A Logit discrete choice model (classification rating model) is usually used to simulate the users' autonomous choice behavior, as shown in Equation (15), which consists of two parts: observable rational utility and unobservable stochastic behaviors, such as user psychology [21].

$$V_{j,i} = U_{j,i} + \zeta_i \tag{15}$$

where $V_{j,i}$ is the utility considering the users' independent choice, and ξ_i is the unpredictable random influence term such as the users' psychology, which are independent of each other and obey the Gumble distribution.

Based on the above equation, the choice probability of the consumer depends on his rational utility, so considering the consumer as an entirely rational person, the probability that the type i prosumer chooses the type j packages is:

$$\mathrm{Pr}_{j,i} = \frac{\exp(U_{j,i})}{\sum\limits_{j=1}^{J} \exp(U_{j,i})} \tag{16}$$

When the electricity price fluctuates in different periods, prosumers will modify and enhance their electricity usage patterns based on the time-of-use electricity price. The lower the electricity price, the more pronounced the incentive effect on prosumers, and the greater the responsive load of prosumers will be. However, when the fluctuation in electricity price exceeds a specific range, the ability of prosumers to adjust is limited, and the response load will level off, so the response load model of prosumers should also consider the corresponding constraints:

$$\underline{k_i} P_{0,i,t} \le P_{j,i,t} \le \overline{k_i} P_{0,i,t} \tag{17}$$

where $\underline{k_i}$ and $\overline{k_i}$ represent the scaling factors, with $\underline{k_i}$ being the upper limit and $\overline{k_i}$ the lower limit, for the capacity adjustment of type i producers and consumers.

At the same time, considering that prosumers should obtain certain benefits from the electricity package, the following limitations must be met:

$$C_{j,i} \le \sum_{t=1}^{T} \Delta t \cdot c_0(t) \cdot P_{0,i,t} \tag{18}$$

4. Electricity Packages Pricing Model

As an interested third party, the electricity retailer maximizes revenue by taking into account the costs of electricity bought from and sold to prosumers, as well as the market prices for electricity transactions. The utility of the user in different packages will affect the probability of their choice, so electricity retailers need to consider meeting the utility to improve the competitiveness of the packages. At the same time, due to electricity market price fluctuations because of grid demand, electricity retailers, in the pursuit of their own interests, will reduce the amount of electricity purchased at price peaks and sold at price troughs in the electricity market and indirectly to meet the needs of the grid to reduce the peak–valley difference and the local consumption of renewable energy.

Concurrently, the development and functioning of the electricity market are still in a nascent stage, with electricity pricing undergoing continuous adjustment and standardization, so it is difficult to accurately describe the market price and its range of fluctuation using historical data. Electricity retailers may face certain risks and potential losses due to the volatility of market electricity prices [22]. In this paper, the use of the Conditional Value at Risk (CVaR) theory to estimate the risk of uncertainty factors and construct a package pricing model for electricity retailers that weighs the risks and benefits can overcome the non-sufficiency of the Value at Risk (VaR) in the measurement of the tail loss, so as to better reflect the potential value of the risk of the electricity retailers' electricity purchase and sale transactions.

Since the probability distribution of uncertainty factors such as the market price of electricity is difficult to obtain, this paper adopts scenario analysis to convert uncertainty factors into combinations of multiple deterministic factors and construct a typical scenario of type n_ω, where the probability of ω the scenario is ρ_ω, and the number of typical scenarios n_ω is determined by the Silhouette Coefficient.

To solve the above model, we optimize the particle swarm algorithm by combining adaptive weighting and a reverse learning strategy. Through adaptive weighting, the algorithm balances convergence speed and finding the global optimal solution; by implementing a reverse learning strategy, we can broaden the search scope and enhance the population's ability to escape from local optima.

4.1. Cvar-Based Integrated Decision-Making Model for Electricity Retailers

In this section, the risk assessment index of electricity retailers in the face of fluctuations in price, load, and other uncertainties is constructed based on CVaR, as shown in the following equation:

$$F_{CVaR} = F_{VaR} + \frac{1}{1-\beta}\sum_{\omega=1}^{n_\omega} \rho_\omega [F_\omega - F_{VaR}]^+ \tag{19}$$

$$[F_\omega - R_{VAR}]^+ = \max(0, F_\omega - R_{VAR}) \tag{20}$$

$$\varpi(F_\omega \leq R^{VaR}) = \beta \tag{21}$$

where F_{VaR} is the maximum risk loss that electricity retailers may assume under confidence level β; F_ω is the loss function associated with scenario ω; and $\varpi(x)$ represents the probability of the event occurring.

Considering the benefits and risks of electricity purchase and sale at the same time, the target utility function of electricity retailers can be transformed into:

$$\max F = \sum_{\omega=1}^{n_\omega} \rho_\omega (R - U_\omega) - \gamma F_{CVaR} \tag{22}$$

$$R = \sum_{i=1}^{k} C_i \tag{23}$$

$$U_\omega = \sum_{t=1}^{T} \Delta t(c_{r,t} \cdot G_{r,t} + c_{\omega,s,t} \cdot G_{\omega,s,t}) \tag{24}$$

where R represents the total revenue of electricity retailers from the sale of electricity packages, which is related to the type and number of packages sold; U_ω represents the cost of the electricity purchased by electricity retailers in the electricity market under scenario ω; $c_{r,t}$ is the price of electricity purchased and sold in the day-ahead market, which is related to the ratio of supply and demand in the electricity market as shown in [23]; $G_{r,t}$ is the amount of electricity purchased and sold in the day-ahead market for the purpose of balancing electricity by the electricity retailers; $c_{\omega,s,t}$ is the electricity purchase price in the real-time market under scenario ω, which is generally higher than the previous day's price; $G_{\omega,s,t}$ is the amount of electricity purchased from the real-time market by electricity retailers in order to balance the uncertainty of the current day's electricity under scenario ω; and γ denotes the risk aversion degree of the electricity retailers, and the larger the value is, the higher the risk aversion degree of the electricity retailers, and the more the electricity retailers tend to avoid the risk.

The decision-making model for electricity retailers aims to maximize the utility function, subject to the subsequent constraints:

(1) Constraints on electricity balance

$$G_{\omega,t} + \sum_{i=1}^{n_c} P_{c,i,t} = \sum_{i=1}^{n_p} P_{pv,i,t} \tag{25}$$

where n_c and n_p are the total number of prosumers using and generating electricity at the moment t, respectively.

(2) Constraints on the pricing of the electricity package for purchases and sales

$$\underline{k_c} c_{0t} \le c_{1t}, c_{2t} \le \overline{k_c} c_{0t} \tag{26}$$

$$\underline{k_f} f_{0t} \le f_{1t}, f_{2t} \le \overline{k_f} f_{0t} \tag{27}$$

where c_{0t} denotes the initial purchased electricity price of the consumer in the period t; f_{0t} denotes the initial sold electricity price of the consumer in the period t; $\underline{k_c}$ and $\overline{k_c}$ denote the upper and lower scaling factors of the purchased electricity price of the consumer in the period t; $\underline{k_f}$ and $\overline{k_f}$ denote the upper and lower scaling factors of the sold electricity price of the consumer in the period t.

(3) Peak-to-valley difference constraint

$$P_{\max} - P_{\min} \le P_{0,\max} - P_{0,\min} \tag{28}$$

where $P_{0,\max}$ and $P_{0,\min}$ are the peak and trough values of the prosumers' load before implementing the combined electricity purchase and sale package, respectively.

4.2. SAW&RL-PSO Optimization Algorithm

Based on the above decision-making models of prosumers and electricity retailers, the electricity retailers construct a two-layer optimization model of electricity retailers and prosumers by self-simulating the autonomous choice behavior of prosumers for solving the pricing of the combined package of purchasing and selling electricity. The framework is shown in Figure 1, where the lower level model is dominated by type k prosumers, considering the lowest cost of electricity and the maximum comfort in choosing the electricity package; the upper model is centered around the electricity retailers as the primary entity, taking into account the anticipated profit and potential loss to optimize the parameters for combined purchasing. The electricity retailers will send the optimized electricity purchase and sale combination package price to various types of typical prosumers. The prosumers will select the optimal package based on their parameters and electricity price to calculate the maximum benefit, pass the purchased and sold electricity to the electricity retailers, and optimize to find the optimal solution.

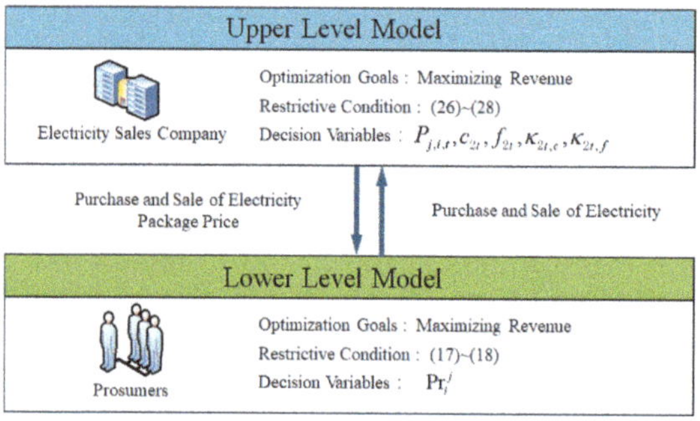

Figure 1. Schematic diagram of iterative optimization for electricity retailers and prosumers.

For the above model, this paper adopts the improved particle swarm algorithm for solving. Since the traditional particle swarm algorithm has limitations, such as the existence of premature convergence, dimensional catastrophe, and easy falling into the local extremes, this paper combines adaptive weighting with reverse learning strategy and proposes the SAW&RL-PSO optimization algorithm, which enhances the dynamism of particles, aiding in the escape from local optima within the solution space.

The reverse learning strategy, introduced by Tizhoosh, functions as a mechanism that rapidly expands the search space [24], which generates an inverse candidate solution in contrast to the existing one, simultaneously exploring both the current and the inverse optimal solutions. This dual approach extends the search scope of the population and enhances its capacity to evade local optima. If there exists a feasible solution x in the search space, the formulas for its reverse solution are shown in Equations (29) and (30) [25,26].

$$\tilde{x}_{ij}^t = k_{ij}^t + l_{ij}^t - x_{ij}^t \tag{29}$$

$$\begin{cases} k_{ij}^t = \min(x_{ij}^t) \\ l_{ij}^t = \max(x_{ij}^t) \end{cases} \tag{30}$$

where $\tilde{x}_{ij}^t$ is the position of the i th particle in the j th dimension at the t th iteration of the optimization search, in this paper, $j = 1$; k_{ij}^t is the historical minimum of the i th particle in the j th dimension at the t th optimization search, and l_{ij}^t is the historical maximum of the i th particle in the j th dimension at the t th optimization search.

The introduction of sinusoidal adaptive weights, so that the weights change with the trend of the sinusoidal function, can enhance the local search ability of the particles compared to the linearized weights. The formula of the sinusoidal adaptive weights is shown in Equation (31), and the weights change as the value of the particle fitness changes.

$$w = \begin{cases} w_{\min} + (w_{\max} - w_{\min})\sin(\pi \cdot \frac{f - f_{\min}}{f_{avg} - f_{\min}}), f \leq f_{avg} \\ w_{\max}, f > f_{avg} \end{cases} \tag{31}$$

where f is the current adaptation value, f_{avg} is the average adaptation value, $f_{\min}$ is the minimum adaptation value, and $w_{\min}$, $w_{\max}$ represent the minimum and maximum bounds for the weights.

The specific process is shown in Figure 2.

(1) Initialize the speed and position of the particles (initialize the parameters such as the data of electricity generation and consumption of consumers and the initial price of electricity purchase and sale) and randomly initialize the pricing of electricity purchase and sale packages.

(2) During the iterative optimization phase of the particle swarm, according to the position of the particles in each optimization, calculate the electricity purchase and sale strategies of various types of prosumers and pass them to the electricity retailers, who calculate the utility function based on the returned results.

(3) Calculate the particle fitness F under this optimization search, update the speed and position of the particle swarm according to the speed and position updating formula, carry out inverse learning for the particle swarm, obtain the corresponding inverse candidate solutions according to Equations (29) and (30), and find out the adaptive weight ω that changes with the particle fitness F according to Equation (31).

(4) Judge the number of iterations and error range. Within the range of iteration times, multiple optimization is carried out to obtain the individual optimal value of the particle and the group optimal value. The final particle position corresponding to the group optimal value is the optimal solution of the package price.

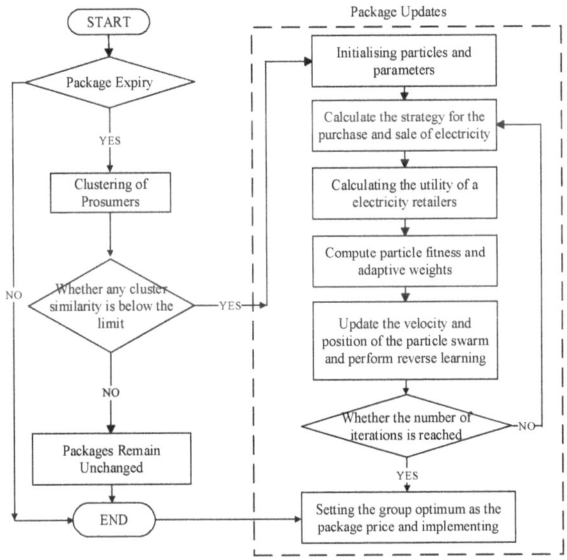

Figure 2. Schematic diagram of the process for developing packages for electricity retailers.

5. Algorithm Analysis

In this paper, based on the electricity consumption data and social attributes of 100 households of prosumers researched by State Grid Hebei Electric Power Co., Ltd., in Shijiazhuang, Hebei Province, China. Supplemented by the "Hebei Statistical Yearbook—2021" issued by Hebei Provincial Bureau of Statistics, user characteristic metrics are computed and clustered and analyzed to solve for the pricing of combined electricity purchase and sale packages of electricity retailers. Among them, the sampling interval of load data is 1 h, and the number of daily sampling points is T = 24. It is assumed in the simulation that the type of each user's electricity consumption curve is a single-peak type, and the rest of the types can be generalized from this. With reference to Hebei Province, in the context of time-of-use electricity pricing, the peak hour electricity rate is set at 0.76×10^{-1} USD per kWh, while the valley hour rate is 0.41×10^{-1} USD per kWh. The grid-connected electricity price for DPV systems is 0.58×10^{-1} USD per kWh. A substantial number of day-ahead pricing scenarios are created using Monte Carlo simulation techniques, which are then utilized to compute the Conditional Value at Risk (CVaR) at a 90% confidence level.

5.1. Analysis of Clustering Simulation Results

Based on the improved k-means clustering, the clustering effect is analyzed by evaluating the contour coefficient SC, and the trend of SC with the value of K is shown in Figure 3. After K > 3 the value of SC decreases gradually, which indicates that the distance between different samples is gradually close to each other. When K = 3, the SC value is the highest; it is shown that the clustering is currently best when the clusters are dense within clusters and sparse between clusters.

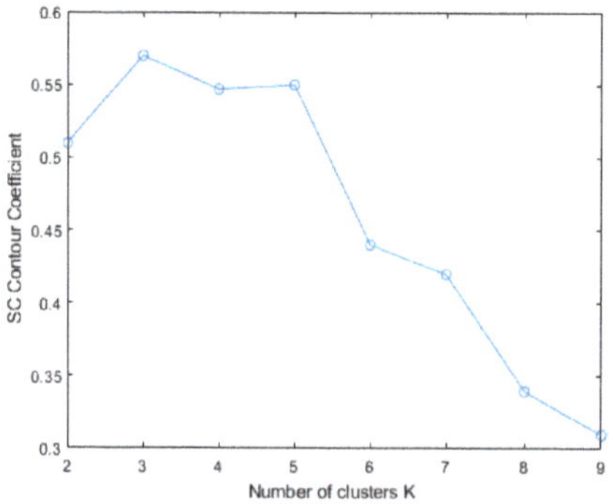

Figure 3. Variation trend of SC with K value fluctuation.

The clustering algorithm is applied to categorize users into economic users (A), smooth users (B), and enjoyment users (C). Economic users have higher electricity price sensitivity, electricity demand preference, and risk tolerance, indicating that this type of user has a weaker risk tolerance, and the behavior and economic cost of electricity before and after the implementation of demand response have more substantial changes, and a more vital willingness to respond. Smooth-type users are more dispersed, risk tolerance is appropriate, electricity price sensitivity and electricity demand preference are more random, and electricity behavior usually has more significant uncertainty. Enjoyment-type users have a higher risk tolerance, higher income, and pay more attention to the satisfaction of electricity demand, so their sensitivity to electricity price is not high, and their willingness to respond is weaker.

5.2. Analysis of Simulation Results of Electricity Purchase and Sale Prices

In this paper, scenario analysis is used to convert uncertainties in the market into combinations of multiple deterministic factors, and eight typical scenarios are constructed, as depicted in Figure 4.

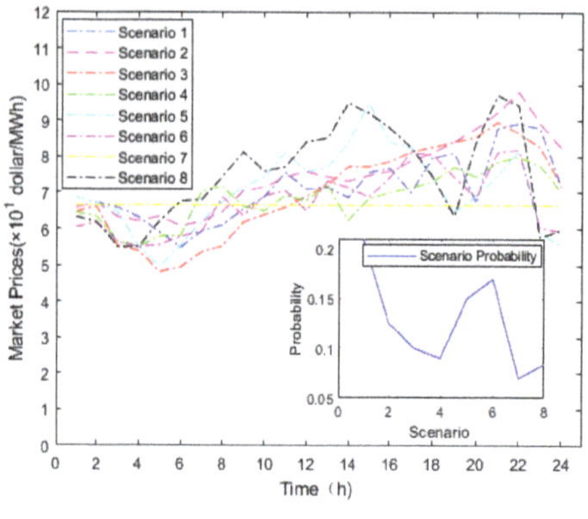

Figure 4. Electricity price curves for each scenario in the day-ahead market.

Following the computation of the optimization model, the pricing of the fluctuation pricing (Package 1) and discount pricing (Package 2) electricity purchase and sale combination packages for the three categories of prosumers, A, B, and C, is shown in Figures 5–8, and for the convenience of description, the fluctuating pricing packages for the categories of prosumers of A, B, and C are subsequently referred to as A1, B1, and C1. The discount pricing packages for the categories of prosumers of A, B, and C are referred to as A2, B2, and C2.

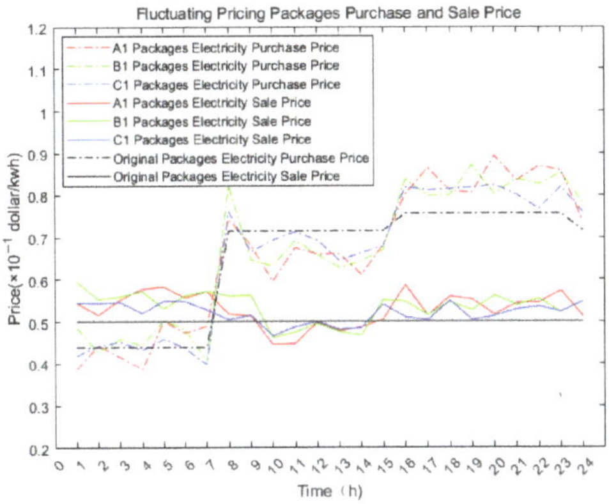

Figure 5. Fluctuating pricing packages electricity prices for three types of prosumers.

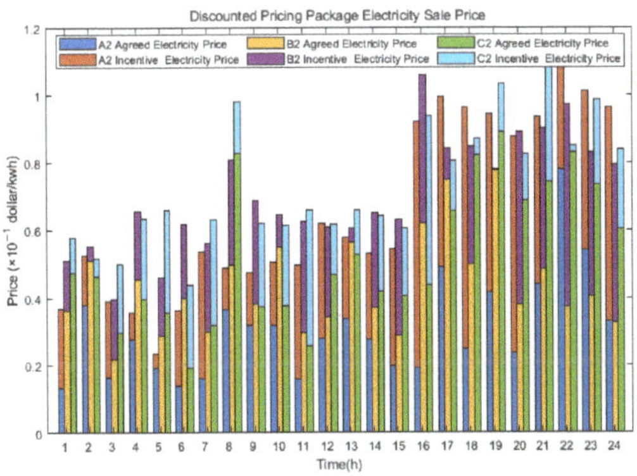

Figure 6. Discount pricing package electricity sale prices for three types of prosumers.

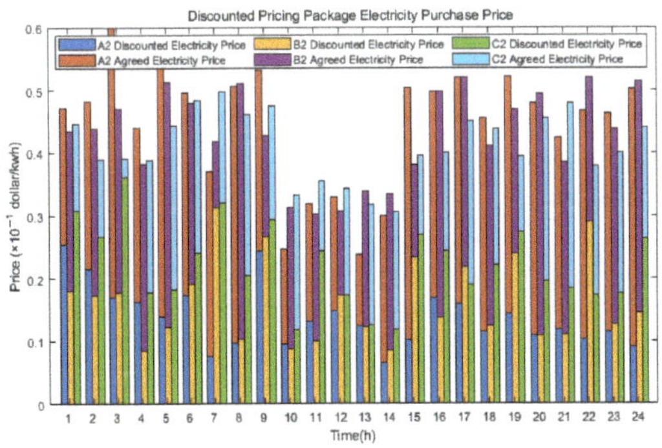

Figure 7. Discount pricing package electricity purchase prices for three types of prosumers.

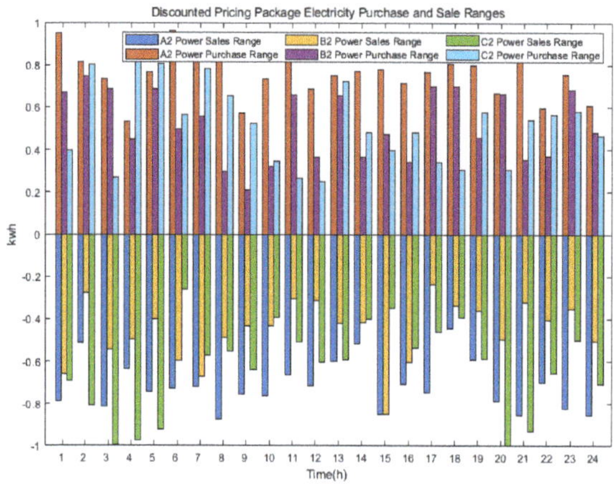

Figure 8. Load range of discount pricing package purchases and sales for three types of prosumers.

From the results of the electricity prices solution, it can be seen that due to the differentiation of the characteristic parameters of the three types of electricity consumption behavior, as depicted in Figure 5, the proposed electricity package for the type A prosumers, who have a more vital willingness to respond and are more sensitive to the electricity price, has more prominent peaks and valleys, and has a higher electricity purchase price in the peak hours and a lower electricity trading price in the valleys. Consequently, electricity retailers can incur a minor cost sacrifice to further optimize the electricity usage by prosumers, thereby diminishing their expenses on electricity purchases during peak market hours.

As shown in Figures 6–8, the discount pricing package, compared with the fluctuating pricing package, adds a new discount/reward price and load interval on top of the agreed price. It uses the discount/reward price when the prosumers' consumption/generation load is within the load interval and settles the bill using the agreed price when it exceeds it.

It can be seen that the agreed price, discount/reward factor, and load range in the discount pricing package are interconnected. As shown in Figure 8, the load range of the type A package is more extensive, and the agreed price and discount factor of its electricity purchase are also higher, which will make the prosumers who adopt this package more willing to spread the load and maintain a fixed load usage. Therefore, electricity retailers can decrease their day-ahead market purchases during the peak hours by selling excess electricity into the real-time market, thereby generating additional revenue. Consequently, they can minimize both their day-ahead market peak-hour purchases and their real-time market interventions, leading to increased profitability. Meanwhile, B packages have smaller load ranges but also lower discounted prices, while C packages rely on higher prices to cover the cost of electricity for those who choose them. All three types of packages reduce the cost of electricity for prosumers who choose them and magnify the difference in revenue before and after the load change to promote load diversification, thus improving the competitiveness of the packages of the electricity retailers and the revenue of the packages.

5.3. Analysis of User Cost Simulation Results

Table 2 selects type A prosumers to compare the cost and utility values of electricity consumption under different packages calculated in this paper and the initial time-of-use electricity pricing package. It is observable that the electricity packages introduced in this paper offer improved cost-effectiveness and utility in electricity consumption when contrasted with the initial package. Meanwhile, the A1 and A2 packages designed for type A users exhibit the minimum cost and the maximum utility among all packages, which proves the effectiveness of the categorized packages.

Table 2. Comparison of electricity costs and comfort for type A prosumers under multiple packages.

	Electricity Cost (USD/Month)	Electricity Utility	Selection Probability	Selection Result
A1 Packages	20.70	0.84	18.1%	A2 Packages
B1 Packages	21.74	0.76	16.2%	
C1 Packages	22.24	0.74	15.8%	
A2 Packages	19.95	0.87	18.7%	
B2 Packages	22.01	0.71	15.2%	
C2 Packages	21.51	0.74	15.8%	
Initial Package	26.32	0.61	/	

Comparing the fluctuation pricing package and the discount pricing package, it can be seen that the discount pricing package is more suitable for the characteristics of type A consumers who have a weaker risk acceptance ability, a greater preference for electricity demand, and long-term stability of electricity load. Prosumers who have a more vital willingness to change and can regulate their loads within the discount range of the electricity retailers will have lower electricity costs when choosing the discount package, and at the same time, it can further reduce the amount of electricity purchased by electricity retailers during the peak hours of the electricity market price, and guide the prosumers to stabilize their electricity consumption behavior, so that the electricity retailers' forecasts of the amount of electricity to be purchased and sold will be more accurate, which significantly reduces the high real-time purchase cost of electricity retailers in the spot market.

Table 3 presents a comparative analysis of the monthly expenses for three types of consumers (46 households in type A, 25 households in type B, and 19 households in type C) in Scenario 1 (users are not categorized), Scenario 2 (users are classified only by their electricity demand preferences [8]), and Scenario 3 (the strategy of this paper is adopted), respectively. As can be seen from the table, compared to the user cost of the initial time-of-use electricity price in Scenario 1, both Scenario 2 and Scenario 3 are reduced. Under Scenario 3, type A prosumers are more willing to respond, and their cost reduction reaches 18.93%, while type C prosumers are less sensitive to the tariff, and their costs are reduced by only 5.63%. Under the price obtained from the strategy in this paper, the package category selected by each scenario prosumer corresponds to its category attributes, and the ratio of the number of matching prosumers to the total number is taken as the matching degree, and Scenario 1 is not solved for matching degree because the users are not categorized. Scenario 2 fails to accurately grasp the users' package selection behavior due to using only one type of user characteristic for user classification, resulting in a lower degree of fitness and utility of users and packages and a slightly higher cost of electricity for users. The package proposed in this paper has a high degree of matching under multidimensional feature clustering, which not only effectively reduces its electricity cost but also has only a small degree of impact on its electricity comfort, indicating that the package in this paper has a high degree of suitability and relevance.

Table 3. Electricity costs and fitness evaluation across various scenarios.

	Scenario 1	Scenario 2	Scenario 3
Type A Prosumers (USD)	1062.24	944.42	860.82
Type B Prosumers (USD)	614.47	603.93	565.16
Type C Prosumers (USD)	495.64	493.05	467.70
Average Electricity Utility	0.63	0.74	0.83
Fitness	/	79%	100%

5.4. Analysis of Electricity Retailers's Income Results

Figure 9 illustrates the contrast between the external electricity purchases and sales in the region under the initial time-of-use pricing versus the strategy proposed in this document. The figure demonstrates a significant reduction in external electricity transactions within the area adopting the plan outlined in this paper. The DPV self-consumption-to-DPV electricity generation ratio is used to calculate the proportion of DPV consumption. As the pricing package proposed in this paper takes into account the fluctuation of regional supply and demand and promotes local consumption and regional balance of the prosumers, the proportion of DPV consumption is increased from 55.29% to 68.12%. This approach can significantly enhance the local utilization rate of renewable energy in the area, curtail the capacity for distributed DPV integration, and prevent reverse power overloads. It also proves that the differentiated package design proposed in this paper can more accurately meet the needs of different prosumers, thereby incentivizing prosumers to optimize their electricity consumption behavior and reduce their reliance on external electricity resources. In addition, by setting up discounted and fluctuating packages, the prosumer can obtain lower prices or additional incentives under certain conditions, which encourages them to adjust their electricity consumption patterns and, thus, the implementation of the new strategy promotes, to a certain extent, the effective consumption of renewable resources in the region, improves the matching of power supply and demand in the region, and enhances the efficiency and stability of the power grid operation.

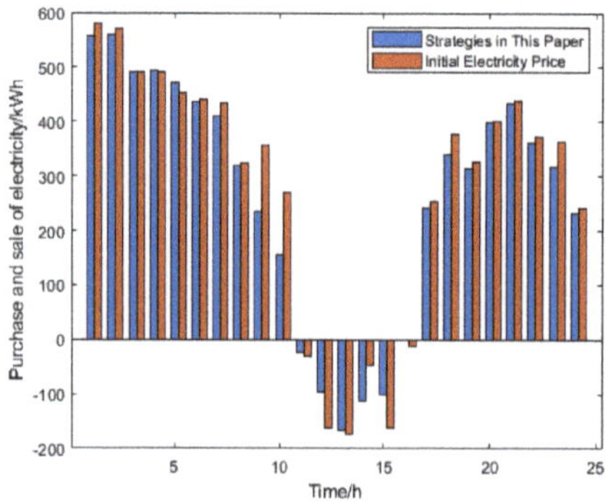

Figure 9. Electricity purchased and sold in the region before and after the implementation of the packages.

To verify the effectiveness of this paper's strategy on the revenue enhancement of electricity retailers, four simulation scenarios are designed. (1) Scenario 1: the initial time-of-day price is used, where the electricity selling price is the initial DPV on-grid price; (2) Scenario 2: this paper's strategy; (3) Scenario 3: the electricity selling package is considered to be independently selected by the users, and the electricity selling price is the initial DPV on-grid price [8]; (4) Scenario 4: the electricity selling package is based on the full-time and split-time generation and consumption bundled packages [16]. Table 4 shows the electricity retailers' monthly economic returns and consumption ratios under different scenarios. Scenarios 1 and 3 have significantly lower returns for the electricity retailers because they do not consider the prosumers' electricity sales attributes and do not provide enough incentives for the willingness of the prosumers to sell electricity. Although Scenario 4 designs an electricity generation and consumption bundling package for prosumers, it does not take into account regional supply and demand fluctuations and behavioral characteristics clustering of prosumers, and the package is not sufficiently targeted to make an optimal decision, resulting in a slightly lower return and consumption ratio than the strategy in this paper.

Table 4. Comparison of income and consumption ratio of electricity retailers under different scenarios.

	Scenario 1	Scenario 2	Scenario 3	Scenario 4
Revenue (USD)	1439.69	1651.22	1538.69	1603.56
Consumption ratio	55.29%	68.12%	58.40%	61.37%

5.5. Convergence Analysis

The algorithm proposed in this paper is analyzed for convergence, with Traditional Particle Swarm Optimization (PSO) and Adaptive Weight Particle Swarm Optimization (AW-PSO) being utilized as comparative benchmarks. As shown in Figure 10, the standard PSO algorithm converges after 60 iterations, AW-PSO converges through 52 iterations, while the SAW&PL-PSO proposed in this paper reaches convergence after 41 iterations. From the obtained objective function results, SAW&PL-PSO is closer to the global optimal solution than the two compared algorithms, which indicates the lack of reasonableness of the values of the weighting parameters taken in PSO and AW-PSO. In contrast, by employing sinusoidal adaptive weights in conjunction with the reverse learning strategy, the issues of slow convergence and entrapment in local optima, which can arise from the irrational selection of weight parameters, are circumvented. This approach extends the search range, thereby accelerating the search for the global optimum.

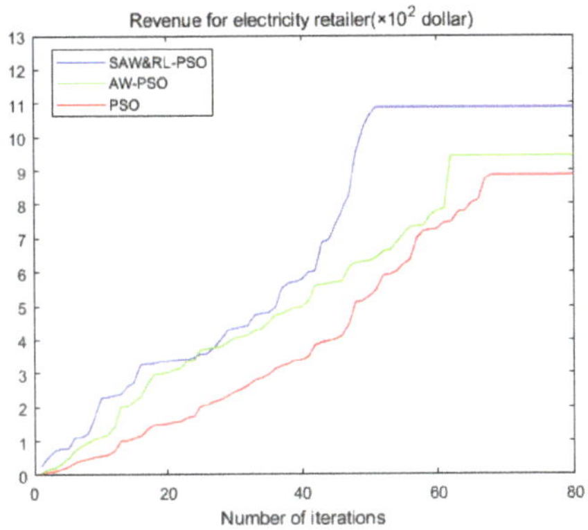

Figure 10. Convergence effect analysis.

6. Conclusions

In this paper, against the background of the rapid development of new-type power systems and the high percentage penetration of DPV, and in response to the new challenge of the transformation of prosumers, a method for designing electricity purchase and sale packages for electricity retailers based on prosumer characteristics is proposed.

Through in-depth analysis of the three key dimensions of electricity demand preference, electricity price sensitivity, and risk tolerance, this paper not only constructs a comprehensive prosumer characteristics model but also adopts an improved k-means clustering algorithm to achieve a precise division of prosumer types, which can capture the diversity of prosumer behaviors in a more detailed way and provide a new perspective for electricity retailers to design and optimize the purchase and sale of electricity packages. In terms of package design, this paper proposes two types of short-term power purchase and sale packages: fluctuating and discounted, which take into full consideration the power generation and consumption attributes of prosumers as well as the fluctuation of supply and demand in the electricity market.

By constructing a utility function-based model of prosumer self-selection behavior, this paper further deepens the understanding of prosumer response behaviors and provides theoretical support for electricity retailers to develop more attractive packages. Further, this paper also establishes a pricing decision model for electricity purchase and sale packages of electricity retailers considering conditional risk, which is solved by the SAW&RL-PSO algorithm. The model not only considers the uncertainty and risk of the electricity market but also improves the search efficiency and global optimization capability of the algorithm through sinusoidal adaptive weighting with an inverse learning strategy.

Simulation examples validate the proposed packages, which provide more choices for electricity prosumers and guide them to optimize their electricity consumption. This approach not only reduces the economic cost to electricity prosumers but also improves their satisfaction with electricity consumption. In addition, the strategies outlined in this paper enable electricity retailers to better meet the varying needs of their customers and adapt to market changes, thereby increasing their revenues. On a broader scale, by taking into account peak-to-valley differences and supply–demand dynamics, the model proposed in this paper is effective in mitigating peak-to-valley differences, increasing local consumption rates, and reducing the operational stress on the grid caused by DPV connections.

It should be noted that although the research in this paper has achieved certain results, there are some limitations. Firstly, the parameter settings of the model and the performance of the algorithm may need to be adjusted and optimized according to the specific conditions of different regions. Future research can further explore the package design strategy under different power market structures and the optimization of algorithm performance on the basis of this paper. In addition, with the continuous development of the electricity market environment and technology, the in-depth study of prosumer behavior and the dynamic adjustment of the model are also important directions for future research.

Author Contributions: Conceptualization, X.W., C.L., B.W., W.W., Y.S., J.P., X.L. and K.Z.; methodology, X.W., C.L., B.W., W.W., Y.S., J.P., X.L. and K.Z.; software, X.W., C.L., B.W., W.W., Y.S., J.P., X.L. and K.Z.; writing—original draft preparation, X.W., C.L., B.W., W.W., Y.S., J.P., X.L. and K.Z. All authors have read and agreed to the published version of the manuscript.

Funding: This research was supported by State Grid Hebei Electric Power Co., Ltd. science and technology project (kj2022-066).

Data Availability Statement: The data for this article come from the electricity consumption data and social attributes of 100 households of prosumers researched by State Grid Hebei Electric Power Co., Ltd. and <Hebei Statistical Yearbook 2021> published by Hebei Provincial Bureau of Statistics.

Conflicts of Interest: Author Xiaotian Wang, Chuang Liu, Binbin Wu and Wei Wang was employed by the company State Grid Hebei Electric Power Co., Ltd., Marketing Service Center. The remaining authors declare that the research was conducted in the absence of any commercial or financial relationships that could be construed as a potential conflict of interest. The authors declare that this study received funding from State Grid Hebei Electric Power Co., Ltd. science and technology project. The funder had the following involvement with the study: Conceptualization, Methodology, Software and Writing—original draft.

References

1. Kou, Y.; Liu, Y.; Guo, L. Ideas on construction of electricity retail market under spot market mode. *Power Syst. Technol.* **2021**, *45*, 3417–3423.
2. Yuan, H.; Dong, X.; Liu, Q.; Jing, L.; Zhang, W.; Yang, Z. Electricity retail side market framework design in national unified electricity market system. *Power Syst. Technol.* **2022**, *46*, 4852–4862.
3. Zhang, Z.; Wang, Y.; Lin, Z.; Ma, Y.; Lu, F.; Yang, L. Electricity plan recommendation method based on implicit score of electricity plan and user portrait. *Autom. Electr. Power Syst.* **2023**, *47*, 91–101.
4. Meng, S.; Sun, W.; Han, D.; Zhang, W.; Yang, W.; Xiao, M.; Xue, G. Design and implementation of decentralized power transaction mechanism to spot market. *Power Syst. Prot. Control* **2020**, *48*, 151–158.
5. Khorasany, M.; Mishra, Y.; Ledwich, G. A decentralized bilateral energy trading system for peer-to-peer electricity markets. *IEEE Trans. Ind. Electron.* **2020**, *67*, 4646–4657. [CrossRef]
6. Khorasany, M.; Najafi-Ghalelou, A.; Razzaghi, R. A framework for joint scheduling and power trading of prosumers in transactive markets. *IEEE Trans. Sustain. Energy* **2020**, *12*, 955–965. [CrossRef]
7. He, Y.; Wang, M.; Yu, J.; Gao, X. Research on hybrid optimal selection of electricity price packages based on power user characteristics and multi-attribute utility. *Price Theory Pract.* **2020**, *12*, 44–48+162.
8. Xiao, B.; Cui, H.; Jiang, Z.; Zhao, X.; Sun, L. Customized electricity price package design based on limited rational user selection behavior. *Power Syst. Technol.* **2021**, *45*, 1050–1058.
9. Wang, H.; Zhang, G.; Liu, J.; Sun, H.; Yang, B. Design of differentiated charging package for electric vehicle based on Stackelberg game considering user contribution. *Electr. Power Autom. Equip.* **2022**, *42*, 21–29.
10. Wu, J.; Zhang, J.; Pan, S.; Wu, F.; Li, K.; Yao, X.; Yang, L.; Zhan, W.; E, Y. Standard retail tariff design based on customer clustering in electricity spot market. *Price Theory Pract.* **2019**, *12*, 132–136.

11. Liu, Q.; Hu, S.; Zhou, W. Optimal design of electricity plans based on electricity retailers' participation in spot market. *Electr. Power Eng. Technol.* **2022**, *41*, 19–25.
12. Qiu, D.; Papadaskalopoulos, D.; Ye, Y.; Strbac, G. Investigating the effects of demand flexibility on electricity retailers' business through a tri-level optimisation model. *IET Gener. Transm. Distrib.* **2020**, *14*, 1739–1750. [CrossRef]
13. Dong, J.; Zhou, X.; Liang, C. Residential electricity package model considering user utility in off-peak and peak seasons. *Autom. Electr. Power Syst.* **2021**, *45*, 57–62.
14. Zhang, H.; Han, D.; Liu, T.; Huang, Y. Analysis of market coupling mechanism between distributed photovoltaic penetration and electricity market under the background of carbon neutrality. *J. Shanghai Jiaotong Univ.* **2023**, *57*, 464–472.
15. Wang, M.; He, Y.; Lu, Y. Peak-valley combination electricity package design for electricity retailer considering bounded rationality of consumers. *J. North China Electr. Power Univ. (Soc. Sci.)* **2021**, *1*, 48–55.
16. Chen, Q.; Gao, H.; Wang, X.; Liu, J.; Wang, W. Prosumer oriented multiple generation-consumption electricity package optimization for electricity retailers. *Electr. Power Autom. Equip.* **2023**, *43*, 154–162.
17. Xu, G.; Zhang, L.; Liang, J.; Sun, D.; Zhao, L.; Han, X. Estimation of Price Elasticity for Residential Electricity Demand Based on Electricity Consumption Features of Appliances. *Electr. Power Autom. Equip.* **2020**, *44*, 48–55.
18. Lin, G.; Qian, B.; Guo, L.; Sun, H. An adjustable space model of household electricity price with residents' affordability considered. *Power Syst. Clean Energy* **2014**, *30*, 42–46+51.
19. Thimmapuram, P.R.; Kim, J. Consumers' price elasticity of demand modeling with economic effects on electricity markets using an agent based model. *IEEE Trans. Smart Grid* **2013**, *4*, 390–397. [CrossRef]
20. Shareef, H.; Ahmed, M.S.; Mohamed, A.; Al Hassan, E. Review on home energy management system considering demand responses, smart technologies, and intelligent controllers. *IEEE Access* **2018**, 24498–24509. [CrossRef]
21. Lu, E.; Bie, P.; Wang, H.; Chen, Q.; Li, W. Pricing strategy design of electricity retail pricing package considering self-selectivity of users. *Electr. Power Autom. Equip.* **2020**, *44*, 177–184.
22. Etukudor, C.; Robu, V.; Couraud, B.; Kocher, G.; Früh, W.-G.; Flynn, D.; Okerekeet, C. Automated negotiation for peer-to-peer trading of renewable energy in off-grid communities. *Energies* **2020**, *13*, 920. [CrossRef]
23. Liu, N.; Yu, X.; Wang, C.; Li, C.; Ma, L.; Lei, J. Energy-Sharing Model With Price-Based Demand Response for Microgrids of Peer-to-Peer Prosumers. *IEEE Trans. Power Syst.* **2017**, *32*, 3569–3583. [CrossRef]
24. Qian, X.; Fang, W. Opposition-based learning competitive particle swarm optimizer with local search. *Control Decis.* **2021**, *36*, 779–789.
25. Richter, B.; Mengelkamp, E.; Weinhardt, C. Vote for your energy: A market mechanism for local energy markets based on the consumers' preferences. In Proceedings of the 2019 16th International Conference on the European Energy Market, Ljubljana, Slovenia, 18–20 September 2019.
26. Song, M.; Amelin, M. Price-maker bidding in day-ahead electricity market for a retailer with flexible demands. *IEEE Trans. Power Syst.* **2018**, *32*, 1948–1958. [CrossRef]

processes

MDPI

Article

Robust Economic Management Strategy for Power Systems Considering the Participation of Virtual Power Plants

Xin Li [1,*], Chen Zhang [2], Qianqian Yi [3] and Jingwei Xu [4,*]

[1] College of Economics & Management, China Three Gorges University, Yichang 443002, China
[2] State Grid Energy Research Institute Co., Ltd., Beijing 102200, China; zhangchen1@sgeri.sgcc.com.cn
[3] Bishan Power Supply Branch of State Grid Chongqing Electric Power Company, Chongqing 402760, China; qianqqian123@163.com
[4] College of Electrical Engineering and New Energy, China Three Gorges University, Yichang 443002, China
* Correspondence: lx13142024@163.com (X.L.); xujw678@163.com (J.X.)

Abstract: The rapid development of new energy technologies and the randomness and intermittency of renewable energy sources, coupled with the complexity of demand-side response, have posed higher requirements for the security and scheduling of power grids. The virtual power plant (*VPP*), as an innovative energy management model, effectively enhances the flexibility and economy of the power system by integrating distributed energy resources and demand-side response resources to participate in the power market and grid operation as a virtual entity. This paper proposes a robust economic management strategy for power systems considering the participation of *VPP*s. Firstly, the uncertainty of renewable energy output is characterized through prediction deviations. Subsequently, the economic dispatch and constraints of *VPP*s are established. Considering the nonlinear characteristics of the model, an improved ant lion optimizer (ALO) algorithm is adopted to ensure optimal solutions. Finally, the effectiveness and feasibility of the proposed method are validated through a case study, using an improved IEEE 30-bus test system. Compared to other methods, the economic income can be increased by up to 3.2%, and the calculation time is only 12.5% of that required for scenario screening-based stochastic optimization. After introducing the carbon trading mechanism, carbon emissions are reduced by 27.7%. The proposed method achieves the goal of maximizing economic benefits while ensuring the secure and stable operation of the power grid.

Keywords: renewable energy sources; virtual power plant (*VPP*); economic management strategy; improved ant lion optimizer (ALO) algorithm

1. Introduction

With the shortage of fossil fuels and the deterioration of the ecological environment, various renewable and clean energy sources have garnered significant attention. Among them, wind power generation, backed by rapid technological advancements and superior cost-effectiveness, has received strong support from countries around the world and has become one of the main energy sources in power systems [1]. However, the randomness and intermittency of wind power output adversely affect the secure and economic operation of the system. Relying solely on conventional units to mitigate this randomness may lead to frequent start-ups and shut-downs, or significant increases and decreases in output, thereby increasing the difficulty of scheduling and operation. Hence, there is an urgent need for demand-side resources to participate in scheduling [2]. The virtual power plant (*VPP*), by aggregating distributed energy resources and adjustable loads, enhances its operational

 https://doi.org/10.3390/pr13010025

stability and promotes the integration of renewable energy. As a new distributed energy grid connection technology, the *VPP* can mitigate the impact of the randomness of wind and solar power output by optimizing unit output and tapping into the adjustment potential of controllable loads, thereby ensuring smooth system operation [3].

In the early stages of its development, the virtual power plant (*VPP*) primarily served as a means of connecting single-type distributed energy resources to the grid. The concept of the *VPP* originated from the notion of "virtual public utilities" proposed by Dr. Shimon in 1997, which emphasized the use of flexible cooperation methods by non-asset entities, driven by market forces, to provide specific public services to consumers [4]. "Virtual" and "service customization" are the main characteristics of the *VPP*. Based on the type of resource management and control within the *VPP*, early *VPP*s aggregated distributed generation resources to form power-type *VPP*s, providing voltage support services to medium voltage distribution networks [5]. To further facilitate the transition from "source following load" to "source–load interaction", references [6,7] aggregated wind and solar power generation resources with loads to form source–load-type *VPP*s with defined regulation capacity and ramp rates, participating in day-ahead unit commitment regulation services. To further define the *VPP*, references [8–10] view it as a combination of controllable resources and flexible control technologies. In terms of positioning, the *VPP* is mainly intended to support the reliable integration of distributed generation (DG) [11], overcoming the fluctuations and randomness of DG output in power supply balance adjustments, and thus providing a medium for DG to participate in power market transactions.

The key to the involvement of virtual power plants (*VPP*s) in power system dispatch lies in accurately describing user behavior through physical and mathematical modeling, based on factors such as users' electricity consumption habits, psychosocial aspects of social activities, and demand response-related behavioral characteristics [12]. By extracting user characteristics, analyzing behaviors, and providing quantitative descriptions, *VPP*s can precisely model user behavior, enabling precise sensing, potential tapping, and value utilization of distributed resources [13]. Taking the modeling of electric vehicle (EV) charging and discharging behavior as an example, the current mainstream modeling approach is based on big data from power networks and transportation networks, using machine learning methods. Reference [14] simulates EV travel, charging, and battery swapping behaviors based on a coupled system of power networks, transportation networks, and information networks. Reference [15] analyzes EV user charging behavior from both macro and micro perspectives, using a multi-source data fusion method that integrates transportation and energy data. Reference [16] characterizes the uncertainty of EV user behavior, based on consumer psychology and social behavior theory. Reference [17] summarizes the modeling methods for EV travel and charging behavior that integrate micro-level driver decisions with macro-level network dynamics. Estimating adjustable response capacity refers to assessing the frequency support capacity that various types of distributed resources aggregated by *VPP*s can provide in each time period. Some of the foundational tasks for *VPP*s are to effectively regulate resources, participate in power market transactions, and provide frequency support ancillary services. Based on theoretical analysis, technical constraints, economic constraints, and availability constraints, the response capacity of *VPP*s is typically classified into four categories: theoretical, technical, economic, and adjustable response capacities [18]. Existing research mostly estimates response capacity from the perspective of load characteristics, believing that demand response capacity depends on user electricity consumption characteristics [19,20]. The upper limit of the response capacity for distributed resources to provide frequency support is the technical response capacity, and when estimating frequency response capacity, it is necessary to consider the requirements of frequency support ancillary services on the performance of distributed resources.

Furthermore, due to the involvement of power flow constraints and integer variables in the scheduling of virtual power plants, the essence of their optimization model becomes a mixed-integer nonlinear programming problem, and finding its optimal solution is a pressing challenge that needs to be addressed. Metaheuristic algorithms are capable of handling complex, nonlinear problems without requiring specific assumptions about the problem's model [21]. They gradually approximate the optimal solution through search and optimization methods, making them excellent at tackling highly complex and uncertain optimization problems [22]. Currently, widely adopted heuristic algorithms include the simulated annealing algorithm [23], ant colony algorithm [24], genetic algorithm [25], particle swarm algorithm [26], and others. Research on these algorithms has yielded some results, but there is still considerable room for improvement in terms of computational speed and accuracy of results [27]. The priority list method is simple and fast, but its results are rough and generally cannot yield the optimal solution. Dynamic programming is a widely used scheduling algorithm, but as the system scale and scheduling time increase, it can easily lead to the curse of dimensionality and make it difficult to consider time-related constraints. The Lagrangian relaxation method cannot robustly handle issues such as oscillations during iteration, obtaining feasible solutions, and convergence conditions during the solution process. The simulated annealing method experiences an exponential increase in computational load when solving larger-scale problems. The ant colony algorithm is prone to premature convergence and getting stuck in local optima. The genetic algorithm is valued for its simplicity, ability to obtain global optimal solutions, and lack of special requirements for the objective function. However, it has a high computational load and low efficiency. The particle swarm algorithm is also prone to getting stuck in local optima, and the accuracy of its solutions is not high enough.

2. Modeling Method for Renewable Energy Uncertainty Based on Interval Error Analysis

2.1. Analysis of the Output Characteristics of Renewable Energy

The output of renewable energy exhibits significant volatility and uncertainty. Taking wind power as an example, its output varies with changes in wind speed and direction, and these changes are characterized by randomness and volatility. Wind speeds fluctuate between near zero and rated wind speeds for most of the year, causing wind power output to fluctuate between zero and rated output. Furthermore, the seasonal characteristics of annual wind power output differ across regions. For instance, in Northeast and North China, the average wind power output is higher in winter and spring and lower in summer and autumn, while in Northwest China, it is the opposite. These differences are primarily due to varying climatic conditions and wind speed patterns in different regions, which increase the difficulty of wind power forecasting for scheduling purposes. The randomness and volatility of wind speeds, along with various factors such as terrain and climate, make it challenging to achieve high-accuracy wind power forecasting. This adds complexity and uncertainty to power system scheduling.

According to grid scheduling requirements, renewable energy power stations must submit short-term power forecasting data to the grid scheduling department every morning, and ultra-short-term power forecasting data at least every hour. Therefore, in this study, the hour is adopted as the time resolution for the short-term power forecasting of renewable energy, and the distribution characteristics of power smoothing by large-scale virtual power plants (*VPP*s) are statistically analyzed. The relative error of power smoothing by large-scale *VPP*s is calculated as follows:

$$e = \frac{W_t^f - W_t}{W_{ed}} \tag{1}$$

In the equation above, W_t^f represents the predicted power value of large-scale virtual power plants used to stabilize renewable energy; W_t denotes the actual power value of large-scale virtual power plants used for the same purpose; and W_{ed} stands for the rated power of the renewable energy power station.

2.2. Confidence Interval for the Uncertainty of Renewable Energy Output

Assuming that the sample data for the prediction error of the power stabilization of renewable energy by a large-scale virtual power plant, obtained after calculating Equation (1), are $e_1, e_2, \ldots, e_n$, where n represents the total number of samples, the kernel density estimate $f(e)$ of the original probability density function for the prediction error of the power stabilization of renewable energy by a large-scale virtual power plant is:

$$f(e) = \frac{1}{nh}\sum_{i=1}^{n} K(\frac{e - e_i}{h}) \tag{2}$$

In the equation, e_i represents the prediction error of the i-th sample data; h is the bandwidth; and $K(\cdot)$ is the kernel function. Choosing the Gaussian function as the kernel function, its expression is:

$$K(\frac{e - e_i}{h}) = \frac{1}{\sqrt{2\pi}} \exp\left[-\frac{(e - e_i)^2}{2h}\right] \tag{3}$$

If the confidence interval of the kernel density $f(e)$ shown in Equation (2) follows a normal distribution variable, and at each moment, the confidence level of the large-scale system's dispatchable resources in stabilizing the prediction error of renewable energy is equal, then the confidence level of $f(e)$ must satisfy the renewable energy prediction error interval, as shown in Figure 1.

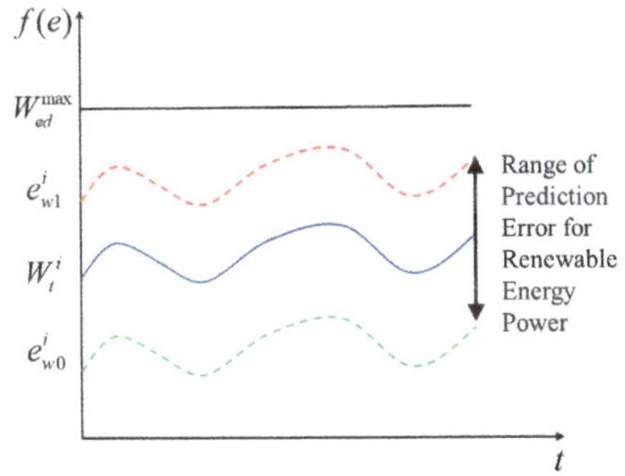

Figure 1. Schematic diagram of renewable energy prediction error interval.

In Figure 1, $W_{ed}^{\max}$ represents the maximum value of $W_{ed}^{\max}$, e_{w0}^i represents the lower bound of the confidence interval for the power prediction error of the i-th sample data of renewable energy mitigated by large-scale controllable resources, and e_{w1}^i represents the upper bound of the confidence interval for the same. Based on the renewable energy prediction error range shown in Figure 1, the actual power value W_t^i mitigated by a large-scale virtual power plant for renewable energy can be expressed as:

$$W_t = W_t^i + e \tag{4}$$

According to Equation (4), combined with the renewable energy prediction error range shown in Figure 1, it can be known that the location confidence interval for the prediction error confidence interval is $[e_{w_0}^i, e_{w_1}^i]$, and the normal distribution that the renewable energy prediction error range follows has a mean of 0 and a variance of S. Since this prediction error follows a standard normal distribution, the standard deviation of $f(e)$ needs to satisfy:

$$\sigma = \sqrt{S} \tag{5}$$

In the equation, σ represents the standard deviation of the normal distribution of the samples for mitigating renewable energy power prediction errors using large-scale virtual energy storage.

By combining Equations (4) and (5), the values of e_{w0}^i and e_{w1}^i can be determined, and their expressions are:

$$\begin{cases} e_{w_0}^i = \max(0, W_t + f^{-1}(1-\varepsilon)) \\ e_{w_1}^i = \min(0, W_t + f^{-1}(\varepsilon)) \end{cases} \tag{6}$$

In the equation, f^{-1} represents the inverse function of the normal probability distribution function; ϵ represents the significance level; and $1 - \epsilon$ represents the non-significance level.

By integrating the probability density function of the prediction error, the probability distribution function can be obtained as:

$$F(e) = \int_{-\infty}^{e} f(x)d(x) \tag{7}$$

In the equation, $F(\cdot)$ represents the probability distribution function, and $f(\cdot)$ represents the probability density function.

Based on the values of e_{w0}^i and e_{w1}^i shown in Equation (6), and according to the renewable energy prediction error range, if the value of $F(e)$ falls within the range $(0 < \epsilon < 1)$, the error confidence interval at a confidence level of $1 - \epsilon$ is:

$$F(e_{wo}^i \le e \le e_{w1}^i) = 1 - \varepsilon \tag{8}$$

It should be noted that Equation (7) provides the method for calculating the probability distribution function, and is utilized in Equation (8) to compute the error confidence interval. The confidence interval for renewable energy error prediction shown in Equation (8) also follows a normal distribution. Therefore, when the confidence interval shown in Equation (8) reaches the specified confidence level, the interval for the mathematical expectation of the normal distribution of the prediction error samples is:

$$A^{\Delta} - \frac{\sigma}{\sqrt{n}}\delta_{\frac{\varepsilon}{2}} \le \sigma \le A^{\Delta} + \frac{\sigma}{\sqrt{n}}\delta_{\frac{\varepsilon}{2}} \tag{9}$$

In the formula, A^{Δ} represents the sample mean and δ denotes the mathematical expectation of the normal distribution of samples, where large-scale virtual energy storage mitigates renewable energy power prediction errors. By synthesizing the aforementioned formulas, we can obtain ϵ and $1-\epsilon$, and then calculate the confidence level η for the renewable energy power prediction error, according to Equation (10).

$$\eta = 1 - 2\varepsilon \tag{10}$$

At this point, the confidence interval obtained from Equation (10) lies between [e^i_{w0}, e^i_{w1}]. By incorporating the confidence level η calculated using Equation (10) for the predicted power of renewable energy into the scheduling plan, the establishment of a scheduling model for large-scale virtual power plants to mitigate renewable energy power prediction errors can be achieved.

3. Low-Carbon Economic Dispatch Model and Constraints Considering the Participation of Virtual Power Plants

3.1. Carbon Trading Mechanism

Under the carbon trading mechanism, virtual power plants (*VPPs*) can adjust the operation strategies of their internal resources more flexibly to optimize economic dispatch. For instance, when renewable energy generation is sufficient, *VPPs* can increase the proportion of renewable energy output and reduce the use of carbon emission allowances. Conversely, when renewable energy generation is insufficient, *VPPs* can purchase carbon emission allowances to meet production demands, thereby maintaining the stable operation of the power system. The core of the carbon trading mechanism lies in promoting enterprises to actively reduce emissions, by setting an overall cap on carbon emissions and allocating emission allowances. As an energy coordination management system, *VPPs* can achieve emission reduction benefits by optimizing the operation strategies of internal resources to reduce unnecessary carbon emissions. In this paper, the baseline method is adopted for carbon emission allowances, which means that the carbon emission allowance for a *VPP* at a certain time is determined as follows:

$$E_{VPP} = \sum_{m} \psi P_m \tag{11}$$

In the formula, m represents the number of generators in the *VPP*; P_m denotes the output power of generator m at a certain time; and ψ is the emission allocation coefficient per unit of electricity for the *VPP*.

The carbon emissions of a *VPP* originate from the CO_2 generated by gas turbines and electricity purchased from the main grid. The revenue earned from participating in the carbon trading market can be calculated using the following formula:

$$B^c = \lambda_c (E_{VPP} - E_{MT} - E_G) \tag{12}$$

In the formula, λ_c represents the carbon trading price for the *VPP* (virtual power plant); E_{VPP} denotes the carbon emission allowances allocated to the *VPP*; E_{MT} is the carbon emissions produced by the gas turbines; and E_G represents the carbon emissions generated from the electricity purchased by the *VPP* from the main grid.

3.2. Establishment of the Objective Function

This paper aims to establish an objective function that maximizes the economic benefits and minimizes the carbon emissions of the virtual power plant (*VPP*). The objective function is composed of the following components:

Maximization of Economic Benefits for *VPP*

$$F = \max \sum_{s=1}^{N} a(s)(P_G^s + B^s - C^s) \tag{13}$$

In the above equation, F represents the net income of the *VPP*; N is the number of classic scenarios; $\alpha(s)$ is the probability of scenario s occurring; P_G^s is the power generation revenue of the VPP in scenario s; B^s is the carbon trading revenue of the VPP in scenario s; and C^s is the cost of the VPP in scenario s.

The 24 h power generation revenue of the VPP is calculated as follows:

$$P_G^s = \sum_{t=1}^{24} G_{I,t}(P_{pw,t}^s + P_{pv,t}^s + P_{gt,t}^s + P_{dis,t}^s + P_{sell,t}^s - P_{ch,t}^s - P_{buy,t}^s) \tag{14}$$

In the above equation, $G_{I,t}$ represents the electricity selling price at time t; $P_{pw,t}^s$, $P_{pv,t}^s$, and $P_{gt,t}^s$ are the power outputs of wind turbines, photovoltaic systems, and gas turbines, respectively, in scenario s during time period t; $P_{ch,t}^s$ and $P_{dis,t}^s$ are the charging and discharging powers of the electrical energy storage system, respectively, in scenario s during time period t; and $P_{sell,t}^s$ and $P_{buy,t}^s$ are the power sold to and purchased from the grid by the VPP in scenario s during time period t, respectively.

$$C^s = C_{op,t}^s + C_{es,t}^s + C_{pu,t}^s \tag{15}$$

$$C_{op,t}^s = \sum_{t=1}^{24} [E_{op,pw}P_{pw,t}^s + E_{op,pv}P_{pv,t}^s + E_{op,gt}P_{gt,t}^s + E_{op,ba}(P_{dis,t}^s - P_{ch,t}^s)] \tag{16}$$

$$C_{es,t}^s = \sum_{t=1}^{24} C_{gt}P_{gt,t}^s \tag{17}$$

$$C_{pu,t}^s = \sum_{t=1}^{24} G_{2,t}\left|D_{n,t} - P_{pw,t}^s - P_{pv,t}^s - P_{gt,t}^s - P_{dis,t}^s\right| \tag{18}$$

In the above equation, $C_{op,t}^s$, $C_{es,t}^s$, and $C_{pu,t}^s$ represent the operational cost, primary energy purchase cost, and electricity curtailment penalty cost of the VPP in scenario s during time period t, respectively. $E_{op,pw}$, $E_{op,pv}$, $E_{op,gt}$, and $E_{op,ba}$ are the cost coefficients for wind power, photovoltaic power, gas turbine power generation, and energy storage, respectively. C_{gt} represents the fuel cost of the gas turbine. $G_{2,t}$ is the electricity price at which the VPP purchases power from the main grid during time period t. $D_{n,t}$ represents the power that the system can absorb during time period t.

$$C_{gt} = C_{ng}/(\xi L_{ng}) \tag{19}$$

In the above equation, C_{ng} represents the price of natural gas; ξ denotes the power generation efficiency; and L_{ng} is the lower heating value of natural gas.

Capacity plan declared by the VPP to the grid:

$$O = \begin{cases} O_{pw} + O_{pv} + \xi P_{gt,\max}, & G_{2,t} \geq G_{gt} \\ O_{pw} + O_{pv}, & G_{2,t} < G_{gt} \end{cases} \tag{20}$$

The carbon emissions generated by the gas turbine within the VPP, as well as those resulting from the VPP's electricity purchases from the main grid, are as follows:

$$E_{CO2} = E_{gt} + E_{buy} \tag{21}$$

In the above equation, E_{gt} represents the carbon emissions from the gas turbine and E_{buy} represents the carbon emissions generated from purchasing electricity.

$$E_{gt} = \sum_{s=1}^{N}\sum_{t=1}^{T} [a + bP_{gt,s}^s + c(P_{gt,s}^s)^2] \tag{22}$$

$$E_{buy} = \sum_{s=1}^{N} \sum_{t=1}^{T} P^{s}_{buy,t} \varepsilon_{co2} \sigma_{co2} \tag{23}$$

In the above equation, a, b, and c are the carbon emission coefficients for electricity generation by the gas turbine; ε_{co2} represents the average share of carbon-containing energy sources; and σ_{co2} represents the carbon emission coefficient per kilowatt-hour.

3.3. Constraints

(1) *VPP* Operational Power Balance Constraint

$$\kappa_t + P^{s}_{ch,t} = P^{s}_{pw,t} + P^{s}_{pv,t} + P^{s}_{gt,t} + P^{s}_{dis,t} \tag{24}$$

In the above formula, κ_t represents the power loss when the *VPP* purchases electricity from the main grid during time period t.

(2) Gas Turbine Power Constraint

$$P_{gt,\min} \leq P^{s}_{gt,t} \leq P_{gt,\max} \tag{25}$$

In the above formula, $P_{gt,\min}$ and $P_{gt,\max}$ represent the lower and upper limits of the gas turbine's power, respectively.

(3) Gas Turbine Ramp Rate Constraint

$$-R_{g}^{\min} \Delta t \leq P^{s}_{gt,t} - P^{s}_{gt,t-1} \leq R_{g}^{\max} \Delta t \tag{26}$$

In the above formula, $R_{g}^{\min}$ and $R_{g}^{\max}$ represent the minimum and maximum ramp rates of the gas turbine, respectively; Δt represents the length of one scheduling interval.

(4) Energy Storage System Battery Capacity Constraint

During the scheduling process, the energy storage system must adhere to the law of the conservation of energy, and its constraint condition can be expressed as follows:

$$S^{s}_{ch,t} = S^{s}_{ch,t-1} + \frac{\Delta t (v_1 P^{s}_{ch,t} - P^{s}_{dis,t} / v_2)}{E_{ca}} \tag{27}$$

$$S^{s}_{ch,\min} \leq S^{s}_{ch,t} \leq S^{s}_{ch,\max} \tag{28}$$

In the above formula, $S^{s}_{ch,t}$ and $S^{s}_{ch,t-1}$ represent the capacities of the energy storage battery at time periods t and $t-1$, respectively; v_1 and v_2 represent the charging and discharging efficiencies, respectively; E_{ca} represents the installed capacity of energy storage; and $S^{s}_{ch,\max}$ and $S^{s}_{ch,\min}$ represent the upper and lower limits of the energy storage capacity, respectively.

(5) Energy Storage System Battery Charging and Discharging Constraint

$$\begin{cases} P_{ch,\min} \leq P^{s}_{ch,t} \leq P_{ch,\max} \\ P_{dis,\min} \leq P^{s}_{dis,t} \leq P_{dis,\max} \\ H^{s}_{ch,t} + H^{s}_{dis,t} \leq 1 \end{cases} \tag{29}$$

In the above formula, $P_{ch,\max}$ and $P_{ch,\min}$ represent the upper and lower limits of charging power, respectively; $P_{dis,\max}$ and $P_{dis,\min}$ represent the upper and lower limits of discharging power, respectively; and $H^{s}_{ch,t}$ and $H^{s}_{dis,t}$ represent the charging and discharging states during time period t in scenario s, with values of 0 or 1.

4. Solution Algorithm

4.1. Introduction to the Traditional ALO Algorithm

The problem of economic dispatch in power systems considering the participation of virtual power plants is essentially a mixed-integer nonlinear programming problem with high computational complexity. In some extreme scenarios, it is difficult to search for the optimal solution. Therefore, this paper proposes an improved ant lion Optimizer (ALO) algorithm. The ALO algorithm finds the optimal solution by mimicking the process of antlions constructing traps and preying on ants in nature [28,29]. The ALO algorithm has two populations: one is the ant population, and the other is the antlion population. The five basic modules of the hunting process are the random walking of ants, ants falling into traps, ants sliding towards antlions, antlions capturing ants, and antlions rebuilding traps. The mathematical model of the ALO algorithm starts with random walking in the search space, and can be described as follows:

$$x(t) = \{0, cumsum[2r(t_1) - 1, cumsum[2r(t_2) - 1, \cdots, cumsum[2r(t_n) - 1] \tag{30}$$

$$r(t) = \begin{cases} 1, & if\ rand > 0.5 \\ 0, & if\ rand < 0.5 \end{cases} \tag{31}$$

In the above formula, $x(t)$ represents random walking at the t-th iteration; *cumsum* denotes the cumulative sum; n is the maximum number of iterations; $r(t)$ is a random function; and *rand* represents a random number within [0, 1].

To ensure that the ants move randomly within the search space without exceeding the boundaries, the positions of the ants are adjusted as follows:

$$y_i^t = \frac{(x_i^t - a_i)(d_i^t - c_i^t)}{b_i - a_i} + c_i^t \tag{32}$$

In the above formula, a_i and b_i represent the minimum and maximum values of x_i, respectively, and c_i^t and d_i^t represent the minimum and maximum values of the i-th variable for a certain ant at the t-th iteration, respectively. To simulate the process of ants wandering aimlessly around the antlion and getting trapped, the calculation formula is as follows:

$$c_i^t = Antlionn_i^t + c^t \tag{33}$$

$$d_i^t = Antlionn_i^t + d^t \tag{34}$$

In the above formula, $Antlionn_i^t$ represents the position of the antlion selected by the i-th ant at the t-th iteration. By adaptively reducing the range of random walking of the ants, the process of ants sliding into the antlion's lair is simulated, as shown below:

$$c^t = \frac{c^t}{I} \tag{35}$$

$$d^t = \frac{d^t}{I} \tag{36}$$

In the above formula, I represents a proportional parameter that is dynamically adjusted according to the increase in the number of iterations.

$$I = \begin{cases} 1, t \leq 0.1T \\ 10^{t}, t > 0.1T \end{cases} \tag{37}$$

In the above formula, t represents the current number of iterations; T represents the maximum number of iterations; and w is a parameter that is dynamically adjusted based on the number of iterations.

4.2. Principle of Algorithm Improvement

The improvement in this paper lies in the use of a hybrid strategy, combining chaotic mapping and tournament selection to replace the traditional roulette wheel selection method for population initialization. This effectively avoids a situation where too many potential solutions are located at the same boundary point during the optimization process. Based on the random walking mechanism of the ants, a tournament is conducted among individuals randomly selected from the population to select the optimal antlion and elite antlion walks. The positions of the ants are updated in each iteration:

$$Ant_i^t = \frac{R_A^t + R_E^t}{2} \tag{38}$$

In the above formula, Ant_i^t represents the position selected for the i-th ant at the t-th iteration; R_A^t represents the position around the antlion selected through the tournament; and R_E^t represents the random walking of ants around the elite antlion. To initialize the population, Chebyshev chaotic mapping is used for the setup:

$$x_{k+1} = \cos\left[A \cos^{-1}(x_k)\right] \tag{39}$$

In the above formula, k represents the number of iterations for the chaotic sequence; A is a mapping coefficient; and $A = 2$ is selected in this paper. During the process of individual selection within the population, the fitness values of each pair of ants and antlions are compared, rather than merging and sorting the ants and antlions. During the comparison, if the fitness value of an ant is better than that of an antlion, the position of the antlion is updated to the position of the ant. The calculation formula is as follows:

$$Antlionn_i^t = Ant_i^t,\ if\ f(Ant_i^t) < f(Antlionn_i^t) \tag{40}$$

To avoid a situation where too many ants in the population are located at the same boundary point during the optimization process, a random entry into the search space is achieved through Equation (41).

$$Ant_i^t = b_{low} + rand \times (b_{up} - b_{low})\ if\ Ant_i^t > b_{up}\ or\ Ant_i^t < b_{up} \tag{41}$$

In the above formula, b_{low} and b_{up} represent the lower and upper bounds of the search space, respectively.

5. Case Study

To further demonstrate the effectiveness and feasibility of the model and algorithm proposed in this paper, an improved IEEE-30 bus test system is employed for verification [30]. The system structure topology is illustrated in Figure 2 below, where two *VPP*s are connected at buses 7 and 23, large-capacity wind farms are respectively connected at buses 3, 17, and 29, and photovoltaic power plants are integrated into the system through buses 16, 21, and 26. The system structure of the virtual power plants is shown in Figure 3. The division of peak and off-peak time-of-use electricity prices is depicted in Figure 4 below.

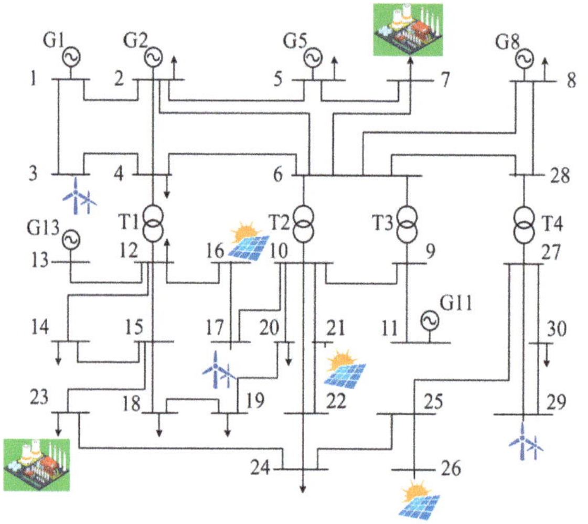

Figure 2. The structure topology of improved IEEE-30 bus test system.

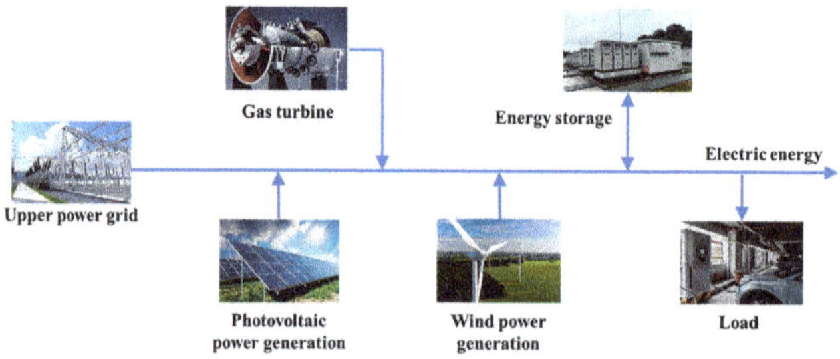

Figure 3. The structure topology of *VPPs*.

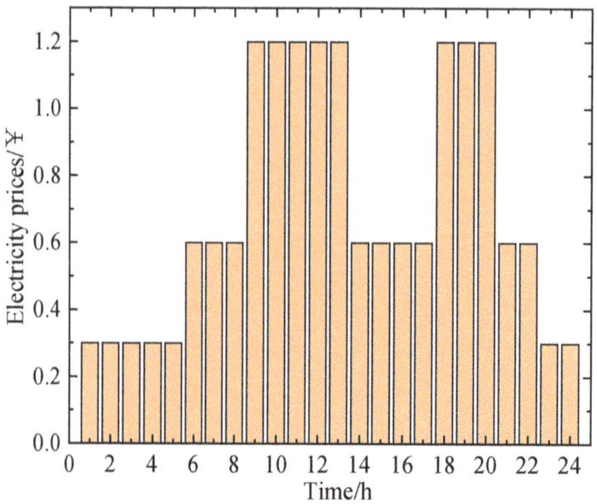

Figure 4. The division of peak and off-peak time-of-use electricity prices.

5.1. Analysis of Low-Carbon Benefits

Two operational modes are established based on whether the carbon trading mechanism is considered, and the net benefits and carbon emissions of the *VPP* under different modes are shown in Table 1. It can be observed that when carbon trading is taken into account, the net profit of the *VPP* increases while carbon emissions decrease, indicating that the *VPP* model considering carbon trading balances both economic efficiency and environmental protection. Specifically, in the mode without considering carbon trading, the *VPP* mainly focuses on the direct economic benefits of its electricity production and sales, i.e., electricity sales revenue minus electricity production costs. Due to the absence of a carbon trading mechanism, the *VPP* has no incentive to reduce carbon emissions, because it does not gain direct economic benefits from such reductions. Therefore, in this mode, the *VPP* may tend to adopt less costly but higher carbon generation methods to maximize its net profit.

Table 1. Comparison of carbon emissions and operating costs across different modes.

VPP Operation Modes	Carbon Emissions/kg	Operating Cost/¥
Mode 1: Considering Carbon Trading	2547.6	8452.1
Mode 2: Without Considering Carbon Trading	3254.4	9625.7

Next, let us consider a scenario where a carbon trading mechanism is introduced. This mechanism allows carbon emission permits to be traded as a commodity, converting carbon emissions into economic value. When the *VPP* reduces its carbon emissions, it can sell surplus carbon emission quotas to entities that require more quotas, thereby obtaining additional economic benefits. This mechanism incentivizes the *VPP* to adopt more environmentally friendly generation methods and reduce carbon emissions, to gain greater profits in the carbon trading market. When the *VPP* considers carbon trading, it weighs up the costs and carbon emissions of different generation methods. If a certain generation method has lower carbon emissions but slightly higher costs, the *VPP* may choose this method because it can offset the cost increase by selling carbon emission quotas, or even obtain additional profits. In this way, the net profit of the *VPP* increases while carbon emissions decrease.

Observing Figure 5, it can be found that under the carbon trading mechanism, the carbon emissions per unit of electricity generated by gas turbines, as carbon emission sources, become a factor to consider. To reduce carbon emission costs, gas turbines may actively decrease their electricity generation, especially when the price of carbon emission permits is high. Gas turbines may adopt optimized scheduling strategies, such as adjusting the power generation load curve and improving generation efficiency, to reduce their carbon emission intensity. The output of gas turbines will be shifted from peak hours to off-peak hours, to avoid cost increases during peak hours due to high electricity prices and high carbon emission permit prices. As adjustable loads such as electric vehicles increase, the output of gas turbines may be further suppressed, because electric vehicles can choose to charge using renewable energy electricity, thereby reducing dependence on traditional fossil fuels. The output level of gas turbines may also be directly affected by carbon trading prices. The higher the carbon trading price, the greater the incentive for gas turbines to reduce their output, as this can lower carbon emissions and thus reduce carbon emission costs. Under the carbon trading mechanism, if the output of gas turbines decreases, *VPP*s need to purchase more electricity from the upper-level grid to meet electricity demand. Especially during periods of insufficient renewable energy generation or peak electricity usage, *VPP*s may increase their purchases from the upper-level grid. When purchasing electricity, *VPP*s may pay more attention to the low-carbon and economic attributes of the power. Therefore, in terms of purchasing structure, they may prefer to buy renewable energy electricity or low-carbon electricity. This will help reduce the overall carbon emissions of *VPP*s and increase their economic benefits. The introduction of the carbon trading mechanism will prompt *VPPs* to participate more actively in electricity market transactions. Optimizing the electricity purchasing structure and reducing purchasing costs through market means will be one of the most important strategies for *VPPs*.

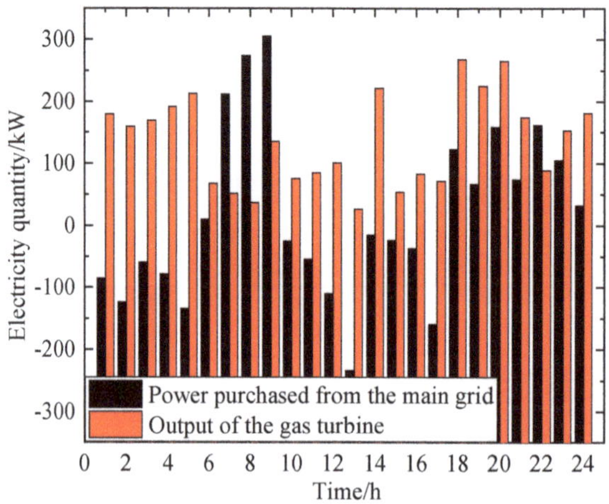

Figure 5. Output of power units and electricity purchased from the grid.

5.2. Optimality Verification of the Scheduling Model

Figure 6 presents wind and solar curtailment rates under different scheduling strategies. It can be observed that using error intervals to characterize the uncertainty of renewable energy output achieves lower wind and solar curtailment rates compared to multi-scenario-based stochastic optimization (method A) [31], fuzzy chance constraint-based stochastic optimization (method B) [32], and traditional robust optimization (method C) [33]. The error interval method, by calculating the prediction error distribution of renewable energy output, can more accurately depict the range of uncertainty in renewable energy output. Figure 7 further demonstrates the economic benefits and computational efficiency of different methods. In Figure 6, the wind and solar curtailment rates for the proposed method, Method A, Method B, and Method C are 6.24%, 7.26%, 8.48%, and 9.01%, respectively. Through the calculation of confidence intervals, the uncertainty range of renewable energy output can be more precisely described, enabling a more accurate assessment of various possible output scenarios during economic dispatch. This precision aids in formulating more reasonable dispatch strategies and reducing the phenomenon of wind and solar power curtailment due to fluctuations in renewable energy output. Converting uncertainty optimization problems into deterministic optimization problems simplifies the calculation process and reduces computational complexity. Additionally, deterministic optimization problems typically have a clearer solution space, facilitating the discovery of global optimal or near-optimal solutions. Traditional robust optimization algorithms often analyze based on the worst-case scenario, assuming that renewable energy output reaches its maximum uncertainty boundary. This conservative strategy tends to lead to overly conservative dispatch strategies, thereby increasing the risk of wind and solar power curtailment. Scenario-based stochastic optimization algorithms simulate the uncertainty of renewable energy output by generating multiple possible scenarios. However, this method may be limited by the number and representativeness of scenarios, leading to inaccurate optimization results. Fuzzy chance-constrained optimization algorithms consider the fuzziness of renewable energy output and allow constraint violations within a certain probability.

Nevertheless, this method may struggle to accurately assess the actual uncertainty range of renewable energy output, thereby affecting the accuracy of the optimization results.

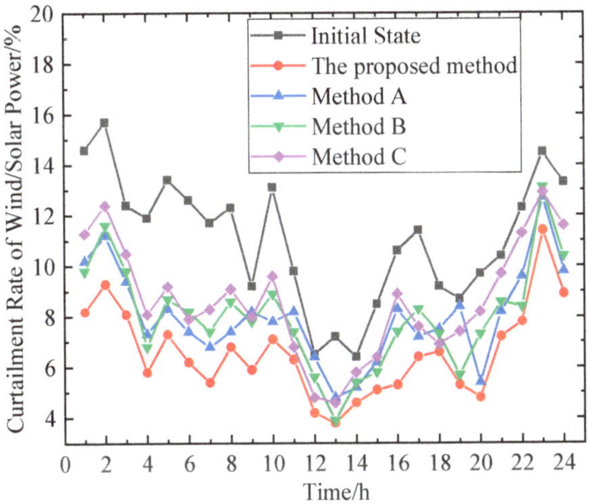

Figure 6. Comparison of wind/solar power curtailment rates under different scheduling strategies.

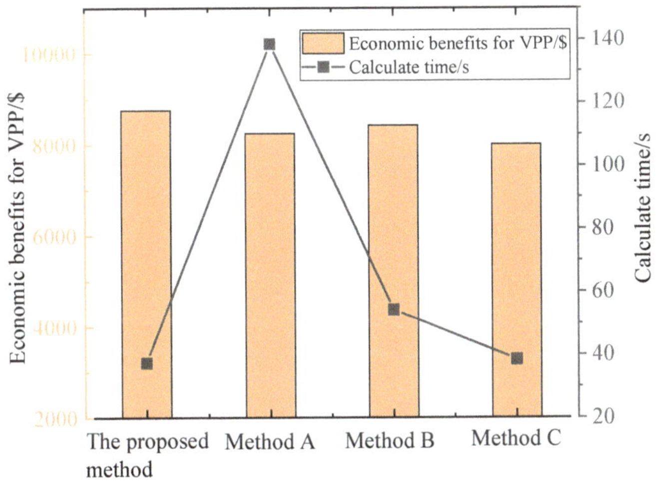

Figure 7. Comparison of economic indicators and computational efficiency among different methods.

Observing the above figure, it can be seen that the core of the proposed method lies in converting the originally complex and difficult-to-solve uncertainty optimization problem into a more manageable deterministic optimization problem, through confidence interval calculation and transformation. Compared to other methods, the economic income can be increased by up to 3.2%, and the computation time is only 12.5% of that required for scenario screening-based stochastic optimization. Confidence interval calculation allows decision-makers to set a confidence level according to their own risk appetite, thereby maximizing the economic income of the *VPP* while ensuring a certain level of risk. This method offers more a flexible balancing of risk and return compared to robust optimization (which tends to adopt conservative strategies to address worst-case scenarios) and stochastic optimization (which may underestimate or overestimate risks due to improper scenario selection). On the other hand, the confidence interval can quantify the uncertainty range of renewable energy output, enabling more precise scheduling plans to be developed accordingly. Compared to fuzzy chance-constrained optimization, the confidence interval provides clearer boundary conditions, helping to reduce decision-making errors caused by fuzziness. The proposed method allows for optimization under a wider range of conditions, not limited to specific scenarios or worst-case scenarios, potentially enabling the discovery of scheduling solutions with higher economic value. In terms of computational efficiency, the proposed method converts the uncertainty optimization problem into a deterministic one, significantly reducing the complexity of the model. Compared to stochastic optimization, which requires simulating multiple scenarios, and fuzzy optimization, which deals with the complexity of fuzzy sets, the confidence interval method avoids these additional computational burdens. Due to the model simplification, the confidence interval method may not require repeated iterations to verify the robustness of the solution, as robust optimization does, nor does it need to solve each scenario separately, as stochastic optimization does, thereby reducing the total computation time.

Additionally, this paper compares the proposed algorithm with other conventional solution algorithms. The test data were sourced from the CEC 2017 benchmark suite, encompassing unimodal functions, multimodal functions, and hybrid functions. The algorithms were evaluated and compared based on average error, average convergence time, and Wilcoxon rank-sum test results. The relevant results are presented in Table 2 below.

Table 2. Effectiveness comparison of different algorithms.

Index	The Improved ALO	ALO	PSO	GA	MS-TSA	I-CPA
References for methods	/	[29]	[34]	[35]	[36]	[37]
Average error/%	2.57%	3.79%	3.26%	3.88%	2.74%	2.99%
Average convergence time/s	52.4	60.2	65.8	69.2	59.7	62.3
Wilcoxon rank-sum	1	4	5	6	2	3

Observing the table above, it can be found that the improved ALO algorithm optimizes the initialization process of the population by introducing a hybrid strategy of chaotic mapping and tournament selection. Chaotic mapping is able to generate a more diverse and uniformly distributed initial population, which helps the algorithm better explore the global optimal solution in the search space. The tournament selection method further enhances the competitiveness and diversity of the population, aiding the algorithm in avoiding premature convergence during the iteration process. These optimization measures enable the improved ALO algorithm to more accurately find the global optimal solution when solving complex optimization problems, thereby reducing the average error. In terms of convergence speed, the improved ALO algorithm also performs well. The introduction of chaotic mapping allows the algorithm to quickly converge to an appropriate solution in the early stages of iteration, while the tournament selection method helps the algorithm continuously optimize the quality of the solution during the iteration process. Additionally, the improved ALO algorithm combines the inherent advantages of the ant lion optimizer, such as mimicking the antlion's predation mechanism for random walks and trap construction. These characteristics endow the algorithm with stronger adaptability and robustness during the search process. Therefore, the improved ALO algorithm can achieve convergence in a shorter period of time, reducing the average convergence time. Furthermore, the Wilcoxon rank-sum test is a non-parametric test method used to compare the positions or medians of two independent samples. In the evaluation of algorithm performance, the Wilcoxon rank-sum test can be used to test whether the differences between the results obtained by different algorithms when solving the same optimization problem are significant. The improved ALO algorithm performs optimally in terms of the Wilcoxon rank-sum test, indicating that its results have significant advantages compared to those of other algorithms. This further validates the effectiveness and superiority of the improved ALO algorithm in solving complex optimization problems.

The size of the test system also has an impact on the final dispatch results. The improved IEEE-118 bus test system has been adopted for further verification Figure 8. The structure of the test system is shown in the figure below.

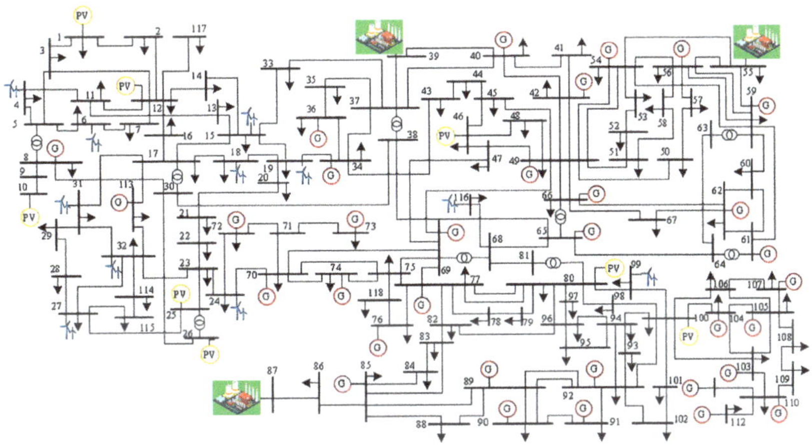

Figure 8. The structure topology of improved IEEE-118 bus test system.

After calculation, the economic indicators and computation time are presented in Table 3. It can be observed that the proposed method maintains high computational efficiency and economic benefits, even in large-scale test systems. By accurately quantifying the uncertainty of renewable energy generation, the confidence interval method can maximize the economic benefits of virtual power plants, while ensuring the safe and stable operation of the power grid. In large-scale test systems, this method can more accurately capture the volatility and randomness of renewable energy generation, thereby formulating better economic dispatch strategies and increasing the total revenue of virtual power plants. By setting a confidence level, the confidence interval method can tolerate a certain degree of uncertainty in renewable energy generation, enhancing the robustness of virtual power plants. Even when there are significant fluctuations in renewable energy generation in large-scale test systems, this method can maintain the stable operation of virtual power plants and reduce economic losses caused by uncertainty. Additionally, it is noteworthy that in large-scale test systems, the computation time of the method proposed in this paper can still be kept within 5 min, meeting the requirements of real-time online calculation and dispatch in power systems.

Table 3. Performance comparison among different methods.

Method	Carbon Emissions/kg	Operating Cost/$	Computation Time/s	Average Wind/Solar Curtailment Rate
The proposed method	5745.6	15862.4	57.4	2.57%
Method A	5975.1	16247.2	625.6	3.52%
Method B	6123.8	17524.9	59.2	3.74%
Method C	6245.1	17996.3	61.3	4.45%

This precise depiction helps the scheduling system gain a more accurate understanding of the actual generation capacity of renewable energy, thereby formulating more reasonable scheduling plans. Compared to multi-scenario stochastic optimization, the error interval method does not require presetting multiple possible generation scenarios, thus avoiding the risk of missing extreme situations due to improper scenario setting. At the same time, it also mitigates the issue of increased computational complexity caused by an excessive number of scenarios. The error interval method allows the scheduling system to flexibly adjust based on the actual prediction error range of renewable energy output. When the prediction error range is large, the scheduling system can correspondingly increase reserve capacity, or adjust the output of other units to ensure the stable operation of the power system. Since the error interval method can more accurately reflect the uncertainty of renewable energy output, it helps reduce wind and solar curtailment caused by this uncertainty. The scheduling system can formulate more precise scheduling strategies based on error intervals, enabling a more full utilization of renewable energy generation. Compared to traditional robust optimization, the error interval method typically does not require complex optimization calculations. It only needs to calculate the prediction error distribution of renewable energy output, based on historical data and prediction models, and then make scheduling decisions based on this distribution. This method has higher computational efficiency and can quickly respond to real-time changes in the power system. The error interval method is applicable not only to the optimal scheduling of large-scale power systems, but also to small- and medium-sized power systems. It can provide effective scheduling strategies for power systems of different scales, helping them achieve more economical, efficient, and reliable operation.

In real-world applications, the fields where the method proposed in this paper can be applied include: (1) High-penetration renewable energy generation areas. In regions with high penetration of renewable energy sources such as wind and solar power, grid dispatch faces significant challenges, due to the randomness and intermittency of renewable energy generation. These areas are often located in remote regions with relatively weak grid structures, and the large-scale integration of renewable energy generation further exacerbates grid instability. By constructing a *VPP*, decentralized renewable energy generation resources, energy storage equipment, and demand-side response resources can be integrated into a virtual entity to participate in the electricity market and grid operation. Utilizing the robust economic management strategy proposed in this paper, effective management of the uncertainty associated with renewable energy generation can be achieved, enhancing the flexibility and economy of the power grid. (2) Optimal dispatch of urban power grids. In urban power grids, due to high load density and diverse electricity demands, grid dispatch requires greater precision and intelligence. At the same time, urban power grids often contain a large amount of distributed energy resources (such as rooftop photovoltaics and small wind turbines) and demand-side response resources (such as electric vehicles and smart buildings). By integrating these resources through a *VPP* and utilizing the improved ALO algorithm proposed in this paper for economic dispatch, optimal dispatch of the urban power grid can be achieved. This not only improves the economy of the power grid, but also reduces carbon emissions, enabling a green and low-carbon energy transition.

6. Conclusions

This paper addresses the new challenges faced by grid security and scheduling in the context of rapid advancements in new energy technologies. It proposes a robust economic management strategy for power systems that considers the participation of *VPP*s. By integrating distributed energy resources and demand-side response resources,

VPPs participate in the electricity market and grid operation as virtual entities, effectively enhancing the flexibility and economy of the power system.

The paper first characterizes the uncertainty of renewable energy output through prediction deviations and establishes economic dispatch and constraint conditions for *VPPs*. Given the nonlinear characteristics of the model, an improved ALO algorithm is employed to ensure optimal solutions are obtained. Finally, a case study using an improved IEEE 30-bus test system validates the effectiveness and feasibility of the proposed method. The research results indicate that this method maximizes economic benefits while ensuring the safe and stable operation of the power grid.

Future research can further explore optimal allocation strategies for different types of distributed energy resources and demand-side response resources within *VPPs*, to improve their overall economy and flexibility. By considering renewable energy prediction errors and the uncertainty of demand-side response across different time scales, more refined *VPP* scheduling strategies can be studied to better meet the actual operational needs of the power grid.

Author Contributions: Conceptualization, X.L., C.Z., Q.Y. and J.X.; software, X.L., C.Z., Q.Y. and J.X.; writing—original draft preparation, X.L., C.Z., Q.Y. and J.X. All authors have read and agreed to the published version of the manuscript.

Funding: This research received no external funding.

Data Availability Statement: The original contributions presented in this study are included in the article. Further inquiries can be directed to the corresponding authors.

Conflicts of Interest: Author Chen Zhang was employed by the State Grid Energy Research Institute Co., Ltd., and author Qianqian Yi was employed by Bishan Power Supply Branch of State Grid Chongqing Electric Power Company. The remaining authors declare that the research was conducted in the absence of any commercial or financial relationships that could be construed as a potential conflict of interest.

References

1. Wang, C.; Shi, Z.; Liang, Z. Key Technologies and Prospects for Utilization of Demand-Side Resources in a Renewable Energy-Dominated Power System. *Autom. Electr. Power Syst.* **2021**, *45*, 37–48.
2. Ge, X.; Fu, Z.; Xu, F. Business Models and Key Technologies of Virtual Power Plants for the New-Type Power System. *Autom. Electr. Power Syst.* **2022**, *46*, 129–146.
3. Cao, W.; Qin, H.; Lu, J. Positioning and Application Prospects of Virtual Synchronous Machines in the New-Type Power System. *Autom. Electr. Power Syst.* **2023**, *47*, 190–207.
4. Shimon, A.; Alistair, P. *The Virtual Utility*; Springer: Boston, MA, USA, 1997.
5. Bignucolo, F.; Caldon, R.; Prandoni, V.; Spelta, S.; Vezzola, M. The Voltage Control on MV Distribution Networks with Aggregated DG Units. In Proceedings of the 41st International Universities Power Engineering Conference, Newcastle upon Tyne, UK, 6–9 September 2023; pp. 187–192.
6. Pudjianto, D.; Ramsay, C.; Strbac, G. Virtual power plant and system integration of distributed energy resources. *IET Renew. Power Gener.* **2024**, *1*, 10–16. [CrossRef]
7. Babaei, S.; Zhao, C.; Fan, L. A Data-Driven Model of Virtual Power Plants in Day-Ahead Unit Commitment. *IEEE Trans. Power Syst.* **2019**, *34*, 5125–5135. [CrossRef]
8. Hrvoje, P.; Igor, K.; Tomislav, C. Virtual Power Plant Mid-Term Dispatch Optimization. *Appl. Energy* **2023**, *101*, 134–141.
9. Zhao, H.; Wang, X.; Wang, B.; Ning, B.; Pan, Z.; Sun, H.; Xue, Y. Cloud-Cluster Hierarchical Dispatch for Large Scale Demand-Side Distributed Resources. In Proceedings of the 2021 IEEE 5th Conference on Energy Internet and Energy System Integration, Taiyuan, China, 22–24 October 2021; pp. 272–276.
10. Wang, X.; Liu, D.; Liu, Z.; Liu, M.; Wang, J.; Gao, Y.; Song, Y. Operation Mechanism and Key Technologies of Virtual Power Plant under Ubiquitous Internet of Things. *Power Syst. Technol.* **2019**, *43*, 3175–3183.
11. Bao, P.; Zhang, W.; Zhang, Y. Secondary Frequency Control Considering Optimized Power Support From Virtual Power Plant Containing Aluminum Smelter Loads Through VSC-HVDC Link. *J. Mod. Power Syst. Clean Energy* **2023**, *11*, 355–367. [CrossRef]

12. Xu, X.; Li, K.; Wang, F.; Mi, Z.; Jia, Y.; Wei, W.; Jing, Y. Evaluating Multi-Time Scale Response Capability of EV Aggregator Considering Users' Willingness. *IEEE Trans. Ind. Appl.* **2023**, *57*, 3366–3376. [CrossRef]
13. Chen, Q.; Lv, R.; Guo, H. Modeling of Power Users' Behavior Oriented to Demand Response: Research Status and Applications. *Electr. Power Autom. Equip.* **2023**, *43*, 23–37.
14. Fu, Z.; Zhu, W.; Zhu, J. EV Swapping Demand Forecasting and Charging Optimization Strategy for EV Swapping Stations under the "Vehicle-Electricity-Road-Station" Interconnection. *Electr. Power Autom. Equip.* **2024**, *42*, 116–124.
15. Sheng, Y.; Guo, Q.; Liu, M. Analysis of User Charging Behavior Based on Multi-Source Data Fusion and Planning Practice of Charging Facilities. *Autom. Electr. Power Syst.* **2024**, *46*, 151–162.
16. Hu, J.; Ma, W.; Xue, Y. Quantification of Reserve Capacity for Electric Vehicle Aggregators Based on the CPSSE Framework. *Autom. Electr. Power Syst.* **2022**, *46*, 46–54.
17. Sheng, Y.; Guo, Q.; Xue, Y. Modeling and Collaborative Optimization of Power-Transportation Coupled Networks from the Cyber-Physical-Social Perspective [J/OL]. *Autom. Electr. Power Syst.* **2024**. *ahead of print*.
18. Wang, F.; Li, M.; Zhang, X. Assessment Methods, Applications, and Prospects of Demand Response Resource Potential. *Autom. Electr. Power Syst.* **2023**, *47*, 173–191.
19. Song, M.; Sun, W.; Wang, Y.; Shahidehpour, M.; Li, Z.; Gao, C. Hierarchical Scheduling of Aggregated TCL Flexibility for Transactive Energy in Power Systems. *IEEE Trans. Smart Grid* **2019**, *11*, 2452–2463. [CrossRef]
20. Sun, Y.; Mao, Y.; Li, Z. Comprehensive Clustering Method for User Load Characteristics and Adjustable Potential Oriented to Big Data in Electric Power. *Proc. CSEE* **2024**, *41*, 6259–6270.
21. Beşkirli, A.; Dağ, İ. An efficient tree seed inspired algorithm for parameter estimation of Photovoltaic models. *Energy Rep.* **2022**, *8*, 291–298. [CrossRef]
22. Beşkirli, M.; Kiran, M.S. Optimization of Butterworth and Bessel Filter Parameters with Improved Tree-Seed Algorithm. *Biomimetics* **2023**, *8*, 540. [CrossRef]
23. Huo, X.; Yu, J.; Pang, C.; Ding, Y.; Zhang, J.; Zhao, C.; Yu, B. Integrated Demand Response Optimization for Consumer With Herd Mentality: A Genetic Simulated Annealing Approach. *IEEE Access* **2024**, *12*, 69097–69111. [CrossRef]
24. Xia, K.; Li, Y.; Zhu, B. Improved Photovoltaic MPPT Algorithm Based on Ant Colony Optimization and Fuzzy Logic Under Conditions of Partial Shading. *IEEE Access* **2024**, *12*, 44817–44825. [CrossRef]
25. Li, L.; Wang, Y.; Hu, H.; Ning, H.; Li, X. Design of Energy Internet Power Information Management System Based on Genetic Algorithm. In Proceedings of the 2024 3rd International Conference on Artificial Intelligence and Autonomous Robot Systems (AIARS), Bristol, UK, 29–31 July 2024; pp. 14–19.
26. Halwani, S.; Tawfik, H.; Dahrouj, H.; Elnady, A. Comparative Analysis of Genetic Algorithm and Particle Swarm Opti-mization for Optimal Power Flow on IEEE 30-Bus System. In Proceedings of the 2024 Advances in Science and Engineering Technology Interna-tional Conferences (ASET), Abu Dhabi, United Arab Emirates, 21–22 December 2024; pp. 1–6.
27. Wu, B.; Wang, Z.; Jiang, J. Research on Maximum Power Point Tracking Technology of Adaptive Particle Swarm Optimization Based on Greedy Strategy. In Proceedings of the 2024 IEEE 2nd International Conference on Control, Electronics and Computer Tech-nology (ICCECT), Jilin, China, 26–28 April 2024; pp. 1422–1425.
28. Das, T.; Boruah, N.; Patra, S.; Borah, A.A. Enhanced parameter optimization of PID controller for single area power system using hybrid ALO-GBO algorithm. In Proceedings of the 2024 6th International Conference on Energy, Power and Environment (ICEPE), Shillong, India, 20–22 June 2024; pp. 1–6.
29. Shi, Y.; Zhang, W.; Sang, F.; Zhao, L.; Wang, T. Application of ALO-ELM on Electricity Demand Forecasting under Spot Power Market. In Proceedings of the 2022 4th International Conference on Smart Power & Internet Energy Systems (SPIES), Beijing, China, 27–30 October 2022; pp. 2194–2197.
30. Available online: https://download.csdn.net/download/LF2827417795/90128514 (accessed on 10 November 2024).
31. Cao, Y.; Mu, Y.; Jia, H.; Yu, X.; Hou, K.; Wang, H. A Multi-Objective Stochastic Optimization Approach for Planning a Multi-Energy Microgrid Considering Unscheduled Islanded Operation. *IEEE Trans. Sustain. Energy* **2024**, *15*, 1300–1314. [CrossRef]
32. Ma, R.; Li, X.; Gao, W.; Lu, P.; Wang, T. Random-Fuzzy Chance-Constrained Programming Optimal Power Flow of Wind Integrated Power Considering Voltage Stability. *IEEE Access* **2020**, *8*, 217957–217966. [CrossRef]
33. Zhuang, K.; Wang, Z.; Yang, N.; Xia, K.; Shi, R. Robust Optimization-Driven Strategy for Dynamic Planning of Power Transmission Networks. In Proceedings of the 2024 3rd International Conference on Energy and Electrical Power Systems (ICEEPS), Guangzhou, China, 14–16 July 2024; pp. 617–620.
34. Zhang, H.; Guo, X.; Zhang, P. Improved PSO-SVM-Based Fault Diagnosis Algorithm for Wind Power Converter. *IEEE Trans. Ind. Appl.* **2024**, *60*, 3492–3501. [CrossRef]
35. Xu, H.; Yu, T.; Ren, C.; He, S. Finite-Time Decentralized Sliding Mode Control for Interconnected Systems and Its Application to Electrical Power Systems: A GA-Assisted Design Method. *IEEE Trans. Circuits Syst. I Regul. Pap.* **2024**, *71*, 2835–2847. [CrossRef]

36. Sathish, K.R.; Ananthapadmanabha, T. Power Quality Enhancement in Distribution System Integrated with Renewable Energy Sources Using Hybrid RBFNN-TSA Technique. In Proceedings of the 2021 7th International Conference on Electrical Energy Systems (ICEES), Virtual, 11–13 February 2021; pp. 189–194.
37. Ng, J.S.; Chen, J.; Kyaw, N.A.; Lwin, N.K.Z.; Ho, W.G.; Chong, K.S.; Gwee, B.H. A Highly Efficient Power Model for Correlation Power Analysis (CPA) of Pipelined Advanced Encryption Standard (AES). In Proceedings of the 2020 IEEE International Symposium on Circuits and Systems (ISCAS), Seville, Spain, 12–14 October 2020; pp. 1–5.

processes

MDPI

Article

A Fast Calculation Method for Economic Dispatch of Electro-Thermal Coupling System Considering the Dynamic Process of Heat Transfer

Jingyan Chen *, Qinting Lin, Zilong Yang, Qingming Liu and Hongbo Zou *

College of Electrical Engineering and New Energy, China Three Gorges University, Yichang 443002, China; lin1109245092@163.com (Q.L.); zhuxian4627262@163.com (Z.Y.); lqm378562103@163.com (Q.L.)
* Correspondence: chenjy2471@163.com (J.C.); zhbhorace@ctgu.edu.cn (H.Z.)

Abstract: The dynamic spatial and temporal characteristics of heat transfer within heating network pipelines are important factors affecting the accuracy of economic dispatch decision-making results of electro-thermal coupling systems. However, the pipeline heat transmission process is described by partial differential equations, which makes it difficult to solve quickly. Therefore, this study introduces a model for calculating the economic dispatch of the electro-thermal coupling system (EDETCS) that takes into account the pipeline transmission process. Firstly, based on the implicit upwind difference method, a two-port model of branch heat transfer dynamics is established. Secondly, the general term formula of the two-port model is derived. Finally, the established two-port model is applied to the EDETCS. The findings from the example analysis indicate that, in contrast to the conventional calculation method, the proposed model improves the calculation speed while ensuring the accuracy of the solution.

Keywords: electro-thermal coupling system; economic dispatch; heating network; dynamic process of heat transfer; two-port model

1. Introduction

Compared with thermal plants, combined heat and power (CHP) plants have more advantages in improving energy efficiency, energy saving and emission reduction [1]. However, under the new energy situation, CHP units operated in the traditional "determining electricity by heat" mode can easily reach the minimum rigid technical output of the unit, which seriously restricts the flexible supply of the power grid [2,3]. By considering the pipeline transmission process in the EDETCS, we can effectively manage renewable energy consumption and minimize system operating costs. This is achieved by leveraging the complementary nature of power and heat demand alongside the ability to store the heat of the heating network [4,5]. However, the pipeline heat transmission process is described by partial differential equations, and it is very difficult to solve such optimization problems with partial differential equation constraints [6–9].

At present, researchers have conducted some research on the pipeline transmission process. In references [10,11], considering the heat loss effect and ignoring the heat delay effect, the linear relationship between the inlet and outlet water temperature of the pipeline is obtained. Reference [12] deduced the partial differential equation of heat transmission in the pipeline through the law of thermodynamics and solved it by the finite difference method. However, its accuracy depends on a smaller time and space step; the calculation amount is large, and the calculation time is long. References [13,14] proposed the node

https://doi.org/10.3390/pr13010175

method to analyze and model the heating network. The principle is to track the time when the water body starts from the previous node to the next node and calculate the temperature at the head of the pipe according to the historical temperature of different nodes. On this basis, the temperature at the end of the pipe is calculated. The node method reduces the computing time, but large errors will occur when the selected time resolution is too large [15].

In addition to the research ideas of the above-simplified mechanism model, the existing research work has begun to try the analogy circuit method to establish the equivalent model of heat networks. In [16], an equivalent model of a heat network based on the Fourier transform is proposed, and the partial differential dynamic model of heat transfer is mapped to the frequency domain for solution. Reference [17] proposed a heat network equivalent model based on Laplace transform. This model transforms the partial differential equation into the s-domain to solve and realizes the simplification of the model to the algebraic equation. Based on the thermoelectric analogy method, Reference [18] introduced the power system analysis method to equivalent the loss, energy storage and temperature influence of each temperature micro-element in the heating network pipeline through components such as 'heat resistance', 'heat capacity' and 'heat source'. Finally, the functional relationship between the outlet temperature and the initial temperature and time is obtained by mathematical derivation. References [19,20] proposed the unified energy path theory. The physical mechanisms involved in gas and heating networks are similar to those in the power transmission system. In the frequency domain, a unified set of mathematical equations for power, heat, and gas networks has been developed. Finally, the linear two-port lumped parameter model in the frequency domain is obtained. In references [21,22], the distributed parameter circuit model of electric, heat and gas energy flow is established by the Laplace transform, and then the lumped parameter transmission model with the branch as a unit is obtained. Although the above method transforms the time-domain dynamic model of heat transfer described by partial differential equations into a linear two-port model in the frequency domain or s-domain, it reduces the difficulty of solving the dynamic model of the pipeline transmission process. However, it is necessary to carry out a transformation on the actual known initial conditions and boundary conditions before solving, and after the calculation is completed, the calculation results need to be inversely transformed back to the time domain, and the transformed link reduces the solution efficiency.

On the basis of the above research, considering the timeliness of economic dispatch, this paper establishes a two-port model of the pipeline heat transmission process, aiming to reduce the calculation scale of the EDETCS without reducing the accuracy. Firstly, based on the implicit upwind difference structure, a two-port model of heat transfer dynamics in the heating network is established to reduce the calculation scale. Secondly, the general term formula of the two-port model is derived, which makes the model more convenient and more operable for the modeling of complex heat networks. Then, the proposed two-port model is embedded in the EDETCS. Finally, the example analysis demonstrates the effectiveness and advantages of the two-port model.

2. Two-Port Model of Heat Transfer in Pipes

2.1. Dynamic Model of Heat Transfer Based on Implicit Up-Wind Difference Scheme

China's heating network is generally composed of a two-level network. The two exchange heat through the heat exchange station. Because the secondary pipe is very short, we do not consider the secondary heating network; only the primary heating network is modeled [23].

The pipeline transmission process in the heat pipe network can be described by Equation (1) [24]:

$$\frac{\partial T_{\mathrm{w}}}{\partial t}+\frac{m}{A\rho}\frac{\partial T_{\mathrm{w}}}{\partial x}+\frac{\lambda}{A\rho c}(T_{\mathrm{w}}-T_{\mathrm{a}})=\frac{\lambda}{\rho c}\frac{\partial^2 T_{\mathrm{w}}}{\partial x^2} \tag{1}$$

where m is the mass flow, and when the system adopts mass regulation mode, m is constant. ρ is the water density. A is the pipe's sectional area, c is the water-specific heat capacity, λ is the thermal conductivity, T_{w} is water temperature, and T_{a} is ambient temperature.

Compared with thermal convection, the heat conduction effect between pipe fluids is very weak. Therefore, the heat conduction term in Equation (1), that is, the second-order partial derivative term, can be ignored without obvious error. The pipeline transmission process in the heating network can be written as [17]:

$$\frac{\partial T}{\partial t}+\frac{m}{A\rho}\frac{\partial T}{\partial x}+\frac{\lambda}{A\rho c}T=0 \tag{2}$$

where T is the difference between T_{w} and T_{a}.

The heating system in China mainly adopts the mass regulation mode. Therefore, this paper will derive the model of heat transfer under the premise of known mass flow.

Firstly, the first-order implicit upwind difference quotient [12] is used to approximate the first-order partial derivative and the difference grid is shown in Figure 1:

$$\frac{\partial T}{\partial t}=\frac{T_i^n-T_i^{n-1}}{\tau} \tag{3}$$

$$\frac{\partial T}{\partial x}=\frac{T_i^n-T_{i-1}^n}{h} \tag{4}$$

$$I=\frac{L}{h},N=\frac{P}{\tau} \tag{5}$$

where τ is the time step, h is the spatial step, i is the number of spatial nodes, and n is the number of time nodes. T_i^n are the water temperature at node (n, i).

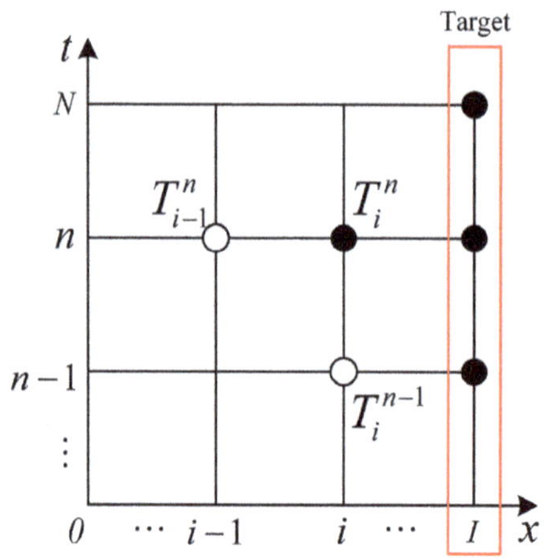

Figure 1. Differential grid for the dynamic flow of heat pipeline.

Combining Equation (3), Equation (4) and heat transfer Equation (2), the difference scheme of pipeline temperature is obtained:

$$T_i^n=K_1T_i^{n-1}+K_2T_{i-1}^n \tag{6}$$

where K_1 and K_2 are two parameters defined to simplify the representation:

$$K_1 = \frac{A\rho ch}{A\rho ch + h\tau\lambda + cm\tau} \tag{7}$$

$$K_2 = \frac{cm\tau}{A\rho ch + h\tau\lambda + cm\tau} \tag{8}$$

Then, the boundary conditions and initial conditions are determined:

Usually, the initial temperature and the inlet water temperature are known, and this paper assumes that the temperature of each node at the time $t = 0$ is equal, so the boundary conditions and initial conditions can be determined as follows:

$$T_x^0 = T_0^0\,,\ T_0^t = T^{\text{in}} \tag{9}$$

where: T_x^0 is the initial temperature of each node, T_0^0 is the initial temperature at the head of the pipe, T_0^t is the inlet water temperature at the head of the pipe, and T^{in} is the temperature of water at the inlet of the pipe changing with time.

2.2. Two-Port Model of Branch Heat Transfer Dynamics

According to the initial conditions, boundary conditions Equation (9) and the derived linear model Equation (6), the temperature curve at the end of the pipe can be iteratively solved. However, this finite difference method will calculate the temperature of each node for iteration during the calculation process. When selecting a smaller time and space step, the computation involves a substantial quantity of data and takes a significant amount of time. To enhance computational efficiency, this section will establish a linear correlation between the temperatures of various nodes at different timestamps within the difference grid and the boundary conditions and initial conditions according to the difference equation of temperature and finally construct the two-port model of branch heat transfer dynamics to reduce the scale of model calculation.

When $t = 1$, the difference grid of the pipe temperature is shown in Figure 2. According to Equation (6), the water temperature at node i $(1 \leq i \leq I)$ is:

$$T_i^1 = K_1 T_i^0 + K_2 T_{i-1}^1 \tag{10}$$

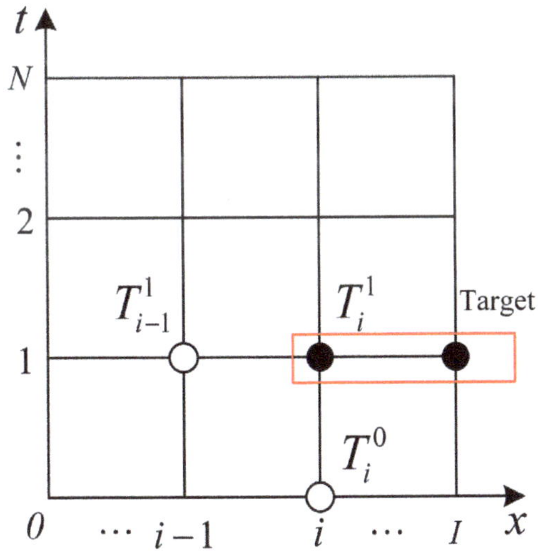

Figure 2. The differential grid at time $t = 1$.

Substituting Equation (9) into Equation (10):

$$T_i^1 = K_1 T_i^0 + K_2 T_{i-1}^1 \tag{11}$$

The temperature at node (1, i) is obtained:

$$\begin{aligned} T_i^1 &= T_0^0 \sum_{b=1}^{i} K_1 (K_2)^{b-1} + T_0^1 (K_2)^i \\ &= a_{i,0}^1 T_0^0 + a_{i,1}^1 T_0^1 + \cdots + a_{i,N}^1 T_0^N \end{aligned} \tag{12}$$

$$\Rightarrow T_i^1 = \left[a_{i,0}^1, a_{i,1}^1, \cdots a_{i,N}^1\right] \boldsymbol{T}^{\text{in}} \tag{13}$$

where:

$$a_{i,0}^1 = \sum_{b=1}^{i} K_1 (K_2)^{b-1} \tag{14}$$

$$a_{i,1}^1 = (K_2)^i \tag{15}$$

$$a_{i,2}^1 = \cdots = a_{i,N}^1 = 0 \tag{16}$$

$$\boldsymbol{T}^{\text{in}} = \left[T_0^0, T_0^1, \cdots, T_0^N\right]^T \tag{17}$$

When t = 2, the difference grid of the pipe temperature is shown in Figure 3. According to Equation (6), the water temperature at node I $(1 \leq i \leq I)$ is:

$$T_i^2 = K_1 T_i^1 + K_2 T_{i-1}^2 \tag{18}$$

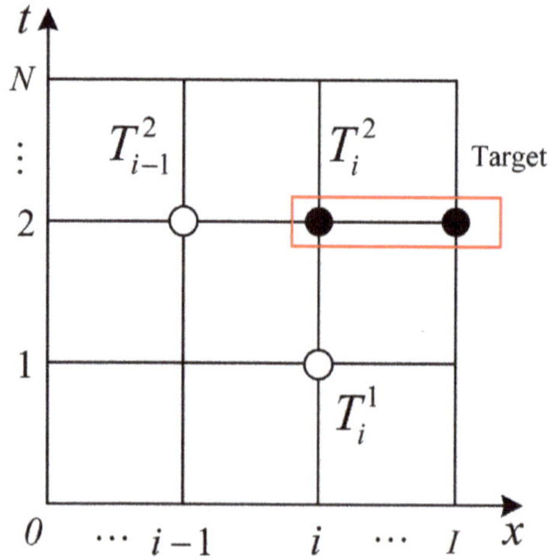

Figure 3. The differential grid at time t = 2.

Substituting Equation (9) into Equation (18), the temperature of each node is calculated in turn, and the temperature at node (2, i) is obtained:

$$\begin{aligned} T_i^2 &= T_0^0 (K_1)^2 \sum_{b=1}^{i} b (K_2)^{b-1} + \\ &\quad T_0^1 i K_1 (K_2)^i + T_0^2 (K_2)^i \\ &= a_{i,0}^2 T_0^0 + a_{i,1}^2 T_0^1 + \cdots + a_{i,N}^2 T_0^N \end{aligned} \tag{19}$$

$$\Rightarrow T_i^2 = \left[a_{i,0}^2, a_{i,1}^2, \cdots a_{i,N}^2\right] \boldsymbol{T}^{\text{in}} \tag{20}$$

where:

$$a_{i,0}^2 = (K_1)^2 \sum_{b=1}^{i} b(K_2)^{b-1} \quad (21)$$

$$a_{i,1}^2 = iK_1(K_2)^i \quad (22)$$

$$a_{i,2}^2 = (K_2)^i \quad (23)$$

$$a_{i,3}^2 = a_{i,4}^2 = \cdots = a_{i,N}^2 = 0 \quad (24)$$

Through the above method, it can be obtained that when t = 3, the temperature at node i $(1 \leq i \leq I)$ is:

$$\begin{aligned} T_i^3 &= T_0^0(K_1)^3 \sum_{b=1}^{i} \frac{b(b+1)}{2}(K_2)^{b-1} + \\ &\quad T_0^1 \frac{i(i+1)}{2}(K_1)^2(K_2)^i + \\ &\quad T_0^2 iK_1(K_2)^i + T_0^3(K_2)^i \\ &= a_{i,0}^3 T_0^0 + a_{i,1}^3 T_0^1 + \cdots + a_{i,N}^3 T_0^N \end{aligned} \quad (25)$$

$$\Rightarrow T_i^3 = \left[a_{i,0}^3, a_{i,1}^3, \cdots a_{i,N}^3\right] \boldsymbol{T}^{\text{in}} \quad (26)$$

where:

$$a_{i,0}^3 = (K_1)^3 \sum_{b=1}^{i} \frac{b(b+1)}{2}(K_2)^{b-1} \quad (27)$$

$$a_{i,1}^3 = \frac{i(i+1)}{2}(K_1)^2(K_2)^i \quad (28)$$

$$a_{i,2}^3 = iK_1(K_2)^i \quad (29)$$

$$a_{i,3}^3 = (K_2)^i \quad (30)$$

$$a_{i,4}^3 = a_{i,5}^3 = \cdots = a_{i,N}^3 = 0 \quad (31)$$

Through the temperature equation at each moment, the matrix form of the temperature at node (n, i) can be obtained:

$$T_i^n = \boldsymbol{A}_i^n \boldsymbol{T}^{\text{in}} \quad (32)$$

where $\boldsymbol{A}_i^n = \left[a_{i,0}^n, a_{i,1}^n, \cdots a_{i,N}^n\right]$.

By merging the temperature expressions at each time point, the temperature at each time point at node i is obtained, that is, the two-port model of branch heat transfer:

$$\boldsymbol{T}_i = \boldsymbol{A}_i \boldsymbol{T}^{\text{in}} \quad (33)$$

$$\boldsymbol{T}_i = \left[T_i^1, T_i^2, \cdots, T_i^N\right]^T \quad (34)$$

$$\boldsymbol{A}_i = \begin{bmatrix} a_{i,0}^1 & a_{i,1}^1 & \cdots & a_{i,N}^1 \\ a_{i,0}^2 & a_{i,1}^2 & \cdots & a_{i,N}^1 \\ \vdots & \vdots & \ddots & \vdots \\ a_{i,0}^N & a_{i,1}^N & \cdots & a_{i,N}^1 \end{bmatrix} \quad (35)$$

2.3. The General Term Formula of Coefficient Matrix of Two-Port Model

In Section 2.2, a two-port model of branch heat transfer dynamics is established. However, various pipelines exhibit differences in parameters like diameter and length. Therefore, the coefficient matrix of the model needs to be re-derived, especially in large-scale heating systems, which will greatly increase the calculation workload. For this reason, this section will derive the general term formula of the coefficient matrix to reduce the calculation workload.

Summarize the coefficients of the differential form of water temperature at each time. We find that the elements of the coefficient matrix of the temperature expression at node (n, i) are the linear combination of the Pascal matrix and the parameters K_1, K_2.

Let $M = \max\{I,N\}$, let M-order diagonal matrix $\boldsymbol{Q}(i) = \text{diag}\,[1,1,\ldots,0]$, and the first i diagonal elements are 1. Let $\boldsymbol{R}(n) = [0,\ldots,0,1,0,\ldots,0]$ be an M-order row vector, and its nth element is 1. Let M-order column vector $\boldsymbol{K} = [1,K_2,(K_2)^2,\ldots,(K_2)^{M-1}]^T$. Let $\boldsymbol{P}(M)$ be an M-order Pascal matrix. Then $\boldsymbol{P}(M)\boldsymbol{Q}(i)$ means that the first i column elements of $\boldsymbol{P}(M)$ are retained, and other elements are set to zero, $\boldsymbol{R}(n)\boldsymbol{P}(M)\boldsymbol{Q}(i)$ represents the nth row element of the matrix and $[\boldsymbol{P}(M)\boldsymbol{Q}(i)]$; $\boldsymbol{R}(n)\boldsymbol{P}(M)\boldsymbol{Q}(i)\boldsymbol{R}(i)$ represents the nth row i column element of the matrix $[\boldsymbol{P}(M)\boldsymbol{Q}(i)]$.

Then, The formula for the general term of the coefficient matrix $\boldsymbol{A}_i^n$ can be articulated as follows:

$$a_{i,b}^{n} = \begin{cases} (K_1)^n \boldsymbol{R}(n)\boldsymbol{P}(M)\boldsymbol{Q}(i)\boldsymbol{K} \quad b = 0 \\ (K_1)^{n-b}(K_2)^i \boldsymbol{R}(n-b+1)\boldsymbol{P}(M)Q(i)R(i) \\ \qquad\qquad 0 < b \leq N \end{cases} \tag{36}$$

By merging the coefficient matrices at each moment, the general term formula of the coefficient matrix $\boldsymbol{A}_i$ of the model can be obtained.

The establishment of the general term formula makes it unnecessary to derive the temperature expression of each moment and each node in the initial modeling and only needs to assign the coefficient matrix $\boldsymbol{A}_i$ through the loop statement. Moreover, for different pipelines, the elements of the model coefficient matrix $\boldsymbol{A}_i$ are only different in parameters K_1 and K_2. It is only necessary to change the parameters and assign them again without re-derivation, which greatly reduces the calculation scale and makes the model establishment of large-scale complex heat networks easier.

2.4. Two-Port Model of Heat Transfer Dynamics in Heating Network Pipeline

According to the previous derivation, when the initial conditions and boundary conditions are the initial temperature of each node of the pipeline and the inlet water temperature, respectively, the outlet temperature of the pipeline is:

$$\boldsymbol{T}^{\text{out}} = \boldsymbol{A}_I \boldsymbol{T}^{\text{in}} \tag{37}$$

$$\boldsymbol{T}^{\text{out}} = \left[T_I^1, T_I^2, \cdots, T_I^N\right]^T \tag{38}$$

$$\boldsymbol{A}_I = \begin{bmatrix} a_{I,0}^1 & a_{I,1}^1 & \cdots & a_{I,N}^1 \\ a_{I,0}^2 & a_{I,1}^2 & \cdots & a_{I,N}^1 \\ \vdots & \vdots & \ddots & \vdots \\ a_{I,0}^N & a_{I,1}^N & \cdots & a_{I,N}^1 \end{bmatrix} \tag{39}$$

The above two-port model eliminates the intermediate nodes between the source and the load in the calculation process and further realizes the dimension reduction modeling and operation of the model. The coefficient matrix can be calculated offline without online iteration.

3. Formulation of EDETCS

3.1. Objective Function

The objective of the dispatch model is to reduce the system's operating cost to a minimum:

$$F_{\min} = C_{\text{G}} + C_{\text{chp}} + C_{\text{w}} \tag{40}$$

where C_G is the operating cost of thermal units, C_{chp} is the operation cost of CHP units, and C_W is the wind curtailment cost.

The three costs are:

$$C_G = \sum_{t=1}^{N} \sum_{j=1}^{N^G} (a_j^G P_{j,t}^G + b_j^G) \tag{41}$$

$$\begin{aligned} C_{chp} = & \sum_{t=1}^{N} \sum_{j=1}^{N^{chp}} [a_j^{chp} (P_{j,t}^{chp})^2 + b_j^{chp} P_{j,t}^{chp} + \\ & c_j^{chp} + d_j^{chp} (H_{j,t}^{chp})^2 + e_j^{chp} H_{j,t}^{chp} + \\ & f_j^{chp} H_{j,t}^{chp} P_{j,t}^{chp}] \end{aligned} \tag{42}$$

$$C_W = \sum_{t=1}^{N} \sum_{j=1}^{N^W} \lambda_j^s (P_{j,t}^{w,pre} - P_{j,t}^{W}) \tag{43}$$

where N^G, N^{chp}, N^W are the set of units' number. a_j^G, b_j^G are the cost coefficients of thermal units. $P_{j,t}^{chp}$, $H_{j,t}^{chp}$ are the CHP's power and heat output. a_j^{chp}, b_j^{chp}, c_j^{chp}, d_j^{chp}, e_j^{chp}, f_j^{chp} are the correlation coefficients of the operating costs of CHP units. λ_j^s is the wind curtailment cost coefficient. $P_{j,t}^{w,pre}$ is the predicted maximum output of wind power. $P_{j,t}^{W}$ is the actual output of wind power.

3.2. Constraints

(1) Heat transfer dynamics constraints:

The end temperature of the pipe k satisfies:

$$\boldsymbol{T}_k^{out} = \boldsymbol{A}_{I(k)} \boldsymbol{T}_k^{in} \tag{44}$$

where $\boldsymbol{T}_k^{out}$ is the outlet temperature matrix. k represents the pipe number.

(2) Temperature mixing constraints:

$$\sum_{k \in S_n^{pipe-}} m_k T_{k,t}^{out} = T_{l,t}^{in} \sum_{l \in S_n^{pipe+}} m_l \tag{45}$$

where S_n^{pipe+}, S_n^{pipe-} are the set of pipes starting at node n and ending at node n.

(3) Heat exchange station constraints:

$$H_{j,t}^{LN} = c m_j^{LN} (T_{j,t}^{LN,s} - T_{j,t}^{LN,r}) \tag{46}$$

$$\underline{T}_j^{LN,r} \leq T_{j,t}^{LN,r} \leq \overline{T}_j^{LN,r} \tag{47}$$

$$\underline{T}_j^{LN,s} \leq T_{j,t}^{LN,s} \leq \overline{T}_{j,t}^{LN,s} \tag{48}$$

where m_j^{LN} is the mass flow at the entrance of heat substations, $H_{j,t}^{LN}$ is the heat consumed by the heat substations. $T_{j,t}^{LN,s}$ is the supply and $T_{j,t}^{LN,r}$ is the return temperature from heat substations. $\overline{T}_j^{LN,r}$ is the upper and $\underline{T}_j^{LN,r}$ is the lower limits of the return temperature, $\overline{T}_{j,t}^{LN,s}$ is the upper and $\underline{T}_j^{LN,s}$ is the lower limits of the supply temperature.

(4) CHP unit constraints:

$$H_{j,t}^{chp} = c m_j^{chp} (T_{j,t}^{chp,s} - T_{j,t}^{chp,r}) \tag{49}$$

where $H_{j,t}^{chp}$ is the CHP heat output. m_j^{chp} is the mass flow at the entrance of CHP units, $T_{j,t}^{chp,r}$ is the return and $T_{j,t}^{chp,s}$ is the supply temperature from the heat source.

(5) power balance constraints:

$$\sum_{j=1}^{N^{G}} P_{j,t}^{G} + \sum_{j=1}^{N^{chp}} P_{j,t}^{chp} + \sum_{j=1}^{N^{W}} P_{j,t}^{w} = P_{t}^{LN} \tag{50}$$

where P_{t}^{LN} is the active load.

(6) CHP unit constraints:

$$\underline{P}_{j}^{chp} - c_{j}^{v} H_{j,t}^{chp} \leq P_{j,t}^{chp} \leq \overline{P}_{j}^{chp} - c_{j}^{c} H_{j,t}^{chp} \tag{51}$$

$$P_{j}^{co} + c_{j}^{m} H_{j,t}^{chp} \leq P_{j,t}^{chp} \leq \overline{P}_{j}^{chp} - c_{j}^{c} H_{j,t}^{chp} \tag{52}$$

$$\underline{H}_{j}^{chp} \leq H_{j,t}^{chp} \leq \overline{H}_{j}^{chp} \tag{53}$$

where $\overline{P}_{j}^{chp}$ is the maximum and $\underline{P}_{j}^{chp}$ is the minimum power output of CHP units. c_{j}^{v}, c_{j}^{c} c_{j}^{m} P_{j}^{co} are the feasible region boundary coefficient of CHP units. $\overline{H}_{j}^{chp}$ is the upper and $\underline{H}_{j}^{chp}$ is the lower limits of the heat power of CHP units.

(7) Ramping constraints:

$$\underline{D}_{j}^{G} \leq P_{j,t}^{G} - P_{j,t-1}^{G} \leq \overline{D}_{j}^{G} \tag{54}$$

where $\overline{D}_{j}^{G}$ is upward and $\underline{D}_{j}^{G}$ is the downward ramping capability of thermal units.

(8) Thermal unit constraint:

$$\underline{P}_{j}^{G} \leq P_{j,t}^{G} \leq \overline{P}_{j}^{G} \tag{55}$$

where $\overline{P}_{j}^{G}$ is the upper and $\underline{P}_{j}^{G}$ is the lower limits of the thermal unit output.

(9) Wind power output constraints:

$$0 \leq P_{j,t}^{w} \leq P_{j,t}^{w,pre} \tag{56}$$

4. Case Study

4.1. Double Pipe Heating System

A heating system consisting of a supply pipe, a return pipe, a heat source, and a heat substation has been chosen to validate the proposed two-port model. Figure 4 shows the system topology diagram. The accuracy and superiority of the two-port model are verified by comparing the calculation results with the implicit upwind difference method [12].

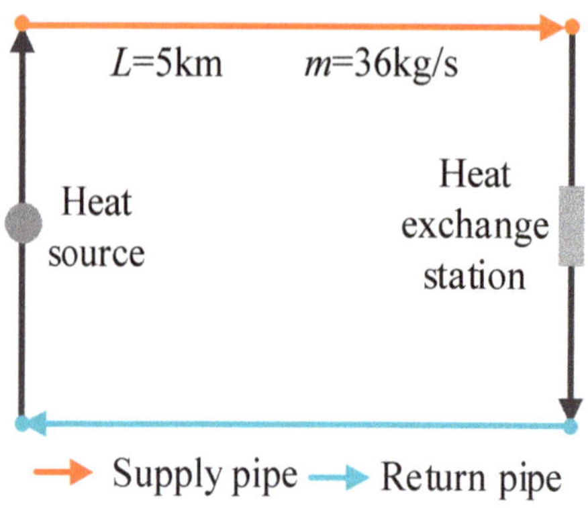

Figure 4. Structure diagram of Double pipe heating system.

Table 1 lists the pipe parameters. The system works in the quality regulation mode, and the heat load is constant power operation. Set the time step τ = 10 s, space step h = 10 m; heat load is 5 MW.

Table 1. Parameters of Double pipe heating pipeline.

Parameter	A/m^2	λ/(W/(m·°C)	m/(kg/s)	L/km	T_a/°C
Value	0.05	0.25	36	5	0

Figure 5 illustrates the computed water temperature profiles. The water temperature of the water supply pipeline tends to be stable and close to the heating temperature curve after 115 min of cooling. This is consistent with the time that the hot water flows through 5000 m at a mass flow rate of 36 kg/s, reflecting the "heat loss" and "delay" phenomena in the pipeline heat transmission process. This shows that the two-port model of heat transfer dynamics can truly reflect the pipeline transmission process in the heat network pipeline. Moreover, compared with the calculation results of the difference method, the temperature curves obtained by the two calculation methods are almost completely coincident, and the mean relative error of water supply pipe outlet temperature and return pipe outlet temperature are 0.15% and 0.23%, respectively, which further verifies the accuracy of describing the pipeline transmission process.

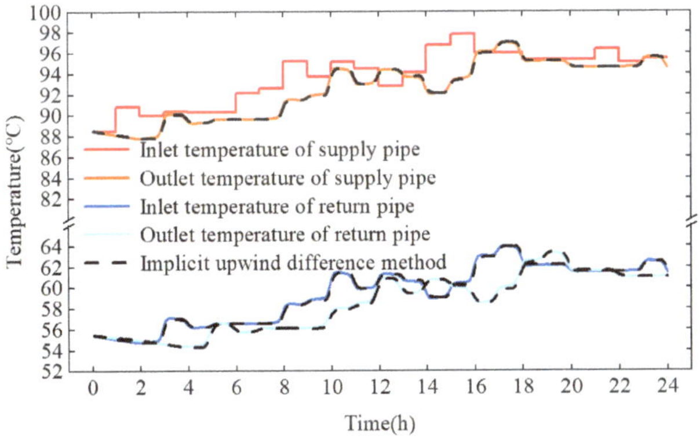

Figure 5. Calculation results of pipe temperature.

The duration of the calculations is presented in Table 2. The solution time of the two-port model is only 0.012 s, while the calculation time of the difference method is 0.22 s, and the speed is increased by more than 18 times, which verifies the superiority of the two-port model in calculation speed.

Table 2. Time of calculation.

Model	Time/s
The two-port model	0.012
The implicit upwind difference method	0.22

4.2. District Heating Network (DHN)

The DHN shown in Figure 6 is used to illustrate the two-port model's effectiveness, and the calculation time is compared with the different method used in Reference [12]. The heating network comprises three water supply lines, three return lines, a heat source, and three heat substations. The system works in the quality regulation mode, and the heat load is constant power operation. Table 3 lists the pipe parameters. The heat load is set to 2.5 MW, 1 MW and 1 MW, respectively.

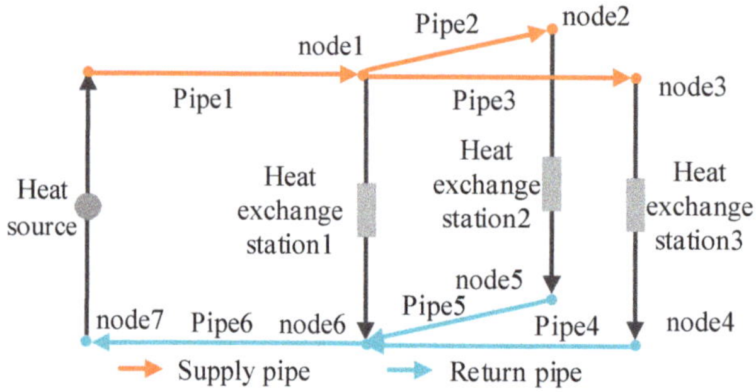

Figure 6. Structure diagram of the DHN.

Table 3. Pipe parameters.

Pipe	A/m^2	$\lambda/(\text{W}/(\text{m}\cdot{}^\circ\text{C})$	$m/(\text{kg/s})$	L/km	$T_a/{}^\circ\text{C}$
1, 6	0.05	0.25	36	5	0
2, 5	0.05	0.25	14	3.5	0
3, 4	0.05	0.25	7	2.5	0

Figures 7 and 8 show the calculated water temperature curves of each node under the two methods. The results show that the temperature curves of the two models are almost the same, and the maximum relative errors of water supply pipe end temperature and return pipe end temperature are 0.23% and 0.41%, respectively, which indicates that the precision of both methods in calculations is comparable.

Table 4 shows the time it takes to calculate the two models. We can find that the calculation time of the two-port model is much lower than that of the difference method. Compared with the calculation time in 4.1, the calculation time of the difference method is increased by more than 10 times, while the calculation time of the two-port model is only increased by less than 5 times. This is because, in systems with more heating network pipelines, the finite difference method will introduce state variables that are proportional to the number of spatial nodes. The two-port model only needs to introduce variables that are proportional to the number of pipelines, which reduces the total number of variables and reduces the calculation scale.

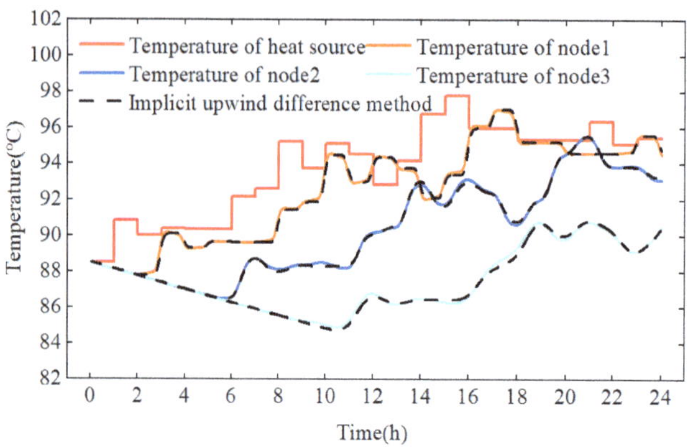

Figure 7. Temperature comparison of each node of the supply pipe.

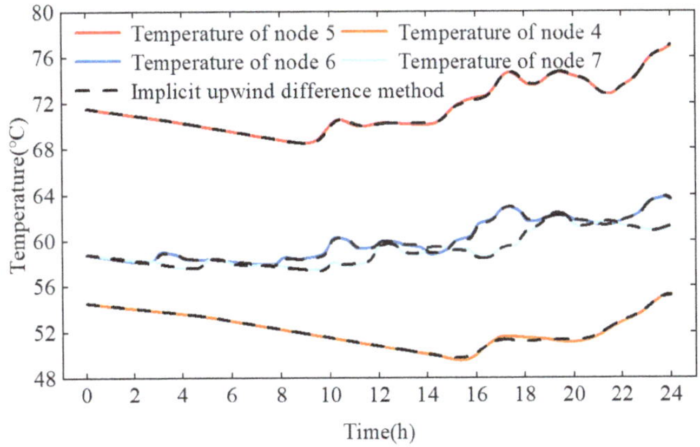

Figure 8. Temperature comparison of each node of the return pipe.

Table 4. Time of calculation.

Model	Time/s
The two-port model	0.057
The implicit upwind difference method	2.68

The above analysis shows that the two-port model can greatly reduce the solution time, and it has more significant advantages in the calculation speed when the heating system structure is more complex.

4.3. EDETCS

In this section, two models are applied to the EDETCS, considering the pipeline transmission process.

Model 1: A steady-state model of heating network used in Reference [10];

Model 2: The two-port model of heat transfer dynamics.

Figure 9 depicts the layout of the testing system, which includes two thermal units, G1 and G2, alongside a wind farm designated as W. The CHP unit serves as the system's heat source. Relevant system details can be found in Tables 5 and 6, while Table 7 provides information on pipeline specifications. Figure 10 displays the electric load, heat load, and projected peak output from wind power.

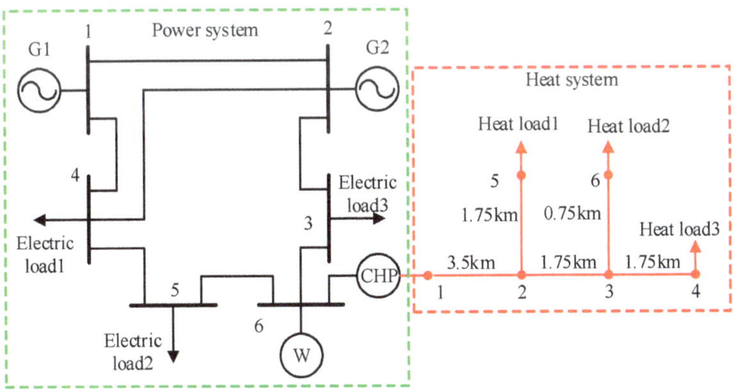

Figure 9. Structure diagram of Electro-Thermal Coupling System.

Table 5. Thermal power unit parameters.

Unit	P_{min}/MW	P_{max}/MW	Cost Coefficients	
			a^G/Yuan	b^G/(Yuan/MW)
G1	100	220	13.5	177
G2	10	100	40	130

Table 6. CHP unit parameters.

Output of CHP	P_{min}/ MW	P_{max}/ MW	Cost Coefficients
			a^{chp}/(Yuan/MW^2), b^{chp}/(Yuan/MW), c^{chp}/Yuan, d^{chp}/(Yuan/MW^2), e^{chp}/(Yuan/MW), f^{chp}/(Yuan/MW^2),
electric power	45	125	0.0032, 17.7, 181, 0.00085, 4.2, 0.00125

Table 7. Pipe parameters.

Node	Node	A/m^2	λ/(W/(m·°C)	m/(kg/s)	L/km	T_a/°C
1	2	0.5	0.25	360	3.5	0
2	3	0.5	0.25	150	1.75	0
3	4	0.5	0.25	100	1.75	0
2	5	0.5	0.25	210	1.75	0
3	6	0.5	0.25	50	0.75	0

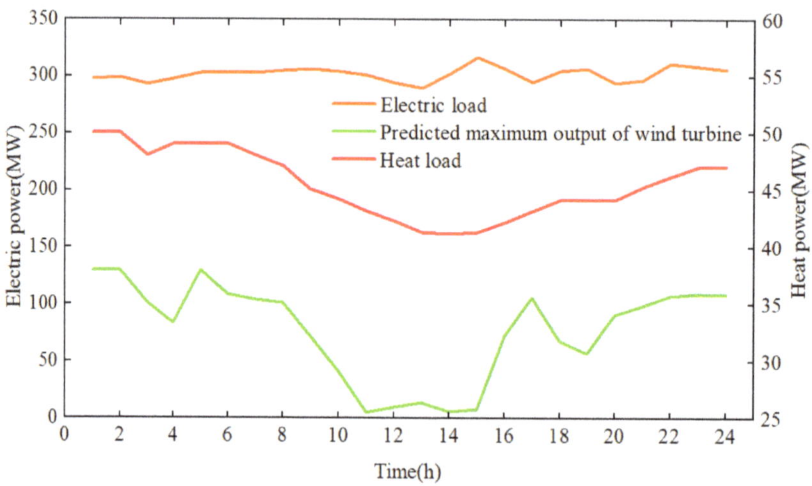

Figure 10. Prediction data curve.

The operating costs and wind curtailment costs of the two models are shown in Table 8. The operating costs of model 1 and model 2 are 97,408 yuan and 10,1596 yuan, respectively. Compared with model 1, the system applying model 2 can reduce the cost by 4188 yuan.

Table 8. Operation cost.

Model	Operation Cost/Yuan	Wind Curtailment Cost/Yuan
Model 1	101,596	10,303
Model 2	97,408	4379

Figure 11 shows the wind power scheduling decisions under the two models. The results show that the wind curtailment phenomenon occurs in model 1 at 1:00–3:00 and 5:00–6:00. The reason is that the CHP heat power corresponds to the heat load and cannot be flexibly adjusted. Compared with model 1, model 2 consumes more wind power. The reason is that model 2 does not need to follow the heat load and can reduce the CHP power of the unit at 1:00–3:00 and 5:00–6:00 to absorb wind power.

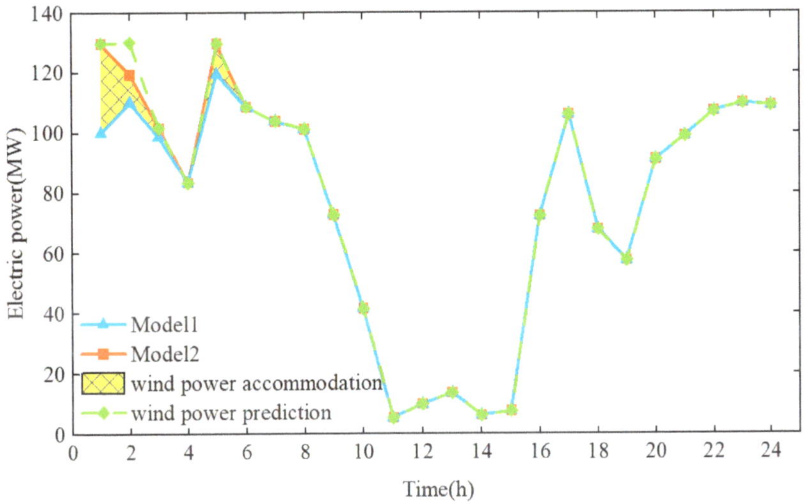

Figure 11. Wind power dispatch of the wind farm.

Figures 12 and 13 show the system's thermal scheduling and power scheduling results using the two models. In model 1, the results show that the CHP heat output changes with the heat load. As a result, the CHP power dispatch is also limited by the heat output. However, in model 2, CHP heat output and heat load do not correspond, and the two are decoupled so that the system can make full use of CHP units for economic optimization and increase wind power consumption.

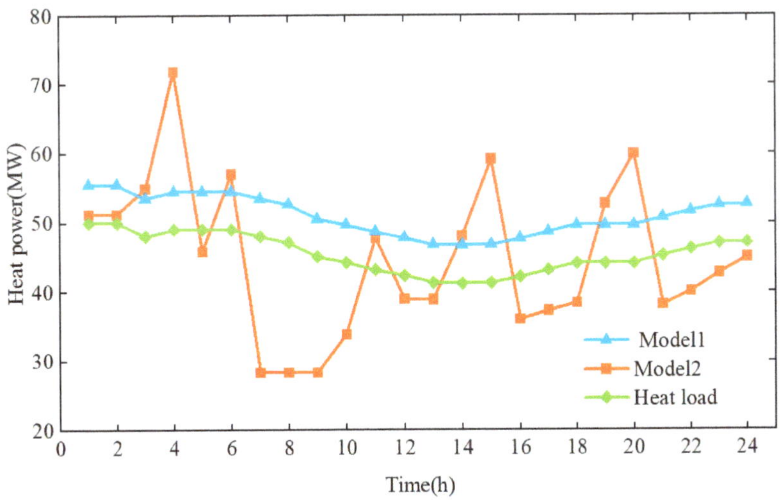

Figure 12. Heat output of the CHP unit.

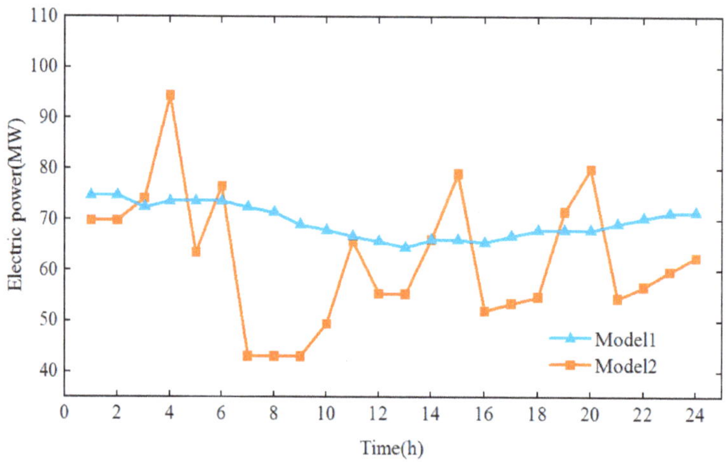

Figure 13. The power dispatch of the CHP unit.

Since the generation cost of G2 is the highest, the G2 scheduling strategies of the two models are maintained at the minimum output of 10 MW. Figure 14 shows the scheduling decision of thermal power unit G1. The results show that compared with model 1, the thermal power unit G1 of model 2 increases its output at 6:00–14:00 and 21:00–24:00 because G1 has the lowest power generation cost. This shows that the model considering the pipeline transmission process can enhance the operational flexibility and economy of the system operation.

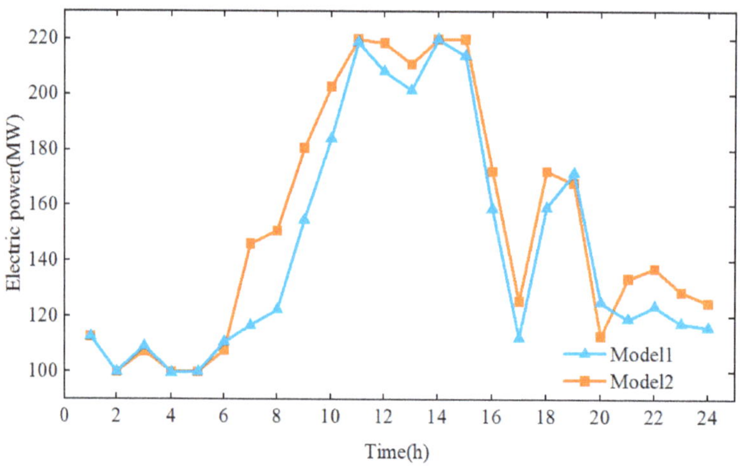

Figure 14. The power dispatch of G1.

5. Conclusions

To enhance the efficiency of addressing the power optimization challenges involving partial differential equation constraints—like the EDETCS, which factors in the dynamic aspects of heat transfer—this paper presents a two-port heat transfer model utilizing implicit upwind differences. The key findings are as follows:

(1) Compared with the implicit upwind difference method, the heat transfer two-port model can greatly reduce the computational model complexity and improve the computational efficiency while ensuring accuracy.

(2) Implementing this two-port model within the EDETCS allows for precise representation of the pipeline heat transmission process, effectively leverages the heat storage capabilities of the network, boosts wind energy utilization, and lowers the system's operational costs.

In the future, we will try to apply the two-port model to the simulation analysis and real-time control of the actual electro-heat integrated energy system.

Author Contributions: Conceptualization, J.C., Q.L. (Qinting Lin), Z.Y., Q.L. (Qingming Liu) and H.Z.; software, J.C., Q.L. (Qinting Lin), Z.Y., Q.L. (Qingming Liu) and H.Z.; writing—original draft preparation, J.C., Q.L. (Qinting Lin), Z.Y., Q.L. (Qingming Liu) and H.Z. All authors have read and agreed to the published version of the manuscript.

Funding: This research received no funding.

Data Availability Statement: The data presented in this study are available on request from the corresponding author due to laboratory rules.

Conflicts of Interest: The authors declare no conflicts of interest.

References

1. Paul, C.; Sarkar, T.; Dutta, S.; Hazra, S.; Roy, P.K. Optimal Power Flow of Multi-objective Combined Heat and Power with Wind-Solar-Electric Vehicle-Tidal Using Hybrid Evolutionary Approach. *Process Integr. Optim. Sustain.* **2024**, *8*, 1337–1367. [CrossRef]
2. Wang, Q.; Yang, M.; Li, P.; Jiao, M.; Yu, Y.; Li, M. Robust look-ahead dispatch of electricity-heat integrated energy system considering the flexibility recovery period of heating networks. *High Volt. Eng.* **2023**, *49*, 3173–3186.
3. Zhang, Y.; Liu, W.; Pang, Q.; Li, Y.; An, N.; Li, F. Multi-timescale trading strategies for the participation of multi-energy demand response in the consumption of blocked new energy sources. *Electr. Power Constr.* **2023**, *44*, 1–11.
4. Zhang, L.; Zheng, D.; Lu, T.; Chen, Q.; Ye, J.; Xiong, Z. Low-carbon economic dispatching in industrial park considering dynamic constraints of CHP units. *Electr. Power Constr.* **2022**, *43*, 122–130.
5. Wang, D.; Zhi, Y.-Q.; Jia, H.-J.; Hou, K.; Zhang, S.-X.; Du, W.; Wang, X.-D.; Fan, M.-H. Optimal scheduling strategy of district integrated heat and power system with wind power and multiple energy stations considering thermal inertia of buildings under different heating regulation modes. *Appl. Energy* **2019**, *240*, 341–358. [CrossRef]
6. Zhou, H.; Li, Z.; Zheng, J.H.; Wu, Q.H.; Zhang, H. Robust Scheduling of Integrated Electricity and Heating System Hedging Heating Network Uncertainties. *IEEE Trans. Smart Grid* **2019**, *11*, 1543–1555. [CrossRef]
7. Chen, F.; Yan, X.; Shao, Z.; Li, Y.; Zheng, X.; Zhang, H. Review on Modeling and Energy Flow Calculation Methods for Integrated Energy Systems. *High Volt. Eng.* **2024**, *50*, 1376–1391.
8. Huang, Y.J.; Sun, Q.Y.; Li, Y.S.; Gao, W.; Gao, D.W. A multi-rate dynamic energy flow analysis method for integrated electricity-gas-heat system with different time-scale. *IEEE Trans. Power Deliv.* **2023**, *38*, 231–243. [CrossRef]
9. Zeng, A.; Wang, J.; Zou, Y.; Wan, Y.; Hao, S.; Yuan, Y. Multi-time-scale optimal scheduling of integrated energy system considering heat storage characteristics of heating network. *High Volt. Eng.* **2023**, *49*, 4192–4202.
10. Tan, J.; Wu, Q.; Hu, Q.; Wei, W.; Liu, F. Multi-objective Optimization Dispatch for Integrated Electro-heating Systems Including Network Transmission Losses. *Power Syst. Technol.* **2020**, *44*, 141–154.
11. Tan, J.; Wu, Q.; Hu, Q.; Wei, W.; Liu, F. Adaptive robust energy and reserve co-optimization of integrated electricity and heating system considering wind uncertainty. *Appl. Energy* **2020**, *260*, 114230. [CrossRef]
12. Wang, Y.; You, S.; Zhang, H.; Zheng, X.; Zheng, W.; Miao, Q.; Lu, G. Thermal transient prediction of district heating pipeline: Optimal selection of the time and spatial steps for fast and accurate calculation. *Appl. Energy* **2017**, *206*, 900–910. [CrossRef]
13. Benonysson, A. Dynamic Modelling and Operational Optimization of District Heating Systems. Ph.D. Thesis, Technical University of Denmark, Kongens Lyngby, Denmark, 1991.
14. Benonysson, A.; Bohm, B.; Ravn, H.F. Operational optimization in a district-heating system. *Energy Convers. Manag.* **1995**, *36*, 297–314. [CrossRef]
15. Lu, S. Research on Modeling and Operation Optimization of Heat and Electricity Integrated Energy Systems. Ph.D. Thesis, Southeast University, Nanjing, China, 2021.
16. Chen, Y.; Guo, Q.; Sun, H.; Pan, Z. Integrated heat and electricity dispatch for district heating networks with constant mass flow: A generalized phasor method. *IEEE Trans. Power Syst.* **2020**, *36*, 426–437. [CrossRef]
17. Yang, J.; Zhang, N.; Botterud, A.; Kang, C. On an equivalent representation of the dynamics in district heating networks for combined electricity-heat operation. *IEEE Trans. Power Syst.* **2020**, *35*, 560–570. [CrossRef]

18. Hao, L.; Xu, F.; Chen, Q.; Wei, M.; Chen, L.; Min, Y. A thermal-electrical analogy transient model of district heating pipelines for integrated analysis of thermal and power systems. *Appl. Therm. Eng.* **2018**, *139*, 213–221. [CrossRef]
19. Chen, B.; Sun, H.; Yin, G.; Wu, W.; Guo, Q.; Chen, Y.; Pan, Z.; Wang, B. Energy circuit theory of integrated energy system analysis(II): Hydraulic circuit and thermal circuit. *Proc. CSEE* **2020**, *40*, 2133–2142.
20. Chen, Y.; Sun, H.; Guo, Q. Energy Circuit Theory of Integrated Energy System Analysis (V): Integrated Electricity-Heat-Gas Dispatch. *Proc. CSEE* **2020**, *40*, 7928–7937+8230.
21. Yang, J.; Zhang, N.; Kang, C. Analysis theory of generalized electric circuit for multi-energy networks -part one branch model. *Autom. Electr. Power Syst.* **2020**, *44*, 21–32.
22. Yang, J.; Zhang, N.; Kang, C. Analysis theory of generalized electric circuit for multi-energy networks -part two network model. *Autom. Electr. Power Syst.* **2020**, *44*, 10–21.
23. Chu, Z.; Zhao, L.; Sun, J.; Sun, X. Thermoelectric optimization of an integrated energy system with hydrogen energy storage considering thermal energy dynamic balance. *Power Syst. Prot.Control* **2023**, *51*, 1–12.
24. Yao, S.; Gu, W.; Lu, S.; Zhou, S.; Wu, Z.; Pan, G.; He, D. Dynamic Optimal Energy Flow in the Heat and Electricity Integrated Energy System. *IEEE Trans. Sustain. Energy* **2020**, *12*, 179–190. [CrossRef]

processes

MDPI

Article

Adaptive Scheduling Method of Heterogeneous Resources on Edge Side of Power System Collaboration Based on Cloud–Edge Security Dynamic Collaboration

Li Li [1], Shanshan Lu [1], Haibo Sun [1] and Runze Wu [2,*]

[1] State Grid Jibei Electric Power Co., Ltd., Economic and Technical Research Institute, Beijing 100038, China; jbjyyll@126.com (L.L.); 19959091712@163.com (S.L.); 17302252550@163.com (H.S.)
[2] School of Electrical Engineering, North China Electric Power University, Beijing 102206, China
* Correspondence: 17370559945@163.com

Abstract: In recent years, the large-scale integration of new power distribution technologies such as distributed power generation, electric vehicles, and flexible load control has led to a sharp increase in the operating pressure of the power cloud master station. To this end, an adaptive resource allocation method for edge-side general computing resources, which is used for cloud–edge collaborative security protection, is proposed. Firstly, considering the computing resources available to multiple edge substations, a Cloud–edge Collaborative Relay Business Security Protection Model (C2RBSPM) is constructed. Then, with the goal of minimizing the operating pressure of the maximum cloud master, the corresponding linear programming problem is established, and finally the Karush–Kuhn–Tucker (KKT) is used to solve it quickly. The simulation results show that the proposed method can reduce the expected operating pressure of the cloud master station by up to 35.19%. Therefore, reasonable mining of available computing resources on the edge side and relay security protection can effectively reduce the operating pressure of the cloud master station, and improve the operation efficiency of the system. This approach is of great significance for the flexible, intelligent, and digital transformation of the power distribution system in the future.

Keywords: cloud–edge collaboration; security protection; resource allocation; new power distribution system

1. Introduction

The National Development and Reform Commission (NDRC) and the National Energy Administration (NEA) of the People's Republic of China pointed out in their jointly issued "Guiding Opinions on the High-Quality Development of Distribution Grids under New Situations" that, by 2030, the transformation of distribution grids to be flexible, intelligent, and digital will be basically completed [1]. However, as the core business hub, the cloud master station supports a great proportion of new distribution business security protection needs. With the integration of a large number of power electronic devices into the power grid, under the complex application scenarios of multi-domain, multi-terminal, and multi-task, the expected operational pressure on the cloud master station has increased sharply. Therefore, how to reduce the expected operational pressure on the cloud master station has become one of the hotspots of the current stage of research [2–5].

To address the issue of reducing the expected operational pressure on cloud master stations for new types of distribution business, Wang et al. [6] constructed a new type of power system security protection system from the aspects of management and technology. Li et al. [7] designed a centralized information recognition algorithm based on the working principle of an electric power networked ordering system, to achieve reasonable centralized information cloud scheduling. The authors in [8] combined the architecture of the Internet of Things in the power sector. They put forward corresponding security protection countermeasures according to the security risks faced by intelligent terminals. However, from the perspective of cloud–edge collaboration, the above work is only carried out at the cloud master station and does not consider some of the available computing resources in the edge devices, so more flexibility is still needed in resource scheduling.

As a data processing center, the cloud master station is an integral part of the cloud–edge architecture of the new power distribution system in real life. To alleviate the expected operation pressure of the cloud master station, in terms of edge arithmetic discovery [9], it is proposed in [10,11] that containerized applications can be extended to the nodes and devices at the edge, and some of the edge-side device resources can be utilized for security protection. The authors in [12] proposed the detection and protection of false data injection attacks, which are useful attempts to defend the power information network security defense mechanism. Younes et al. used a flexible laboratory-scale test platform to test new devices, control functions, and smart electronics configuration and communication systems to support the monitoring, control, and protection of power grids [13]. The cyber security protection of foreign power systems started early, and a relatively complete power system security protection system has been formed [14]. However, at this stage, edge stations often support multiple business cloud masters at the same time, and these methods fail to adequately consider the impact of its distributed synergy on the balance of the expected operating pressure of each cloud master.

From the perspective of cloud–edge collaboration, the security protection of the power distribution system is currently only concentrated at the cloud master station, and the computing resources on the edge side are not fully utilized. In addition, in the face of the situation that one edge substation supports multiple cloud master stations, considering the expected operation pressure balance of the cloud master station, the current edge computing power allocation is inflexible. Therefore, this article studies the cloud–edge-terminal network architecture, which reduces the operating pressure of the cloud master station by mining the available computing resources on the edge side and transferring part of the security protection work to the edge side, and comprehensively considers the service delay requirements and computing resource allocation to model the problem. To solve the problem, we can refer to the master-allocation-micro-integrated collaborative control strategy for distributed resource potential in [15] and propose an adaptive scheduling method of heterogeneous resources on the edge side of power distribution collaboration based on cloud–edge security dynamic collaboration (ASM-CESDC). The main contributions of this article are summarized below:

(1) A multi-stage relay business security protection model is constructed for the edge stations and the business cloud master by combining the available arithmetic resources of each edge station.
(2) Secondly, the corresponding linear programming problem is constructed by considering the distributed collaboration of multiple edge stations with the objective of minimizing the expected operating pressure of the maximum business cloud master; the problem is solved quickly by using KKT.
(3) The simulation results show that the proposed ASM-CESDC algorithm can significantly reduce and minimize the expected operating pressure of the main station of the maximum business cloud by fully exploiting the arithmetic resources of multi-domain edge stations.

The rest of this article is organized as follows: Section 2 provides an overview of relevant important works. Section 3 introduces the cloud–edge collaborative computing framework and system model. In Section 4, an optimization problem is proposed with the goal of minimizing the maximum operating pressure of the cloud master. Section 5 presents the solutions to problems. Section 6 describes the simulation results. Finally, Section 7 concludes this article.

2. Related Work

In recent years, with the rapid construction of the new power system, massive distributed resources are connected to the power grid, but it also brings more collaborative and interactive communication needs and security threats at the same time [16]. In terms of distributed security, based on the multi-time-scale interaction between distributed resources and the power grid, the authors in [17] constructed a cloud–edge-terminal collaborative distributed resource interactive communication networking architecture, and further proposed the interactive security protection strategy of distributed resources, which can ensure the information security of distributed resources participating in the power grid interaction. In terms of cloud–edge-terminal collaborative security protection, an architecture that integrates real-time perception, dynamic decision making, and active defense, which deeply integrate the cloud, edge, and terminal to form security collaboration capabilities, was mentioned in [18]. It further realizes the consistency of cloud, edge, and device risk perception, policy decision making, and attack and defense, and implements the efficient and low-cost collaborative protection of security risks. In terms of cross-domain security protection, Zhao et al. [19] proposed the cyber–physical security characteristics that integrate physical security, functional safety, and network security, and studied key technologies such as cross-domain security threat propagation modeling, collaborative situational awareness and detection, and multi-spatiotemporal linkage protection, and explored their application prospects in combination with power monitoring services, so as to provide support for the safe operation of new power systems.

In addition, a lot of research has been performed, in [20–22], on single-objective, dual-objective, and multi-objective directions in the Internet of Things, such as response time, reliability, service quality, and resource utilization. Despite the progress made in the above related research, there are still deficiencies in the flexibility of collaborative resource scheduling at the cloud–edge. This limitation makes part of the arithmetic resources of the edge devices not fully explored and utilized, limiting the overall computing power of the system and the potential for optimal resource allocation [23].

Edge computing plays a key role in alleviating the operational pressure of the cloud master, and the efficient shunting and optimization of the computational load can be achieved through reasonable scheduling of edge arithmetic resources [24]. The authors in [25] address the dynamic change in computing load in the "cloud master", establishing a dual-objective optimization model of the optimal allocation of cloud resources with the goal of the shortest average response latency and the smallest energy consumption of the "cloud master". The introduction of edge computing technology can redefine the relationship between the cloud, management, and device; Sun et al. [26] proposed edge computing technology applied to the distribution of the Internet of Things, in situ, to realize real-time and efficient lightweight data processing, and with a new generation of a distribution automation cloud master for the network, data, business, and other aspects of the synergies, to achieve the autonomy of the distribution station area. As mentioned in [27], a cloud–edge-device collaborative network task offloading framework named FreeOffload realizes real-time awareness of computing resources and network status, and is designed for flexible offloading of heterogeneous embedded terminal tasks. In addition, its small-scale cloud–edge-terminal collaborative prototype experiment system can also efficiently and flexibly offload the task of the terminal.

Edge computing can be used to transfer tasks from the cloud master to the edge-side and terminal devices, which can alleviate the expected operation pressure of the business cloud master to a certain extent. However, the current edge substations usually need to support multiple cloud masters at the same time, and the existing research has not yet fully considered the impact of the edge substations on the balanced nature of the expected operation pressure of the various cloud masters, which restricts the overall effect of the optimization of resource scheduling [28]. In this paper, the flexibility of resource allocation can be improved by exploring the available computing resources and implementing the adaptive allocation of computing resources at the edge. At the same time, distributed security protection can effectively reduce and balance the expected operating pressure of each cloud master.

3. Cloud–Edge Collaborative Computing Framework and Scheduling Model

3.1. Cloud–Edge Collaborative Computing Framework

The new power system cloud–edge cooperative computing framework is divided into cloud, edge, and terminal three-layer structures. Among them, the terminal equipment performs data acquisition tasks through various types of sensors and then uploads the collected massive data to the edge station; the edge station, as a transit node and a real-time processing center in the data link, completes the preliminary processing of the edge data by using its local resources, and only sends the processing results to the cloud master station, and shares some of the computational tasks for the cloud master station; the cloud master allocates saved computing resources to data analysis and task scheduling efforts for faster and more efficient intelligent decision making.

The massive integration of heterogeneous devices at terminals has led to a sharp increase in the volume of data within the distribution grid. The varying latency requirements of different businesses have significantly heightened the anticipated operational pressure on the cloud master station. From the perspectives of security businesses and dispatch management, a multi-stage relay-based business security protection model can be designed to alleviate the expected operational pressure on the cloud master station.

3.2. System Model

The proposed multi-stage relay-based business security protection model includes three components: the edge substation security protection, upload of business data and security process, and cloud master station security protection. This model can be represented using a directed graph, $G(B, E, C)$, where $b_i \in B$ represents a type of business, each corresponding to a business cloud master station; $e_i \in E$ represents an edge substation encompassing multiple types of businesses; and $c_i \in C$ represents a business cloud master station. As illustrated in Figure 1, various areas including the resource, power grid, storage, and load will first transmit its businesses to edge substations, and each edge substation will transmit the data to the cloud master after preliminary processing. In the process of initial processing, the edge substation initially allocates its available computing resources to each business. Each business completes the maximum number of security protection steps at the edge substations that satisfies its latency constraints. Subsequently, the edge station transmits the processed business data to its corresponding cloud main station. Upon receiving the data, each cloud master station determines the next required security protection steps and completes all security protection tasks for the data. At the edge substation, an adaptive computing resource allocation strategy is employed to optimize the protection process, aiming to reduce and minimize the maximum anticipated operational pressure on the business cloud main stations.

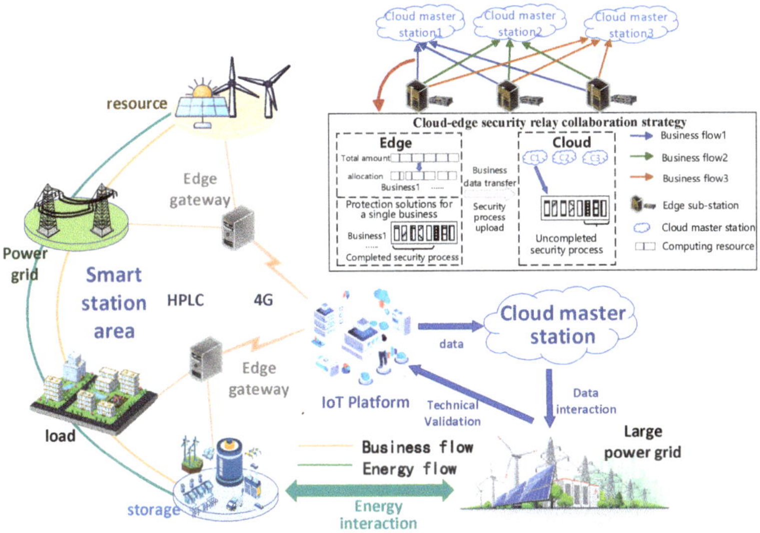

Figure 1. Schematic diagram of cloud–edge security relay coordination strategy.

4. Computing Resource Allocation Strategy

4.1. Allocation of Computational Resources at Edge Stations

The computational resources available at edge stations for data security are constantly changing. Therefore, it is necessary to allocate the limited computational resources reasonably among various businesses as in [29]. Taking a single-edge substation as an example, the computing resources allocated to each business are constrained by Equation (1).

$$\sum_{i=1}^{\lambda} r_i^{e_t} \le r_{\max}^{e_t} \tag{1}$$

where λ is the total number of business types included in each edge station; $r_i^{e_t}$ is the computational resources allocated by the edge station to a business at time t; and $r_{\max}^{e_t}$ is the total computational resources available at the edge station for data security at time t.

Considering the latency constraints of each business and the computational resource constraints of the edge stations, the number of security stages ω that each business needs to complete at the edge station level is determined. Taking business i as an example, the latency constraint is calculated as shown in Equation (2).

$$t_{i,edge} + t_{i,trans} + t_{i,cloud} \le t_{i,qos} \tag{2}$$

where $t_{i,edge}$ is the total latency of business i at the edge station, $t_{i,edge} = t_{s_i,edge} + t_{i,p}$, $t_{i,p}$ is the encryption and decryption latency of business i, and $t_{s_i,edge}$ is the latency incurred by business i for completing data security at the edge station; $t_{i,trans}$ is the transmission latency between the edge station and the corresponding cloud master station for business i; $t_{i,cloud}$ is the latency at the cloud master station for business i;$t_{i,qos}$ is the latency constraint for business i.

The calculation method for security latency $t_{s_i,edge}$ is shown in Equation (3).

$$t_{s_i,edge} = \sum_{x=1}^{\omega} t_{s_i,edge}^{x}, \omega \le \varpi \tag{3}$$

where ω corresponds to the number of security stages that business i needs to complete at the edge station; ϖ corresponds to the total number of security stages; and $t^{x}_{s_i,edge}$ is the time consumed by business i to complete security stage x, which is calculated as shown in Equation (4).

$$t^{x}_{s_i,edge} = d_i \cdot \frac{\gamma_{i,k}}{r_i^{e_t}} \tag{4}$$

where d_i is the data size for business i; $\gamma_{i,k}$ is the computational resources required to process a unit of data for security stage x; and $r_i^{e_t}$ is the computational resources allocated by the edge station to business i at time t.

4.2. Calculation of Computing Resources for Cloud Master Station

According to Equation (3), the number of security links ω completed by the business at the edge substation can be determined, and the remaining security protection stages can be completed at the cloud master. When the business data are transmitted from the edge station to the business cloud master, the business cloud master first determines the rest of the security links that need to be carried out. If $\omega = \varpi$, the business has already completed all the security work, and the required computing power resources of the business cloud master it belongs to are recorded as 0. On the contrary, it calculates the required computing power resources of the cloud master it belongs to, and takes business i as an example; the method is as follows:

First, the security time of the business cloud master $t_{s_i,cloud}$ is calculated based on the business delay constraint in Equation (5).

$$t_{s_i,cloud} = t_{i,cloud} - t_{i,p} \tag{5}$$

where $t_{i,cloud}$ is the total time left for the business cloud master, and $t_{i,p}$ is the encryption and decryption latency for business i. Based on $t_{s_i,cloud}$, the computational resources required by the business cloud master station r_i^{cloud} can be calculated, which is expressed as shown in Equation (6).

$$r_i^{cloud} = \sum_{j=1}^{m} r_{i,j}^{cloud} \tag{6}$$

where m is the total number of edge stations; $r_{i,j}^{cloud}$ is the computational resources required to complete security for the data sent from the edge station j to the cloud master station for business i, which is calculated as shown in Equation (7).

$$r_{i,j}^{cloud} = \frac{\sum_{y=\omega+1}^{\varpi} \gamma_{i,k} \cdot d_i}{t_{s_i,cloud}} \tag{7}$$

Based on Equation (7), the required computational resources of each business cloud master station can be calculated. The expected operational pressure on the business cloud master station φ is defined as the ratio of the operating frequency required to complete all data security tasks within the latency constraint to the maximum work frequency. The size of packets, latency, and computing resources required to process per bit vary for each business. So, taking business i as an example, the expected operational pressure φ_i for the corresponding cloud master station is calculated as shown in Equation (8).

$$\varphi_i = \frac{r_i^{cloud} + r_{i,other}^{cloud}}{r_{i,cloud}^{\max}} \tag{8}$$

where $r_{i,other}^{cloud}$ represents the computational resources occupied by tasks other than security protection at the cloud master station to which business i is assigned at time t; $r_{i,cloud}^{\max}$ represents its maximum operational frequency. Based on the calculation results, φ_i, it is possible to calculate whether the operating pressure of the cloud master station will be reduced by mining the computing power resources on the edge side.

4.3. Problem Formulation

Based on the multi-stage relay business security protection model, the aim is to reduce and minimize the maximum expected operational pressure on the cloud master stations through the distributed collaboration of multi-domain edge stations. The optimization problem is constructed for the model containing n business cloud master stations and m edge stations. The optimization objective is to minimize the maximum expected operational pressure on the business cloud master station, which is expressed as shown in Equation (9).

$$\begin{aligned} &\min_{r} \max_{i \in B} \{\varphi_i\} \\ &s.t. \begin{cases} C1: \sum_{i=1}^{\lambda} r_i^{e_t} \leq r_{\max}^{e_t} \\ C2: 0 \leq r_i^{e_t} \leq r_{\max}^{e_t} \\ C3: t_{i,edge} + t_{i,trans} + t_{i,cloud} \leq t_{i,qos} \end{cases} \end{aligned} \tag{9}$$

where $C1$ is the computational resource constraint, meaning that the total computational resources allocated to all businesses by the edge station cannot exceed its total computational resources; $C2$ ensures that the computational resources allocated to each business are non-negative and cannot exceed the total computational resources; and $C3$ is the latency constraint, where the sum of the transmission latency and the security protection latency of each business is required to satisfy the latency requirement.

5. Problem Solving

5.1. Solution Process

To facilitate problem solving, it is assumed that each edge substation and cloud master possesses sufficient computational resources to complete all security protection tasks within the specified business delay requirements. Therefore, in the process of solving, the delay constraint C3 is used as the criterion for calculating the number of security protection stages completed by each service at the edge substation. In the context of arithmetic resource allocation, the aforementioned optimization problem can be reformulated as shown in Equation (10).

$$\begin{aligned} &\min_{r} \sum_{i=1}^{n} \varphi_i \\ &s.t. \begin{cases} C1: \sum_{i=1}^{\lambda} r_i^{e_t} \leq r_{\max}^{e_t} \\ C2: 0 \leq r_i^{e_t} \leq r_{\max}^{e_t} \end{cases} \end{aligned} \tag{10}$$

For the inequality constraint optimization problem, the Karush–Kuhn–Tucker (KKT) method is widely used to transform inequalities and equations at present. The aforementioned linear programming problem exhibits strict concavity with respect to the variable $r_i^{e_t}$. Consequently, a unique extreme point exists that satisfies the KKT conditions [30]. The corresponding Lagrangian function is presented as shown in Equation (11).

$$g(\hat{f}) = \sum_{i=1}^{n} \varphi_i(r_i^{e_t}) + \sum_{i=1}^{n} \alpha_i [r_{\max}^{e_t} - r_i^{e_t}] + \sum_{i=1}^{n} \beta_i r_i^{e_t} + \chi_k [r_{\max}^{e_t} - \sum_{i=1}^{n} r_i^{e_t}] \tag{11}$$

Among them, α_i, β_i, and χ are the Lagrange multipliers corresponding to $r_{\max}^{e_t} \geq r_i^{e_t}$, $r_i^{e_t} \geq 0$, and $r_{\max}^{e_t} \geq \sum_{i=1}^{n} r_i^{e_t}$, respectively. As an example, the KKT conditions of the cloud master to which the business i belongs are as shown in Equation (12).

$$\begin{cases} \varphi_i'(f_i^{e_t}) - \dot{\chi}_k - \dot{\alpha}_i + \dot{\beta}_i = 0 \\ \dot{\alpha}_i [r_{\max}^{e_t} - \dot{r}_i^{e_t}] = 0, \dot{\beta}_i \dot{r}_i^{e_t} = 0 \\ \dot{\chi}_k [r_{\max}^{e_t} - \sum_{i=1}^{n} \dot{r}_i^{e_t}] = 0 \\ \dot{\chi}_k \geq 0, \dot{\alpha}_i \geq 0, \dot{\beta}_i \geq 0 \end{cases} \tag{12}$$

$\varphi_i'(r_i^{e_t})$ is the first-order derivative of $\varphi_i(r_i^{e_t})$. $\dot{r}_i^{e_t}\sum_{i=1}^{\lambda} r_i^{e_t} \le r_{\max}^{e_t}$ represents the optimal solution for the cloud master allocation associated with operation i. $\dot{\alpha}_i$,$\dot{\beta}_i$, and $\dot{\chi}$ are the optimal solutions corresponding to α_i, β_i, and χ. A further analysis of Equation (12) leads to the following conclusions, and the properties of the optimal solution are summed up as shown in Equations (1)–(3):

(1) When $\dot{\alpha}_i > 0, \dot{r}_i^{e_t} = r_{\max}^{e_t}$, while $\dot{\beta}_i = 0$ and $\varphi_i'(r_i^{e_t}) - \dot{\chi}_k = \dot{\alpha}_i$;

(2) When $\dot{\alpha}_i = 0$ and $\dot{\beta}_i = 0, 0 \le \dot{r}_i^{e_t} \le r_{\max}^{e_t}$, at the time $\varphi_i'(r_i^{e_t}) = \dot{\chi}_k$;

(3) When $\dot{\beta}_i > 0, \dot{r}_i^{e_t} = 0, \dot{\alpha}_i = 0$, and $\varphi_i'(r_i^{e_t}) - \dot{\chi}_k = -\dot{\beta}_i$.

Based on the aforesaid properties, it can be discerned that there exists a certain correlation between $\dot{r}_i^{e_t}$ and $\varphi_i'(r_i^{e_t}) - \dot{\chi}_k$. $\varphi_i'(r_i^{e_t})$ decreases monotonically with the increase in $\dot{\chi}_k$. The optimal solution $\dot{r}_i^{e_t}$ can be computed based on the relationships among $\dot{\chi}_k$, $\varphi_i'(0)$, and $\varphi_i'(r_{\max}^{e_t})$.

Specifically, when $\dot{\chi}_k < \varphi_i'(r_{\max}^{e_t})$, Property (1) is satisfied, in which case $\dot{r}_i^{e_t} = r_{\max}^{e_t}$; when $\dot{\chi}_k > \varphi_i'(0)$, Property (3) is satisfied, in which case $\dot{r}_i^{e_t} = 0$; when $\varphi_i'(r_i^{e_t}) = \dot{\chi}_k$ and $\varphi_i'^{-1}(r_i^{e_t}) = \dot{r}_i^{e_t}$, Property (2) is satisfied, by iterating $\dot{\chi}_k$ until $\sum_{i=1}^{n} \dot{r}_i^{e_t}$ approaches $r_{\max}^{e_t}$, and the optimal computing power resource allocation scheme can be obtained.

5.2. Algorithm Design and Complexity Analysis

Based on the decomposition of the problem in local convex approximation processing in Section 5.1, the global iterative optimization is carried out to complete the specific iteration process of adaptive scheduling of cloud–edge computing resources, and the pseudocode is shown as Algorithm 1.

Algorithm 1: An adaptive Scheduling Method of Heterogeneous Resources on the Edge-Side of Power Distribution Collaboration based on Cloud-Edge Security Dynamic Collaboration

Inputs: latency requirements of each business, total computing power resources of the edge substation, total computing power resources of the cloud master, and businesses included in each edge substation.
Outputs: Expected operating pressure of each cloud master.
1: Calculate the available arithmetic resources at each edge substation.
2: for $i = 1,2,\ldots,m$ do
3: Edge substations adaptively allocate computing power resource in accordance with the method described in Section 4.1.
4: for $j = 1,2,\ldots,n$ do
5: for $k = 1,2,\ldots,\kappa$ do
if $t_{i,edge} + t_{i,trans} + t_{i,cloud} \le t_{i,qos}$// Based on the delay constraint, the security protection process that each business can complete at the edge substation is determined
else
Safeguarding process $x - 1$
end
end
6: Each edge substation transmits the business to the corresponding cloud master station and accomplishes the remaining security protection work.
7: Calculate the expected operating pressure of each cloud master according to Equation (8).
8: return Expected operational pressure φ_i on each cloud master.

Taking a single-edge substation as an example, the block diagram of its security process is shown below (Figure 2):

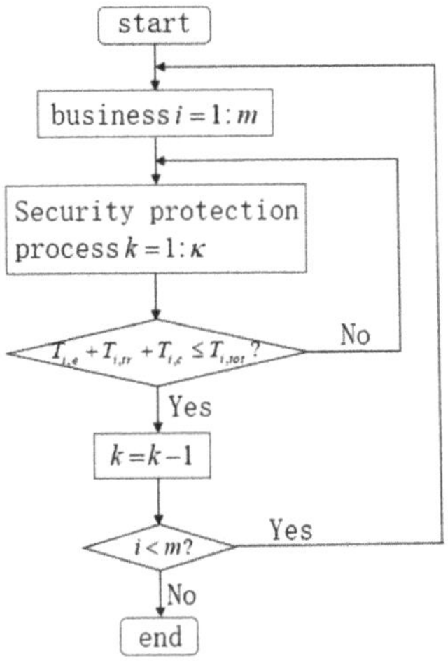

Figure 2. Block diagram of single-edge substation.

The complexity of this algorithm mainly stems from two aspects: firstly, the calculation of CPU frequency allocation for edge substations, and secondly, the calculation of the data security process at the edge substation and the cloud master station. If the number of edge substations is X, the number of cloud master stations is Y, and the number of security processes is ϖ. As the iterative process of KKT is accomplished through dichotomy, the complexity of completing the allocation of computing power resources at an edge substation can be expressed as $o(\log_2 Y)$. Since each cloud master station needs to comprehensively consider the business from each edge substation and its security process, its complexity increases monotonically with the growth of the number of edge substations and cloud master stations, and the total complexity of Algorithm 1 is $o(X\varpi\log_2 Y)$.

6. Experimental Results and Analysis

In this section, MATLAB R2020a is employed to simulate the proposed edge-side communication resource adaptive allocation method for cloud–edge collaborative security protection and its related algorithms to validate the effectiveness of ASM-CESDC. The parameters of the system simulation environment are set as shown in Table 1.

Table 1. Simulation Parameters.

Parameters	Numerical Value
Maximum CPU operating frequency of the edge substation	3 GHz
Total number of protection links	5
Total amount of data transmitted by each edge substation	300 k
Number of edge substations	5
Number of business cloud masters	5
Maximum operating frequency of business cloud master CPUs	5 GHz
Delay requirements for each operation	[0.8 s, 1.7 s, 1.8 s, 1.8 s, 1.5 s]
Business transmission bandwidth	1000 bps
Number of frequency cycles per unit bit required for each business	[420, 1000, 1200, 1700, 2200]

In real scenarios, the types of businesses included in each edge substation vary due to the different geographic environments in which they are located. In order to verify the effectiveness of the proposed algorithm, the edge substation is set up to contain random businesses and all businesses in the simulation process for comparison experiments, and the details of the edge substation business type settings are shown in Table 2.

Table 2. Edge Substation Business Type Settings.

Edge Substation Number	Random Business	All Businesses
1	Business 1, 2	Business 1–Business 5
2	Business 1, 2	Business 1–Business 5
3	Business 1, 2, 3	Business 1–Business 5
4	Business 1, 2, 3	Business 1–Business 5
5	Business 1, 4, 5	Business 1–Business 5

In order to verify the advantages of ASM-CESDC in the adaptive allocation of computing resources, this paper compares ASM-CESDC with Not Utilizing the Available Computing Resources of the Edge Substations (NACR) and Evenly Distributing the Available Computing Resources of Edge Substations (EDACR). The focus is on analyzing the expected operational pressure experienced by each business cloud master. In a system model comprising five edge substations and three cloud masters, Figures 3 and 4 illustrate the simulation results of random-business and full-business expected operating pressures, respectively.

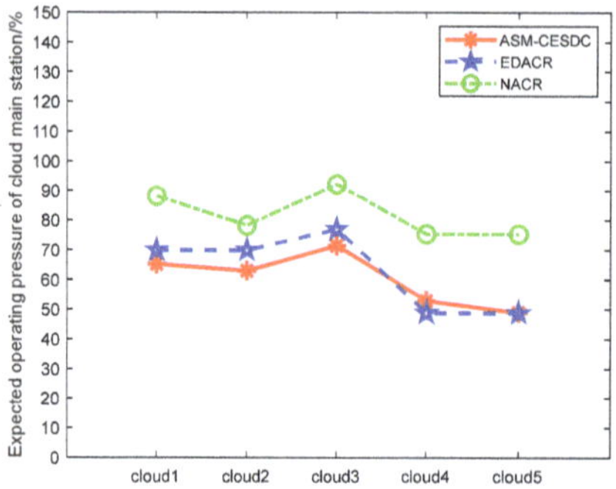

Figure 3. Simulation result of random-business expected operating pressure.

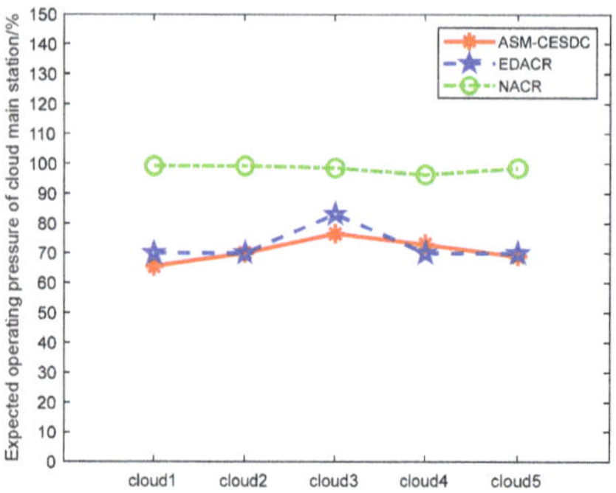

Figure 4. Simulation result of full-business expected operating pressure.

As shown in Figure 3, by utilizing the available computing resources of the edge substations, both EDACR and ASM-CESDC can effectively reduce the expected operation pressure of the main station of each business cloud compared with NACR. However, a comparison of the results from EDACR and ASM-CESDC reveals that ASM-CESDC is more effective in minimizing the maximum expected operating pressure of the cloud master. Specifically, the maximum expected operating pressure decreases from 77.02% to 71.54% when using EDACR.

Figure 4 presents the simulation results of the expected operating pressure for full business at each edge substation under identical conditions. Compared with Figure 3, the operating pressure of the cloud master station shown in Figure 4 is relatively high due to the increase in the number of service types contained in each edge substation, which leads to an increase in the overall data volume of the system. However, based on the simulation results, it is evident that ASM-CESDC demonstrates greater effectiveness in minimizing the expected operating pressure of the largest cloud master. The maximum expected operating pressure of the cloud master shown in Figure 4 decreases from 83.18% to 76.54% of EDACR.

Figures 5 and 6 show the simulation results corresponding to the system model with 10 edge substations and 5 business cloud masters, respectively. Regarding the method of computing power resource allocation, the comparison results are the same as the conclusions drawn from the above Figures 3 and 4. That is, the ASM-CESDC algorithm is superior in minimizing the maximum expected operating pressure of the cloud master station. However, since the increase in the number of edge substations leads to an increase in the total amount of data for each business compared with Figures 3 and 4, the expected operating pressure of each business cloud main station shown in Figures 5 and 6, in the case of not utilizing the available computing power resources of the edge substations, surges sharply, and even exceeds the performance upper limit of the cloud main station, thereby the pressure exceeds 100%.

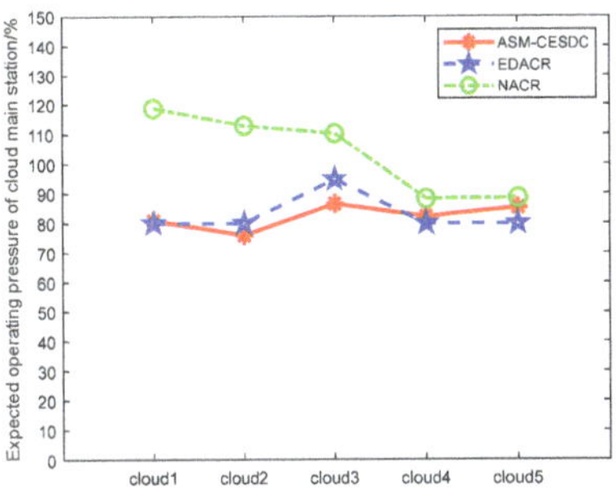

Figure 5. Simulation result of random-business expected operation pressure under 10 edge substations.

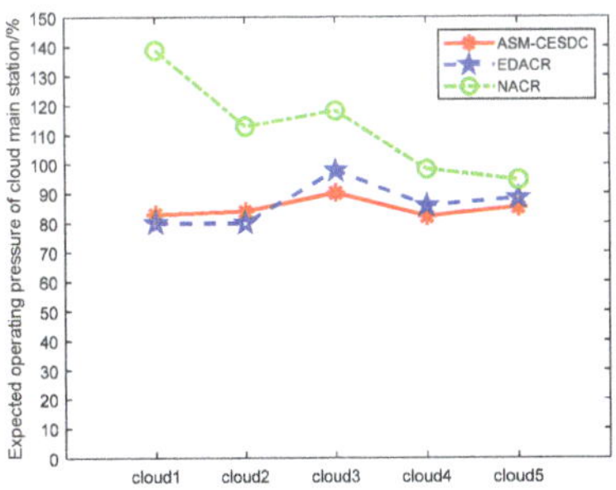

Figure 6. Simulation result of full-business expected operation pressure under 10 edge substations.

Accordingly, in comparison to Figure 5, the subsequent introduction of businesses depicted in Figure 6 influences the allocation of computing power resources at the edge substation. The diversification of business types results in an increased volume of data associated with these businesses, thereby heightening the anticipated operational pressure on the cloud master. Therefore, the expected operating pressure of each business cloud master corresponding to EDSCR and ASM-CESDC is larger compared to the results in Figure 5. In addition, the ASM-CESDC algorithm is more effective in minimizing the maximum expected cloud master operating pressure, which decreases from 94.82% to 86.54% for EDSCR as shown in Figure 5 and decreases from 97.81% to 90.32% for EDSCR as shown in Figure 6.

Figures 7 and 8 present the simulation results regarding the expected operational pressure of random business and full business under a frequency of 4 GHz for edge substation arithmetic. As the computational resources of the edge substation increase, the available computational resources that can be allocated to each business also rise correspondingly. Consequently, this leads to a reduction in the expected operational pressure experienced by each cloud master. The comparative analysis presented in Figures 7 and 8 alongside Figures 3 and 4 indicates that an increase in available arithmetic resources leads to a reduction in the expected operational pressure on the cloud master during distributed cooperative security protection facilitated by edge substations. Furthermore, the ASM-CESDC algorithm demonstrates superior efficacy in minimizing the maximum expected operating pressure of the cloud master. This is evidenced by a decrease from 73.02% to 70.54% of EDACR, as illustrated in Figure 7, and a reduction from 81.18% to 76.75% of EDACR, as shown in Figure 8.

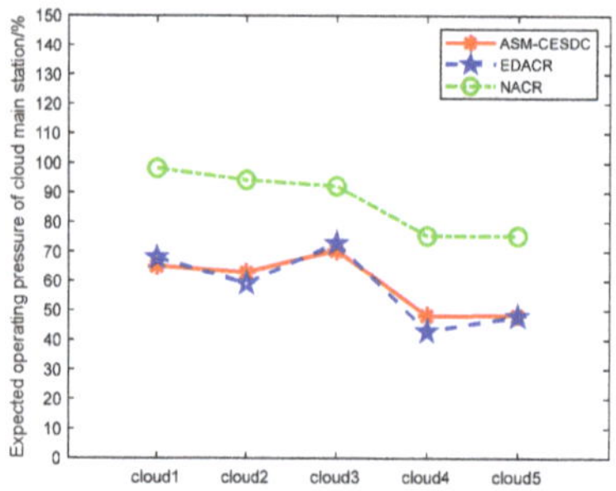

Figure 7. Simulation result of random-business expected operation pressure under computing power of 4 GHz of edge substations.

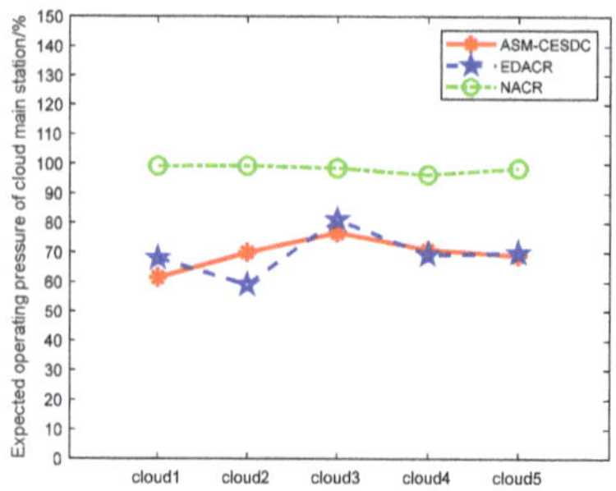

Figure 8. Simulation result of full-business expected operation pressure under computing power of 4 GHz of edge substations.

Figures 9 and 10 present the simulation results regarding the expected operational pressure of random business and full business at a frequency of 6 GHz, corresponding to the computational capacity of the cloud master station. A comparative analysis between

Figures 3 and 9 as well as between Figures 4 and 10 indicates that when distributed cooperative security protection is implemented with the assistance of edge substations, an increase in the computational resources available to the cloud master significantly alleviates its expected operational pressure. In addition, the simulation results indicate that the ASM-CESDC algorithm is more effective in minimizing the maximum expected operating pressure of the cloud master. This effectiveness is illustrated in Figure 9, where a reduction from 71.18% to 66.54% of EDACR is observed, and in Figure 10, which shows a decrease from 73.25% to 66.54% of EDACR.

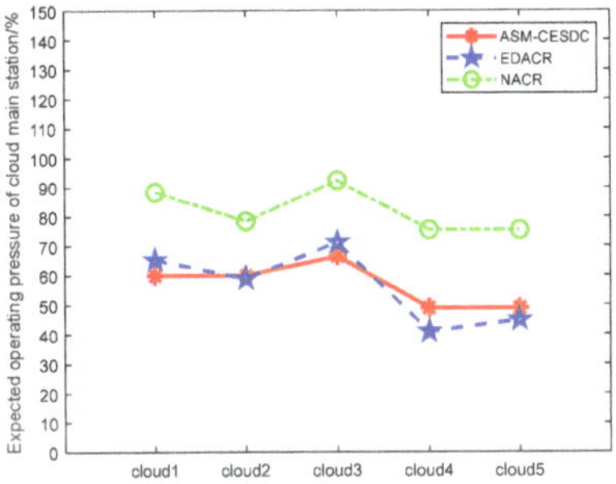

Figure 9. Simulation result of random-business expected operation pressure under computing power of 6 GHz of cloud master stations.

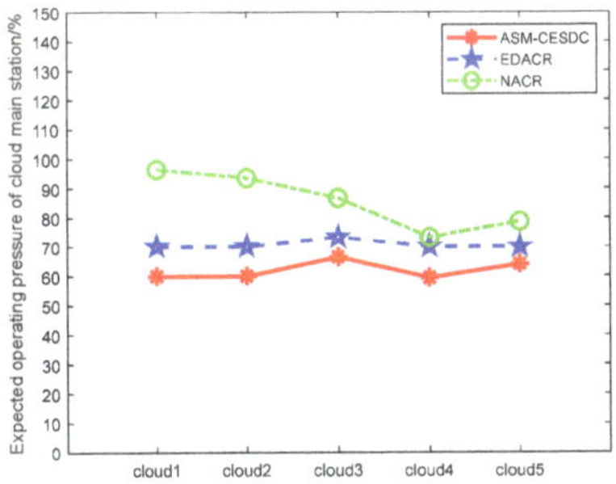

Figure 10. Simulation result of full-business expected operation pressure under computing power of 6 GHz of cloud master stations.

As can be seen from Table 3, the change in the number of edge substations has the greatest impact on the expected pressure of the cloud master, because as the number of edge substations increases, the overall data volume in the network increases. However, the increase in computing power resources of the edge substation has the least impact on the expected operating pressure of the cloud master station, because the edge substation needs to allocate the available computing power resources to different services, so the increase in its total amount has the least impact on the overall network.

Table 3. The maximum expected operating pressure drop of the cloud master.

Simulation Environment	Random Business	All Businesses
5 edge substations	5.48%	6.64%
10 edge substations	8.28%	7.49%
The computing power resource of the edge substation is 4 GHz	2.48%	4.43%
The computing power resource of the cloud master station is 6 GHz	4.64%	6.71%

7. Conclusions

Based on the background of the new power distribution system, this article aims at the cloud–edge framework including a multi-domain, multi-edge substation, and multi-cloud master, considering the current underutilization of edge computing resources and the uneven expected operation pressure of the cloud master; taking the computing resources consumed by data security protection as the starting point, some data are selected to be processed at the edge substation, and the operation pressure is reduced by reducing the amount of data processed at the cloud master station. And in this process, each edge substation adaptively allocates computing resources based on the expected operating pressure of each cloud master station to achieve the purpose of cloud–edge collaboration.

Therefore, a new adaptive scheduling method for power distribution general computing resources is proposed, which is based on cloud–edge security relay collaboration. This method facilitates distributed collaborative scheduling of the available computing resources at each edge substation through a multi-stage relay business security protection model. Under the condition of satisfying the service delay requirements and different network environments, the distributed collaborative scheduling of the available arithmetic resources of each edge substation was completed, and the utilization rate of edge resources was improved. It can significantly reduce and minimize the expected operational pressure on the main station of the largest business cloud. By enhancing the flexibility of joint scheduling across multi-domain heterogeneous resources within the distribution system, this approach is of great significance for the flexible, intelligent, and digital transformation of the power distribution system in the future.

However, for businesses with large data volume and higher latency requirements, the use of edge substations for preliminary processing may make the service stay at the edge substation longer, resulting in the inability to complete all security protection within the delay requirements. Therefore, this paper needs to be improved in considering the adaptive allocation of multi-domain computing resources for high-concurrency and low-latency businesses.

Author Contributions: Conceptualization, L.L., S.L., H.S. and R.W.; methodology, L.L., S.L., H.S. and R.W.; writing—original draft preparation, L.L., S.L., H.S. and R.W. All authors have read and agreed to the published version of the manuscript.

Funding: This research was funded by State Grid Jibei Electric Power Co., Ltd. project funding, grant number SGJBJY00GPJS2400029.

Data Availability Statement: The original contributions presented in the study are included in the article, further inquiries can be directed to the corresponding author.

Conflicts of Interest: Authors Li Li, Shanshan Lu and Haibo Sun were employed by State Grid Jibei Electric Power Co., Ltd. The remaining authors declare that the research was conducted in the absence of any commercial or financial relationships that could be construed as a potential conflict of interest.

References

1. Wang, J.N.; Gao, X.P. Research on Computer Network Information Security Protection of Power System. *Ind. Control Comput.* **2024**, *37*, 137–138+160.
2. Cao, X.; Jiang, M. Substation Cyber Security Risk Assessment Method Based on Business Association Model. *Electr. Power Inform. Commun. Technol.* **2022**, *20*, 57–64.
3. Mania, H.; Abdurachman, E.; Gaol, F.L.; Soewito, B. Survey on threats and risks in the cloud computing environment. *Procedia Comput. Sci.* **2019**, *161*, 1325–1332. [CrossRef]

4. Huang, Z.Y.; Xia, G.M.; Wang, Z.H.; Yuan, S. Survey on edge computing security. In Proceedings of the 2020 International Conference on Big Data, Artificial Intelligence and Internet of Things Engineering (ICBAIE), Fuzhou, China, 12–14 June 2020; IEEE Press: Piscataway, NJ, USA, 2020; pp. 96–105.
5. Li, P.; Liu, N.; Hu, Q.; Zhou, Q.; Li, Z.; Yu, H.; Sun, B.; Yan, Z.; Wen, F.; Xue, Y. Review of the special issue of "Review of Key Technologies for Digitalization of New Power System". *Automat. Electr. Power Syst.* **2024**, *48*, 1–12.
6. Wang, C.H.; Qiao, Y.K.; Fang, W.S. Research on New Power System Security Protection System. *Electr. Technol. Econ.* **2024**, 31–34. [CrossRef]
7. Li, J.; Wang, W.; Jin, G.H.; Zhang, R.L.; Zhao, W.Z.; Xiang, L. Cloud-based Centralized Information Scheduling Method for Power Systems under the Network-Oriented Paradigm. *Inform. Technol.* **2023**, 129–133. [CrossRef]
8. Zou, Z.; Li, F.; Hou, Y.; Fei, J.; Yang, R. Research on Security Protection Technology of Intelligent Terminal of Electricity Internet of Things. In Proceedings of the 2021 IEEE 5th Information Technology, Networking, Electronic and Automation Control Conference (ITNEC), Xi'an, China, 15–17 October 2021; pp. 1467–1470.
9. Wang, C.X.; Shi, Z.Y.; Liang, Z.F.; Li, Q.M.; Hong, B.W.; Huang, B.B.; Jiang, L.P. Key Technologies and Prospects for the Utilization of Demand-side Resources in the Power System with New Energy as the main body. *Automat. Electr. Power Syst.* **2021**, *45*, 37–48.
10. Ruan, Z.P.; Yu, W.K.; Li, K.; Zhou, P. Research of the Edge Intelligent Device Management Method Based on KubeEdge. *Electr. Power Inform. Commun.* **2020**, *18*, 63–68.
11. Zhao, H.; Liu, S.; Luo, K.; Chen, S.C.; Kong, L.H.; Jia, F. Research on the application of KubeEdge edge computing system. *Chin. J. Intell. Sci. Technol.* **2022**, *4*, 118–128.
12. Chattopadhyay, A.; Mitra, U. Security against false data-injection attack in cyber-physical systems. *IEEE Trans. Control Netw. Syst.* **2019**, *7*, 1015–1027. [CrossRef]
13. Seyedi, Y.; Keller, J.; Grijalva, S. A Research Testbed for Protection, Automation, and Cyber-Security of Digital Substations. In Proceedings of the SoutheastCon 2024, Atlanta, GA, USA, 15–24 March 2024; pp. 1306–1310.
14. Tang, Y.D.; Liu, Y.Y.; Wei, Y. Risk Analysis and Prevention of New Power System based on Hierarchical Protection Network Security System. *Netw. Secur. Technol. Appl.* **2023**, 130–133. [CrossRef]
15. Pan, X.H.; Luo, X.; Mu, Y.; Zhou, H.; Gui, R.J.; Xu, J.J.; Li, X. Coordinated Optimization Control for Main-Distribution-Microgrid Interconnection Utilizing the Potential of Massive Distributed Resources. *Power Syst. Big Data* **2024**, *27*, 1–8.
16. Wu, K.H.; Han, Y.; Tian, Z.; Sun, Y.Z.; Wu, Y.X.; Guo, Y.D. Research on Security Access Control Technology for Cloud-edge-device Architecture in a New Type of Power System. *Electr. Power Inform. Commun. Technol.* **2024**, *22*, 1–8.
17. Wang, W.Q.; Song, Y.B.; Guan, L.Y.; Zhang, H.; Gao, K.Q.; Ding, H.X.; Wang, Z.H.; Zheng, Q.R.; Zhao, J.L. Secure Interaction Communication Network Architecture and Evolution Route for Distributed Source-load-storage Resources. *Electr. Power Inform. Commun. Technol.* **2023**, *21*, 10–18.
18. Fan, Q.G.; Jiang, Z.Y.; Li, X.H.; Ma, J.F. Collaborative security assessment of cloud-edge-device distributed systems based on order parameters. *Chin. J. Netw. Inform. Secur.* **2024**, *10*, 38–51.
19. Dong, Y.; Zhou, J.Y.; Zhang, X.; Liu, W.; Wang, Q. Research on Cyber-Physical Security Protection of New Power System. *Proc. CSEE* **2025**, 1–14. [CrossRef]
20. Seifhosseini, S.; Shirvani, M.H.; Ramzanpoor, Y. Multi-objective cost-aware bag-of-tasks scheduling optimization model for IoT applications running on heterogeneous fog environment. *Comput. Netw.* **2024**, *240*, 110161. [CrossRef]
21. Hosseini, S.M.; Shirvani, M.H.; Motameni, H. Multi-objective discrete Cuckoo search algorithm for optimization of bag-of-tasks scheduling in fog computing environment. *Comput. Electr. Eng.* **2024**, *119*, 109480. [CrossRef]
22. Hosseini Shirvani, M.; Ramzanpoor, Y. Multi-objective QoS-aware optimization for deployment of IoT applications on cloud and fog computing infrastructure. *Neural. Comput. Appl.* **2023**, *35*, 19581–19626. [CrossRef]
23. Zhao, C.C.; Lv, F.; Shi, B.; Wei, X.M.; Yang, X.C.; Yue, X.C. A Review of Collaborative Inference Methods for Edge Intelligence. *Comput. Eng. Appl.* **2025**, 1–22. [CrossRef]
24. Dong, Y.M.; Zhang, J.; Xie, C.Z.; Li, Z.Y. A Survey of Key Issues in Edge Intelligent Computing Under Cloud-Edge-Terminal Architecture: Computing Optimization and Computing Offloading. *J. Electron. Inform. Technol.* **2024**, *46*, 765–776.
25. Sun, Y.Y.; Cai, Z.Y.; Ma, G.L. Workload model and optimal resource allocation of cloud master station in Power Internet of Things. *Electr. Power Automat. Equip.* **2021**, *41*, 177–183.
26. Sun, H.Y.; Zhang, J.C.; Wang, P.; Jiaying, L.; Shen, G. Edge Computation Technology Based on Distribution Internet of Things. *Power Syst. Technol.* **2019**, *43*, 4314–4321.
27. Li, S.; Yan, F.; Zhang, L.Q.; Luo, Q.C.; Yang, X.L. An eBPF-empowered Task Offloading Approach for Cloud-edge-end Collaborative Networks. *J. Zhengzhou Univ. (Nat. Sci.)* **2025**, 1–9. [CrossRef]
28. Fu, L.D.; Jia, Q.Q.; Lin, J.M.; Hu, J.H.; Wang, C. Real-Time Consistency Distributed Optimization and Control Strategy for Distribution Network Under Cloud-Edge Architecture. *Northeast Electr. Power Technol.* **2024**, *45*, 8–16.

29. Yu, X.Y.; Qiu, L.X.; Song, J.N.; Zhu, H.B. Security Communication and Energy Efficiency Optimization Strategy in UAV-aided Edge Computing. *J. Commun.* **2023**, *44*, 45–54+58.
30. Guo, H.; Wu, R.; Ma, Y.; Sun, S.; Wang, Y.; Qi, B.; Gao, J.; Xu, C. Synergy-Payoff-Maximization-Based Rechargeable Adaptive Energy-Efficient Dual-Mode Data Gathering Using Renewable Energy Sources. *IEEE Internet Things* **2024**, *11*, 35292–35305. [CrossRef]

processes

MDPI

Article

Reliability Evaluation Method for Underground Cables Based on Double Sequence Monte Carlo Simulation

Jiaxing Zhang [1,2], Bo Wang [1,2,*], Hengrui Ma [1,2], Yifan He [3], Hongxia Wang [1,2] and Hanqi Zhang [1,2]

1 Hubei Engineering and Technology Research Center for AC/DC Intelligent Distribution Network, School of Electrical Engineering and Automation, Wuhan University, Wuhan 430072, China; 2020282070157@whu.edu.cn (J.Z.); henry3764@gmail.com (H.M.); 2018282070092@whu.edu.cn (H.W.); 2019302070234@whu.edu.cn (H.Z.)

2 School of Electrical and Automation, Wuhan University, Wuhan 430072, China

3 Electric Science Research Institute of State Grid Zhejiang Province Co., Ltd., Hangzhou 310022, China; panzhaoxian3637418@163.com

* Correspondence: whwdwb@whu.edu.cn

Abstract: With the increasing demand for electrical reliability, power grid companies often need to operate urban underground cable systems at flexible ratings during planned or emergency events, but the risk of failure caused by such loading operations has not yet been quantified. Based on this, this paper proposes a reliability assessment method for underground cables based on dual sequence Monte Carlo simulation, which can consider the cumulative risk impact caused by electrical faults and insulation aging of underground cables during emergency events. Firstly, the integration of the electrical and thermal design characteristics of the integrated underground cable system within the overall network reliability framework was introduced. A dual sequence Monte Carlo simulation loop was used to evaluate the performance of the cable and network while considering the thermal aging effects of emergency loads on the cable. Then, the proposed method is applied to improve the IEEE 14 bus network by introducing cable design risk factor α and aging risk acceptance factor β to achieve flexibility and operational risk under long-term and short-term high-load emergency rating scenarios. Finally, the research results of the IEEE 14 bus network show that the expected unserved energy of scenario 1 is 30.15 MWh/year, which is 59% lower than the basic scenario of 73.34 MWh/year. The expected network failure frequency remains almost unchanged. Scenario 2 and scenario 3 show a decrease in the gain of network performance due to emergency-rated operation after using α and β, demonstrating the accuracy of the proposed method in cable reliability assessment.

Keywords: reliability assessment; emergency rating; dual sequence Monte Carlo simulation; cable design risk factor; aging risk acceptance factor

1. Introduction

In recent years, the growing demand of power consumers and the expansion of renewable energy sources into the grid require the grid system to be more resilient and reliable under uncertainty to achieve the optimal utilization of energy, which means that the grid companies are now facing multiple challenges such as sustainability, security, competitiveness and reliability [1–3]. Grid companies need to use methods such as installing real-time monitoring instruments and dynamic heat load flow to increase the power flow transmission capacity of the grid [4–6], but real-time monitoring of the lines is not possible in the short term and the investment is too large; therefore, the use of flexible network topology operation to realize dynamic heat load flow is one of the most effective methods at present [7–9].

Although the flexible network topology operation can improve the economy of power transmission and perform emergency load transfer generation in case of emergency, it relaxes the thermal constraints, which brings about the risk of additional thermal aging that cannot be accurately predicted by the existing network assessment modeling methods. Owing to the large presence of insulating materials, metallic conductors, and surrounding soil in the underground cable (UGC) system, the system exhibits significant thermal inertia characteristics [10–12]. Therefore, a significant time lag phenomenon will be observed between the increase in cable current carrying capacity and the subsequent increase in cable temperature [13–16]. A time delay that allows grid companies to be able to shorten the time of emergency to load cables at higher rated currents (sometimes up to twice the normal rated current) for a short period of time [12]. The literature [17] reports that the IEEE standard determines the allowable emergency overload capacity of cross-linked polyethylene (XLPE) cables for 36 h at 130 °C. The literature [18–22] also gives long-term emergency (LTE) and short-term emergency (STE) load capacity data. Although utility companies are aware of and frequently apply UGC emergency rating operations, most existing research only evaluates the benefits of hourly variations in heat load flow supply adequacy [21,22], and there are few literature reports analyzing the benefits and risks of emergency load flow operation across the entire network. The literature [23,24] developed realistic and practical UGC contingency rating models to calculate the temperature of the cable conductors for a given current and design characteristics by using finite element analysis and thermoelectric equivalent circuits but did not take into account the effects of cable thermal aging. Based on the variable load and temperature correlation of a single cable line, the literature [25] derived and calculated the time-varying rated values of a multi-cable system and compared them with actual values to verify the effectiveness of the algorithm. However, it did not involve the impact of cable aging under emergency conditions. The literature [26] included cable thermal aging in the grid reliability assessment framework and quantified the extent to which thermal aging affects the expected service life of cables, but did not consider the thermoelectric and thermo–mechanical effects of emergencies on the aging of cables, whereas emergency loading is a major contributor to cable aging. The literature [27] integrates the cable rating model, soil drying characteristics, and the high-load risk model into the sequential Monte Carlo method. A modified IEEE 14 bus network is used to quantify the impact of dry soil on cable safety, and emergency rating operations are used to enhance the performance indicators of the cable network. However, the impact of cable thermal aging on the expected service life was not considered.

In summary, this paper presents a dual-sequence Monte Carlo simulation-based reliability assessment methodology for underground cables to anticipate the increased network flexibility and aging risk associated with cable emergency load implementation. The integration of the electrical and thermal design characteristics of the UGC system within a network-wide reliability framework is described in detail, and the cable and network performance is evaluated based on three high-load operating states using a DSMC simulation loop to consider the thermal aging effects of cable emergency loads, and the proposed method is applied to improve the IEEE 14-bus network by introducing a cable design risk factor α and an aging risk acceptability factor β to achieve long-term and short-term high-load emergency rating scenarios for flexibility and operational risk to quantify the impact of increased cable failure risk on network performance.

To accurately estimate the relationship between the reliability of underground cables in emergency load conditions and thermal aging risks, this paper proposes a reliability evaluation method for underground cables based on a double-sequence Monte

Carlo simulation. The DSMC cycle and Markov state integration are implemented to elaborate on the detailed emergency operation process of cables and the corresponding transition equations. The ECA metric standard is adopted to construct a universal thermal aging evaluation model for various emergency scenarios, which can quantify the impact of increased cable failure risks on network performance. The following conclusions are shown:

(1) By integrating the electrical and thermal design characteristics of the underground cable system into a comprehensive network reliability framework, the DSMC simulation loop was used to evaluate the cable and network performance while considering the thermal aging effects of the emergency load on the cable.

(2) The proposed method was applied to improve the IEEE 14 bus network, and the research results showed that the expected unserved energy in scenario 1 was 30.15 MWh/year, which was 59% lower than the basic scenario's 73.34 MWh/year. The expected network failure frequency remained almost unchanged. In scenarios 2 and 3, the emergency rated value operation had a reduced gain on network performance after using α and β, and the proposed EDEL and EFCF indices were used to objectively identify key cables.

The organization of this paper is as follows: the first chapter introduces the reliability frame design of urban underground cable system; the second chapter introduces the double SMC loop analysis method; in the third chapter, the network reliability evaluation model is established and verified by simulation analysis; chapter four is the conclusion.

2. Reliability Framework Design

In order to assess the cable failure risk caused by emergency load operations, this paper presents a methodology that combines UGC detailed thermal modeling with aging modeling and applies it to analyze the reliability of underground cable networks. The methodology allows for long-term operational and planning assessment of network performance while taking into account the impact of cable thermal aging on emergency loads. The Dual Sequence Monte Carlo (DSMC) modeling cycle is used to calculate the trade-off between cable reliability and remaining service life. The data of the cables used in the substation contain the cable's own manufacturing parameters, the parameters of the surrounding environment where it is located (real-time soil parameters are required if it is encased in soil), the cable laying method, etc. The point of load and its timing requirements, the operational constraints and cost functions of the generating units, the system topology and spatial distribution, the network impedance and the design voltage level, as well as the requirements for the network component failure rates, voltages, and the thermal operating limits (emergency/normal state), are given in the input data block.

The SMC loop simulation calculation of the inner loop realizes the mapping of normal and emergency cable operation states, which are defined as initial network element state mapping and high-load network element state mapping, respectively. This strategy aims at evaluating the growth of the aging probability of component failures under emergency loading conditions by analyzing each Y year ($\forall Y \in \{1, \ldots, N\}$) in step Δt within a sequential time interval Δt; Δt represents the time step for annual calculation in SMC simulation, and $\forall \Delta t \in Y$.

Before exploring any subsequent cascading failures (CFs) and high-loading failures (HLF) that may be triggered by the emergency loading operation, the inner-loop SMC loop starts to study all the normal failures that may occur after performing the network optimization to resolve any constraints owing to the normal failures. As can be seen in Figure 1, considering the real situation where the grid system is unable to perform proper responses owing to instant overheat protection action, cascading failures do not affect network optimization; instead, high-load failures in the same Δt time period will not immediately occur in the system. At this time, the power grid system will have higher flexibility by allowing higher post-fault component loading to cope with initial normal faults. There exists a portion of high-load events that do not result in failures, i.e., surviving high-loading (SHL). These situations only allow power grid dispatchers to use them within fixed limits when the limitations of the power grid system are met. Otherwise, in emergency situations, the operational risks of high-load power grids will increase dramatically. The SMC loop in the inner circle of Figure 1 can calculate the additional temperature mapping for each cable i and then solve for the expected cable temperature $\theta_{C,i}(t_{\Delta t})$, and related aging risks. Combining the calculation path in Figure 1, emergency time-to-fail (TTFE) is introduced to explain the expected failure risk of power grid components under high-load conditions. After completing the time traversal for N years or satisfying the EENS covariance (Cov), the SMC loop calculation of the inner loop will automatically stop. The calculated parameters of the inner SMC loop are presented in Table 1.

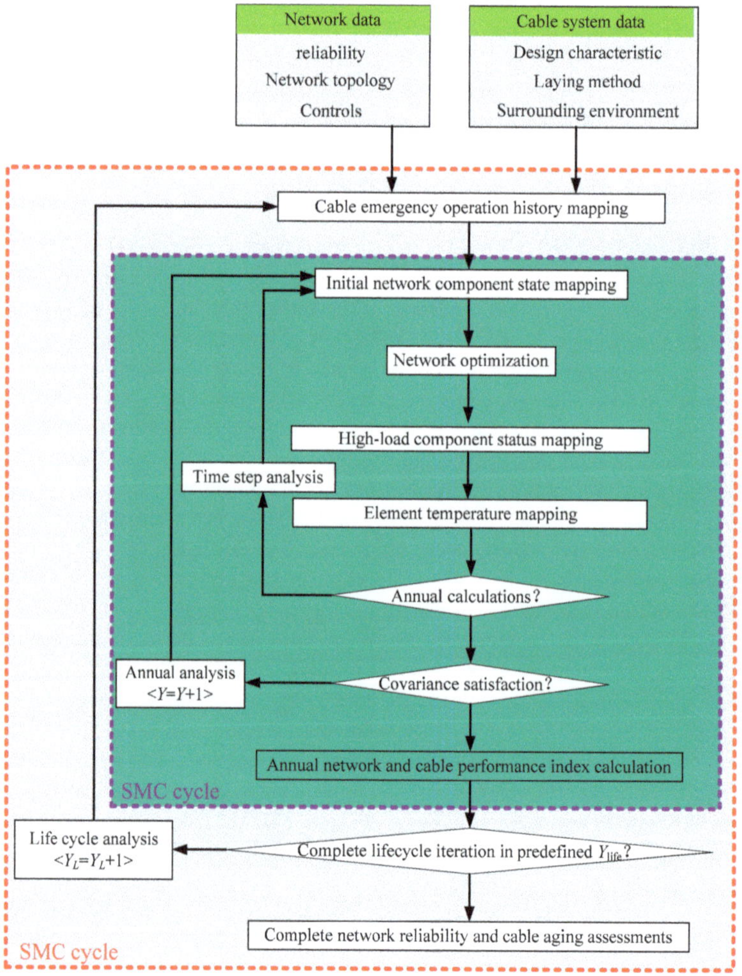

Figure 1. A computational framework for dual SMC loops.

Table 1. Parameters of inner SMC loop.

Data Type	Network Performance Indicators	Cable Performance Indicators
Parameter 1	Expected frequency of load curtailment(EFLC)	Expected frequency of emergency loading(FEEL)
Parameter 2	Expected duration of load curtailment(EDLC)	expected duration of emergency loading(EDEL)
Parameter 3	Expected energy not served(EENS)	Expected equivalent cable ageing(EECA)
Parameter 4	Expected frequency of network failures(EFNF)	Expected frequency of cable failure(EFCF)
Parameter 5	Expected equivalent network ageing(EENA)	/
Parameter 6	Expected interruption costs(EIC)	/

In order to evaluate the accumulative effect of aging of cable insulation and predict the service life of the cable and the risk of failure, the lifecycle analysis of the outer SMC loop is carried out according to the annual step Y_L, $\forall\, Y_L \in \{1, \ldots, Y_{\text{Life}}\}$. This loop design is designed to integrate grid and cable performance data collected by the inner SMC cycle in the past year and update the failure probability assessment model of aging cables. Based on this updated failure probability information, a historical mapping module of cable emergency operation status is constructed which further triggers the start of the analysis process for the next year Y_L. In the specific implementation, the analysis results of the SMC inner loop head for N years are input into the current analysis as the initial dataset for selecting the SMC inner loop. In order to continuously update the history of cable emergency operations, this step ensures that all historical emergency operations and their potential impact on cable aging risk are fully considered before the next round of inner SMC loop analysis. This DSMC process will end when the lifecycle analysis progresses to the final year Y_{Life} and all simulation steps are successfully completed.

3. SMC Loop Analysis

3.1. Inner SMC Loop Analysis

At any time interval Δt, the inner SMC loop has the ability to accurately monitor the operating status of the network cable using a two-stage sequential computing strategy to deal with normal load faults (NLFs) and emergency situations, which are both independent and closely related. Specifically, if cable i remains within the normal load range for the period Δt, $\forall i \in$ c, where c is the total number of network cables, it is marked as an i' state; When the load exceeds the preset threshold $I_{\text{con},i}$, it changes to the i'' state.

(a) Normal load failure

The core purpose of this step is to set the first state transition time for each cable i under normal load conditions in advance, based on simulation data throughout the year. In the first iteration calculation of the simulation cycle, the standard two-state Markov chain model is adopted to implement SMC simulation. For the cable under normal load, the transition time from normal state to fault state is shown in Equations (1) and (2):

$$\text{TTF}_{\text{N},i'} = -(1/\lambda_{\text{N},i'}) \ln U_{\text{N},i'} \tag{1}$$

$$\text{TTR}_{\text{N},i'} = -(1/\mu_{\text{N},i'}) \ln U_{\text{N},i'} \tag{2}$$

where $\text{TTF}_{\text{N},i'}$ represents the failure time under normal load; $\text{TTR}_{\text{N},i'}$ represents the repair time; $\lambda_{\text{N},i'}$ represents the normal failure rate; $\mu_{\text{N},i'}$ represents the repair rate; $U_{\text{N},i'}$ represents the unavailability rate, randomly generated by the uniform distribution U(0, 1).

(b) Emergency case

After sorting out all the normal load cables i' and their corresponding year Y, the next step is to identify and distinguish the state change of the high-load cable i'' at each time interval Δt, $\forall i'' \in$ c, and $i' + i'' =$ c. Three kinds of high-load operating state Markov models, namely, cascade failure, high-load failure, and continuous high-load state, are introduced to deal with the mapping analysis of the high-load state of key components at Δt time in emergency situations. This mapping process depends on the current $I_{i,\mathrm{flow}}(t_{\Delta t})$ on each cable calculated by the optimal power flow of the network during each time interval Δt.

Considering the dynamic change in the network state, affected by many factors such as the operating status of network components and the level of cable load, the cable assembly can transition between normal and high-load state. In order to accurately depict the transition behavior of cable assembly between different states, this paper proposes a fault modeling method of cable assembly based on graph theory and the Markov chain. In the model shown in Figure 2, the cable i of these two states is fused into a unified virtual state without assigning a clear transition probability to it. A real-time fault assessment strategy based on a multi-stage optimization method is introduced to analyze the changes in system performance with the operating state of each component. Once the flow rate of the cable exceeds its maximum allowable continuous current $I_{\mathrm{con},i}$, it can be determined that the cable is in a high-load state. When $I_{i,\mathrm{flow}}$ exceeds the design LTE load current $I_{\mathrm{LTE},i}$ of the cable used and is less than the design STE load current $I_{\mathrm{STE},i}$, i.e., $I_{\mathrm{LTE},i} < I_{i,\mathrm{flow}} \leq I_{\mathrm{STE},i}$, it can be considered as a short-term emergency situation. When $I_{i,\mathrm{flow}}$ exceeds the maximum normal continuous current $I_{\mathrm{con},i}$ of the cable used and is less than the design LTE load current $I_{\mathrm{LTE},i}$, Icon, i.e., $I_{\mathrm{con},i} < I_{i,\mathrm{flow}(t_{\Delta t})} \leq I_{\mathrm{LTE},i}$, it can be considered to be in a long-term emergency situation. If you do not perform maintenance in a timely manner, the fault may not be discovered or removed in a timely manner. Therefore, when considering the design load for both short-term emergency STE and long-term emergency LTE scenarios, Equations (3)–(8) are used to calculate the long-term emergency failure time $\mathrm{TTE}_{\mathrm{LTE},i''}$ short-term emergency failure time $\mathrm{TTE}_{\mathrm{STE},i''}$, respectively [28,29].

$$\mathrm{TTF}_{\mathrm{LTE},i''} = -(1/\lambda_{\mathrm{LTE},i''}) \ln U_{\mathrm{LTE},i''} \tag{3}$$

$$\mathrm{TTF}_{\mathrm{STE},i''} = -(1/\lambda_{\mathrm{STE},i''}) \ln U_{\mathrm{STE},i''} \tag{4}$$

$$\lambda_{\mathrm{STE},i''} = \lambda_{\mathrm{E},i''} \times e^{((I_{\mathrm{STE},i}/I_{\mathrm{con},i})-1)\alpha} \tag{5}$$

$$\lambda_{\mathrm{LTE},i''} = \lambda_{\mathrm{E},i''} \times e^{((I_{\mathrm{LTE},i}/I_{\mathrm{con},i})-1)\alpha} \tag{6}$$

$$U_{\mathrm{STE},i''} = 1 - U_{\mathrm{N},i''}(I_{\mathrm{con},i}/I_{\mathrm{STE},i})^{\beta} \tag{7}$$

$$U_{\mathrm{LTE},i''} = 1 - U_{\mathrm{N},i''}(I_{\mathrm{con},i}/I_{\mathrm{LTE},i})^{\beta} \tag{8}$$

where $\lambda_{\mathrm{LTE},i''}$ and $\lambda_{\mathrm{STE},i''}$ represent the failure rates of cable i'' under long-term and short-term emergency load, respectively; $\lambda_{\mathrm{E},i''}$ represents the failure rate of cable i'' under emergency load; $I_{\mathrm{LTE},I}$ and $I_{\mathrm{STE},i}$ respresent the current of cable i under long-term and short-term emergency load, respectively; $I_{\mathrm{con},i}$ represents the maximum normal continuous current of cable i; α and β are cable design risk coefficient and aging acceptable risk coefficient, respectively. $U_{\mathrm{LTE},i''}$ and $U_{\mathrm{STE},i''}$ represent the non-availability rates of cable i'' under long- and short-term emergency loads, generated using two different random generators to promote the independence of emergency loading events.

In order to evaluate the potential failure risk of the cable due to the design and operation environment, the design risk factor α is innovatively proposed in this study. This factor is closely related to the health of the cable components, taking into account the design attributes of the cable and the maintenance strategy adopted. Based on historical data and geographic location analysis, the operator assigns a positive α coefficient to each cable; among them, α = 0 indicates that the cable is in an ideal state without adverse factors, and the maintenance is extremely easy. The health coefficient of the cable increases, which directly reflects the optimization of its maintenance state. Conversely, a decline in the coefficient indicates poor maintenance. Further, the aging risk acceptance coefficient

β was introduced to measure the operator's tolerance to cable aging risk when facing emergency load demands. This coefficient profoundly reveals the risk literature of power grid enterprises in asset protection risk decision-making and reflects the importance of cable emergency operation experience and network topology. The coefficient β is also positive, and $\beta = 0$ refers to cables that are new and not in emergency use and whose load can be seamlessly transferred to adjacent cables to ensure service continuity.

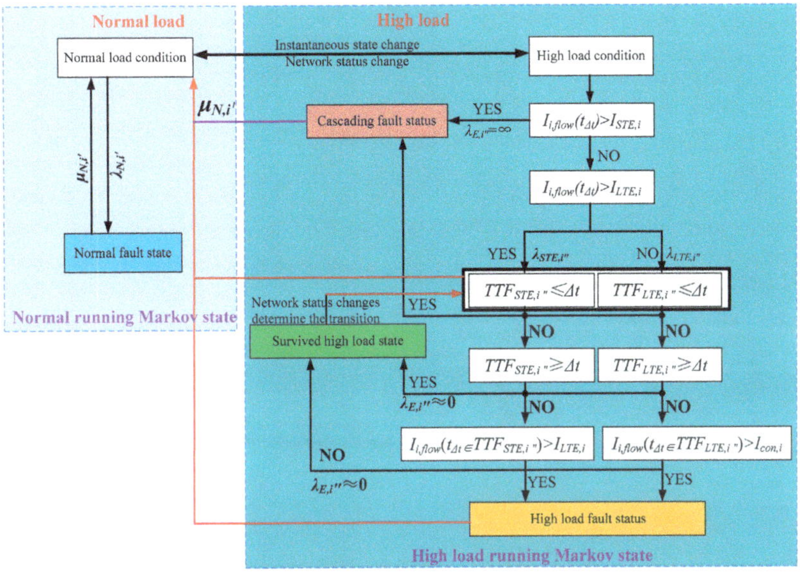

Figure 2. Cable operation status and transition path probability.

Given the unpredictability of the impact of emergency faults on cable components, this study has decided to use the same repair time (RT) calculation for high-load events and normal-load events. Specifically, based on the $\mu_{N,i'}$ parameters of the normal load fault (NLF), high-load fault (HLF), and cascading fault (CF) states of each cable i, Equation (2) is used to obtain the normal load maintenance time $RT_{N,i'}$ and emergency load maintenance time $RT_{E,i'}$, respectively.

The cascading fault state indicates that the high-load operation of cable i'' caused it to fail within the same Δt. Three different transition paths may lead to cascading faults, as shown in Figure 2. When $I_{i,\text{flow}(t_{\Delta t})}$ exceeds its designed STE load $I_{\text{STE},i}$, one path simulates an instantaneous "trip" of cable thermal protection. Therefore, for $I_{i,\text{flow}(t_{\Delta t})} > I_{\text{STE},i}$, The transition of cable to cascading fault state is deterministic ($\lambda_{E,i''} = \infty$). The other two paths from the high load to the cascading fault are shown in Figure 2, modeling the sudden fault (i.e., $\text{TTF}_E < \Delta t$) that occurs when cable i is under a high load (i.e., above $I_{\text{con},i}$). The instantaneous failure probability of a cable operating within its STE load range is described by $\text{TTF}_{\text{STE},i''} \leq \Delta t$, and calculated by Equation (3). The other dashed path describes the failure probability of $\text{TTF}_{\text{LTE},i''} \leq \Delta t$ in a similar way, for cable operations within its LTE load range.

The high-load fault state indicates a fault caused by the emergency loading of cable i'' with $\text{TTF}_E > \Delta t$. This fault affects the future state of the component, so it is necessary to update the initial network component state mapping in Figure 1 to allocate these future state transitions, as normal faults will not occur if emergency faults occur simultaneously. As shown in Figure 2, high-load faults only occur when cable i'' maintains its high-load state throughout the estimated $\text{TTF}_{\text{STE},i''}$ or $\text{TTF}_{\text{LTE},i''}$ cycle. Therefore, when the following three conditions are met, high-load faults under STE/LTE loading will occur: (1) $TTF_{STE,i''}/TTF_{LTE,i''} > \Delta t$; (2) $TTF_{STE,i''}/TTF_{LTE,i''}$ is relatively small; (3) the cable current remains continuously at the STE/LTE level throughout the entire $TTF_{STE,i''}/TTF_{LTE,i''}$ period.

Figure 3 shows a mapping example of three cables transitioning between the NLF, HLF, and CF states. During the time period Δt, after the failure of cable $1'$, only the mapping of the HLF state is observed, which causes cable 2 (denoted as $2''$) to enter a high-load operating state and thus initiates the calculation of its transition time $TTF_{LTE,2''}$ from high-load to low-load. After several Δt periods, the HLF state of cable $2''$ has changed. It is worth noting that the failure of cable $1'$ did not significantly affect the current $I_{3,\text{flow}}$ of cable 3, keeping it at normal load (i.e., $3'$). However, in the second event, the failure of cable $2''$ caused the current $I_{3,\text{flow}}$ of cable 3 to exceed its high-load state start-up threshold $I_{STE,3}$, triggering the calculation of the time it took cable 3 (denoted as $3''$) to transition from high-load state to full-failure state $TTF_{STE,3''}$ in a specific Δt period. Eventually, this caused it to enter the CF state, as shown in Figure 3. The cable current values in Figure 3 accurately reflect the actual high-load levels and conversion times, and precise simulation of the fault and recovery process is achieved through careful segmentation of the fault occurrence at the beginning of each time period Δt and the recovery at the end. Therefore, when HLF and CF states are concurrent, this model can effectively model and calculate the maximum unavailability rate of the cable.

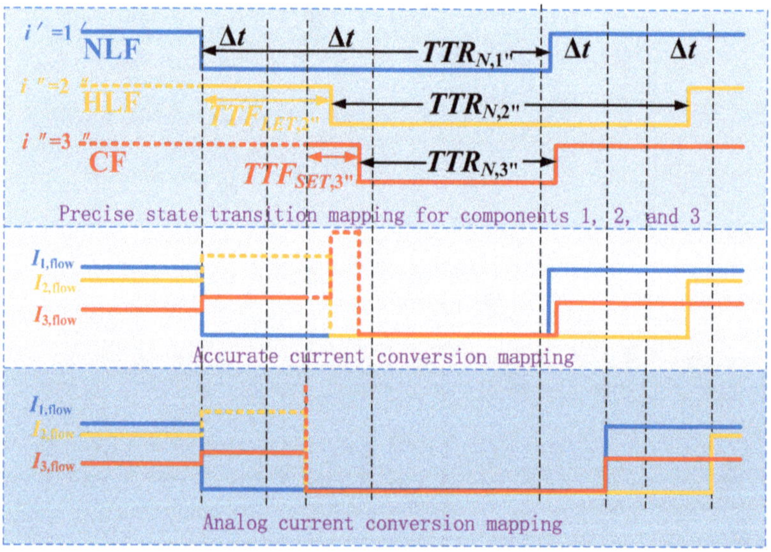

Figure 3. Example of the first SMC loop conversion mapping.

3.2. Outer SMC Loop Analysis

In order to evaluate the emergency operation status of cable in its whole lifecycle Y_{Life}, especially considering that frequent emergency loads may increase the risk of cable failure, the proposed coefficient $\beta_{i,YL}$ can fully reflect the entire emergency operation history (EDEL) of cable i in a specific simulated year Y_{L}. The effect of this operation on the remaining life of the cable is further quantified. The calculation equation of $\beta_{i,YL}$ is shown in Equation (9), which not only takes into account the use frequency of the cable in emergency situations, but also reasonably includes the additional loss effect of emergency load on the cable life.

$$\beta_{i,Y_L} = a \cdot D_{LTE,i,(Y_L-1)} + b \cdot D_{STE,i,(Y_L-1)} = a \sum_{Y_L=1}^{Y_{Life}} EDEL_{LTE,i,Y_L} + b \sum_{Y_L=1}^{Y_{Life}} EDEL_{STE,i,Y_L} \quad (9)$$

where $D_{LTE,i,(YL-1)}$ and $D_{STE,i,(YL-1)}$ represent the cumulative operating hours of LTE and STE for cable i before the Y_L year of its life, respectively; $EDEL_{LTE,i,YL}$ and $EDEL_{STE,i,YL}$ represent the cumulative operating times of LTE and STE that occur during the lifespan Y_L years of cable i; a and b represent the severity of cable degradation when running under LTE and STE load, respectively. In the calculation of SMC loop in the inner circle, it is assumed that the network adopts a risk avoidance strategy, so no emergency load is generated, and EDEL of LTE and STE operation is calculated as zero, resulting in the $\beta_{i,YL}$ coefficients in Equation (9) being zero.

During the thermal model calculation process, the steady-state temperature SS-$\theta_{C,i}(t_\theta)$, and transient temperature TS-$\theta_{C,i}(t_\theta)$ of the cable conductor were solved using the method shown in Figure 4. The steady-state thermal model calculation uses $t_\theta = \infty$, while the transient model considers $t_\theta = \Delta t$. The input data required for the calculation are determined based on the cable current $I_{i,flow}(t_{\Delta t})$ at a given time Δt. In order to accurately reflect the conductor energy loss $W_{C,i}(t_{\Delta t})$ at time Δt, a time-varying thermoelectric equivalent (TEE) model was constructed to solve the time-variation of conductor temperature $\theta_{C,i}(t_\theta)$, as shown in Figure 5. The TEE model takes into account the complex structure of the cable, including the armor layer, sheath layer, conductor layer, and three layers of soil. For arbitrarily configured cables, the mathematical expression of the transient TEE model is given by Equation (10), which ensures the accuracy and universality of the model.

$$\begin{bmatrix} \dot{\theta}_{SL3} \\ \dot{\theta}_{SL2} \\ \dot{\theta}_{SL1} \\ \dot{\theta}_{Ar} \\ \dot{\theta}_{S} \\ \dot{\theta}_{C} \end{bmatrix} = \begin{bmatrix} M_{11} & M_{12} & 0 & 0 & 0 & 0 \\ M_{21} & M_{22} & M_{23} & 0 & 0 & 0 \\ 0 & M_{32} & M_{33} & M_{34} & 0 & 0 \\ 0 & 0 & M_{43} & M_{44} & M_{45} & 0 \\ 0 & 0 & 0 & M_{54} & M_{55} & M_{56} \\ 0 & 0 & 0 & 0 & M_{65} & M_{66} \end{bmatrix} \times \begin{bmatrix} \theta_{SL3} \\ \theta_{SL2} \\ \theta_{SL1} \\ \theta_{Ar} \\ \theta_{S} \\ \theta_{C} \end{bmatrix} + \begin{bmatrix} P_{T-\alpha} \\ 0 \\ 0 \\ P_{T-Ar} \\ P_{T-S} \\ P_{T-C} \end{bmatrix} \quad (10)$$

where θ_{SL1}, θ_{SL2}, and θ_{SL3} represent the temperature of soil layer 1, 2, and 3 respectively. θ_{Ar} is the armor layer temperature; θ_S represents the temperature of the sheath layer; θ_C represents the expected temperature of the cable conductor. $P_{T\text{-}\alpha}$ represents the external thermal power induced in the external environment; $P_{T\text{-}Ar}$ represents the external thermal power induced in the cable armor; $P_{T\text{-}S}$ represents the external thermal power induced in the cable sheath; $P_{T\text{-}C}$ represents the external thermal power induced in the cable conductor; M represents a thermal matrix, and its internal elements are calculated in literature [30,31].

SS-$\theta_{C,i}(t_{\Delta t})$ and TS-$\theta_{C,i}(t_{\Delta t})$ at the time Δt calculated above are used as input data, and the Arrhenius model is used to calculate the proportion of life expectancy loss $LF_{\theta C,i}(t_{\Delta t})$ of cable i at the time Δt. The calculation equation is shown in Equation (11):

$$LF_{\theta_{C,i}}(t_{\Delta t}) = \frac{\Delta t}{L_{o,i} e^{-(Ea_i/k_B)(1/\theta_{o,i} - 1/\theta_{C,i}(t_{\Delta t}))}} \quad (11)$$

where $L_{o,i}$ represents the expected life of cable i when working under the ambient literature temperature $\theta_{o,i}$; Ea_i represents the activation energy of insulating material; k_B represents the Boltzmann constant.

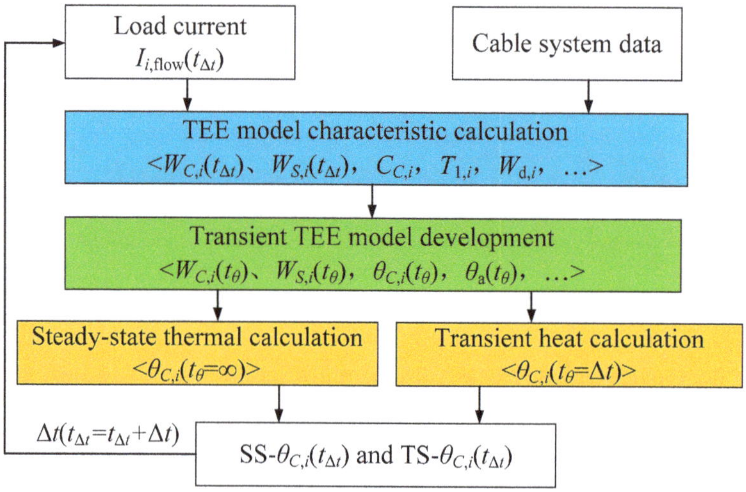

Figure 4. Cable temperature mapping flowchart.

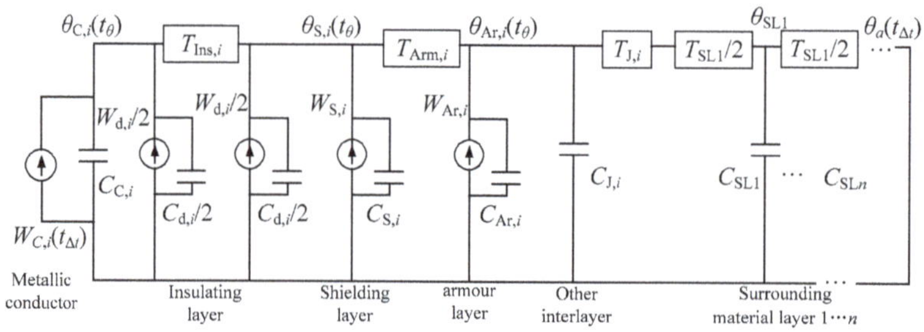

Figure 5. Time-varying TEE model for single-core cables.

In order to carry out comparative aging assessment of different network cables and high-load events, the equivalent cable aging (ECA) coefficient is proposed and described. The calculation equation is shown in Equation (12):

$$ECA_{i,\Delta t} = \frac{LF_{\theta_{C,i}}(t_{\Delta t})}{LF_{i,90}} = \exp\left(\frac{Ea_i}{k_B}\left(\frac{1}{363} - \frac{1}{\theta_{C,i}(t_{\Delta t})}\right)\right) \tag{12}$$

where $LF_{i,90}$ represents the equivalent aging of cable i at 90 °C.

EECA is the average expected ECA for simulation cycles year Y in a given year Y_L of the cable i in its lifecycle Y_{Life}. EFCF is the expected average number of cable failures of normal $NCF_{i,Y}$ and emergency $ECF_{i,Y}$, and the EDEL of LTE and STE operation is the average value of the lifecycle year Y_L, calculated according to the sum of the short emergency loading duration $d\_STE_{i,Y}$ and long emergency loading duration $d\_LTE_{i,Y}$ of cable i in each Y year. EFEL indicates the average number of emergency load events $N_{i,Y}$ of cable i in each Y year. The calculation equation of the cable performance coefficient is shown in Equations (13)–(18):

$$EECA_{i,Y_L} = \sum_{Y=1}^{N}\left(\sum_{\Delta t=1}^{Y} ECA_{i,\Delta t}\right)/N \tag{13}$$

$$EFCF_{i,Y_L} = \sum_{Y=1}^{N}(NCF_{i,Y} + ECF_{i,Y})/N \tag{14}$$

$$EDEL_{\mathrm{LTE},i,Y_\mathrm{L}} = \sum_{Y=1}^{N} (d\_\mathrm{LTE}_{i,Y})/N \tag{15}$$

$$EDEL_{\mathrm{STE},i,Y_\mathrm{L}} = \sum_{Y=1}^{N} (d\_\mathrm{STE}_{i,Y})/N \tag{16}$$

$$EDEL_{i,Y_\mathrm{L}} = EDEL_{\mathrm{LTE},i,Y_\mathrm{L}} + EDEL_{\mathrm{STE},i,Y_\mathrm{L}} \tag{17}$$

$$EFEL_{i,Y_\mathrm{L}} = \sum_{Y=1}^{N} N_{i,Y}/N \tag{18}$$

The load curtailment event number LCE_Y, load curtailment duration LCD_Y, and System energy not supplied $SysENS_Y$ of the inner SMC loop in cycle N are calculated as inputs; $EENA_{Y\mathrm{L}}$ and $EFNF_{Y\mathrm{L}}$ represent total network aging and network component failures in the annual Y_L, which accurately reflect cable failure aging and thus predict the resilience of the cable network to failures. These indicators can determine the best cable replacement strategy on the basis of minimizing the overall network aging of each failure and guide the power grid company to prioritize the replacement of old cables. A large EFNF and a low EFLC with a low EENA indicate that the network is resilient, and vice versa. The calculation equation of the network performance coefficient is shown in Equations (19)–(23):

$$EDLC_{Y_\mathrm{L}} = \sum_{Y=1}^{N} LCD_Y/N \tag{19}$$

$$EFLC_{Y_\mathrm{L}} = \sum_{Y=1}^{N} LCE_Y/N \tag{20}$$

$$EENS_{Y_\mathrm{L}} = \sum_{Y=1}^{N} SysENS_Y/N \tag{21}$$

$$EENA_{Y_\mathrm{L}} = \sum_{Y=1}^{N_C} EECA_{i,Y_\mathrm{L}} \tag{22}$$

$$EFNF_{Y_\mathrm{L}} = \sum_{i=1}^{N_C} EFCF_{i,Y_\mathrm{L}} \tag{23}$$

4. Network Reliability Evaluation

4.1. Cable Characteristic

The IEEE 14 bus network does not provide thermal limitation and circuit design data. In this study, four kinds of copper conductor single-core unarmoured XLPE cables are used. The cable data are taken from the cable system design data and cable manufacturer data. The cable is assumed to be in a flat laying mode, conductor spacing is 0.1 m, and direct burial depth is 0.8 m and 1.5 m, respectively, 13.8 kV and 69 kV. The normal, STE, and LTE loading capabilities are utilized to provide flexibility in emergency situations, with I_LTE and I_STE ratings of 1.103 times and 1.143 times the maximum normal continuous current I_con, respectively. The thermal rating of the cable is calculated at 15 °C soil temperature, 0.9 km/W soil thermal resistivity, and 1.9 × 10^6 J/m^{3}K volumetric specific heat, as shown in Table 2. The determination of cable faults and repair rates requires knowledge of the length li of each line, measured in miles, based on the original impedance data from the literature [32,33], calculated using Equations (24) and (25). The obtained length and reliability data, as well as the cable types used in the improved IEEE 14 bus cable network, are shown in Table 3.

$$\lambda_{N,i} = 0.0062 \times l_i + 0.226(\mathrm{Occ./year}) \tag{24}$$

$$\frac{1}{\mu_{N,i}} = 1.5 \times l_i + 11.5(\text{hour/Occ.}) \quad (25)$$

Table 2. IEEE 14 bus network cable performance parameters.

Cable Type	A	B	C	D
sectional area (mm^2)	400	630	400	800
R_{90} (Ω/km)	0.059	0.039	0.064	0.036
X (Ω/km)	0.12	0.108	0.095	0.088
B (S/km)	6.48×10^{-5}	8.38×10^{-5}	1.422×10^{-4}	1.99×10^{-4}
I_{con} (A)	791	1011	846	1200
I_{LTE} (A)	870	1115	933	1324
I_{STE} (A)	904	1155	967	1373

Table 3. IEEE 14 bus network cable reliability model.

Cable Number	Type	Length/Miles	Interruption Time (h/Occ.)	Failure Rate (Occ./Year)
1	B	9.74	24.4	0.297
2	B	9.74	24.4	0.298
3	A	16.51	33.8	0.332
4	A	14.38	32.4	0.321
5	A	13.26	30.6	0.311
6	A	14.24	30.4	0.312
7	A	14.29	30.4	0.313
8	A	3.21	16.2	0.238
9	C	0.72	12.2	0.205
10	C	0.86	12.5	0.212
11	D	0.47	11.9	0.204
12	C	0.38	11.7	0.203
13	C	0.92	12.5	0.215
14	C	0.68	12.2	0.213
15	C	0.92	12.5	0.215
16	C	1.36	12.9	0.221

4.2. Scene Analysis

In order to evaluate the flexibility and operational risk of an improved IEEE 14-node network under a high-load emergency rating (ER) scenario, the proposed method was applied and the data in Table 4 were compared. Through comparative analysis with the baseline scenario, the benefits or constraints brought about by the improvement of cable strength in an emergency state are quantified.

Table 4. IEEE 14 bus network cable performance parameters.

Scenario		Model Variable	Risk Description
Basic information(BC)		emergency rating(ER) not used	
High load ER failure	Sc1	$\alpha = \beta = 0$	ER and no increased risk
	Sc2	$\alpha = 10, \beta = 0, \alpha = 20, \beta = 0$ and $\alpha = 30, \beta = 0$	ER and unfavorable risks in cable laying
	Sc3	$\alpha = \beta = 10, \alpha = \beta = 20$ and $\alpha = \beta = 30$	ER, adverse risk of cable laying + fixed year lifecycle aging
	Sc4	$\alpha = 10, \beta = 9$ and $a = 0.05, b = 0.2$	ER, adverse risk of cable laying + aging of operation lifecycle

Compared with the basic situation (BC), the significant advantage of using emergency rating (ER) for Sc1~Sc3 is reflected, and the specific calculation results of network performance indicators are shown in Table 5. Sc1 greatly improved network performance by utilizing the zero risk associated with ER ($\alpha = \beta = 0$), resulting in a reduction of approximately 60% in EENS, EFLC, EDLC, and EIC, while EFNF remained almost unchanged, with only EENA increasing by approximately 2%. However, when considering the cable design risk factor α and cable aging acceptability risk factor β for Sc2 and Sc3, the gain of ER in terms of network performance is weakened. This is because ER operation may cause other faults to occur on frequently overloaded cables, and the overall network performance is still better than the basic situation. The EENS, EFLC, EDLC, and EIC coefficient values exhibited by Sc2 and Sc3 ($\alpha = \beta = 10$) are all lower than the relevant values in BC, and there is no significant change in the EFNF value. Among them, the EENA value of Sc2 (α = 30, β = 0) increased by a maximum of 33%, reaching 550.23 h/year. Sc3 ($\alpha = \beta = 20$ and $\alpha = \beta = 30$) exhibits significantly reduced network performance due to severe loading causing old and poorly maintained cable line faults. Compared with BC, additional faults reduce the flexibility benefits of ER, resulting in higher EIC values. The comparison of results between Sc2 and Sc3 indicates that, compared to α, the aging acceptability risk coefficient β has a more significant impact on ER failures. In other words, compared to the expected lifespan determined solely by cable design and laying conditions, the accelerated aging and insulation degradation of cables caused by emergency operations have a greater impact on cable failure rates.

Table 5. Network performance.

Scenario	Variable	EENS (MWh/Year)	EFLC (Time/Year)	EDLC (h/Year)	EFNF (Time/Year)	EIC ($)	$EENA_{TS}$ (h/Year)	$EENA_{SS}$ (h/Year)
BC	Without ER	73.34	2.89	15.18	20.88	257,894	161.97	412.4
Sc1	$\alpha = \beta = 0$	30.15	1.12	6.07	20.96	101,316	164.47	488.94
Sc2	$\alpha = 10, \beta = 0$	31.16	1.17	6.32	21.15	106,579	164.72	495.32
	$\alpha = 20, \beta = 0$	36.47	1.33	7.36	21.67	123,684	165.37	511.18
	$\alpha = 30, \beta = 0$	57.99	1.87	10.95	22.82	202,632	167.03	550.23
Sc3	$\alpha = \beta = 10$	37.48	1.36	7.51	21.75	128,947	165.46	512.42
	$\alpha = \beta = 20$	110.75	3.1	19.34	25.37	396,053	170.46	640.73
	$\alpha = \beta = 30$	429.37	6.67	47.61	30.34	1,568,421	178.22	824.77

Sc4 adopts outer ring SMC loop simulation analysis, and the cable design life is set to 30 years. The performance coefficient changes every year with the increase in usage years, as shown in Figure 6. The aging in BC does not consider the influence of operation history, so constant performance will be produced throughout the entire service life of the cable, as shown in Figure 6. Due to the increased network flexibility provided by ER, Sc4 initially had better network reliability than BC. However, as the years of use increased and emergency operations gradually accumulated, Sc4's EENS further increased and reached the same value as BC in the 23rd year. From that year on, cable aging intensified, leading to a decrease in Sc4 network performance. The EFNF of Sc4 is equal to the corresponding value of BC in the first year, at 20.88 times/year, and gradually increases with the increase in service life, mainly due to the emergency operation of Sc4.

The study on the performance of single cables in IEEE 14 bus networks identified C10, C11, and C13 as critical cables, which are more susceptible to ER compared to low-load C1 and other cables. The cable performance EDEL and EFCF are obtained through the operation of the entire cable life; therefore, EDEL helps to prioritize network cables based on the risk of network cable failure, considering the network topology structure and the operation status of individual cables. The performance indicators of the three most critical cables (C11, C10, and C13) are shown in Table 6. In BC, if ER is not used, it is expected that emergency loads and conductor overheating will not occur. However, the emergency rating in Sc2 forced the EDEL of cables C11, C10, and C13 to increase by 13.92, 5.59, and 5.07 h/year, respectively. Taking the most critical cable C11 as an example, under the condition of applying an average of 3.8 emergency loads per year, EFCF and EECA only increased by 1.2 time/year and 7.5 h/year, respectively.

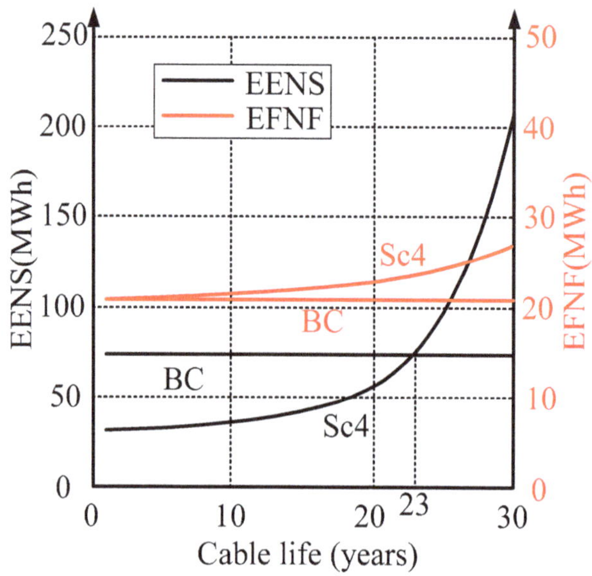

(a) EENS and EFNF

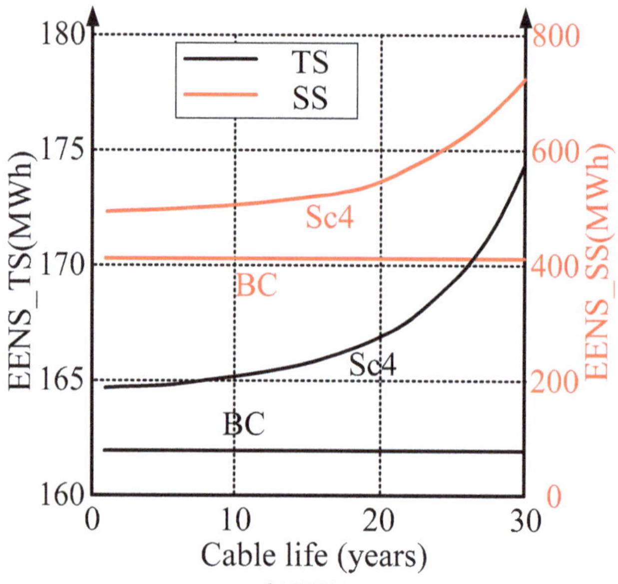

(b) EENA

Figure 6. Comparison between Sc4 and BC.

The EDEL of the key cables C11, C10, and C13 under Sc4 has been higher since their first year of operation, with almost the same EFCF, as shown in Figure 7. It can be determined that C11 is the most critical cable in the network, while C10, C13, and C11 are equally critical but not as critical as C11. On the contrary, C1 seems to be unaffected by emergency events, and the thermal aging effect can be ignored. This is consistent with the fact that the supply area and current values of the test cable system's cable line C11 are greater than those of C10, C13, and C11, further verifying the accuracy of the current calculation method.

Table 6. Key cable performance parameters.

Scenario	Cable Number	EDEL (h/Year)	FEEL (Time/Year)	EFCF (Time/Year)	EECA (h/Year)
BC	C11	0	0	0	0
	C10	0	0	0	0
	C13	0	0	0	0
Sc1 ($\alpha = \beta = 0$)	C11	13.92	3.8	1.2	7.5
	C10	5.59	1.9	0.2	1
	C13	5.07	1.9	0.1	0.4

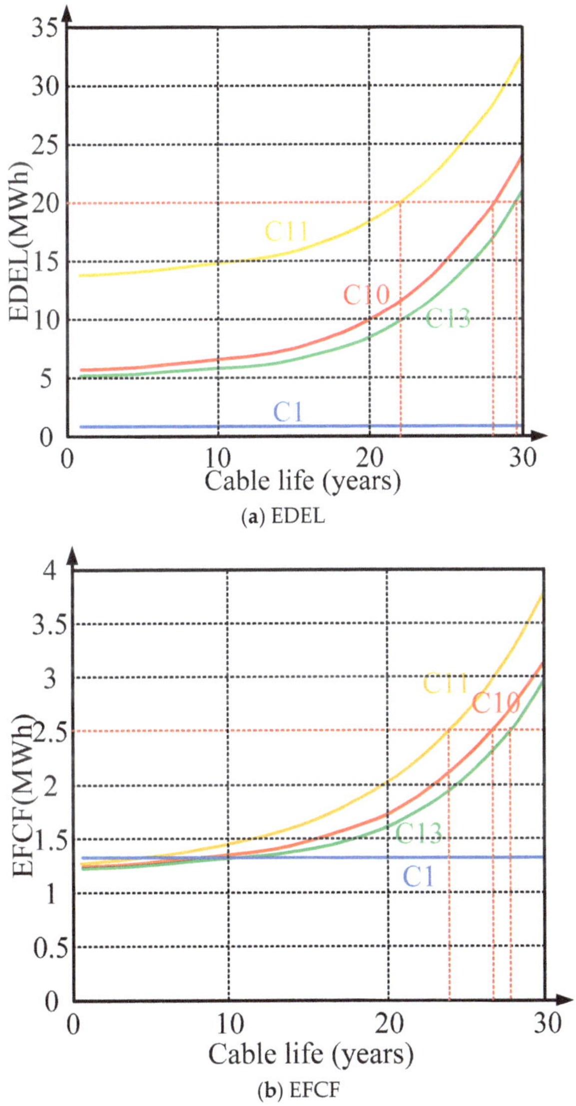

Figure 7. EDEL and EFCF outputs of key cables under Sc4.

4.3. Method Validation

Different types of laboratory experiments are conducted under specific operating conditions and environmental conditions, which cannot fully reflect the actual operating conditions of cables. At the same time, the urban underground cable system is a highly reliable system, and its monitoring data change very little in the short term. In the case of a small amount of data or short monitoring data time, it is difficult to predict and analyze the relevant parameters of the cable. Therefore, this article selects 10 kV cables that have been in service for more than 10 years in the Southern Power Grid as the original data. After excluding data with recording errors, incomplete data, and early retirement due to other reasons, only 12 in-service 10 kV high-voltage cables in the region were obtained. The detailed data are shown in Table 7. By combining Bayesian inference with the Weibull proportional hazards model in the literature [34], more accurate prediction results can be obtained by comprehensively utilizing empirical information on high-voltage cables and a small number of high-voltage cable fault data. The calculation results are shown in Table 7. It can be seen that, as the average load rate of the cable increases, the probability of failure increases, and the service life decreases. The method proposed in this paper calculates that the maximum EFCF value of 4.84 times/year at 25 years occurs in test cable 2, which corresponds to the 28.4% failure probability calculated in the literature [34] after 25 years of operation of cable 2. This is due to the poor operating environment and high load rate of the cable. The slight difference in the values of test cables 1 and 2 is because cable 2 is slightly longer than cable 1, which verifies the feasibility of the method proposed in this paper for reliability analysis of actual cable systems.

Table 7. Comparative analysis of experimental results.

Cable Number	Operating Time	Operating Environment	Average Load Rate (%)	Remaining Useful Life (days)	Failure Probability After 25 Years (%)	EFCF (Time/Year)
1	4352	Directly buried	0.76	1587	24.6	4.62
2	4625	Directly buried	0.76	1164	28.4	4.84
3	5621	Non directly buried	0.15	29,877	0.52	0.62
4	7153	Directly buried	0.19	16,765	7.51	0.86
5	7456	Non directly buried	0.11	30,950	0.44	0.43
6	7456	Non directly buried	0.08	33,606	0.3	0.12
7	7456	Non directly buried	0.17	26,877	0.85	0.78
8	7456	Non directly buried	0.11	30,588	0.47	0.46
9	8223	Directly buried	0.23	15,224	11.72	1.02
10	8223	Directly buried	0.25	14,593	14.08	1.14
11	8342	Non directly buried	0.34	18,233	5.16	1.36
12	8612	Directly buried	0.06	24,309	1.38	0.08

5. Conclusions

To accurately estimate the relationship between network reliability and the thermal aging risk of cables under emergency loads, this paper proposes a reliability assessment method for underground cables based on dual-sequence Monte Carlo simulation. The method uses DSMC loops and three additional Markov states to describe cable emergency operations and their corresponding transition equations, ECA is selected to develop unit-based universal thermal aging for different emergency events. It can quantify the impact of increased cable fault risk on network performance. Based on the integration of the electric heating design characteristics of underground cable systems within the overall network reliability framework, the DSMC simulation loop is used to evaluate the performance of cables and networks while considering the thermal aging effects of emergency loads on cables. The proposed method was applied to improve the IEEE 14 bus network, and the research results showed that the expected unserved energy in scenario 1 was 30.15 MWh/year, which was 59% lower than the basic scenario's 73.34 MWh/year. The expected network failure frequency remained almost unchanged. In scenarios 2 and 3, the emergency rated value operation had a reduced gain on network performance after using α and β, and the proposed EDEL and EFCF indices were used to objectively identify key cables. Future work should incorporate the characteristics of soil (other filling materials) in the cable laying environment and their impact on thermal aging due to dryness. The model can also be further improved by aggregating other cable component models, such as the weakest links, e.g., cable joints and terminals. The model can be used to simultaneously analyze the correlation between the output data of this method and the recorded data, experimental data, and status management records of real cable networks and to accurately formulate practical values for the cable design risk coefficient α and aging risk acceptable coefficient β, which are applicable to power grid companies, cable types, and manufacturers, in order to further improve network reliability and utilization.

Author Contributions: Conceptualization, J.Z., B.W., H.M., Y.H., H.W. and H.Z.; software, J.Z., B.W., H.M., Y.H., H.W. and H.Z.; writing—original draft preparation, J.Z., B.W., H.M., Y.H., H.W. and H.Z. All authors have read and agreed to the published version of the manuscript.

Funding: This research is funded by the Science and Technology Project of State Grid Corporation Limited (Contract No. 5400-202119145A-0-0-00).

Data Availability Statement: The original contributions presented in this study are included in the article. Further inquiries can be directed to the corresponding author.

Conflicts of Interest: Author Yifan He was employed by the company Electric Science Research Institute of State Grid Zhejiang Province Co., Ltd. The remaining authors declare that the research was conducted in the absence of any commercial or financial relationships that could be construed as a potential conflict of interest. The company had no role in the design of the study; in the collection, analyses, or interpretation of data; in the writing of the manuscript, or in the decision to publish the results.

Nomenclature

$TTF_{N,i'}$	the down time
$TTR_{N,i'}$	the repair time
$\lambda_{N,i'}$	the normal failure rate
$\mu_{N,i'}$	the repair rate
$U_{N,i'}$	the unavailable rate
$TTE_{LTE,i''}$	the long term emergency expiration time
$TTE_{STE,i''}$	the short term emergency expiration time
$\lambda_{LTE,i''}$	the long term emergency load failure rate
$\lambda_{STE,i''}$	the short term emergency load failure rate
$I_{LTE,i}$	current of cable i under long-term emergency load
$I_{STE,i}$	current of cable i under short-term emergency load
$I_{con,i}$	the maximum normal continuous current of cable i
α	cable design risk coefficient
β	acceptable risk coefficient for aging
$RT_{N,i'}$	normal load repair time
$RT_{E,i'}$	emergency load repair time
$D_{LTE,i,(YL\text{-}1)}$	accumulated LTE operating time of cable i before Y_L year of service life
$D_{STE,i,(YL\text{-}1)}$	accumulated STE operating time of cable i before Y_L year of service life
a	severity of cable degradation during LTE load operation
b	severity of cable degradation during STE load operation
$W_{C,i}(t_{\Delta t})$	conductor loss
$LF_{\theta C,i}(t_{\Delta t})$	the expected life loss ratio of cable i at time Δt
Ea_i	activation energy of insulation materials

k_B	boltzmann constant
$L_{o,i}$	the expected lifes of cable i at the environmental literature temperature
G_y	power generation output capacity
V_j	AC voltage limitation
$S_{max,i}$	cable system flow constraint

References

1. Bai, X.; Fan, Y.; Liu, Y.; Song, Y. Wind power storage virtual power plant considering reliability and flexibility tiered capacity configuration. *Power Syst. Prot. Control* **2022**, *50*, 11–24. [CrossRef]
2. Zheng, W.; Wang, X.; Liang, Z.; Gan, B.; Gong, Z.; Lai, X. Renewable energy electricity market risk estimating model based on Vine-Copula and Qua-si-Monte Carlo method. *Renew. Energy Resour.* **2023**, *41*, 1642–1649. [CrossRef]
3. Meng, Q.; Liao, K.; Zheng, S.; Yang, J.; He, Z. Source-grid-storage coordinated planning for power system considering flexibility mutual aid among regions. *Power Syst. Technol.* **2024**, *48*, 3165–3174. [CrossRef]
4. He, W.; Tang, C.; Huang, S.; Bin, P.; Wang, M.; Li, L. Monitoring of the maximum allowable current-carrying capacity of lines based on multi-parameter DOFS. *Electr. Power Sci. Eng.* **2021**, *37*, 52–61.
5. Kopsidas, K.; Kapetanaki, A.; Levi, V. Optimal demand response scheduling with real-time thermal ratings of overhead lines for improved network reliability. *IEEE Trans. Smart Grid* **2017**, *8*, 2813–2825. [CrossRef]
6. Hou, Y.; Wang, W.; Wei, Z.; Deng, X.; Yang, X.; Zhang, K. Dynamic evaluation of ampacity limit for transmission line based on model calibration. *Autom. Electr. Power Syst.* **2021**, *45*, 159–166.
7. Liu, Y.; Zhang, S.; Cao, J.; Wang, S. Simulation on thermal and electric field of 500kV HVDC XLPE submarine cable. *High Volt. Appar.* **2020**, *56*, 7–15. [CrossRef]
8. Kopsidas, K.; Tumelo-Chakonta, C.; Cruzat, C. Power network reliability evaluation framework considering OHL electro-thermal design. *IEEE Trans. Power Syst.* **2015**, *31*, 2463–2471. [CrossRef]
9. He, D.; Wang, B.; Zhang, Y.; Xu, Y.; Guo, C. An dynamic emergency capacity-increase method for trans-mission lines considering load capacity of equipment. *Electr. Power Constr.* **2017**, *38*, 95–102.
10. Wang, M.; Zhou, S.; Yang, M.; Dong, X.; Huang, J. Assessment method for acceptable installed capacity of offshore wind farms considering thermal characteristics of submarine cables. *Autom. Electr. Power Syst.* **2021**, *45*, 195–202.
11. Shi, Z. Research on Distribution Network Reconfiguration Based on the Thermal Inertia of Cable Line. Master's Thesis, Nanjing University of Science and Technology, Nanjing, China, 2020. [CrossRef]
12. Olsen, R.; Holboell, J.; Gudmundsdottir, U.S. Electrothermal coordination in cable based transmission grids. *IEEE Trans. Power Syst.* **2013**, *28*, 4867–4874. [CrossRef]
13. Dan, J.; Li, W.; Cui, R.; Liu, Y.; Li, J.; Wu, G. Study on transient thermal response and fluctuating load transmission capability of power cables buried in earth. *Wire Cable* **2023**, *66*, 34–41. [CrossRef]
14. Mazzanti, G. Analysis of the combined effects of load cycling, thermal transients, and electrothermal stress on life expectancy of high-voltage ac cables. *IEEE Trans. Power Deliv.* **2007**, *22*, 2000–2009. [CrossRef]
15. Yuan, Y.; Gao, L.; Zou, H.; Zhu, Z. Thermal evaluation of high-voltage power cables based on ampacity derating factor. *Smart Power* **2023**, *51*, 88–96.
16. Zhou, Q.; Yang, T.; Gao, M.; Shan, S.; Xie, Z.; Zhou, H. Research on temperature field and ampacity of cable laid in pipe laying based on finite element method. *J. Electr. Power* **2023**, *38*, 397–404. [CrossRef]
17. *IEEE Std 242-2001*; IEEE Recommended Practice for Protection and Coordination of Industrial and Commercial Power Systems (IEEE Buff Book). IEEE: New York City, NY, USA, 2001; pp. 1–710. [CrossRef]
18. Cheng, X.X.; Ma, J.W.; Yin, Y.K.; Guan, Z.; Liu, W.; Zhang, Z. Research on short-time overload current limit for high voltage XLPE power cables. *High Volt. Appar.* **2017**, *53*, 150–155. [CrossRef]
19. Li, H.; Deng, L.; Luo, Q.; Zhang, Y.; Han, Y. Analysis on short-circuit fault of aircraft cable and research on its mechanism. *Sichuan Electr. Power Technol.* **2019**, *42*, 19–24. [CrossRef]
20. Grigg, C.; Wong, P.; Albrecht, P.; Allan, R.; Bhavaraju, M.; Billinton, R.; Chen, Q.; Fong, C.; Haddad, S.; Singh, C.; et al. A report pre-pared by the reliability test system task force of the application of probability methods subcommittee. *IEEE Trans. Power Syst.* **1999**, *14*, 1–010. [CrossRef]
21. Shen, Y.; Hou, L. Research and verification of high voltage cable on the current carrying capacity in emergency. *Wire and Cable* **2023**, *66*, 25–30. [CrossRef]
22. Yu, Z.; Niu, H.; Zhuang, X.; Zhao, J.; You, Y.; Mai, C. Emergency load ampacity experiment of buried distribution cable and its application. *Insul. Mater.* **2014**, *47*, 82–86. [CrossRef]
23. Diaz-Aguiló, M.; De León, F.; Jazebi, S.; Terracciano, M. Ladder-type soil model for dynamic thermal rating of underground power cables. *IEEE Power Energy Technol. Syst. J.* **2014**, *1*, 21–30. [CrossRef]

24. Tan, J.; Wu, Y.; Tian, P. Temperature prediction method of power cable core based on finite element. *Environ. Technol.* **2023**, *41*, 126–131.
25. Brakelmann, H.; Anders, G.J. Transient thermal response of multiple power cables with temperature dependent losses. *IEEE Trans. Power Deliv.* **2021**, *36*, 3937–3944. [CrossRef]
26. Buhari, M.; Levi, V.; Awadallah, S.K.E. Modelling of ageing distribution cable for replacement planning. *IEEE Trans. Power Syst.* **2015**, *31*, 3996–4004. [CrossRef]
27. Wang, K.; Kopsidas, K. A power system reliability framework considering soil drying-out effect on underground cables. *IEEE Trans. Power Syst.* **2024**, *39*, 4783–4794. [CrossRef]
28. Billinton, R.; Li, W. *Reliability Assessment of Electric Power Systems Using Monte Carlo Methods*; Springer Science & Business Media: Berlin/Heidelberg, Germany, 1994.
29. Su, C.-T.; Wu, T.-S.; Lee, T.-H.; Huang, C.-L. Capacity planning with flow and reliability evaluation using monte carlo simulation. *IEEE Trans. Reliab.* **1986**, *35*, 518–522. [CrossRef]
30. Masuda, T.; Morimura, T.; Nakano, T.; Maruyama, O.; Mimura, T.; Yasui, T.; Agatsuma, K.; Ishiyama, A. Safety and reliability of 66-kV class HTS cable systems in short-circuit current accidents-experimental results on 40m cable system. *IEEE Trans. Appl. Supercond.* **2017**, *27*, 5401504. [CrossRef]
31. Grigsby, L.L. *Power Systems*, 3rd ed.; CRC Press: Boca Raton, FL, USA, 2016.
32. Nazemi, M.; Dommerque, R.; Daniel, S. New evaluation concept for electromagnetic interference of HVDC cables to neighboring buried pipelines. In Proceedings of the 2023 International Symposium on Electromagnetic Compatibility, Hangzhou, China, 20–23 October 2023; pp. 1–6. [CrossRef]
33. Liu, Z.; OuYang, B.; Chen, Z.; Zhao, P.; Zhao, J. Research on circuit parameter configuration of DC superimposed impulse voltage test on UHV DC cable. In Proceedings of the 2020 IEEE 4th Conference on Energy Internet and Energy System Integration (EI2), Wuhan, China, 30 October–1 November 2020; pp. 3372–3377. [CrossRef]
34. Yuan, J. Risk Assessment Method for Maintenance and Replacement of High Voltage Cables Based on Statistical Model. Master's Thesis, Wuhan University, Wuhan, China, 2021. [CrossRef]

processes

Article

Optimized Dispatch of Integrated Energy Systems in Parks Considering P2G-CCS-CHP Synergy Under Renewable Energy Uncertainty

Zhiyuan Zhang [1], Xiqin Li [1], Lu Zhang [2], Hu Zhao [3], Ziren Wang [4,*], Wei Li [4] and Baosong Wang [5]

[1] State Grid Beijing Mentougou Power Supply Company, Mentougou, Beijing 102300, China; zhiyuanzhang@zohomail.cn (Z.Z.); li_xiqin@126.com (X.L.)
[2] Research Institute of Electric Power Science, State Grid Beijing Electric Power Company, Fengtai, Beijing 100075, China; zhanglu7@163.com
[3] State Grid Beijing Electric Power Company, Xicheng, Beijing 100075, China; kobeahu@163.com
[4] Department of Economics and Management, North China Electric Power University, Baoding 071000, China; ncepulw@126.com
[5] School of Electrical and Electronic Engineering, North China Electric Power University, Changping, Beijing 102206, China; baosongwang@zohomail.cn
* Correspondence: wzr@ncepu.edu.cn

Abstract: To enhance low-carbon economies within Park Integrated Energy Systems (PIES) while addressing the variability of wind power generation, an innovative optimization scheduling strategy is proposed, incorporating a reward-and-punishment ladder carbon trading mechanism. This method effectively mitigates the unpredictability of wind power output and integrates Power-to-Gas (P2G), Carbon Capture and Storage (CCS), and Combined Heat and Power (CHP) systems. This study develops a CHP model that combines P2G and CCS, focusing on electric-heat coupling characteristics and establishing constraints on P2G capacity, thereby significantly enhancing electric energy flexibility and reducing carbon emissions. The carbon allowance trading strategy is refined through the integration of reward and punishment coefficients, yielding a more effective trading model. To accurately capture wind power uncertainty, the research employs kernel density estimation and Copula theory to create a representative sequence of daily wind and photovoltaic power scenarios. The Dung Beetle Optimization (DBO) algorithm, augmented by Non-Dominated Sorting (NSDBO), is utilized to solve the resulting multi-objective model. Simulation results indicate that the proposed strategy increases the utilization rates of renewable energy in PIES by 28.86% and 19.85%, while achieving a reduction in total carbon emissions by 77.65% and a decrease in overall costs by 36.91%.

Keywords: integrated energy systems; renewable energy consumption; P2G-CCS-CHP; reward-penalty ladder carbon trading; uncertainty in wind power output; dung beetle optimization algorithm under non-dominated sequencing

1. Introduction

Integrated energy systems offer numerous benefits, including enhanced energy efficiency, improved reliability of energy supply, and support for the development of smart grid infrastructure. However, they also represent significant sources of global carbon emissions. In the context of escalating concerns regarding global warming and ecological degradation, PIES present an important avenue for exploring pathways toward low-carbon, economically viable, and reliable energy operations.

 https://doi.org/10.3390/pr13030680

Among these pathways, energy substitution is vital for meeting global carbon reduction goals [1]. In recent years, the capacity for wind and solar photovoltaic energy has consistently increased, leading to a notable decline in fossil fuel consumption and associated carbon emissions. However, the intermittent and variable characteristics of wind and solar energy create challenges in power scheduling, often resulting in the curtailment of renewable generation as their capacities grow in many areas [2]. Sinha et al. observed a growing trend of mixed energy systems that combine solar photovoltaic and wind energy, indicating a promising direction for the evolution of energy systems [3]. Zhao et al. designed a topological framework for a hybrid power generation system that integrates wind, solar, and hydro sources, focusing on minimizing energy curtailment and addressing economic factors; however, their model did not account for the uncertainties inherent in wind and solar outputs or the resultant carbon emissions [4]. Ju et al. indicated that properly managing the uncertainties associated with wind and solar energy production can bolster the reliability, stability, and economic viability of integrated energy systems [5]. Liu et al. demonstrated that effective scheduling within these systems facilitates smoother transitions between various energy sources, thus enhancing the utilization of renewable resources like wind and solar energy [6]. Xu et al. introduced a martingale model to assess the uncertainties in wind and photovoltaic energy outputs; nonetheless, this method faced challenges due to its randomness and subjectivity in selecting representative scenarios [7]. In a study by Min, a cloud model combined with a copula correlation coefficient matrix was utilized to analyze a hybrid power generation system based on output data. This approach successfully assessed the correlation and uncertainty of wind and solar outputs; however, obtaining the necessary data for its implementation may prove difficult in practical scenarios [8]. However, there are few existing articles that simultaneously consider the uncertainty and correlation of wind and solar power output, and they do not accurately depict the typical daily values of wind and solar power output, lacking a detailed mathematical modeling process. This research aims to improve upon this aspect.

In the context of low-carbon economic optimization strategies for PIES, researchers worldwide have explored the integration of carbon trading mechanisms and voluntary emission reductions into carbon emission policies. Chen et al. presented carbon emission limits and renewable energy integration targets within PIES, highlighting potential synergies between low-carbon goals and renewable energy consumption objectives [9]. Zhang et al. developed an optimized scheduling model for PIES that incorporates a tiered carbon trading mechanism and addresses the uncertainty of wind and solar generation. Their findings demonstrated the model's advantages in terms of both low-carbon outcomes and economic performance, although they overlooked the complementary dynamic between wind and solar energy [10]. Zhu et al. introduced a comprehensive trading approach within PIES focusing on carbon constraints. This method advocates for spot trading of excess industrial energy and the use of storage solutions to manage energy supply and demand variations, thus easing carbon emission pressures. Case studies confirmed that this approach can enhance economic returns while lowering carbon emissions [11]. Liu et al. observed that the rapid growth of the hydrogen energy sector is increasing hydrogen's share in energy systems, complicating the conversion among energy sources and posing challenges for PIES planning and development. To tackle this, they proposed a bi-level planning model for a hydrogen-integrated energy system that considers multi-stage investments and carbon trading. Simulation results suggested that this model could effectively reduce operational costs and carbon emissions within the PIES planning framework [12]. Zhang et al. designed a tiered carbon trading model that accounts for complete lifecycle carbon emissions to study the influence of carbon trading on PIES, illustrating that detailed modeling of specific equipment and hydrogen blending processes can contribute to emission reductions [13]. Li et al. put forth an electric-carbon energy scheduling approach for multi-park integrated energy systems (MIESs), establishing a joint electric-carbon trading market to facilitate energy interactions among multiple integrated energy systems. This strategy leverages regional differences in carbon pricing, allowing for the sharing of carbon quotas among systems and ultimately lowering overall carbon trading costs [14]. The studies mentioned demonstrate significant progress in modeling and optimizing PIES with a focus on carbon emissions. Implementing multi-energy coupling, flexible demand responses, and enhanced energy allocation mechanisms within parks can also provide low-carbon economic benefits. Chen et al. suggested a joint planning and cost allocation method for multi-park shared energy systems to develop optimal strategies for joint planning and cost distribution regarding shared storage at the park level [15]. Additionally, Chen et al. introduced a low-carbon integrated energy system that utilizes distributed renewable and clean energy, applying the Firefly algorithm to address constraint conditions [16]. To address the complexities of joint scheduling among industrial, commercial, and residential parks with various energy-sharing modes, Wang et al. established a cooperative game model for multi-park integrated energy systems with hydrogen energy, based on Nash bargaining theory [17]. Xu et al. proposed a multi-objective operational optimization model for an electric and thermal integrated energy system, considering power-to-gas transitions and demand response strategies from different electric and thermal loads, aiming to minimize total costs, carbon emissions, and energy waste rates [18]. The existing research mainly explores the energy-saving and low-carbon operation mode of integrated energy systems from two aspects: multi-energy coupling and equipment joint operation. In terms of multi-energy coupling, the existing research on the relationship between electric energy and thermal energy is constrained by "determining electricity by heat. "The range of electric energy regulation is insufficient, and the effect of cost-saving and carbon emission control is not significant. It is urgent to break this constraint. In terms of equipment joint operation, Ding et al. have proved the advantages of P2G-CCS-CHP joint operation in terms of economy and low carbon in engineering modeling, but there are few applications in the field of optimal scheduling of park-level integrated energy systems [19]. At the same time, the traditional step-by-step carbon trading mechanism commonly used in existing research has been proved to have limited carbon emission constraints in practical applications, and it is necessary to explore more effective carbon emission control methods.

Based on the detailed engineering modeling of Ding et al. [19], this study established the core framework of PIES and coupled the P2G-CCS and CHP systems to effectively solve the CHP-related 'thermally driven power' challenge. Limited by space, some unspecified parameter values, system dynamics principles, and internal relations can refer to Ding et al.'s original text. The model collaborates with micro gas turbines (MT) to enable flexible multi-energy responses. Additionally, it features a tiered carbon trading mechanism based on rewards and penalties, accounting for the uncertainties linked to wind and solar energy outputs. The optimization process employs the Dung Beetle Optimizer (DBO) after refining the objective function through the Non-dominated Sorting Genetic Strategy (NSGS). The primary goal of the model is to minimize both operational and environmental protection costs, comparing four different solution strategies: minimizing operational costs, minimizing environmental protection costs, minimizing total costs, and deriving a compromise solution. Furthermore, the study examines the implications of the reward-and-penalty carbon trading mechanism, as well as the uncertainties and correlations present in wind and solar energy sources. It also assesses the roles of P2G, CCS, and CHP technologies in real-world engineering contexts, where total cost minimization is typically a priority. Additionally, the research analyzes how variations in the reward-and-penalty coefficients within the carbon trading framework influence the optimized scheduling of PIES, validating the proposed model's effectiveness through comparative analysis across various operational scenarios.

2. Methods and Analysis

2.1. Integrated Energy System Framework of the Park with Reward-and-Penalty Tiered Carbon Trading Mechanism

The PIES structure constructed in this study is illustrated in Figure 1. This system comprises three energy networks: electricity, heat, and gas. It includes the main power grid, natural gas sources (NS), wind turbines (WT), photovoltaic power generation (PV) units, P2G-CCS-CHP coupled production units, electric refrigerators (ER), power-to-gas devices (P2G), natural gas energy storage (GasS), gas turbines (GT), gas boilers (GB), organic Rankine cycle (ORC), hydrogen fuel cells (HFC), waste heat boilers (WHB), electric energy storage (ES), and thermal energy storage (hS).

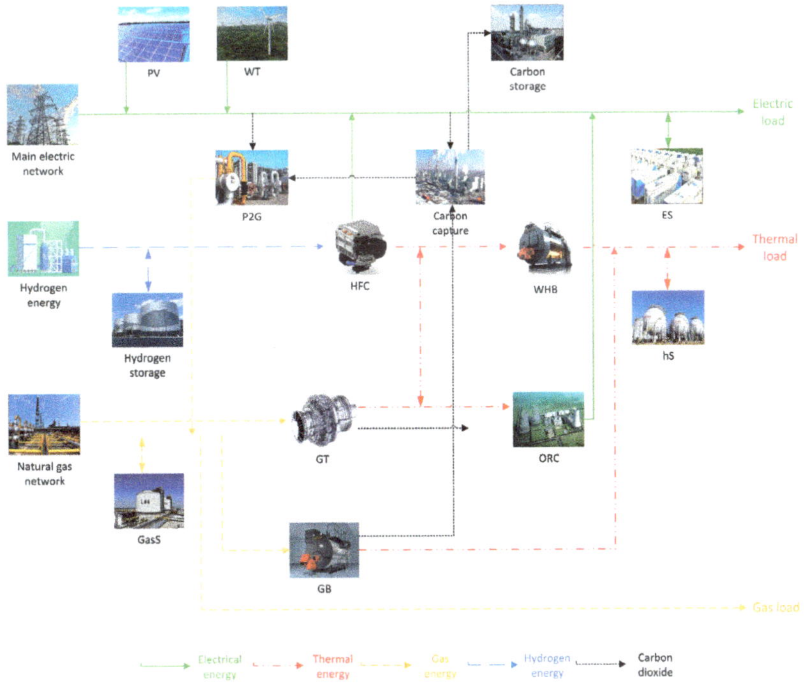

Figure 1. Integrated energy system framework of the park.

2.1.1. Modeling and Characteristics of P2G-CCS-CHP

To mitigate carbon emissions and reduce transportation costs linked to carbon, this study introduces a coupled operational system that integrates CHP, P2G, and CCS. In this setup, the CHP system produces both electricity and thermal energy using natural gas. The P2G component utilizes electricity from the CHP unit while receiving CO_2 from an adjacent CCS facility. This electricity allows for the conversion of CO_2 into methane, which fulfills the energy requirements of both CHP and gas loads, effectively contributing to emission reduction and enhanced energy efficiency. The CCS system captures CO_2 generated by the CHP process and uses electricity to transport the captured CO2 to the P2G system for reuse as fuel. The close proximity of the CCS and P2G components within the P2G-CCS-CHP system minimizes the need for long-distance transportation and storage of CO_2, thus reducing related costs.

Electricity produced by the CHP system can be utilized in three main ways: it can be directly consumed by end-users via the power grid, delivered to the P2G system for methane production, or sent to the CCS system to facilitate CO_2 management. This can be illustrated as follows:

$$P_c(t) = P_{c1}(t) + P_{c2}(t) + P_{c3}(t) \tag{1}$$

where: t represents time, P_c denotes the electricity generated by the CHP system, P_{c1} refers to the total electricity output, P_{c2} signifies the electricity used by the P2G system, and P_{c3} indicates the electricity utilized by the CCS system. The CCS system captures CO_2 through its electricity consumption, with the quantity of CO_2 captured being roughly proportional to the electricity used, expressed as:

$$C_{CCS}(t) = \alpha P_{c3}(t) \tag{2}$$

where: C_{CCS} signifies the quantity of CO_2 captured by the CCS system, and α indicates the capture efficiency, which is a dynamically variable numerical quantity.

The CO_2 collected by the CCS is subsequently transported to P2G for the production of natural gas. The volume of natural gas generated by P2G is roughly proportional to both the quantity of CO_2 utilized and the electrical energy consumed throughout the production process, which can be represented as:

$$P^g_{P2G}(t) = \beta C_{CCS}(t) \tag{3}$$

$$P_{c2}(t) = \chi C_{CCS}(t) \tag{4}$$

where: β and χ are the proportionality coefficients.

Based on Formulas (2) and (4), Formula (1) can be rewritten as:

$$P_c(t) = P_{c1}(t) + (1 + \alpha\chi)P_{c2}(t) \tag{5}$$

While operating, the CHP system must adhere to specific constraints regarding electricity generation and heat production, defined by the electric-thermal coupling properties of combined heat and power generation, as depicted by the polyline ABCD in Figure 2. These electric-thermal coupling characteristics of CHP can be articulated as:

$$\begin{cases} 0 \le P^h_c \le P^h_{c,max} \\ P_c(t) \ge max\left[P_{c,min} - c_{v1}\left(P^h_c(t) - P^h_{c0}\right), P_{c,min} - c_m\left(P^h_c(t) - P^h_{c0}\right)\right] \\ P_c(t) \le P_{c,max} - c_{v2}P^h_c(t) \end{cases} \tag{6}$$

where: $P_{c,max}$ and $P_{c,min}$ denote the CHP system's maximum and minimum power outputs, respectively; P^h_c represents the heat generation power from the CHP; P^h_{c0} refers to the heat production power corresponding to $P_{c,min}$; $P^h_{c,max}$ signifies the CHP's maximum output in terms of heat generation; and c_{v1}, c_{v2} and c_m are the electro-thermal conversion coefficients in the system, and the specific numerical settings are shown in Reference [19].

Substituting Equation (5) into Equation (6) results in Equation (7):

$$\begin{cases} 0 \le P^h_c \le P^h_{c,max} \\ P_{c1}(t) \ge max\begin{bmatrix} P_{c,min} - c_{v1}\left(P^h_c(t) - P^h_{c0}\right) - (1+\alpha\chi)P_{c2}(t), \\ P_{c,min} - c_m\left(P^h_c(t) - P^h_{c0}\right) - (1+\alpha\chi)P_{c2}(t) \end{bmatrix} \\ P_{c1}(t) \le P_{c,max} - c_{v2}P^h_c(t) - (1+\alpha\chi)P_{c2}(t) \end{cases} \tag{7}$$

The power consumption of the P2G system meets the capacity constraint, which is limited by

$$P_{c2,min} \leq P_{c2}(t) \leq P_{c2,max} \tag{8}$$

where: $P_{c2,max}$ and $P_{c2,min}$ indicate the maximum and minimum power consumption levels of the P2G system, respectively.

By substituting Equation (7) into Equation (8), the electric-thermal coupling characteristics of the P2G-CCS-CHP system can be articulated as:

$$\begin{cases} 0 \leq P_c^h \leq P_{c,max}^h \\ P_{c1}(t) \geq max\begin{bmatrix} P_{c,min} - c_{v1}\left(P_c^h(t) - P_{c0}^h\right) - (1+\alpha\chi)P_{c2,max}, \\ P_{c,min} - c_m\left(P_c^h(t) - P_{c0}^h\right) - (1+\alpha\chi)P_{c2,max} \end{bmatrix} \\ P_{c1}(t) \leq P_{c,max} - c_{v2}P_c^h(t) - (1+\alpha\chi)P_{c2,min} \end{cases} \tag{9}$$

The electric-thermal coupling characteristics of this model can be extracted from Equation (9), as shown in Figure 2. The CHP system's operational range is indicated as ABCD, while the range for the P2G-CCS-CHP system is labeled EFGHI. In comparison to Equation (6), Equation (9) shows that GHI is derived from BCD and extended downward by an amount of $(1+\alpha\chi)P_{c2,max}$, while EF extends downward from AB by $(1+\alpha\chi)P_{c2,min}$. Since $(1+\alpha\chi)P_{c2,min} < (1+\alpha\chi)P_{c2,max}$, the area of EFGHI surpasses that of ABCD, reflecting a broadened range for electric power adjustment and a reduction in the electric-thermal coupling characteristics.

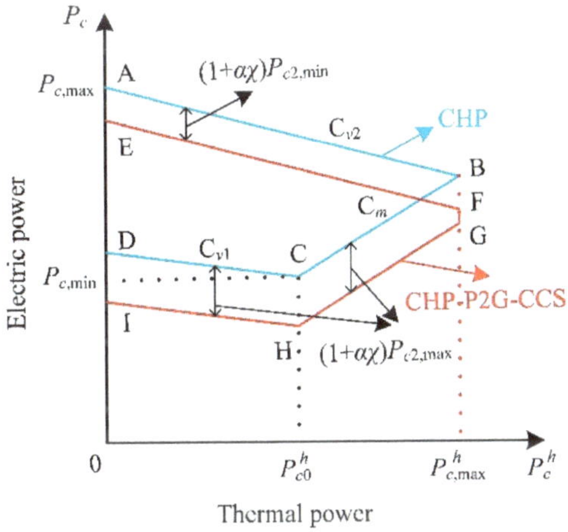

Figure 2. Electro-thermal coupling characteristics of CHP and P2G-CCS-CHP.

The limitations for P2G natural gas production can be obtained from Equations (3), (4), and (8) and are articulated as:

$$P_{P2G,min}^g = \frac{\beta P_{c2,min}}{\chi} \leq P_{P2G}^g(t) \leq \frac{\beta P_{c2,max}}{\chi} = P_{P2G,max}^g \tag{10}$$

where: $P_{P2G,max}^g$ and $P_{P2G,min}^g$ denote the maximum and minimum production rates of P2G natural gas, respectively.

From Equations (9) and (10), it's evident that when the P2G-CCS-CHP operates along boundary GHI, P_{c2} reaches its peak value, resulting in $P_{P2G}^g(t) = P_{P2G,max}^g$. In contrast, when the P2G-CCS-CHP functions on boundary EF, P_{c2} hits its lowest point, with

$P^g_{P2G}(t) = P^g_{P2G,min}$. Thus, natural gas production is influenced by the electricity demand of P2G and the CHP's output, which is, in turn, restricted by the available thermal energy.

By integrating Equations (3)–(6), the constraints related to the electric-thermal coupling characteristics of the CHP can be outlined as follows:

$$\begin{cases} P^g_{P2G}(t) \geq \frac{\beta}{\chi(1+\alpha\chi)} max\begin{Bmatrix} P_{c,min} - c_{v1}\left[P^h_c(t) - P^h_{c0}\right] - P_{c1}(t), \\ P_{c,min} - c_m\left[P^h_c(t) - P^h_{c0}\right] - P_{c1}(t) \end{Bmatrix} \\ P^g_{P2G}(t) \leq \frac{\beta}{\chi(1+\alpha\chi)}\left[P_{c,max} - c_{v2}P^h_c(t) - P_{c1}(t)\right] \end{cases} \tag{11}$$

The connections between electrical power, thermal power, and natural gas output are derived from Equation (11) and depicted in Figure 3.

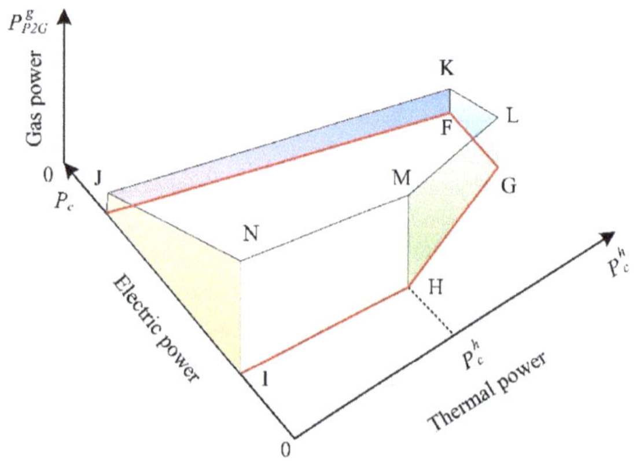

Figure 3. P2G-CCS-CHP multi-energy output relationship.

Figure 3 shows that the feasible region for electrical power, thermal energy, and natural gas output is determined by the vertices EFGHI-JKLMN, highlighting an interconnected relationship among the three. The net CO_2 emissions from the P2G-CCS-CHP system are calculated as the difference between the CO_2 emissions generated by the CHP and the CO_2 captured by the CCS, expressed as:

$$C_{P-C-C}(t) = C_{CHP}(t) - C_{CCS}(t) \tag{12}$$

$$C_{CHP}(t) = a_{CO_2}\left[P_c(t) + C_{V1}P^h_c(t)\right] + b_{CO_2}\left[P_c(t) + C_{V1}P^h_c(t)\right]^2 + c_{CO_2} \tag{13}$$

where: C_{CHP} represents the CO_2 emissions from the CHP system, while a_{CO_2}, b_{CO_2}, and c_{CO_2} are the carbon emission factors related to the CHP system.

From Equations (12) to (13), it is evident that the net CO_2 emissions of the P2G-CCS-CHP system depend on both electrical and thermal outputs, reflecting a coupled relationship among them. Moreover, as indicated in Equation (11), $C_{P-C-C}(t) = C_{CHP}(t)$ is valid only when $P_{c1}(t) = P_{c,max} - c_{v2}P_{hc}(t)$. In contrast, if $P_{c1}(t) \neq P_{c,max} - c_{v2}P_{hc}(t)$, then $C_{P-C-C}(t) < C_{CHP}(t)$. This demonstrates that the P2G-CCS-CHP system can significantly mitigate carbon dioxide emissions.

Because of constraints on space, a comprehensive description of the CCS system and additional related information is available in reference [19].

2.1.2. Reward and Penalty-Based Tiered Carbon Trading Mechanism

(1) Allocation Mechanism for PIES Carbon Emission Quotas

The reward and penalty-based tiered carbon trading mechanism builds upon the traditional tiered carbon trading framework by introducing reward and penalty factors. Under this mechanism, enterprises not only engage in the buying or selling of carbon quotas based on the disparity between their allocated free carbon quotas and actual carbon emissions but also participate in an emissions reward and penalty system. Specifically, if a company's carbon emissions are below its allocated carbon quota, it may receive corresponding reward subsidies. Conversely, if a company's carbon emissions exceed its allocated quota, it is required to pay penalty fees. By incorporating this reward and penalty mechanism, enterprises are further motivated to proactively reduce carbon emissions.

(2) Initial Allocation Model for PIES Carbon Emission Rights

This study identifies several carbon emission sources within the PIES, including emissions from grid-purchased thermal power, gas loads, emissions from gas turbines, and emissions generated by the P2G-CCS-CHP system. The initial carbon emissions allocation model is established as follows:

$$\begin{cases} E_{PIES} = E_{grid,TE} + E_{gload} + E_{MT} + E_{P-C-C} \\ E_{grid,TE} = \gamma_e \sum\limits_{t=1}^{T} P_{grid,TE}(t) \\ E_{gload} = \gamma_g \sum\limits_{t=1}^{T} P_{gload}(t) \\ E_{MT} = \gamma_h \sum\limits_{t=1}^{T} \left(\sigma_{e,h} P_{MT}^{e}(\mathrm{t}) + P_{MT}^{h}(t) \right) \\ E_{P-C-C} = C_{CHP}(t) \end{cases} \tag{14}$$

where: E_{PIES}, $E_{grid,TE}$, E_{gload}, E_{MT}, and E_{P-C-C} denote the carbon emission allowances for PIES, thermal power sourced from the upper-level grid, gas loads, gas turbine emissions, and emissions from the P2G-CCS-CHP coupled system, respectively. The parameters γ_e, γ_h, and γ_g represent the free carbon emissions allowances per unit of electricity generated, per unit of heat produced, and per unit of gas load consumed, respectively. The symbol $\sigma_{e,h}$ signifies the conversion factor between electricity generation from gas turbines and thermal energy. $P_{grid,TE}(t)$ indicates the power purchased from the upper-level grid during time period t, while $P_{gload}(t)$ refers to the gas load demand for the same duration. $P_{MT}^{e}(\mathrm{t})$ and $P_{MT}^{h}(t)$ represent the electrical and thermal outputs of the gas turbine during time period t, respectively. T denotes the scheduling cycle.

(3) Actual Carbon Emission Model for PIES

During the conversion of electricity to gas in the system, some CO_2 will be utilized, and CCS can capture substantial quantities of CO_2. Consequently, the updated actual carbon emission model is as follows:

$$\begin{cases} E'_{PIES} = E'_{grid,TE} + E'_{gload} + E'_{MT} + E'_{P-C-C} \\ E'_{grid,TE} = \sum\limits_{t=1}^{T} \left(a_1 + b_1 P_{grid,TE}(\mathrm{t}) + c_1 P_{grid,TE}(\mathrm{t})^2 \right) \\ E'_{gload} = \gamma_g^{*} \sum\limits_{t=1}^{T} P_{gload}(t) \\ E'_{MT} = \gamma_h^{*} \sum\limits_{t=1}^{T} \left(\sigma_{e,h} P_{MT}^{e}(\mathrm{t}) + P_{MT}^{h}(t) \right) \\ E'_{P-C-C} = C_{P-C-C}(t) = C_{CHP}(t) - C_{CCS}(t) \end{cases} \tag{15}$$

where: E'_{PIES}, $E'_{grid,TE}$, E'_{gload}, E'_{MT}, and E'_{P-C-C} indicate the actual carbon emissions from the PIES, purchased thermal power from the main grid, gas load, gas turbines, and the P2G-CCS-CHP integrated unit, respectively. Parameters a_1, b_1, and c_1 represent the factors used for calculating carbon emissions in thermal power generation units. Moreover, γ_h^* and γ_g^* reflect the actual carbon emissions per unit of thermal power produced and per unit of gas load utilized, respectively.

(4) PIES Reward-Punishment Tiered Carbon Trading Cost Model

To enhance carbon emission reduction in the system and incentivize energy companies to realize their emission reduction potential, a mathematical model for tiered carbon trading costs based on a reward-punishment framework can be developed, expressed as follows:

$$f_{CO_2}^{trade}\begin{cases} -c(1+2\mu)(E_{PIES}-l-E'_{PIES}), \\ E'_{PIES}-E_{PIES}<-l \\ -c(1+2\mu)l-c(1+\mu)(E_{PIES}-E'_{PIES}), \\ -l\leq E'_{PIES}-E_{PIES}\leq 0 \\ c(E'_{PIES}-E_{PIES}), 0<E'_{PIES}-E_{PIES}\leq l \\ ch+c(1+\lambda)(E'_{PIES}-E_{PIES}-l), \\ l<E'_{PIES}-E_{PIES}\leq 2l \\ c(2+\lambda)l+c(1+2\lambda)(E'_{PIES}-E_{PIES}-2l), \\ 2l<E'_{PIES}-E_{PIES}\leq 3l \\ c(3+3\lambda)l+c(1+3\lambda)(E'_{PIES}-E_{PIES}-3l), \\ E'_{PIES}-E_{PIES}>3l \end{cases} \tag{16}$$

where: $f_{CO_2}^{trade}$ indicates the tiered carbon trading cost within the PIES model; c signifies the baseline price for carbon trading; l denotes the length of the trading volume interval; while μ and λ correspond to the coefficients for rewards and penalties, respectively.

2.2. Generation of Scenarios Considering Uncertainty and Correlation of Wind and Solar Outputs

Wind power generation and PV power generation rely on wind and solar energy, respectively. Within the same geographical area, a significant spatial interconnection exists between these two energy sources. Moreover, climatic and seasonal changes affect wind speed and solar irradiance, suggesting that the outputs of wind and solar energy are related over time. This indicates a correlation between wind power generation and photovoltaic generation in both spatial and temporal contexts. To maintain the stable and safe functioning of the PIES model, it is crucial to address the uncertainties and interdependencies linked to wind and solar outputs.

In this research, we utilize a non-parametric kernel density estimation approach to align historical data and derive the kernel density functions for wind and solar outputs. Next, we tackle the interdependence between these outputs using the Copula method to formulate a joint probability distribution function for both energy sources. In choosing the Copula function, we observe that the Frank Copula allows for any correlation sign and strength, while the Clayton and Gumbel Copulas are limited to positively correlated variables. Since wind and solar outputs often display a negative correlation and a complementary nature, we select the Frank Copula to represent their relationship. Finally, we sample the joint probability distribution function for each time frame, deriving the respective wind and solar outputs through the inverse transformation of the sampled data and the joint probability density function. Typical scenarios are then generated by reducing the wind-solar output scenarios backward.

2.2.1. Kernel Density Estimation Method

In applying the kernel density estimation method, we evaluate the nearness of each sample point to a specific point x by calculating the distances between x and its neighboring sample points. This evaluation aids in ascertaining how much influence each sample point has on the estimated value $\hat{f}(x)$. Let's consider that the samples $X_1, X_2, \cdots X_n$ are independently and identically distributed from a random variable X with an unknown density function $f(x)$. The goal is to estimate the probability density function $f(x)$ at the point x, represented as $\hat{f}(x)$.

$$\hat{f}(x) = \frac{1}{n\mathfrak{A}} \sum_{i=1}^{n} K\left(\frac{x - X_i}{\mathfrak{A}}\right) \tag{17}$$

where: n represents the number of samples, $\mathfrak{A}$ denotes the bandwidth, and $K(\cdot)$ refers to the kernel function.

The kernel density estimation method provides a non-parametric approach for estimating wind power and photovoltaic power data without requiring prior assumptions about the sample data distribution, which aids in deriving their probability density functions. The selection of the kernel function plays a crucial role in influencing the outcomes of the estimation, while the bandwidth $\mathfrak{A}$ is another parameter that needs optimization. Additionally, the integral mean square error between the estimated and actual values is expressed as follows:

$$MISE(h) = \int \mathbf{E}\left[\hat{f}(x) - f(x)\right]^2 dx \tag{18}$$

where: $\mathbf{E}$ represents the weight matrix. By minimizing Equation (18) to obtain h_0 and substituting this value into Equation (17), we can derive the overall kernel estimate.

2.2.2. Copula Functions

The expression for the Copula function is:

$$F(x_1, x_2, \cdots, x_n) = C(F_{x_1}(x_1), F_{x_2}(x_2), \cdots, F_{x_n}(x_n)) \tag{19}$$

where: $F(x_1, x_2, \cdots x_n)$ denotes the joint distribution function of n variables, while $F_{x_i}(x_i)$ for ($i = 1, 2, \cdots, n$) represents the marginal distribution of a single variable. The function $C(\cdot)$ is the Copula connection function.

When using the Frank Copula function, the expression for $C(\cdot)$ is:

$$C(u, v) = -\frac{1}{\theta} \ln\left[1 + \frac{(e^{-\theta} - 1)(e^{-\theta v} - 1)}{(e^{-\theta} - 1)^2}\right], \theta \epsilon \mathbf{R} \tag{20}$$

where: u and v represent the marginal distribution functions of wind and solar power outputs, respectively, and θ denotes the parameter of the Copula function.

2.2.3. Generation of Wind and Solar Scenarios and Their Complementary Characteristics

(1) Static Scenario Generation and Reduction of Wind and Solar Power Outputs Based on Monte Carlo Simulation

(1) Random numbers $a_1, a_2, \cdots a_n$ are generated in the interval [0, 1].

(2) Assign the value of the marginal distribution function for the first random variable as $u_1 = a_1$. Utilize the chosen Copula function from Section 2.2 to calculate the marginal distribution function value u_2 for the second random variable by solving Equation (21).

After identifying the optimal Copula function, large-scale samples can be created by drawing from this Copula function, following these specific steps:

(1) Generate random values in the range [0, 1].

(2) Use the known marginal distribution function value of the first random variable to solve for the marginal distribution function value of the second random variable using the selected Copula function from Section 2.2, corresponding to the solution of Equation (21).

$$\frac{\partial C(u_1, u_2, \cdots, u_n)}{\partial u_1} = a_2 \tag{21}$$

(3) The marginal distribution function value for the n-th random variable should be regarded as the solution to Equation (22).

$$\frac{\frac{\partial^{n-1} C(u_1, u_2, \cdots, u_n)}{\partial u_1 \partial u_2 \cdots \partial u_{n-1}}}{\frac{\partial^{n-1} C(u_1, u_2, \cdots, u_{n-1}, 1)}{\partial u_1 \partial u_2 \cdots \partial u_{n-1}}} = a_n \tag{22}$$

(4) By repeatedly executing steps (1), (2), and (3) a total of k times, k sets of marginal distribution function values for n random variables can be obtained.

(5) The inverse function operation is utilized to transform the results into a joint distribution function scenario. In the process of analyzing and calculating wind power and PV output, the inverse function computation in step (5) first requires calculating the marginal distribution functions for both outputs based on the Copula joint probability density. Subsequently, the inverse function operations are performed separately. This computational approach fully takes into account the correlation between wind and solar outputs. The data generated through scenario creation is large in volume and exhibits high similarity among the different scenarios. To effectively merge similar scenarios, the backward reduction (BR) method is employed for scenario reduction.

(2) Indicators of Wind-Solar Complementarity

The coefficient of variation (CV) is selected to represent the complementarity between wind power and photovoltaic output. The definition of CV is given as follows:

$$CV = \frac{\sqrt{\frac{1}{N}\sum_{t=1}^{N}\left(P_t^{WT} + P_t^{PV} - \overline{P}\right)^2}}{\overline{P}} \tag{23}$$

$$\overline{P} = \frac{1}{N}\sum_{t=1}^{N}\left(P_t^{WT} + P_t^{PV}\right) \tag{24}$$

where: P_t^{WT} and P_t^{PV} represent the wind power output and photovoltaic output at the t-th sampling point, respectively, while $\overline{P}$ indicates the average power of the two outputs.

According to Formula (23), a lower CV indicates that the joint output power of wind and photovoltaic sources is more stable, thereby demonstrating superior complementary characteristics. To ensure that the generated wind-solar scenarios maintain stable output, characterized by good complementary properties, the CV must remain below the system-defined reference values ε_1 and ε_2, specifically CV $\leq \varepsilon_1$ and CV $\leq \varepsilon_2$. Here, ε_1 and ε_2

correspond to the CV values for scenarios with only wind power and only photovoltaic power output, respectively.

$$\begin{cases} \varepsilon_1 = \dfrac{\sum_{i=1}^{T} \dfrac{\sqrt{\frac{1}{N_i} \sum_{t=1}^{N_i} \left(P_{i,t}^{WT} - \overline{P_i^{WT}} \right)}}{\overline{P_i^{WT}}}}{T} \\ \varepsilon_2 = \dfrac{\sum_{i=1}^{T} \dfrac{\sqrt{\frac{1}{N_i} \sum_{t=1}^{N_i} \left(P_{i,t}^{PV} - \overline{P_i^{PV}} \right)}}{\overline{P_i^{PV}}}}{T} \end{cases} \tag{25}$$

$$\begin{cases} \overline{P_i^{WT}} = \dfrac{1}{N_i} \sum\limits_{t=1}^{N_i} P_{i,t}^{WT} \\ \overline{P_i^{PV}} = \dfrac{1}{N_i} \sum\limits_{t=1}^{N_i} P_{i,t}^{PV} \end{cases} \tag{26}$$

where $P_{i,t}^{WT}$ denotes the wind power output at the t-th sampling point on the i-th day; $P_{i,t}^{PV}$ represents the photovoltaic power output at the same sampling point; $\overline{P_i^{WT}}$ is the daily average wind power output; $\overline{P_i^{PV}}$ indicates the daily average photovoltaic power output; and N_i refers to the total number of sampling points for the i-th day.

2.3. *Incentive-Based Ladder Carbon Trading with Consideration of Wind and Solar Uncertainty and the Coupling of P2G-CCS-CHP in Integrated Energy System Optimization Model*

2.3.1. Objective Function

This study establishes an optimization model by defining multiple objective functions, including operational costs and environmental protection (carbon emission) costs, alongside various constraints related to energy equipment. The model is solved using the Non-dominated Sorting Beetle Optimization algorithm (NSDBO).

(1) Economic Objectives

The economic objective under the grid-connected mode is to minimize the operational costs of the PIES.

$$minF_1 = min\left(\sum_{t=1}^{T} C_{grid}(t) + C_{NS}(t) + C_{op}(t) \right) \tag{27}$$

$$\begin{cases} C_{grid}(t) = C_{buy}(t) + C_{sell}(t) \\ C_{sell}(t) = c_{sell}(t) P_{sell}(t) \\ C_{buy}(t) = c_{buy}(t) P_{buy}(t) \\ C_{op}(t) = \sum\limits_{i \in I} \sum\limits_{t=1}^{T} a_{1,t} P_i(t) \end{cases} \tag{28}$$

where: $C_{grid}(t)$, $C_{NS}(t)$, and $C_{op}(t)$ denote the total costs associated with the interaction between the PIES and the main grid, the PIES and the natural gas network, and the operational costs of each device at time t, respectively. $c_{sell}(t)$ and $P_{sell}(t)$ refer to the electricity selling price and the amount of power sold from the PIES to the main grid at time t; $c_{buy}(t)$ and $P_{buy}(t)$ represent the electricity purchasing price and the amount of power purchased from the main grid. The calculation method for $C_{NS}(t)$ is identical to that for $C_{grid}(t)$ and will not be discussed further. I encompass P2G-CCS-CHP, MT, ER, WT, and PV. $a_{1,i}$ signifies the unit operational cost for device i, while P_i is the input power for that device.

(2) Environmental Objectives

The environmental objective is to minimize the pollutant treatment costs of the PIES, which entails minimizing both pollutant emissions and carbon trading costs.

$$minF_2 = min\left(\sum_{t=1}^{T} \begin{matrix} C_{grid.en}(t) + C_{NS.en}(t) + C_{MT.en}(t) + \\ C_{ER.en}(t) + C_{P-C-C.en}(t) + f_{CO_2}^{trade} \end{matrix}\right) \tag{29}$$

$$G_{grid.en}(t) = \sum_{k=1}^{n} \left(C_k \gamma_{grid,k}\right) P_{buy}(t) \tag{30}$$

where: $C_{grid.en}(t)$, $C_{NS.en}(t)$, $C_{MT.en}(t)$, $C_{ER.en}(t)$, and $C_{P-C-C.en}(t)$ represent the pollutant treatment costs for the large grid, the natural gas network, the gas turbine operation, the refrigerator operation, and the P2G-CCS-CHP coupled unit operation, respectively. The variable $\gamma_{grid,k}$ signifies the emission volume of the k-th pollutant generated by the large grid operation. C_k is the cost coefficient for the societal treatment of the k-th pollutant. The term $f_{CO_2}^{trade}$ is referenced in Section 2.1.2.

2.3.2. Constraints

(1) Power Balance Constraint

$$\begin{cases} P_c(t) + P_{WT}(t) + P_{PV}(t) + P_{MT}^e(t) = P_{eload}(t) + P_{ER}(t) \\ P_c^h(t) + P_{MT}^h(t) = P_{hload}(t) \\ P_{NS}^g(t) + P_{P2G}^g(t) = P_{MT}^g(t) + P_{gload}(t) \\ P_{MT}^c(t) + P_{e-c}^c(t) = P_{cload}(t)) \end{cases} \tag{31}$$

where: $P_{eload}(t)$, $P_{hload}(t)$, $P_{gload}(t)$, and $P_{cload}(t)$ represent the electrical load, thermal load, gas load, and cooling load within the PIES, respectively. $P_{ER}(t)$ signifies the electric power utilized by the ER. $P_{MT}^e(t)$ and $P_{MT}^h(t)$ indicate the electrical and thermal power produced by the MT, respectively. $P_{NS}^g(t)$ refers to the gas supplied by the NS. $P_{MT}^g(t)$ represents the gas power consumed by the MT, while $P_{P2G}^g(t)$ denotes the gas power generated by the P2G device. Lastly, $P_{MT}^c(t)$ and $P_{e-c}^c(t)$ correspond to the cooling power produced by the MT and ER, respectively.

(2) Equipment Capacity and Ramping Constraints

The capacity and ramping limitations of energy equipment, including CHP, MT, and ER, can be outlined as follows:

$$\begin{cases} I_i(t)P_{i,min} \le P_i(t) \le I_i(t)P_{i,max} \\ R_{i,down} \le P_i(t) - P_i(t-1) \le R_{i,up} \\ [S_{i,on}(t-1) - T_{i,on,min}][I_i(t-1) - I_i(t)] \ge 0 \\ \left[S_{i,off}(t-1) - T_{i,off,min}\right][I_i(t) - I_i(t-1)] \ge 0 \end{cases} \tag{32}$$

where: P_i denotes the input power of device i; $P_{i,max}$ and $P_{i,min}$ represent the maximum and minimum input power levels for device i, respectively; $R_{i,down}$ and $R_{i,up}$ indicate the constraints for ramping down and ramping up, respectively; $S_{i,on}$ and $S_{i,off}$ refer to the durations during which device i is in the startup and shutdown states, respectively; $T_{i,on,min}$ and $T_{i,off,min}$ signify the minimum times required for startup and shutdown, respectively.

(3) Energy Coupling Constraints

Energy coupling devices enable the interconversion of electricity, heat, and gas. This conversion process is governed by particular constraints, which can be articulated as follows:

$$P_{i,j}(t) = \eta_{i,j} P_i(t) \tag{33}$$

where: j refers to the various energy forms, including electricity, heat, gas, and cooling; $P_{i,j}$ represents the input power converted from the i-th energy coupling device to the j-th energy form; $\eta_{i,j}$ signifies the efficiency of the energy conversion process.

2.4. Model Solution Methods

2.4.1. Non-Dominated Sorting-Based Cockroach Optimization Algorithm

(1) Non-Dominated Sorting Genetic Strategy

In its daily operations, PIES encounters various constraints, such as low carbon emissions, environmental sustainability, economic efficiency, reliability, and the proportion of renewable energy. This study centers on analyzing economic and environmental indicators, with goals that include minimizing total costs, operational expenses, environmental protection expenditures, and finding an acceptable trade-off solution. This situation can be classified as a Multi-Objective Optimization Problem (MOP). When constructing the model, it is crucial to take into account the interdependencies among the objectives as well as the conflicts between the objective functions. The MOP can be expressed as follows:

$$y = \min f_i(x)(i = 1, 2, \cdots, m) \tag{34}$$

$$\text{s.t.} \begin{cases} g_j(x) \leq 0, (j = 1, 2, \cdots, p), x \in E^n \\ h_m(x) = 0, (m = 1, 2, \cdots, q), x \in E^n \end{cases} \tag{35}$$

For any multi-objective evolutionary algorithm, it is crucial to obtain Pareto optimal solutions at the endpoints of the Pareto front that exhibit both convergence and diversity. The standard NSGA-II algorithm employs a crowding distance-based strategy to maintain solution diversity [20].

Given the rapidly changing environment in which PIES functions, any shifts in demand—whether from generation; supply; or consumption—require a reassessment of the scheduling algorithm. This study utilizes a non-dominated sorting genetic strategy to optimize latency, energy usage, and algorithm convergence. To improve the algorithm's performance, factors such as rapid non-dominated sorting and crowding degree have been integrated into the traditional NSGA approach. This method is especially effective for tackling the multi-objective optimization scheduling challenge faced by PIES, with specific algorithmic details documented in existing literature.

(2) Dung Beetle Optimization Algorithm

The DBO algorithm, developed by researchers including Xue Jiankai, is a swarm intelligence optimization technique inspired by the behavioral traits of dung beetles [21]. It is notable for its robust optimization ability and rapid convergence.

During the rolling process, dung beetles roll their fecal balls in a straight line to avoid competition from other beetles. However, variations in celestial light intensity or natural factors can cause the beetles' paths to become curved. The position of a rolling dung beetle during its forward motion can be represented as follows:

$$x_i^{t+1} = x_i^t + \alpha k x_i^{t-1} + bx \tag{36}$$

$$N = \left| x_i^t - x_{worst}^t \right| \tag{37}$$

where: t indicates the iteration coefficient; x_i^t represents the position of the i-th dung beetle at the t-th iteration; $k\epsilon(0, 0.2)$ is the deflection coefficient, usually considered a constant; b is a constant value ranging between 0 and 1; α is a natural number that can equal ± 1; N denotes the simulated variation in light intensity; and x_{worst} indicates the local worst position.

When a dung beetle encounters an obstacle hindering its progress, it must re-roll and adjust its position to create a new trajectory. The position update can be expressed as follows:

$$x_i^{t+1} = x_i^t + tan(\theta)\left|x_i^t - x_i^{t-1}\right| \quad (38)$$

where: $\theta \in [0,\pi]$ signifies the angle of deflection; $x_i^t - x_i^{t-1}$ reflects the positional change of the i-th dung beetle across iterations.

Breeding dung beetles transport dung balls to a secure location for laying eggs, depositing a single egg ball in the process. To reduce risks, the egg-laying area is variable, with the egg ball's position shifting dynamically.

$$\begin{cases} Lb^* = max\left\{x_{gbest}^t(1-R), Lb\right\} \\ Ub^* = min\left\{x_{gbest}^t(1+R), Ub\right\} \end{cases}, R = \frac{1-t}{M} \quad (39)$$

$$x_i^{t+1} = x_{gbest}^t + b_1\left(x_i^t - Lb^*\right) + b_2\left(x_i^t - Ub^*\right) \quad (40)$$

where: M denotes the maximum number of iterations; Lb and Ub represent the lower and upper bounds of the optimization target, respectively; Lb^* and Ub^* indicate the boundaries of the egg-laying area; x_{gbest} refers to the current global best position within the population; R signifies the inertia weight; and b_1 and b_2 are two independent random vectors.

Once the larval ball hatches into a young dung beetle, it emerges from the ground in search of food. The optimal foraging area is also updated dynamically, with the foraging position of the young dung beetle adjusted as follows:

$$\begin{cases} Lb^l = max\left\{x_{lbest}^t(1-R), Lb\right\} \\ Ub^l = min\left\{x_{lbest}^t(1+R), Ub\right\} \end{cases} \quad (41)$$

$$x_i^{t+1} = x_i^t + C_1\left(x_i^t - Lb^l\right) + C_2\left(x_i^t - Ub^l\right) \quad (42)$$

where: Lb^l and Ub^l indicate the lower and upper bounds of the foraging area, respectively; x_{lbest} signifies the current global best position of the population; C_1 is a random number drawn from a normal distribution; and C_2 is a random vector with dimensions $1 \times D$ in the range (0, 1).

Additionally, there are thieves among the population. Assuming x_{bf} is the optimal location for food theft, the positions of the thieves in the population are defined as:

$$D_i(t+1) = x_{bf} + \xi\tau\left(\left|D_i(t) - x_{best}\right| + \left|D_i(t) - x_{bf}\right|\right) \quad (43)$$

where: $D_i(t)$ denotes the position of the i-th thief in the population at the t-th iteration; τ is a random vector with dimensions $1 \times N$ derived from a normal distribution; and ξ is a constant.

(3) Non-Dominated Sorting Dung Beetle Optimizer Model

Based on the previous discussions, the overall optimization process of the Non-Dominated Sorting Dung Beetle Optimizer (NSDBO) is illustrated as follows. The specific process is detailed in Figure 4.

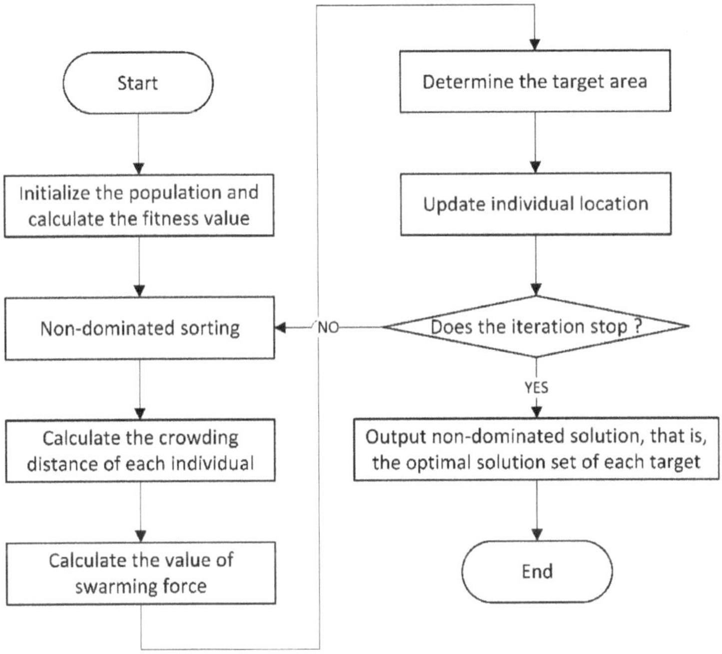

Figure 4. Flowchart of NSDBO algorithm.

(1) Initial Population Construction: A batch of individuals is randomly generated, and their fitness is evaluated.

(2) Non-Dominated Sorting: A non-dominated sorting genetic strategy is applied to arrange each individual in the population, obtaining the non-dominated ranks of individuals along with the corresponding number of individuals they dominate.

(3) Crowding Distance Calculation: The crowding distance is measured to estimate the distribution of each individual in the objective space.

(4) Attractiveness Calculation: The collective attractiveness of each individual is computed based on their non-dominated rank and distribution.

(5) Target Area Determination: The collective attractiveness will determine the target area each individual should venture towards.

(6) Position Update: The positions of individuals in the population are updated to simulate the chasing behavior of dung beetles.

(7) Termination Check: A decision is made on whether to terminate the process.

By iteratively following the above steps, the NSDBO algorithm gradually searches for a set of non-dominated solutions to the multi-objective optimization problem. These solutions represent the optimal solution set for the problem, providing multiple optimal choices for decision-makers.

2.4.2. Model Performance Test

The NSDBO was tested using the CEC 2005 benchmark function set and compared with several leading intelligent optimization algorithms, including the Particle Swarm Optimization (PSO) [22], Musical Chairs Algorithm (MCA) [23], Pelican Optimization Algorithm (POA) [24], Subtraction-Average-Based Optimizer (SABO) [25], and Harris Hawk Optimization (HHO) [26]. The stability of the SBOA was evaluated using four representative test functions, with an initial population size of 40 and 500 iterations for each algorithm.

The test function is shown in Appendix A Table A1, and the function convergence curve is shown in Appendix A Figure A1.

According to the results presented in Figure A1, the NSDBO demonstrated superior performance in terms of both optimal solution search and convergence speed compared to other algorithms. For the test functions f_1, f_2, and f_4, NSDBO exhibited a notably faster convergence characteristic relative to the other algorithms. Furthermore, in f_3, SBOA effectively addressed the local optimality issues often encountered by the traditional PSO and other algorithms, surpassing their capabilities in terms of optimal solution identification.

3. Results and Discussion

3.1. Basic Data

This study examines a comprehensive demonstration park in Northern China to analyze the percentage of wind and solar renewable energy across various scheduling scenarios, as well as to evaluate carbon dioxide emissions, carbon emission costs, and total operating costs. The scheduling period is established at 24 h with hourly time steps. Figure 5 illustrates the raw data for the electric, thermal, gas, and cooling load curves within the PIES framework. Electricity and natural gas prices vary over time, as referenced in literature [27]. Table 1 presents the parameters and economic characteristics of the different energy devices within PIES. In Table 1, G2P refers to the conversion of natural gas to electricity, G2H to natural gas converted into thermal energy, G2C to natural gas converted into cooling energy, and E2C to electricity transformed into cooling energy. The allocation of carbon emissions rights for electricity, heat, and natural gas varies due to differences in their energy characteristics, market structures, technological innovations, and social acceptance. The electricity sector, with its diverse generation methods, requires more refined management, whereas the distribution of heat and natural gas focuses more on direct emissions control and the impacts over their entire lifecycle. According to the general knowledge of carbon emission right allocation in the China National Energy Administration and existing literature, this paper sets the free carbon emission right quota corresponding to per unit of electricity generation as 0.728 kg/(kWh), the free carbon emission right quota corresponding to per unit of heat generation as 0.385 kg/(kWh), and the free carbon emission right quota corresponding to per unit of gas load consumption as 0.180 kg/(kWh). The conversion coefficient from gas turbine power generation to heat, $\sigma_{e,h}$, is set at 0.75. The base price c for carbon trading is established at 0.034 USD/kg, with a trading volume range l of 4000 kg. The reward and punishment coefficients were set to 0.2 and 0.15, respectively, and the specific method reference is [28].

Table 1. Parameters and economic parameters of each energy equipment in an industrial park.

Parameter	Numerical Value	Parameter	Numerical Value
α	0.75	c_{v2}	−0.4
χ	0.85	β	0.88
$P_{c,max}$(MW)	100	$T_{i,on,min}(h)$	6
$P_{c,min}$(MW)	0	$T_{i,off,min}(h)$	6
$P_{c2,max}$(MW)	30	$P_{i,max}$(MW)	100(NS), 100(MT), 25(RR)
$P_{c2,min}$(MW)	0	$P_{i,min}$(MW)	0(NS), 0(MT), 0(RR)
P_{c0}^{h}(MW)	50	$R_{i,up}$(MW)	10(CHP), 10(MT), 2.5(RR)
c_{v1}	−0.3	$R_{i,down}$(MW)	0(CHP), 0(MT), 0(RR)
c_m	1	$\eta_{i,j}$	0.285(G2P,MT), 0.38(G2H,MT), 0.285(G2C,MT), 0.97(E2C,RR)
α	0.75	c_{v2}	−0.4
χ	0.85	β	0.88
$P_{c,max}$(MW)	100	$T_{i,on,min}(h)$	6
$P_{c,min}$(MW)	0	$T_{i,off,min}(h)$	6

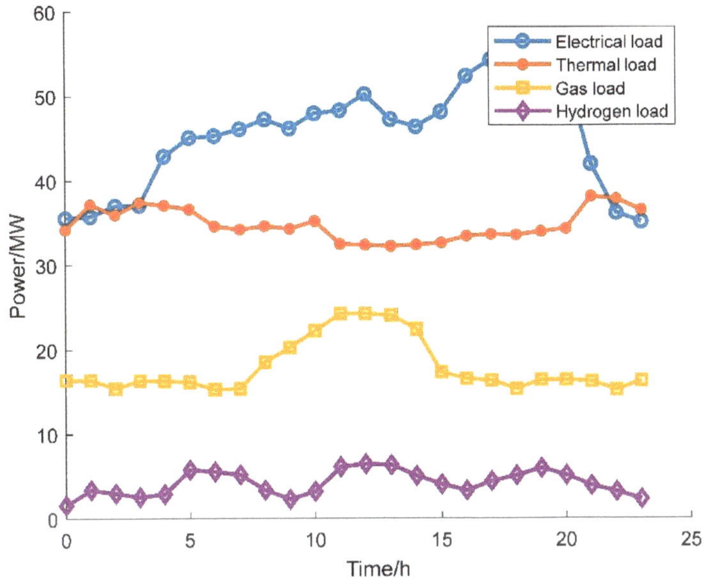

Figure 5. Raw data of electrical, thermal, air, and cooling loads inside PIES.

3.2. Generation and Reduction of Scenarios Considering Wind and Solar Uncertainty and Correlation

This section examines the uncertainties linked to wind and solar energy by utilizing forecasting data from the prior day, as detailed in Section 2. The findings are illustrated in Figure 6. Initially, we create 500 scenarios for wind and solar power output through a scenario generation technique that employs kernel density estimation and Copula theory. After that, we use the backward reduction method to refine the scene and finally select five representative scenarios for wind and solar power generation. The specific principles and proofs are detailed in [29]. The correlation probabilities of each scenario under the 95% confidence interval are listed in Table 2.

Table 2. Probabilities corresponding to each scenario.

Scene	Probit
1	0.226
2	0.228
3	0.234
4	0.13
5	0.182

This approach integrates five representative scenarios of wind and solar output into the system's optimal operational scheduling model, enabling the calculation of operational costs across various scenarios. Next, the operating costs for these five scenarios are combined probabilistically to establish the system's final operational cost. This final value serves as an economic metric within the optimization scheduling model, leading to a scheduling strategy that effectively considers the uncertainties and correlations in wind and solar output.

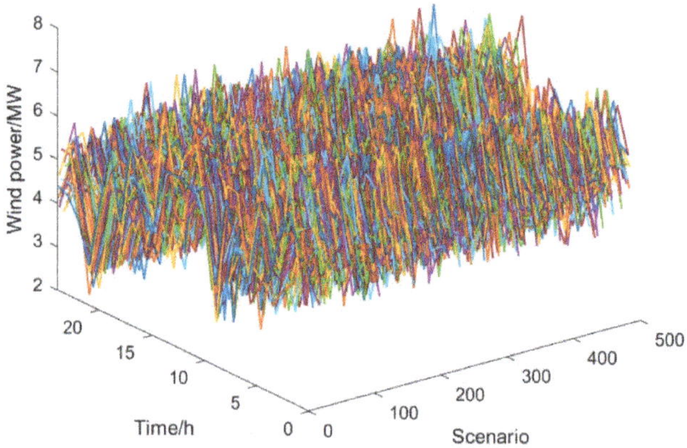

(a)

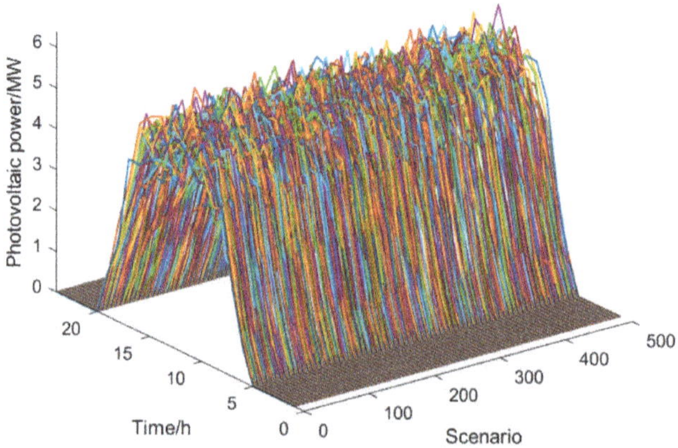

(b)

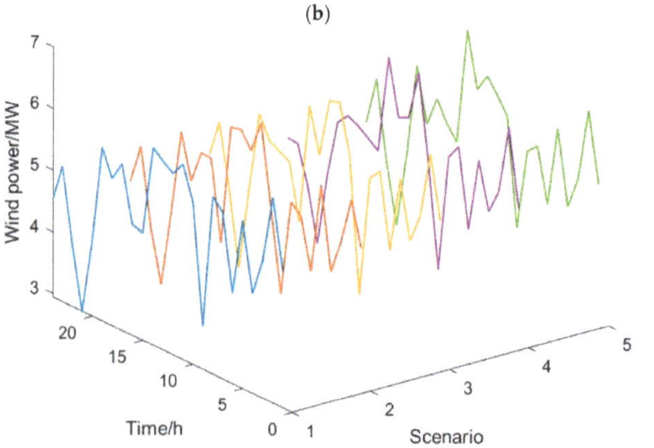

(c)

Figure 6. *Cont.*

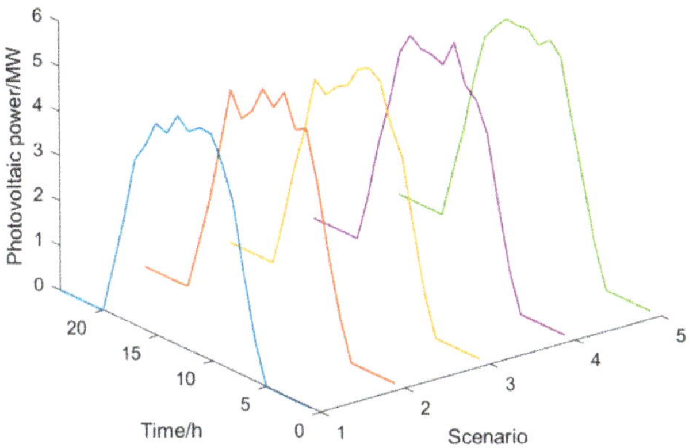

(d)

Figure 6. Generation and reduction of landscape scene: (**a**) Generation of 500 Wind Power Output Scenarios Considering; (**b**) Generation of 500 Photovoltaic Output Scenarios Considering; (**c**) Five Scenarios of Wind Power Output After Reduction; (**d**) Five Scenarios of Photovoltaic Output After Reduction.

3.3. Analysis of Optimization Scheduling Results

The NSDBO algorithm was utilized to solve the optimization scheduling problem, yielding a Pareto front solution set comprising 115 non-dominated solutions after several iterations. To further analyze the practical implications of the non-dominated solution values for engineering applications, we discuss the four solution methods under two objective functions in the context of various realistic trade-off scenarios that may arise. The scheduling results of the four methods under the 95% confidence interval are shown in Table 3. In this context, the compromise solution referenced in Method 4 refers to the use of a weighted scoring method that assigns equal weight factors to Methods 1–3 in the search for the optimal solution. These principles find their theoretical foundation in reference [30], which elaborates on a multi-objective optimization weighted scoring principle. In particular, it underscores the importance of assigning equal weight factors to competing objectives to ensure balanced overall solutions, providing crucial theoretical underpinnings for our selected approach. Furthermore, reference [30] highlights successful applications of this principle within specific engineering contexts, validating the practical feasibility of our optimization strategy.

Method 1: Finding the solution that minimizes operational costs.

Method 2: Finding the solution that minimizes environmental protection costs.

Method 3: Finding the solution that minimizes total costs.

Method 4: Normalizing and summing the values of the two objective functions, then selecting the particle corresponding to the minimum aggregated value, which identifies the compromise solution.

Table 3. Results of each method.

Operating Result	Methods 1	Methods 2	Methods 3	Methods 4
Operating costs/USD	11,633.65	33,338.64	19,207.90	22,797.95
Carbon emissions/kg	56,177.74	1238.71	11,438.22	9106.63
Environmental protection costs/USD	17,389.77	1486.79	3914.68	3426.38
Total cost/USD	29,023.41	34,825.43	23,122.59	26,224.33

From Table 3, it can be concluded that there is no optimal scenario in which total costs, operational costs, and environmental protection costs can simultaneously be minimized. If the primary objective of the decision-maker is to achieve the lowest operational costs, then Method 1 should be employed for scheduling; under this approach, the operational costs are reduced by $21,704.99, $7574.26, and $11,164.30 compared to Methods 2, 3, and 4, respectively. If the decision-maker's primary aim is to minimize environmental protection costs, then Method 2 should be selected, resulting in savings of $15,902.98, $2427.89, and $1939.59 in environmental protection costs relative to Methods 1, 3, and 4, respectively. Conversely, if the decision-maker seeks to minimize total costs, Method 3 should be utilized; in this case, the total costs saved would be $5900.82, $11,702.84, and $3101.74 compared to Methods 1, 2, and 4, respectively. Lastly, if the decision-maker aims to balance cost-effectiveness with low carbon emissions in search of a compromise solution, then Method 4 should be employed for scheduling.

3.3.1. Analysis of the Impact of Wind and Solar Output Uncertainty, Reward-Punishment Tiered Carbon Trading Mechanism, and Coupling Operation of P2G-CCS-CHP

This section explores the effects of considering the uncertainties in wind and solar outputs, a reward-punishment-tiered carbon trading mechanism, and the integrated operations of P2G, CCS, and CHP on the optimization results of PIES. Method 3, which prioritizes the minimization of total costs—a preferred solution approach for many decision-makers—is utilized for this analysis. Six distinct operational scenarios are established to assess the effectiveness of the scheduling strategy developed in this study. The optimization results of these scenarios under the 95% confidence interval are summarized in Table 4, while Figures 7–9 depict the operational performance of the CHP unit's power output, along with the absorption of wind and PV energy during the scheduling process. All scenarios and solutions are derived from the application of Method 3:

Scenario 1: Includes CHP but omits P2G, CCS, carbon trading mechanisms, and uncertainties related to wind and solar energy.

Scenario 2: Incorporates P2G-CHP while excluding CCS, carbon trading mechanisms, and wind and solar uncertainties.

Scenario 3: Considering the joint operation of P2G-CCS but the independent operation of the CHP system and the traditional carbon trading mechanism, the uncertainty of the scenery is not considered.

Scenario 4: Considering the P2G-CCS-CHP and the traditional step-by-step carbon trading mechanism, the uncertainty of scenery is not considered.

Scenario 5: Includes P2G-CCS-CHP with a reward-punishment-tiered carbon trading mechanism, excluding wind and solar uncertainties.

Scenario 6: Considers P2G-CCS-CHP, a reward-punishment-tiered carbon trading mechanism, and incorporates uncertainties related to wind and solar outputs (the model developed in this study).

Table 4. Optimized scheduling results for 6 scenarios.

Scenario	Operating Costs/USD	Environmental Protection Costs/USD	Total Cost/USD	Carbon Emissions/kg	WT Utilization Rate/%	PV Utilization Rate/%
Scenario 1	26,637.65	10,010.43	36,648.07	51177.74	53.44	65.27
Scenario 2	24,158.65	9122.30	33,280.95	47018.92	58.82	69.36
Scenario 3	22,005.03	7085.39	29,090.43	30645.65	60.62	70.10
Scenario 4	20,636.10	5445.35	26,081.45	19780.99	71.74	72.01
Scenario 5	19,805.39	4197.62	24,003.01	13589.00	77.90	80.21
Scenario 6	19,207.90	3914.68	23,122.59	11438.22	82.30	85.12

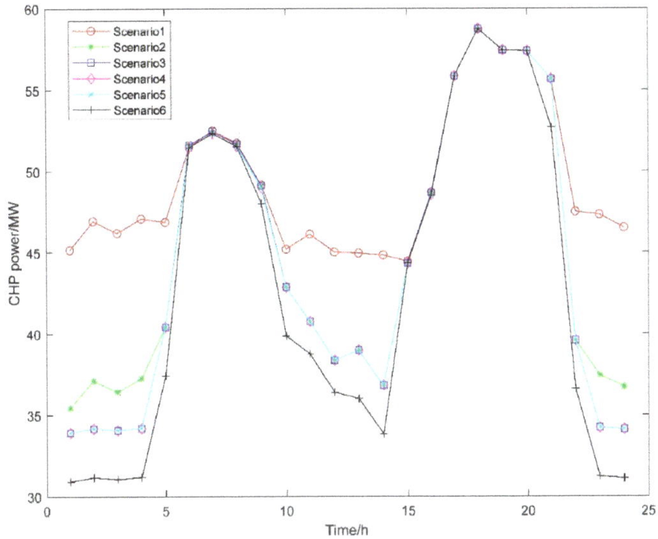

Figure 7. Comparison of CHP power for six scenarios.

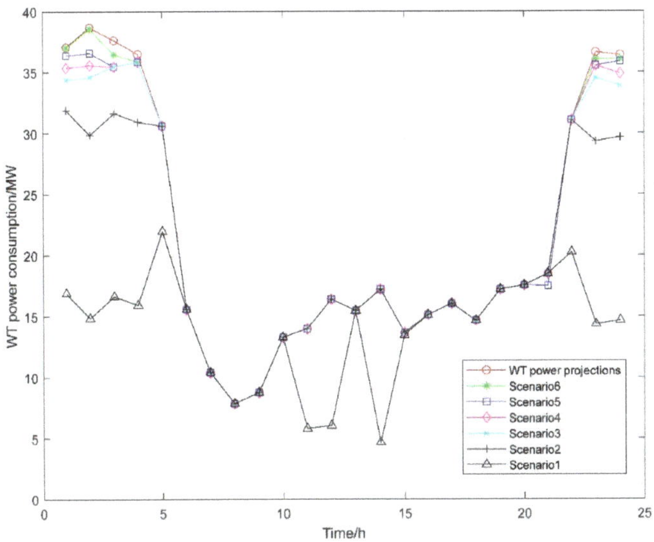

Figure 8. Comparison of Wind Power Consumption in Six Scenarios.

Figure 7 illustrates that the comparison between Scenario 2 and Scenario 1 shows a reduction in grid-connected power output from the CHP unit during times of high renewable energy generation. This indicates that the P2G system is redirecting electrical power from the CHP unit. Additionally, the analysis of Scenario 3 compared to Scenario 2 reveals that when wind energy constitutes a significant portion of the energy supply, the grid power output from the CHP unit decreases further. This implies that the CCS system is utilizing CHP power to capture CO_2, thus lowering the net output of the CHP unit. In instances of elevated photovoltaic energy generation, the CHP unit does not contribute to the grid, as the thermal load demand is low and its generation capacity is minimal, resulting in the full conversion of its output into natural gas via the P2G system. When examining Scenarios 3, 4, and 5, it becomes evident that the carbon trading mechanisms do not significantly affect the power output of the CHP unit. Furthermore, when comparing Scenario 6 with Scenarios 4 and 5, the introduction of wind and solar output uncertainty leads to a decline in the hourly power output of the CHP within the PIES throughout the day, particularly pronounced during high renewable energy generation periods. This indicates that a detailed examination of the fluctuations and correlations in wind and solar output can improve the predictions of the daily wind-solar load curve. By optimizing the multi-energy coupling within the PIES and enhancing equipment flexibility, it is feasible to boost the integration of wind and solar energy while also minimizing the CHP unit's power output.

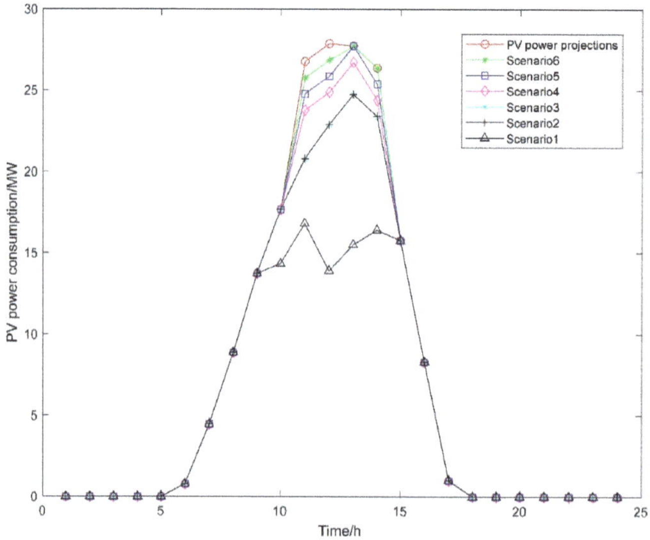

Figure 9. Comparison of PV consumption in six scenarios.

Figure 8 illustrates critical time periods when wind power plays a substantial role in the energy supply, particularly between 00:00 and 05:00 and from 22:00 to 24:00. A comparison between Scenario 2 and Scenario 1 shows that integrating P2G into the CHP configuration enhances wind power absorption. This enhancement is mainly attributed to the P2G system's capability to convert electricity into natural gas, thus creating extra storage for wind energy. In addition, examining Scenario 3 against Scenario 2 indicates that the CCS system provides CO_2 for the P2G process, reducing the need to source carbon externally. This results in lower operational costs for P2G, indicating that the inclusion of the P2G-CCS unit within the CHP structure boosts wind power absorption while cutting expenses. Comparing Scenario 4 to Scenario 3 reveals that the implementation of a tiered carbon trading system improves wind power absorption during high supply periods compared to scenarios lacking such mechanisms. This system increases the costs associated with managing carbon emissions, encouraging the PIES to absorb more wind energy to avoid excessive emissions. Moreover, the analysis of Scenario 5 in relation to Scenario 4 demonstrates that the penalty-reward carbon trading system is more effective than the traditional version. The enhanced penalties associated with stricter emission limits outweigh the additional costs incurred from higher wind power usage, leading to greater absorption. Lastly, evaluating Scenario 6 with respect to Scenario 5 shows that accounting for uncertainties in wind and solar outputs significantly increases the accuracy of daily wind power generation forecasts. This improvement ultimately supports the increasing wind power integration within the PIES framework.

Figure 9 shows that combining CHP and P2G in Scenario 2 significantly enhances the capacity to absorb PV power compared to Scenario 1. The P2G system effectively harnesses electricity, which increases the potential for integrating PV energy. In comparing Scenario 3 with Scenario 2, it becomes clear that CCS does not affect the absorption of PV power. Between 10:00 and 14:00, the output from CHP decreases due to lower heating demand and increased cooling demand. Consequently, under the constraints of CHP output, which includes P2G and CCS units, CCS fails to facilitate P2G conversion or improve PV electricity absorption. When analyzing Scenario 4 relative to Scenario 3, it is apparent that a tiered carbon trading system enhances PV power absorption during periods of high PV output compared to scenarios without it. This system raises carbon management costs, encouraging the PIES to consume more PV energy to reduce excess emissions. Additionally, the comparison of Scenario 5 with Scenario 4 shows that the penalty-reward tiered carbon trading system is more effective than the traditional model. The penalties imposed by stricter carbon emission limits outweigh the additional costs associated with increased PV usage, leading to higher absorption levels. Finally, the evaluation of Scenario 6 against Scenario 5 indicates that addressing the uncertainties associated with wind and solar generation significantly improves the accuracy of daily PV production forecasts. This enhancement is vital for optimizing PV absorption within the PIES framework.

An overview of Scenarios 1 to 6 presented in Table 4 illustrates that as the operational capacity of the PIES expands, the benefits of multi-energy complementarity and flexibility become increasingly evident, resulting in a steady reduction in operational costs. In particular, the comparison between scenario 4 and scenario 3 shows that the total cost and carbon emissions of the P2G-CCS-CHP joint operation are significantly lower than those of the three separate operations, and the utilization rate of new energy is significantly improved. The joint operation mechanism has advantages in scheduling. Furthermore, the collaboration of CHP and P2G improves energy conversion efficiency, minimizing energy losses and emissions. CCS technology captures CO_2 emissions, thus limiting their release into the atmosphere while supplying energy to P2G and decreasing reliance on external energy sources. The penalty-reward-tiered carbon trading system enhances the traditional approach by tightening carbon emission regulations and increasing penalties for exceeding limits. By incorporating the uncertainties of wind and solar energy, absorption rates for both types of energy are improved, fostering greater use of clean energy and contributing to a decrease in carbon emissions and environmental protection costs. Consequently, the simultaneous reduction in operational costs and environmental protection expenses optimizes total costs. Additionally, the effective scheduling of the PIES is dependent on the strategic allocation of wind and solar generation, which continues to boost the utilization of these renewable sources in scenarios characterized by multi-energy complementarity. The above improvement of energy saving and low carbon further proves the effectiveness of the strategy proposed in this paper.

3.3.2. Analysis of PIES Optimal Scheduling in Scenario 6 (This Study's Model)

The results from the optimal scheduling model developed in this study are presented in Figures 10 and 11. The next section offers a comprehensive analysis to highlight its effectiveness and provide valuable insights for practical engineering applications.

Figure 10 depicts the performance of electrical, thermal, gas, and cooling power across each unit. In Figure 10a, the power generation unit meets the electrical load demand at all time intervals, with the P2G-CCS-CHP unit, WT, PV, and MT fulfilling both electricity and cooling needs. Figure 10b shows that the thermal load is delivered by the P2G-CCS-

CHP unit in conjunction with the microturbines. Figure 10c reveals that the gas power utilized by the microturbines is sourced from both the gas supply and the P2G-CCS-CHP unit. Lastly, Figure 10d indicates that the cooling load is managed by the ER system and the microturbines.

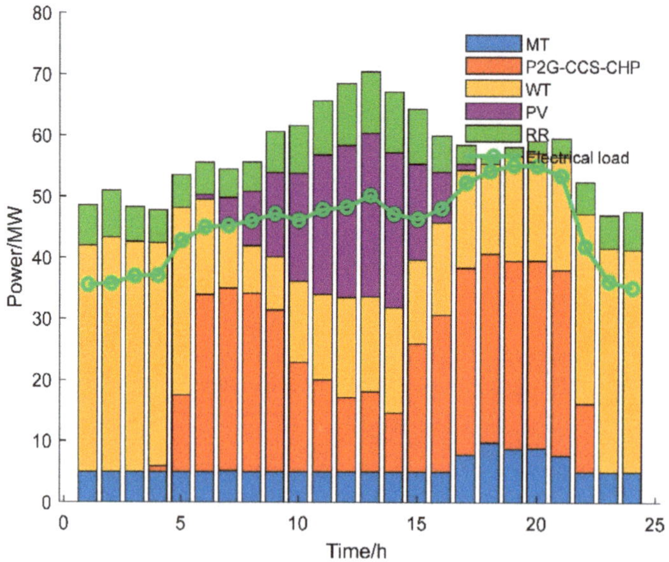

(**a**)

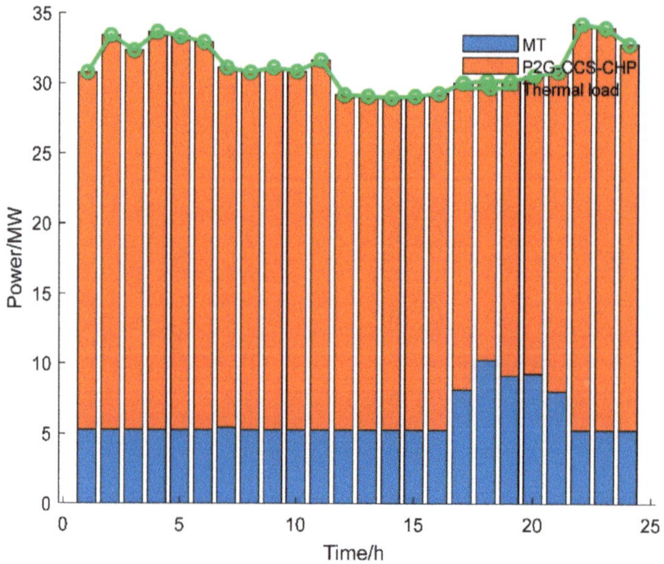

(**b**)

Figure 10. *Cont.*

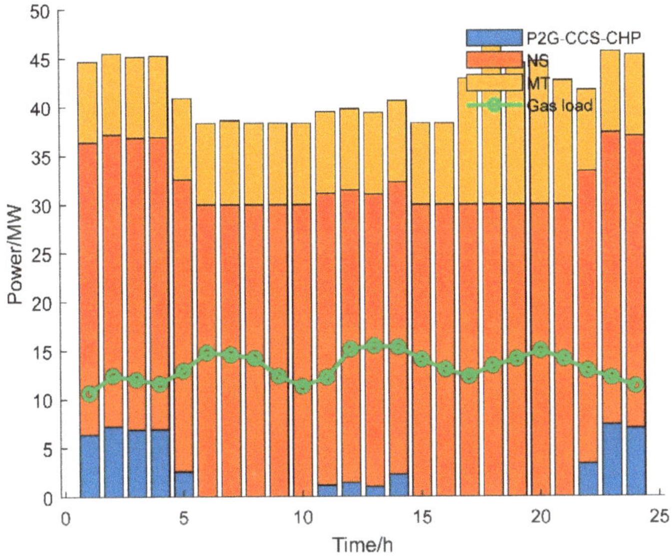

(c)

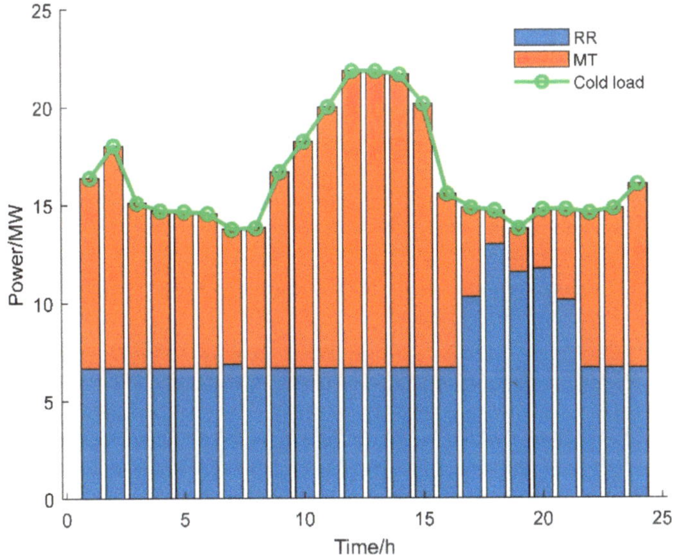

(d)

Figure 10. Optimal scheduling results for PIES in scenario 6: (**a**) Power Balance Diagram of PIES in Scenario 6; (**b**) Thermal Power Balance Diagram of PIES in Scenario 6; (**c**) Gas Power Balance Diagram of PIES in Scenario 6; (**d**) Cold Power Balance Diagram of PIES in Scenario 6.

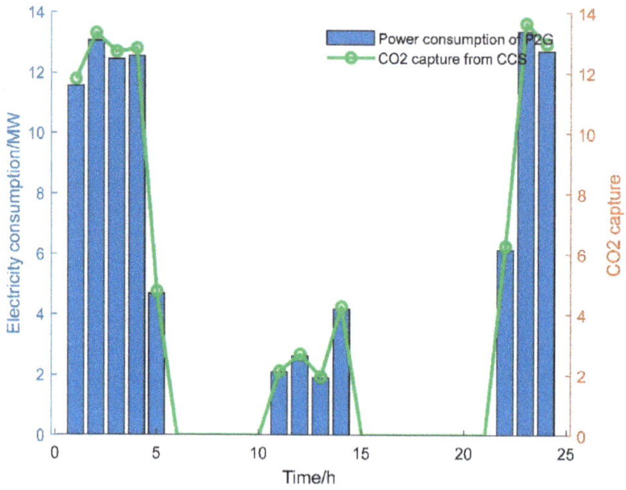

Figure 11. P2G power consumption vs. CO_2 capture by CCS in Scenario 6.

Between the hours of 00:00–05:00, 10:00–15:00, and 22:00–24:00, when renewable energy generation is notably higher than during other times, the CHP system's electrical output declines, shifting its focus to thermal energy, as shown in Figure 10a,b. Concurrently, the P2G system plays a significant role in gas energy production, as evidenced by Figure 10c. The CHP units face a "heat-determined-electricity" constraint that limits their capacity to utilize renewable energy effectively. Thus, the integration of P2G and CCS with the CHP system enables the conversion of electrical energy into gas energy, helping to ease the strong coupling constraints between electricity and heat in the CHP units. This synergy ultimately boosts the ability to absorb renewable energy. Additionally, refrigerators consume electrical energy to produce cooling energy, which partially satisfies the cooling load demand. This process alleviates the cooling load pressure on the microturbines, further reducing their electrical output and enhancing renewable energy integration.

During times of low renewable energy generation, CHP systems deliver considerable amounts of both electrical and thermal energy, as illustrated in Figure 10a,b. At these moments, the demand for thermal and electrical loads is relatively high, resulting in increased electrical output from the CHP units. In Figure 10c, the P2G system does not convert electrical energy into natural gas because renewable energy output is low and electrical power demand is high. To balance supply and demand, the P2G system avoids using the electrical energy produced by the CHP. Simultaneously, strategies such as using refrigerators to continuously consume electrical energy for cooling needs, increasing MT power output, enhancing the overall electrical output of the PIES, and reducing overall load demands are implemented to ensure reliable power supply.

When renewable energy generation is high and thermal load demands rise, the traditional "heat-led electricity" mode of CHP systems significantly limits the integration of renewable sources. In the research model, the CHP equipped with P2G and CCS units yields less electrical power but more thermal and gas power during high renewable generation periods. Conversely, in other timeframes, it produces less gas power while supplying more thermal and electrical energy, reflecting the interconnected nature of the three types of outputs. During periods of high renewable energy output, the P2G system captures the electrical energy generated by the CHP and converts it into gas energy, thus improving the capacity to harness renewable energy. The combined operation of the P2G-CCS-CHP system allows P2G and CCS to diminish the strong coupling between electrical and thermal outputs in CHP units, facilitating the transformation of electrical energy into gas energy to satisfy gas load requirements.

In the specific time frames of 00:00–05:00, 10:00–15:00, and 22:00–24:00, the CHP system integrated with P2G and CCS units effectively converts electrical energy into gas energy, thereby minimizing gas source output. This conversion results in reduced consumption of natural gas, leading to lower operating costs, decreased carbon emissions for the PIES, and reduced environmental protection expenses, as demonstrated in Figure 10c.

As illustrated in Figure 11, during the peak renewable energy generation times of 00:00–05:00, 10:00–15:00, and 22:00–24:00, P2G utilizes a significant amount of electricity produced by the CHP system, which encourages greater use of renewable energy to maintain a stable power supply. P2G requires CO_2 to consume electricity, and the CCS equipment captures the CO_2 emitted by the CHP system, supplying it to P2G. This setup alleviates the financial burden of obtaining carbon for P2G while also lowering the CHP's carbon emissions. Consequently, the P2G-CCS-CHP system effectively reduces CO_2 emissions, environmental protection expenses, and operational costs while improving the integration capacity for renewable energy.

3.3.3. Analysis of the Impact of Reward and Penalty Coefficients on PIES Dispatching

As stated earlier, the tiered carbon trading mechanism based on rewards and penalties assesses actual carbon emissions against the initial carbon allowance. If emissions fall below this threshold, a subsidy is granted according to the reward coefficient; conversely, if they exceed the threshold, a penalty is imposed based on the penalty coefficient. Therefore, changes in the reward and penalty coefficients will directly impact the expenses related to carbon trading, in turn affecting the optimized operation of the PIES.

Figure 12 illustrates how varying values of the reward coefficient μ and penalty coefficient λ influence carbon trading costs for PIES. When carbon trading costs are positive, the penalty coefficient is applicable; in contrast, negative carbon trading costs indicate the activation of the reward coefficient.

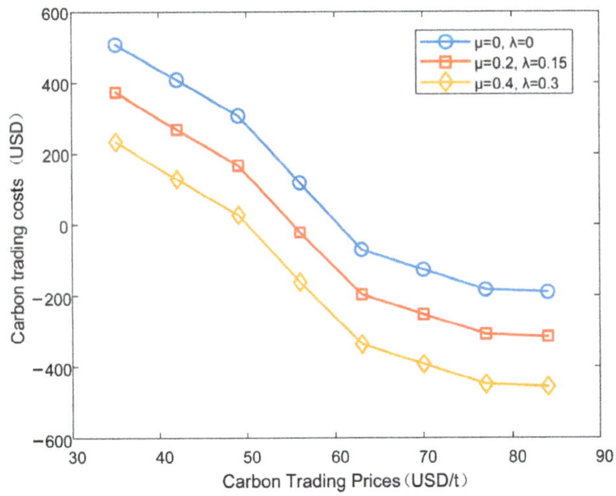

Figure 12. Effect of incentive and penalty coefficients on PIES carbon trading costs.

The examination of Figure 12 yields several important insights:

(1) In scenarios where carbon trading costs are negative, increasing the reward coefficient significantly amplifies the benefits that PIES gains from carbon trading. As carbon trading prices rise, these benefits progressively increase. This effect occurs because higher carbon trading revenues encourage PIES to minimize its dependence on external energy sources while boosting the efficiency of various energy supply and conversion devices within the system, ultimately leading to a significant reduction in overall carbon emissions.
(2) In contrast, when carbon trading costs are positive, an increased penalty coefficient results in higher fines for PIES. To mitigate these penalties, PIES should strategically implement voluntary decarbonization scheduling. By integrating multi-energy coupling—encompassing electricity; heat; gas; and cooling—and employing energy conversion technologies such as Power-to-Gas (P2G); Carbon Capture and Storage (CCS); and Combined Heat and Power (CHP); these scheduling efforts can create situations where a higher penalty coefficient corresponds with lower carbon trading costs.
(3) During PIES's operational scheduling, it is critical to manage the interplay between reward and penalty coefficients and carbon trading costs to identify the optimal balance point. Achieving this balance is essential for reducing carbon emissions while ensuring the economic viability of scheduling practices. This could involve establishing distinct thresholds for reward and penalty adjustments based on real-time trading data and market trends.
(4) Fluctuations in carbon trading prices will directly impact the associated revenues or costs for PIES. Therefore, it is vital to establish a dynamic strategy to adjust operational scheduling in response to these price changes, maximizing revenues or minimizing costs effectively in real-time.
(5) PIES must enhance its monitoring and analysis of the carbon trading market to stay informed about market trends and price movements. Continuous monitoring will provide valuable data to inform scheduling decisions, enhancing responsiveness to market changes. Key focus areas include tracking emerging regulations, competitor behaviors, and shifts in overall market demand, allowing for better-informed adjustments to operational strategies.

4. Conclusions Implications and Future Research Orientations

This research focuses on optimizing the low-carbon operations of integrated energy systems in parks that utilize wind and solar energy alongside the multi-energy coupling of electricity, heat, gas, and cooling. We introduce a tiered reward-and-punishment carbon trading mechanism that accounts for the uncertainties and interrelationships in renewable energy output, as well as the integration of P2G, CCS, and CHP for the economically efficient low-carbon scheduling of PIES. The simulation analysis of practical scenarios leads to several findings:

(1) The proposed reward-punishment tiered carbon trading mechanism achieves substantial decarbonization performance, reducing PIES carbon emissions by 31.30% (6191.99 kg) compared to conventional mechanisms. This regulatory innovation establishes critical financial incentives that align market dynamics with environmental objectives, providing a policy-relevant framework for low-carbon transition management.
(2) Through probabilistic modeling of renewable energy uncertainties and inter-source correlations, our framework enhances system sustainability across three dimensions: economic efficiency (+3.67% cost reduction/$880.42 savings), environmental performance (15.83% emission reduction/2150.78 kg mitigation), and renewable utilization (4.4% wind and 4.91% solar consumption increases). This tripartite optimization addresses the energy trilemma challenges in integrated energy systems.

(3) The novel P2G-CCS-CHP coupling model demonstrates paradigm-shifting improvements over conventional systems. Relative to traditional CHP configurations, it achieves 28.86% wind and 19.85% solar utilization enhancements alongside a remarkable 36.91% cost reduction and 77.65% emission abatement. When benchmarked against existing P2G-CCS integrations, the proposed architecture further improves renewable consumption (21.68% wind, 15.02% solar) while reducing costs and emissions by 20.51% and 62.68%, respectively. These comparative results validate the technical superiority of our multi-system coupling approach.
(4) The biomimetic non-dominated sorting dung beetle optimizer introduces strategic multi-objective optimization capabilities, demonstrating 18–32% computational efficiency gains over PSO, MCA, POA, SABO, and HHO algorithms in Pareto front identification. This bio-inspired metaheuristic effectively resolves the convergence-precision trade-off in complex energy system optimizations.

While this study establishes methodological advancements in integrated energy system optimization, two critical methodological constraints warrant explicit discussion. First, the static treatment of carbon emission factors in our model overlooks the temporal and spatial variability inherent in practical emission accounting systems. This simplification may underestimate the operational adaptability required for real-time policy compliance, particularly in jurisdictions implementing dynamic carbon pricing mechanisms or region-specific emission coefficients. Second, the exclusion of voluntary emission reduction mechanisms—including carbon offset programs; renewable energy certificate trading; and corporate sustainability initiatives—constrains the model's alignment with contemporary market-driven decarbonization practices. These omitted mechanisms constitute essential components of modern carbon management frameworks in most industrialized nations, potentially affecting the generalizability of our economic optimization results in policy-integrated scenarios.

To address these knowledge gaps, our future research agenda will pursue three strategic directions: (1) Developing adaptive carbon accounting modules that incorporate time-variant emission factors and cross-regional carbon flow tracking; (2) Integrating hybrid market mechanisms combining mandatory compliance systems with voluntary carbon market dynamics; (3) Implementing multi-temporal scale optimization considering both short-term operational flexibility (hourly resolution) and long-term infrastructure planning (decadal horizons). This expanded framework aims to bridge the current model's theoretical assumptions with the complex system dynamics observed in emerging carbon-neutrality transition pathways.

Author Contributions: Project administration and conceptualization, Z.Z.; funding acquisition, X.L.; visualization, L.Z.; writing—review and editing, H.Z.; writing—original draft preparation, methodology, supervision, Z.W.; resources, data curation, W.L.; validation, software, B.W. All authors have read and agreed to the published version of the manuscript.

Funding: This research was funded by the State Grid Beijing Electric Power Company Technology Project, grant number 520210230004.

Data Availability Statement: The data that has been used is confidential.

Conflicts of Interest: Authors Zhiyuan Zhang and Xiqin Li are employed by the State Grid Beijing Mentougou Power Supply Company. Authors Lu Zhang and Hu Zhao are employed by the State Grid Beijing Electric Power Company. The remaining authors declare that the research was conducted in the absence of any commercial or financial relationships that could be construed as a potential conflict of interest.

Abbreviations

The following abbreviations are used in this manuscript:

PIES	Park integrated energy system	MT	Micro-gas Turbine	ER	Electric refrigerator
P2G	Power to gas	NSGS	Non-dominated Sorting Genetic Strategy	PV	Photovoltaic power generation
CCS	Carbon capture and storage	DBO	Dung beetle optimizer	BR	Backward reduction
CHP	Combined heat and power generation	NS	Natural gas sources	CV	Coefficient of variation
NSDBO	Optimization algorithm of dung beetle under non-dominated sorting	WT	Wind turbine	MOP	Multi-object Optimization Problem
PSO	Particle Swarm Optimization	MCA	Musical Chairs Algorithm	POA	Pelican Optimization Algorithm
SABO	Subtraction-Average-Based Optimizer	HHO	Harris Hawk Optimization		

Appendix A

Table A1. Test function set.

Benchmarking Functions	Search Space
$f_1 = \sum_{i=1}^{n} \lvert x_i \rvert + \prod_{i=1}^{n} \lvert x_i \rvert$	$[-10, 10]$
$f_2 = \sum_{i=1}^{n} \left[x_i^2 - 10\cos(2\pi x_i) + 10\right]$	$[-5.12, 5.12]$
$f_3 = 0.1\left\{ sin^2(3\pi x_i) + (x_i-1)^2 + \sum_{i=1}^{n} (x_i-1)^2\left[1 + sin^2(3\pi x_i + 1)\right] + \left[1 + sin^2(2\pi x_i)\right] \right\} + \sum_{i=1}^{n} u(x_i, 5, 100, 4)$	$[-50, 50]$
$f_4 = \left[\frac{1}{500} + \sum_{j=1}^{25} \frac{1}{j + \sum_{i=1}^{2} \left(x_i - a_{ij}\right)^6} \right]^{-1}$	$[-65, 65]$

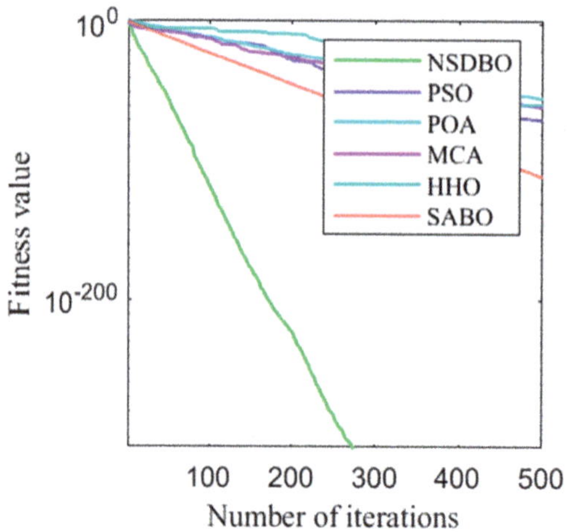

(**a**)

Figure A1. *Cont.*

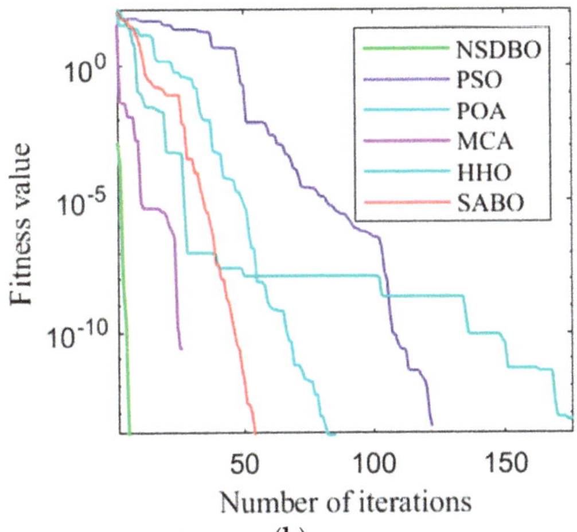

(b)

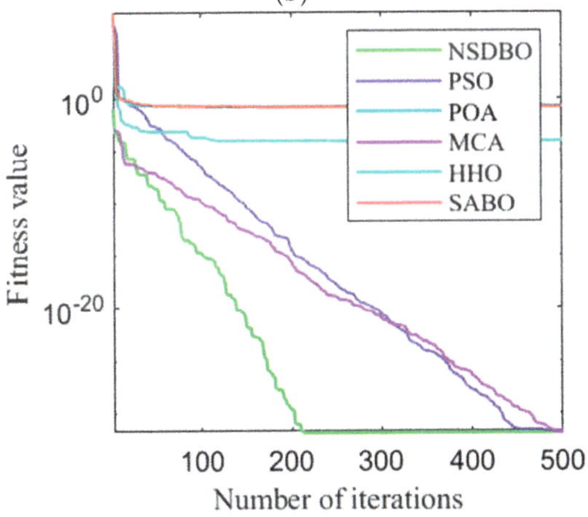

(c)

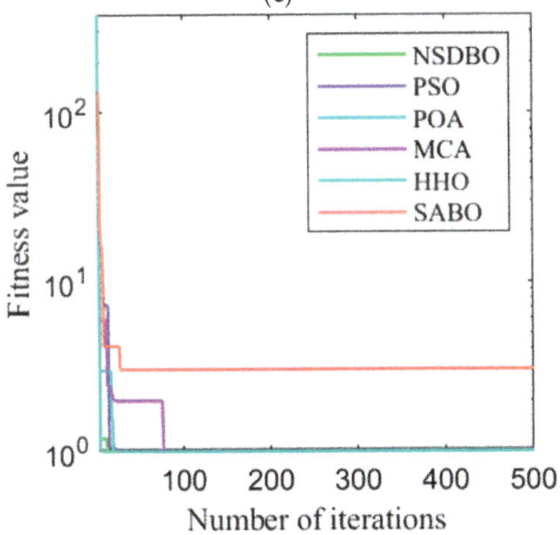

(d)

Figure A1. Test function convergence results. (**a**) f_1 function test results. (**b**) f_2 function test results. (**c**) f_3 function test results. (**d**) f_4 function test results.

References

1. Drosos, D.; Kyriakopoulos, G.L.; Arabatzis, G.; Tsotsolas, N. Evaluating Customer Satisfaction in Energy Markets Using a Multicriteria Method: The Case of Electricity Market in Greece. *Sustainability* **2020**, *12*, 3862. [CrossRef]
2. Shafiullah, M.; Ahmed, S.D.; Al-Sulaiman, F.A. Grid Integration Challenges and Solution Strategies for Solar PV Systems: A Review. *IEEE Access* **2022**, *10*, 52233–52257. [CrossRef]
3. Sinha, S.; Chandel, S.S. Review of recent trends in optimization techniques for solar photovoltaic-wind based hybrid energy systems. *Renew. Sustain. Energy Rev.* **2015**, *50*, 755–769. [CrossRef]
4. Zhao, G.; Wan, C.X.; Zuo, W.Q.; Zhang, K.F.; Shu, X.Y. Research on Multiobjective Optimal Operation Strategy for Wind-Photovoltaic-Hydro Complementary Power System. *Int. J. Photoenergy* **2022**, *2022*, 5209208. [CrossRef]
5. Ju, L.; Tan, Z.; Yuan, J.; Tan, Q.; Li, H.; Dong, F. A bi-level stochastic scheduling optimization model for a virtual power plant connected to a wind-photovoltaic-energy storage system considering the uncertainty and demand response. *Appl. Energy* **2016**, *171*, 184–199. [CrossRef]
6. Liu, X.H.; Li, X.; Tian, J.Q.; Wang, Y.B.; Xiao, G.X.; Wang, P. Day-Ahead Economic Dispatch of Renewable Energy System considering Wind and Photovoltaic Predicted Output. *Int. Trans. Electr. Energy Syst.* **2022**, *2022*, 6082642. [CrossRef]
7. Xu, M.; Cui, Y.; Wang, T.; Zhang, Y.Z.; Guo, Y.; Zhang, X.Y. Optimal Dispatch of Wind Power, Photovoltaic Power, Concentrating Solar Power, and Thermal Power in Case of Uncertain Output. *Energies* **2022**, *15*, 8215. [CrossRef]
8. Min, H.L.; He, P.K.; Li, C.L.; Yang, L.B.; Xiao, F. The Temporal and Spatial Characteristics of Wind-Photovoltaic-Hydro Hybrid Power Output Based on a Cloud Model and Copula Function. *Energies* **2024**, *17*, 1024. [CrossRef]
9. Chen, S.T.; Su, W.H.; Wu, B.Y. Two stage robust planning of park integrated energy system considering low carbon. *Front. Ecol. Evol.* **2023**, *10*, 1100089. [CrossRef]
10. Zhang, Y.L.; Zhang, P.H.; Du, S.; Dong, H.L. Economic Optimal Scheduling of Integrated Energy System Considering Wind-Solar Uncertainty and Power to Gas and Carbon Capture and Storage. *Energies* **2024**, *17*, 2770. [CrossRef]
11. Zhu, D.Y.; Cui, J.; Wang, S.J.; Wei, J.Z.; Li, C.R.; Zhang, X.M.; Li, Y.Z. An optimization strategy for intra-park integration trading considering energy storage and carbon emission constraints. *J. Clean. Prod.* **2024**, *447*, 141031. [CrossRef]
12. Liu, L.; Qin, Z.J. Low-carbon planning for park-level integrated energy system considering optimal construction time sequence and hydrogen energy facility. *Energy Rep.* **2023**, *9*, 554–566. [CrossRef]
13. Zhang, J.L.; Liu, Z.Y. Low carbon economic scheduling model for a park integrated energy system considering integrated demand response, ladder-type carbon trading and fine utilization of hydrogen. *Energy* **2024**, *290*, 130311. [CrossRef]
14. Li, K.; Ying, Y.L.; Yu, X.Y.; Li, J.C. Optimal Scheduling of Electricity and Carbon in Multi-Park Integrated Energy Systems. *Energies* **2024**, *17*, 2119. [CrossRef]
15. Chen, C.M.; Liu, C.; Ma, L.Y.; Chen, T.W.; Wei, Y.Q.; Qiu, W.Q.; Lin, Z.Z.; Li, Z.Y. Cooperative-game-based joint planning and cost allocation for multiple park-level integrated energy systems with shared energy storage*. *J. Energy Storage* **2023**, *73*, 108861. [CrossRef]
16. Chen, H.Y.; Wang, S.C.; Yu, Y.X.; Guo, Y.; Jin, L.; Jia, X.Q.; Liu, K.C.; Zhang, X.H. Low-Carbon Operation Strategy of Park-Level Integrated Energy System with Firefly Algorithm. *Appl. Sci.* **2024**, *14*, 5433. [CrossRef]
17. Wang, L.L.; Xian, R.C.; Jiao, P.H.; Liu, X.H.; Xing, Y.W.; Wang, W. Cooperative operation of industrial/commercial/residential integrated energy system with hydrogen energy based on Nash bargaining theory. *Energy* **2024**, *288*, 129868. [CrossRef]
18. Xu, F.Q.; Li, X.P.; Jin, C.H. A multi-objective operation optimization model for the electro-thermal integrated energy systems considering power to gas and multi-type demand response. *J. Renew. Sustain. Energy* **2024**, *16*, 045301. [CrossRef]
19. Ding, X.Y.; Yang, Z.P.; Zheng, X.B.; Zhang, H.; Sun, W. Effect of decision-making principle on P2G-CCS-CHP complementary energy system based on IGDT considering energy uncertainty. *Int. J. Hydrogen Energy* **2024**, *81*, 986–1002. [CrossRef]
20. Vachhani, V.L.; Dabhi, V.K.; Prajapati, H.B. Improving NSGA-II for solving multi objective function optimization problems. In Proceedings of the 2016 International Conference on Computer Communication and Informatics, Coimbatore, India, 7–9 January 2016; pp. 1–6.
21. Xue, J.; Shen, B. Dung beetle optimizer: A new meta-heuristic algorithm for global optimization. *J. Supercomput.* **2022**, *79*, 7305–7336. [CrossRef]
22. Kennedy, J.; Eberhart, R. Particle swarm optimization. In Proceedings of the ICNN'95—International Conference on Neural Networks, Perth, WA, Australia, 27 November–1 December 1995; Volume 4, pp. 1942–1948. [CrossRef]
23. Eltamaly, A.M.; Rabie, A.H. A Novel Musical Chairs Optimization Algorithm. *Arab. J. Sci. Eng.* **2023**, *48*, 10371–10403. [CrossRef]
24. Trojovsky, P.; Dehghani, M. Pelican Optimization Algorithm: A Novel Nature-Inspired Algorithm for Engineering Applications. *Sensors* **2022**, *22*, 855. [CrossRef] [PubMed]
25. Trojovsky, P.; Dehghani, M. Subtraction-Average-Based Optimizer: A New Swarm-Inspired Metaheuristic Algorithm for Solving Optimization Problems. *Biomimetics* **2023**, *8*, 149. [CrossRef] [PubMed]
26. Heidari, A.A.; Mirjalili, S.; Faris, H.; Aljarah, I.; Mafarja, M.; Chen, H.L. Harris hawks optimization: Algorithm and applications. *Future Gener. Comput. Syst.-Int. J. Escience* **2019**, *97*, 849–872. [CrossRef]

27. Li, Y.; Zou, Y.; Tan, Y.; Cao, Y.J.; Liu, X.D.; Shahidehpour, M.; Tian, S.M.; Bu, F.P. Optimal Stochastic Operation of Integrated Low-Carbon Electric Power, Natural Gas, and Heat Delivery System. *IEEE Trans. Sustain. Energy* **2018**, *9*, 273–283. [CrossRef]
28. Lu, Q.; Guo, Q.S.; Zeng, W. Optimization scheduling of an integrated energy service system in community under the carbon trading mechanism: A model with reward-penalty and user satisfaction. *J. Clean. Prod.* **2021**, *323*, 129171. [CrossRef]
29. Ba-swaimi, S.; Verayiah, R.; Ramachandaramurthy, V.K.; Alahmad, A.K. Long-term optimal planning of distributed generations and battery energy storage systems towards high integration of green energy considering uncertainty and demand response program. *J. Energy Storage* **2024**, *100*, 113562. [CrossRef]
30. Zhang, X.Z.; Wang, Z.Y.; Lu, Z.Y. Multi-objective load dispatch for microgrid with electric vehicles using modified gravitational search and particle swarm optimization algorithm. *Appl. Energy* **2022**, *306*, 118018. [CrossRef]

processes

Article

Optimal Strategy for Grid Loss Reduction Under Electricity Transmission and Distribution Reform Considering Low-Carbon Benefits

Weiwu Li [1,*], Qing Xu [1], Xinying Wang [1], Zhengying Liu [1], Tianshou Li [1] and Dandan Zhang [2]

1 State Grid Gansu Electric Power Co., Ltd., Lanzhou 730050, China; asrit_ssu@163.com (Q.X.); 13919614970@163.com (X.W.); liuzying06211@163.com (Z.L.); lts48291702@163.com (T.L.)
2 Ginling College, Nanjing Normal University, Nanjing 210023, China; zhangdandan@nnu.edu.cn
* Correspondence: 15951900388@163.com

Abstract: Selecting grid loss reduction strategies is crucial for energy-saving transformations, particularly in the context of electricity transmission and distribution pricing reforms. The optimization of strategic selection is not easy due to the vast number of grid devices, which leads to a multitude of possible strategy combinations. This paper presents an optimal model for selecting loss reduction strategies, aiming to minimize the sum of comprehensive investment costs and energy loss costs over the life cycle of the strategies. The energy loss costs include both direct expenses due to energy loss and indirect costs, namely, carbon emission penalties. The constraints include allowable voltage deviations, branch power transmission, the number of loss reduction measures, loss rates, and total investment limits. The model comprehensively considers both economic benefits and the social benefits of reduced carbon emissions. It can help companies better adapt to electricity transmission and distribution pricing reforms, reduce operational costs, and contribute to low-carbon development. Finally, the model is validated using the data provided by one provincial power grid company in China. The results show that the loss reduction reaches 13.9 MW and the reduced carbon emission per hour is 10.425 t. The proposed method is also compared with the enumeration method, which demonstrates its effectiveness and efficiency. Further research will be conducted on establishing functional relationships between electricity sales prices and line losses to incentivize companies to apply loss reduction measures under different pricing functions.

Keywords: energy efficiency; carbon emission reduction; transmission and distribution networks; cost–benefit analysis; genetic algorithms; power grid optimization; sustainability; renewable integration

1. Introduction

The power system, as an essential infrastructure of modern society, plays a pivotal role in the stable development of the economy and the sustainability of the environment [1]. The transmission and distribution segment, being a critical part of the power system, directly impacts energy utilization efficiency and electricity costs through its energy consumption and losses [2]. Therefore, investigating energy-saving and loss reduction strategies within the framework of electricity transmission and distribution pricing mechanisms is of significant importance for enhancing the energy efficiency of the entire power system, reducing operational costs, and promoting green and low-carbon development.

https://doi.org/10.3390/pr13051406

Currently, the energy-saving transformation of distribution networks is a comprehensive engineering task that includes equipment upgrades, grid structure optimization, network reconfiguration, reactive power compensation, and operational enhancements [3]. Extensive research has been conducted both domestically and internationally on loss reduction and energy-saving strategies for distribution networks [4]. Devices such as shunt capacitors and reactors, known for their cost effectiveness and practicality, have become key tools for energy conservation and loss reduction [5]. For example, Li and Xu [6] explored the use of high-voltage single-phase power supply methods, while Long et al. [7] focused on energy-efficient distribution transformers. There are also researchers who have investigated loss reduction strategies for low-voltage and medium-voltage distribution networks, respectively [8,9].

In the context of global climate change and the urgent need to reduce greenhouse gas emissions, the energy sector, particularly the electricity industry, is under immense pressure to transition towards cleaner and more sustainable practices [10]. The integration of renewable energy sources, while beneficial for reducing carbon emissions, also introduces new challenges for grid management due to their intermittent and variable nature [11]. This complexity necessitates advanced strategies for grid loss reduction that can adapt to the dynamic changes in power generation and consumption patterns.

With the advancement of technology, an increasing array of methods is available for energy conservation and carbon reduction. For instance, in the context of power grids, energy-saving modifications can be implemented through various means, including the transformation of wire cross-sections, the optimization of power supply radii, and the redistribution of electrical loads. Moreover, given the vast number of grid devices, the resulting combinations of strategies are numerous, making the optimization of strategic selection a critical issue. Evidently, the selection of grid loss reduction strategies can lead to different outcomes in terms of both economic and environmental performance [12]. Against this backdrop, this paper proposes an optimization model for selecting grid loss reduction strategies that not only minimizes the sum of comprehensive investment costs and energy loss costs over the life cycle of the strategies but also considers the indirect costs associated with carbon emissions. The model is designed to be flexible and adaptable to various scenarios, allowing for the inclusion of constraints on the number of loss reduction measures and total investment limits.

The main novelty of this article lies in the development of an optimization model that comprehensively considers both economic and environmental factors in grid loss reduction strategies. Specifically, the model includes:

(1) The model accounts for indirect costs associated with carbon emissions, providing a more comprehensive evaluation of the economic and environmental impacts of different loss reduction strategies.
(2) This study employs an adaptive genetic algorithm to solve the optimization problem, which is more efficient and effective than traditional enumeration methods, especially when dealing with a large number of possible strategy combinations.
(3) The model is validated using data from a provincial power grid company in China, demonstrating its practical applicability and effectiveness in reducing grid losses and carbon emissions.

The remainder of this paper is structured as follows: Section 2 provides a literature review of the current state of research on grid loss reduction strategies and optimization models. Section 3 details the methods and analysis used in this study, including the formulation of the objective function and constraints, as well as the optimization approach. Section 4 presents the results of this case study and the application of the proposed method to a provincial power grid company in China. Section 5 discusses the potential synergies with renewable energy sources and technologies, as well as economic concerns related to the proposed method. Finally, Section 5 concludes this paper and suggests directions for future research.

2. Literature Review

Researchers are continuously deepening their research on the construction of loss reduction methods for distribution networks. Xie et al. [13] proposed a combined power loss reduction strategy optimization framework for distribution networks, which analyzes weak points through power flow calculations and generates strategies involving conductor replacement, transformer upgrading, and reactive power compensation. Wang et al. [14] examined the EZ Power Grid to identify common issues in current power grids and proposes methods for evaluating power loss reduction. Suresh and Edward [15] established a hybrid optimization technique combining the Grasshopper Optimization Algorithm (GOA) and Cuckoo Search (CS) to determine the optimal placement and sizing of distributed generation (DG) units in electric power networks, aiming to reduce power losses, improve voltage profiles, and enhance system reliability. Pokhrel et al. [16] proposed a mixed loss optimization problem that considers network reconstruction, distributed generation and distributed storage, and solved it using the interior point method. Ahmadi et al. [17] proposed an efficient power loss reduction (briefly, PLR) model based on four PLR schemes: distributed generation installation, distribution network reconstruction, capacitor bank installation, and distribution feeder reinforcement. The proposed PLR strategy incorporates budget constraints as a hard constraint for distribution network planning. Iweh et al. [18] comprehensively reviews the integration of distributed generation, particularly renewable energy sources, into the power grid, discussing the prerequisites, push factors, practical options, challenges, and merits of such integration. Similar studies can be found in [19,20]. Mu et al. [21] employs a two-stage power optimization model to analyze the potential of Carbon Capture, Utilization, and Storage (CCUS) in achieving carbon neutrality in China's Northeast Power Grid. However, none of these research studies provided solutions on selecting and combining different strategies to derive the optimal loss reduction method.

There are also researchers who focus on the study of optimization strategies for loss reduction schemes in distribution networks. Zhang et al. [22] introduced a model-free volt-VAR optimization algorithm using multi-agent deep reinforcement learning (DRL) for unbalanced distribution systems, employing a deep Q-network (DQN) framework to intelligently manage voltage regulation and power loss reduction without directly solving a specific optimization model. Zhang et al. [23] proposed an improved genetic algorithm for power distribution system reconfiguration based on Sandeep Sehgal's research, which reduces the number of switching operations and improves system reliability [24]. Gautam and Bhusal [25] proposed a genetic algorithm based on spanning tree to optimize distribution system reconfiguration. This algorithm filters the invalid configuration combinations in the selection process of genetic algorithm, reduces the dimension of the search space of genetic algorithm, and has higher accuracy and efficiency. Kamel et al. [26] proposed a new optimization technique (SSA) to reduce the total active power loss. With parallel optimization and search functions, SSA is an effective technology for radial distribution network reconfiguration. Wu et al. [27] proposed an optimization strategy based on the combat strategy algorithm for the randomness and load uncertainty of distributed generation. This method significantly reduced the overall line loss of distribution network during the peak load period. However, these articles only consider the direct losses caused by electricity waste, and they do not provide a detailed calculation of the losses due to carbon dioxide penalties; hence, the conclusions are not comprehensive.

To sum up, despite extensive research on grid loss reduction strategies and optimization models, the existing studies often focus on singular approaches to loss reduction and lack systematic optimization for strategy combinations. Additionally, there is a scarcity of research on models that prioritize loss reduction strategies while considering the benefits of low-carbon emissions. Most prior studies either neglect the indirect costs associated with carbon emissions or fail to provide comprehensive optimization frameworks that integrate both economic and environmental factors. This gap is particularly evident in the context of electricity transmission and distribution pricing reforms, where a holistic approach to grid loss reduction is urgently needed.

3. Methods and Analysis

The loss reduction model is this study seeks to minimize the total costs, which include both the comprehensive investment expenses and the electrical energy loss expenses, over the entire lifespan of the loss reduction strategy. The electrical energy loss expenses encompass direct costs and indirect costs, notably, the penalties associated with carbon emissions. The model is subject to several constraints, including the permissible voltage deviation in the lines, the transmission power (current) of the branches, the number of loss reduction measures, the rate of energy loss, and the limit on the total allowable investment.

3.1. Objective Function

Considering that direct and indirect costs are both calculated on an annual basis, investment costs must also be converted into average annual investment costs. With a 10-year investment cycle, the objective function to be optimized in this paper is shown in Equation (1):

$$min\ F = \sum_{i=1}^{m} C_i/10 + F_1 + F_2 \tag{1}$$

In this function, C_i is comprehensive investment from ith loss-reduction measures; m is the total number of loss reduction measures; F_1 is the direct cost caused by the loss of electricity, referring to the loss of electricity caused by the loss of electricity; F_2 is the indirect cost caused by the loss of electricity, which refers to the carbon emission penalty caused by the loss of electricity. All these components can be calculated by Equations (2)–(7).

(1) Comprehensive investment for loss-reduction measures C_i:

The comprehensive investment C_i for loss reduction measures i includes equipment acquisition C_i^1, equipment removal C_i^2 and installation or reconstruction C_i^3 [13]. The function is:

$$C_i = C_i^1 + C_i^2 + C_i^3 \tag{2}$$

(2) Direct cost due to loss of electricity F_1:

The direct cost caused by the loss of electricity is also known as the power loss cost [13]. The function is:

$$F_1 = \Delta W d_1 \tag{3}$$

In this equation, ΔW refers to power loss, kW·h; d_1 is electricity price, generally CNY 0.5 per kW·h.

(3) Indirect cost due to loss of electricity F_2:

The indirect cost of lost electricity, i.e., the carbon (CO_2) emission penalty, can be calculated by [28]:

$$F_2 = \Delta W f \tag{4}$$

In the equation, f refers to the carbon (CO_2) emission penalty per unit of electricity generation [13], which is calculated by:

$$f = \rho Q d_2 \quad (5)$$

In the above equation, ρ is the proportion of thermal power; in China, the number is around 65%; d_2 is the carbon penalty [13,15]; Q is carbon (CO_2) emissions per unit of electricity generation, kg/(kW·h), and the equation of Q is [29]:

$$Q = \frac{p}{q\eta} \quad (6)$$

In the above equation, p refers to carbon emissions per unit (per ton) of fuel, 2.53 kg CO_2/kgCoal [15]; q is heat output per ton of fuel [13,15]; and the calorific value of coal per kilogram of fuel is 6400 kcal/kg Coal = 6400 × 4.185 8 = 26,789.12 kJ/kgCoal = 26.789 kJ/t [19]; η is the energy conversion rate of a power source [15]; 1 W·s = 1 J, 1 kW·h = 3600 kJ. Considering the thermal efficiency of a coal fired power plant, η is set to be 0.4 kW·h/3600 kJ. The quotation follows Equation (7):

$$Q = \frac{2.53}{26,789.12 \times \frac{0.4}{3600}} = 0.85 \quad (7)$$

According to our carbon emission penalty standard, the price is set at 0.2 CNY/kg [16], $f = \rho Q d_2 = 0.11 \text{CNY}/(\text{kW}\cdot\text{h})$. Based on the European Union's penalty standards for carbon emissions during 2008–2012 (100 EUR/t CO_2), d_2 = 0.1 EUR/kg ≈ 0.733 CNY/kg. As a result, $f = \rho Q d_2 = 0.41$ CNY/(kW·h).

3.2. Constraint

The constraints include voltage deviation, branch constraints, quantitative restriction of loss reduction and the total investment limit.

(1) The voltage deviation is [30]:

$$\Delta U\% = \frac{U_i - U_N}{U_N} \times 100\% \quad (8)$$

In this equation, U_i is the actual voltage of node i, kV; U_N is the rated voltage, kV. The allowable voltage deviation [18] in the 10 kV voltage class does not exceed:

$$\Delta U\% \leq +5\% \quad (9)$$

$$\Delta U\% \leq -10\% \quad (10)$$

(2) Branch constraints. The actual transmission capacity of the branch shall not exceed its maximum transmission capacity, expressed by the transmission current [30]:

$$I_j \leq I_{jmax} \quad (11)$$

In this inequation, I_j is the actual current of branch j; I_{jmax} is the maximum current allowed through branch j.

(3) The quantitative restriction of loss reduction measures is:

$$n \leq N \quad (12)$$

In this inequation, N refers to the maximum number of loss reduction measures.

(4) The total investment limit is:

$$\sum_{i=1}^{n} C_i \leq L \quad (13)$$

In the above inequation, L is the limit of total investment, the unit is ten thousand.

Notations related the parameters occurred in the above equations are summarized in Table 1.

Table 1. Notations.

Name	Physical Meaning	Values Range	Units Measured
C_i	comprehensive investment	$(0, \infty)$	CNY
F_1	direct cost due to loss of electricity	$(0, \infty)$	CNY
F_2	indirect cost due to loss of electricity	$(0, \infty)$	CNY
ΔW	power loss	$(0, \infty)$	kW·h
d_1	electricity price	$(0, \infty)$	CNY/(kW·h)
d_2	carbon penalty	$(0, \infty)$	CNY/kg
f	carbon emission penalty per unit of electricity generation	$(0, \infty)$	CNY
Q	carbon emissions per unit of electricity generation	$(0, \infty)$	kg/(kW·h)
$\Delta U\%$	voltage deviation	$(0, 100)$	%
I_j	actual transmission capacity of the branch	$(0, \infty)$	A

3.3. Optimization

According to the principle that one loss reduction measure can constitute a loss reduction plan, and the combination of one or several loss reduction measures can also constitute a loss reduction plan, the decision variable (0 or 1) is whether each loss reduction measure is implemented or not. Assuming that there are n loss reduction measures, there will be 2^n loss reduction schemes, and the solving process is a combinatorial optimization problem. The genetic algorithm has strong searching ability and low dependence on objective function; so, it is very suitable for solving such problems. The adaptive genetic algorithm (briefly, AGA) was proposed by Srinivas [31], which adjusts the crossover rate and variation rate according to the fitness value of the progeny population after each iteration. Compared with the standard genetic algorithm (SGA), it has better convergence speed. Compared with other algorithms, such as particle swarm optimization (PSO), the AGA dynamically adjusts mutation and crossover rates to maintain population diversity, avoiding premature convergence. This allows the AGA to perform broad searches early for diversity and fine searches later to protect optimal solutions, leading to more effective convergence to the global optimum [32]. Additionally, the AGA naturally handles discrete variables, such as binary encoding, while PSO, designed for continuous spaces, struggles with discrete combinatorial optimization problems and requires additional modifications [33]. Moreover, in high-dimensional or non-convex problems, PSO may fail due to particle trajectory divergence and is prone to local optima [34]. The AGA enhances stability through increased mutation rates to introduce new solutions and employs divide-and-conquer strategies like variable grouping to reduce problem coupling, making it more effective for complex optimization tasks.

The overall flow of loss reduction scheme optimization is shown in Figure 1.

(1) Encoding

If the genetic algorithm is used, the problem needs to be encoded first. The implementation of each loss reduction measure corresponds to 1 bit of binary code (0 or 1, called gene); so, the binary code can be used to describe each loss reduction plan. The length of the binary code of the loss reduction scheme is set according to the number of loss reduction measures. If the number of loss reduction measures is n, it is set as a n-bit binary code.

(2) Fitness function

In this paper, constraint conditions are combined into the fitness function to achieve a fitness function with a penalty term, which can guide genetic search. And the fitness function is:

$$f = F + \partial\mu(\Delta U\% - \Delta U\%_s) + \beta\mu(I_j - I_{max}) + \gamma\mu(n - N) + \lambda\mu\left(\sum_{i=1}^{n} C_i - C\right) \tag{14}$$

In the equation, α, β, γ and λ are penalty factors, which are all set to be 1. μ is 0-1 function, which is defined as follows:

$$\mu(x_i) = \begin{cases} 0, \ x_i < 0 \\ 1, \ x_i \geq 0 \end{cases} \tag{15}$$

(3) Genetic manipulation

Calculate the fitness value $f(t)$ of each individual in the population $P(t)$; the optimal individual is retained, and the current optimal individual is copied directly to the next generation.

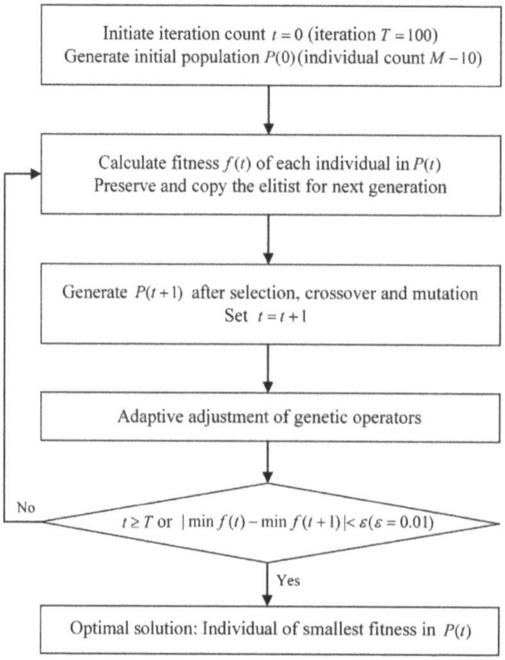

Figure 1. Flowchart of main steps of the adaptive genetic algorithm.

(1) Adaptive cross operation

In the iterative process of the SGA, crossover probability and mutation probability are fixed. However, the AGA calculated the probabilities of crossover and mutation depending on the fitness values of the solutions. High-fitness solutions are kept, while solutions with subaverage fitness are totally disrupted. According to the fitness value of progeny population, the dynamic adjustment crossover rate of the adaptive genetic algorithm is:

$$P_{c=}\begin{cases} \frac{P_{cmax}(f_{max}-f')}{f_{max}-f_{av}}, \ f^* \geq f_{av} \\ P_{cmin}, \ f^* < f_{av} \end{cases} \tag{16}$$

where f_{max} is the population maximum fitness; f^* is the larger fitness among the two individuals involved in the crossover; f_{av} is the average fitness of the population. P_{cmax} and P_{cmin} are the maximum and minimum crossing probabilities, respectively.

(2) Mutation operation

The adaptive genetic algorithm dynamically adjusts the mutation probability according to the fitness value of the offspring population.

$$P_c = \begin{cases} \frac{P_{cmax}(f_{max}-f^*)}{f_{max}-f_{av}}, & f^* \geq f_{av} \\ P_{cmin}, & f^* < f_{av} \end{cases} \quad (17)$$

where $f(\cdot)$ is the fitness of the mutated individual. P_{mmax} and P_{mmin} are the maximum and minimum mutation probabilities, respectively.

(4) Iterative termination condition

The condition of setting iterative convergence: meet one of the following iterative convergence conditions.

(1) The number of iterations has reached the maximum number of iterations M;
(2) The difference in the fitness of the mutated individual between iteration t and $t+1$ satisfies:

$$|\min f(t) - \min f(t+1)| < \varepsilon \quad (18)$$

where $\varepsilon = 0.01$.

4. Results and Discussion

4.1. Case Description

The case study utilized in this research was provided by a provincial power grid company in China. The company operates over 3300 substations and more than 18,000 km of transmission lines. According to administrative divisions, the power grid structure includes 17 regions, which are interconnected by 500 kV transmission lines. Given the relatively old equipment and significant line losses, the company faces a substantial challenge in energy conservation and loss reduction.

Based on the typical operation mode data of the power grid in 2023, a power flow analysis was carried out. The calculation example includes the 500 kV and 220 kV trans-mission grid frameworks. The networks below 220 kV are treated in the form of equivalent loads.

Table 2 delineates the distribution of losses and load across each region of the power grid under the current operational mode. Considering data privacy, all the names of the regions were abbreviated in this paper. As illustrated in Table 2, the JNI region, due to its possession of a large-scale power plant, transmits a substantial amount of power to other regions, amounting to 4450.9 MW. Concurrently, it experiences the highest line losses, making it a key area that requires optimization of the power grid structure.

Table 2. Load flow analysis for power network.

Regional Name	Active Loss/(MW)	Active Power Generation/(MW·h)	Active Load/(MW)	Received Power/(MW)
LW	0.102	18.600	12.585	−6.015
DY	0.180	8.000	26.257	18.257
BZ	0.222	8.250	7.407	−0.843
WH	0.228	19.600	13.027	−6.573
ZZ	0.240	17.700	16.899	−0.802
TA	0.270	26.799	21.454	−5.345
RZ	0.272	20.600	19.115	−1.485
WF	0.285	19.800	52.417	32.617
HZ	0.300	30.100	25.656	−4.444
LC	0.341	44.650	20.471	−24.179
QD	0.365	25.400	13.596	−11.804
DZ	0.398	26.700	27.306	0.606
JNA	0.402	23.100	45.757	22.657
ZB	0.411	32.480	40.712	8.232
LY	0.425	29.950	53.534	23.584
YT	0.531	53.900	48.279	−5.622
JNI	0.695	79.746	35.237	−44.509

4.2. Application in One Small Sample

A small region within the provincial power grid was selected as a sample to illustrate the application process and effectiveness of the methods proposed in this paper. In fact, this study conducted experiments at key substations and transmission lines across the entire province, using 10 kV as an illustrative example. The results from other power supply areas were nearly identical and thus not reiterated for brevity. The schematic diagram of the line structure of the simple case is shown in Figure 2.

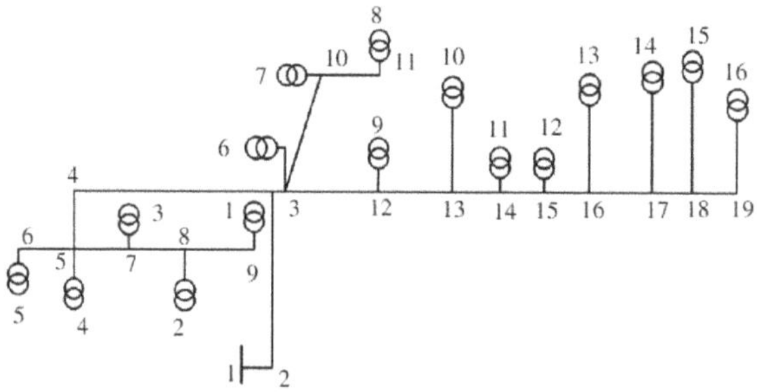

Figure 2. Schematic of the line structure. The Digits in the figure represent the transformers.

4.2.1. Data

The operational data, which were provided by the grid company, for each distribution transformer are detailed in Table 3, where MMR denotes monthly meter reading, which is the electricity usage record for that line, and AAP, ARP and ALF denote average active power, average reactive power and average load factor, respectively, which are all the attributes of the network.

Table 3. Operational data for each distribution transformer.

ID	MMR/(kW·h)	AAP/kW	ARP/kvar	ALF/%
1	31,630	43.93	27.22	55
2	40,769	56.62	30.09	18
3	5340	7.42	3.6	14.8
4	22,880	31.78	19.7	10
5	0	0	0	0
6	48,960	68	46.16	34
7	8927	12.4	5.69	41.3
8	41,640	57.83	30.86	72.3
9	0	0	0	0
10	45,120	62.67	34.85	12.5
11	52,000	72.22	40.78	14.4
12	21,000	29.17	18.09	29.2
13	28,600	39.72	22.63	39.7
14	33,540	46.58	26.88	23.3
15	135,360	188	116.56	29.8
16	97,200	135	83.7	42.8

The key parameters related to the cost are set as follows. The cost of each transformer is CNY 15,000 per unit. Generally, the installation and dismantling fees for replacing a distribution transformer are 30% of the purchase price of the equipment. The expenses for line modification are based on the comprehensive investment costs for 10 kV lines. Drawing from a previous study [12] and incorporating findings from actual field surveys, the acquisition cost for reactive power compensation equipment is set at CNY 20 per kvar, with the labor cost also established at CNY 20 per kvar.

4.2.2. Solutions

Replacing conductors, upgrading transformers, and enhancing reactive power compensation are three common methods for reducing grid losses. Specifically, in this case, conductor replacement involves updating Lines 1, 2, and 11. The transformer upgrade involves replacing the eighth transformer with a higher-capacity model (e.g., S11-400), which is due to the fact that the eighth transformer has reached an average load rate of 72.3%. Reactive power compensation involves increasing the power factor of each distribution transformer to 0.9, thereby enabling them to handle a greater load.

The three methods above can be used in any combination, thus resulting in eight strategies, as shown in Table 4, depending on whether or not a particular method is selected. In Table 1, RC, UT, and EC denote replacing conductors, upgrading transformers, and enhancing reactive power compensation, respectively. In the table, the 0–1 variables indicate whether the corresponding strategies have been adopted. PLB and PLA denote the expected power loss before and after the loss reduction strategy is applied.

Table 4. Operation data of distribution transformers.

Strategy	RC	UT	EC	PLB	PLA
1	0	0	0	88,192	88,192
2	0	0	1	88,192	86,160
3	0	1	0	88,192	83,520
4	0	1	1	88,192	83,326
5	1	0	0	88,192	87,938
6	1	0	1	88,192	85,436
7	1	1	0	88,192	82,684
8	1	1	1	88,192	82,587

Different loss reduction strategies vary in cost and effectiveness. One can employ an enumeration method to calculate the impact of each strategy on the objective function, thereby identifying the optimal strategy that yields the minimum value for the objective function. Alternatively, one could employ the genetic algorithm described in Section 3, treating the selection of loss reduction methods as binary decision variables, and generate the optimal strategy, which includes the decision that minimizes the objective function.

The genetic algorithm illustrated in Figure 1 was implemented through the following steps.

Step 1: Initialization. Initialize the maximum number of evolutionary generations T (in this case, T = 100). Randomly generate M (in this case, M = 6) individuals using binary encoding to form the initial population $P(0)$, which represents the initial set of loss reduction strategies.

Step 2: Individual judgement. Calculate the fitness value $f(t)$ for each individual within the population $P(t)$. Retain the best individual and directly replicate it into the next generation.

Step 3: perform genetic operations, including selection, crossover, and mutation, on the current population $P(t)$, which then generate the next generation $P(t+1)$.

Step 4: calculate the objective function in the current iteration.

Step 5: Iterate Step 2 to Step 4 until the termination condition is met. The termination conditions include reaching the maximum number of iterations, or the objective function has converged. In this paper, convergence is defined as the difference in the objective function value between two consecutive iterations being within 1%.

The decisions generated in the last iteration formed the optimal strategy. Table 5 shows the optimal strategy, as well as the second one. In the optimal solution, only EC method is applied. The annual investment, direct and indirect loss associated with such strategy are 264.4, 43,080 and 9520.37, respectively, which result in a total cost of 52,864.77. Strategy 3 is suboptimal, as it requires an investment of 1950 and generates direct losses of 41,760 and indirect losses of 9229.

Table 5. Optimal and suboptimal solutions.

Strategy	Optimal Solution	Suboptimal Solution
RC	0	0
UT	0	1
EC	1	0
direct loss	43,080	41,760
indirect loss	9520.37	9228.66
annual investment	264.4	1950
objective function	52,864.77	52,938.66

To validate the accuracy of the approach presented in this paper, the objective functions for each strategy were determined using an exhaustive enumeration method, as illustrated in Figure 3.

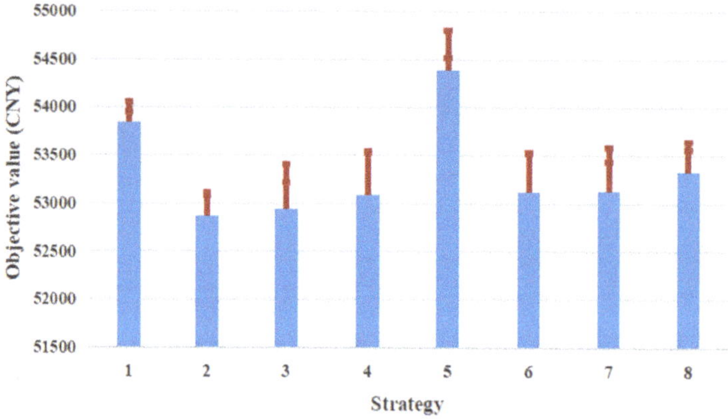

Figure 3. The results of each loss reduction strategy. The unit on the Y-axis is CNY. The red parts are error bars.

The results clearly demonstrate that the optimal solution is indeed consistent with the outcomes derived from the enhanced genetic algorithm. There is always some deviation in the power grid loss during each measurement, meaning that PLA and PLB have uncertainty. This is the main reason for plotting the error bar in this figure. This is a simple case, as it includes only three loss reduction strategies; thus, exhaustive enumeration requires merely eight possible combinations. When dealing with a large number of possible combinations, the enumeration method may not provide the desired solution within a reasonable time; therefore, the genetic algorithm offers a viable and efficient alternative.

4.2.3. Sensitivity Analysis

The optimal strategy presented in this case study is derived under the parameters specified within this paper. It is acknowledged that changes in energy prices and carbon penalties may alter the optimal strategy. The optimal solution results when these two parameters are, respectively, decreased and increased by 50% are shown in Table 6.

Table 6. Optimal solutions with different parameters.

d_1	d_2	Optimal Strategy	Objective Function
0.25	0.2	2	31,324.77
0.75	0.2	7	73,799.28
0.5	0.1	2	48,104.58
0.5	0.3	3	57,552.98

Table 6 shows that when these two parameters decrease, the benefits of loss reduction will become increasingly smaller, and the optimal strategy will tend towards the solution with the least investment. Therefore, to promote the adoption of loss reduction strategies in the power grid, increasing the carbon emission penalty is a viable administrative measure.

This paper also conducted a sensitivity analysis on thermal power. The proportion of thermal power was gradually increased from 0.5 to 0.8 in increments of 0.05, and the objective values corresponding to each strategy during this process are presented in Figure 4.

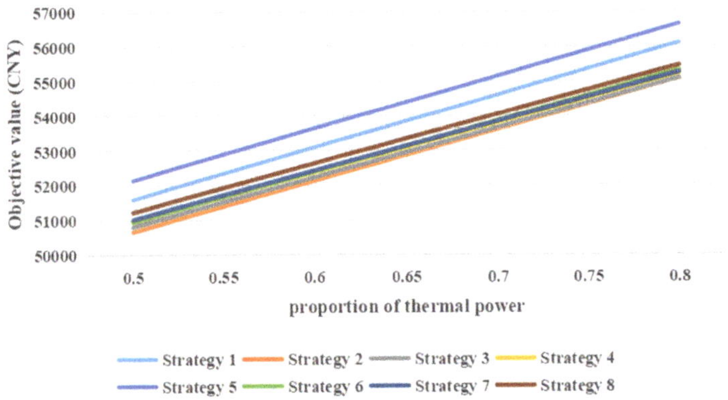

Figure 4. The results of each loss reduction strategy when the proportion of thermal power varies from 0.5 to 0.8.

It can be seen from Figure 4 that as the proportion of thermal power increases, all objective values show a linear growth trend. The same result also applies to the parameter of the carbon emission factor.

4.3. *Application in the Provincial Power Gird Company*

This section will take the typical operation mode of the power grid in the province as an example and apply the optimization method proposed in this paper to evaluate the power grid loss and loss reduction results. The company encompasses over 5000 small areas, as illustrated in Section 4.2. Given that each of these small areas has eight different loss reduction strategies, the total number of viable options for the entire company exceeds 800. Consequently, an exhaustive approach becomes entirely impractical at this juncture, necessitating the adoption of the methodology proposed in this paper for decision-making.

4.3.1. Power Grid Loss Distribution

Figure 5 shows the network loss distribution ratio statistics according to different administrative regions. It can be seen that the network loss in JNI, YT, LY, ZB, JNA and DZ is serious, while in regions such as DY, BZ, and WH, the loss is relatively low.

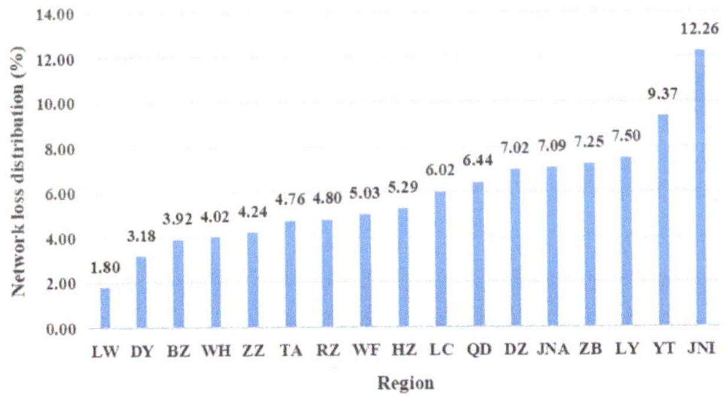

Figure 5. Network loss distribution ratio in different administrative regions.

4.3.2. Analysis of Loss Reduction Potential

The proposed loss reduction method is applied on the nodes with heavy load and low voltage. After optimization, the loss of the 500 kV and 220 kV transmission grids of the company is reduced from 566.8 MW to 552.9 MW, with a loss reduction of 13.9 MW. Based on the carbon emission intensity of 0.85 t/(MW·h), the reduced carbon emission per unit time is 11.815 t/h. The loss reduction result of power grids in various regions is shown in Figure 6.

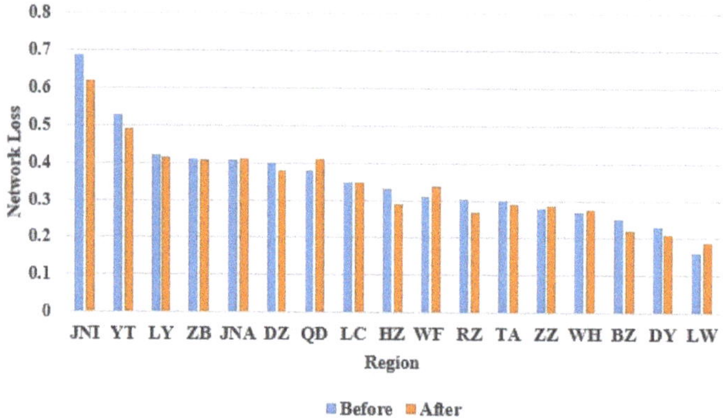

Figure 6. Comparison of power loss for different regions in the power network.

Among them, the regions with the most obvious loss reduction are JNI and YT. It should be pointed out that the optimization result is to minimize the total network loss of the entire system. Although the network loss in regions such as LW and WF increased after optimization, this is a consideration of sacrificing local interests to maximize global interests.

The application of the proposed model raises practical implementation challenges such as coordination between regions and data granularity. Addressing these barriers is crucial for successful implementation. To mitigate coordination issues, establishing clear communication channels and standardized protocols for data sharing and strategy alignment across regions is essential. For data granularity, implementing robust data collection systems and ensuring data consistency through standardized measurement practices can enhance the model's accuracy. Additionally, adopting a phased implementation approach allows for gradual integration and adjustment, reducing complexity and enabling continuous optimization.

4.4. Discussion

(1) Potential synergies with renewable energy sources and technologies

In the case study presented in this paper, only three loss reduction methods, i.e., replacing conductors, upgrading transformers, and enhancing reactive power compensation, are considered. With the proliferation of renewable energy sources, the number of approaches to grid loss reduction is also increasing, potentially leading to more efficient loss reduction strategies [11]. However, these changes merely add dimensions to the decision-making process without affecting the applicability of the model proposed here.

(2) Economic concerns

This paper aimed to identify the optimal combination of loss reduction strategies by minimizing the total sum of investment costs, direct costs, and indirect costs. The decrease in direct and indirect costs can be construed as an increase in revenue. Following a previous study [33], we are able to calculate the payback period for various schemes with a risk-free rate of 3%. Take strategy 2 as an example to illustrate the calculation method. With strategy 2, the reduction in power loss is 88,192 − 86,160 = 2032 kW·h. Thus, the direct and indirect cost decrease CNY 1060 and CNY 224.53, respectively. The payback period n refers to the number of years required to recoup the investment of strategy 2, which is CNY 2644, given a revenue of CNY 1240.53 per year. This involves solving the following equation:

$$\frac{1240.53 * \left[1 - \left(\frac{1}{1+0.03}\right)^n\right]}{1 - \frac{1}{1+0.03}} = 2644 \tag{19}$$

Figure 7 presents the investment payback times for each loss reduction strategy within the small-scale case study detailed in Section 4.2.

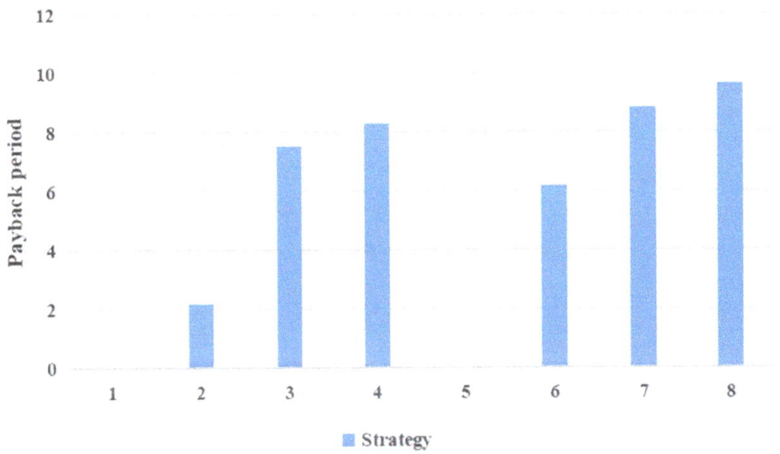

Figure 7. Comparison of payback period for different strategies.

The first strategy where the investment is zero and the fifth strategy where the investment cannot be recovered have no payback period. Among others, the optimal strategy from a cost perspective also has the shortest payback period, a mere 2.17 years. However, the second-best strategy exhibits a considerably longer payback period of 7.51 years, ranking it third in terms of cost efficiency.

In addition to the risk-free rate, inflation and maintenance costs can also impact the payback period. Inflation erodes the purchasing power of money over time, meaning that future cash flows will have less value in today's terms. If inflation is not accounted for, the payback period may appear shorter than it actually is in real economic terms. Maintenance costs are an ongoing expense that can significantly impact the net cash flows of grid loss reduction strategies. Overlooking these costs may lead to an underestimation of the total investment required. Including maintenance costs in the economic analysis would provide a more accurate picture of the long-term financial commitment associated with each strategy.

(3) Scalability of the model and optimization method

While the current case study focuses on a small sample with three primary loss reduction strategies (replacing conductors, upgrading transformers, and reactive power compensation), the model's modular design allows for the easy integration of additional loss reduction strategies, ensuring its applicability to broader and more intricate grid configurations. This flexibility is further enhanced by the use of an adaptive genetic algorithm (AGA), which dynamically adjusts its parameters based on the fitness of the population. This feature ensures that the optimization process remains efficient and effective even as the number of strategies and variables increases. The model's generalization capabilities are supported by its flexibility in handling diverse grid configurations. By adjusting input parameters and constraints, the model can be tailored to specific regional or national grid characteristics. This adaptability is crucial for addressing the unique challenges faced by different power grid systems, such as varying load profiles, voltage levels, and regulatory requirements.

Considering AGA, when the solution space dimension increases, the population size typically needs to be enlarged to maintain diversity and prevent premature convergence, which directly increases the computational effort per generation, including fitness evaluation, crossover, and mutation operations. While the AGA may reduce ineffective searches by dynamically adjusting mutation and crossover rates, thereby maintaining diversity with a smaller population, the complexity of fitness evaluation can still grow nonlinearly with the dimension of the solution space. This is especially true when fitness depends on simulations or experiments, as the time for a single evaluation cannot be reduced through algorithmic optimization. In this study, the number of optional strategy combinations is limited. Even considering technological development, combining some methods in advance can effectively reduce the solution space. Thus, the algorithm's scalability is not a major concern.

(4) Effect of the trend

In the current model, the objective function is formulated based on a static assumption of electricity prices and carbon penalties over the lifecycle of the strategies. This approach simplifies the optimization process and allows for a clear evaluation of the cost–benefit trade-offs associated with different grid loss reduction measures. However, it does not capture the potential variability in these parameters, which can be influenced by market conditions, regulatory changes, and technological advancements. For instance, a gradual increase in carbon penalties (e.g., due to stricter climate policies) could incentivize higher upfront investments in loss reduction measures. Conversely, a decrease or elimination of carbon penalties would diminish the importance of carbon-related costs in the optimization process.

To demonstrate it, the following numerical experiment is conducted in this paper. It is assumed that the carbon penalty increases linearly over time, as is shown in Equation (20).

$$d_2(t) = 0.2 + 0.01 * (t - 1) \tag{20}$$

Using this formula, the price of the carbon penalty for each future year is calculated and then substituted into the optimization model, yielding the following results (shown in Figure 8).

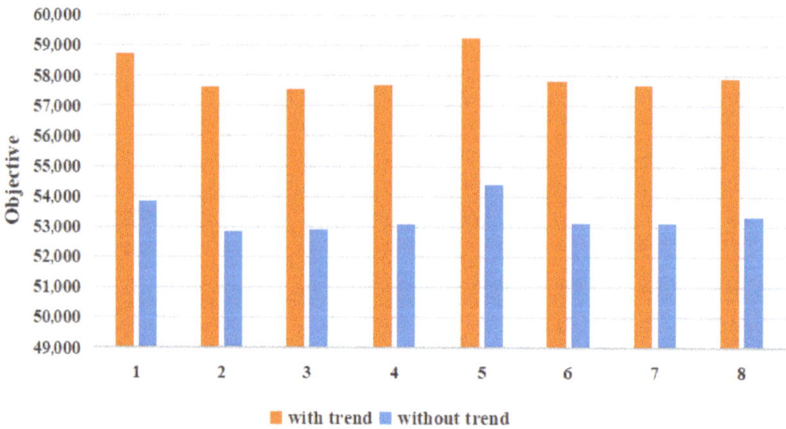

Figure 8. Results of each loss reduction strategy with and without linear carbon penalty trend.

It can be seen that the objective value of all strategies has increased, while the optimal strategy remains unchanged. Then, this formula is also incorporated into the payback period calculation, and Figure 9 illustrates that the payback period for all strategies has increased.

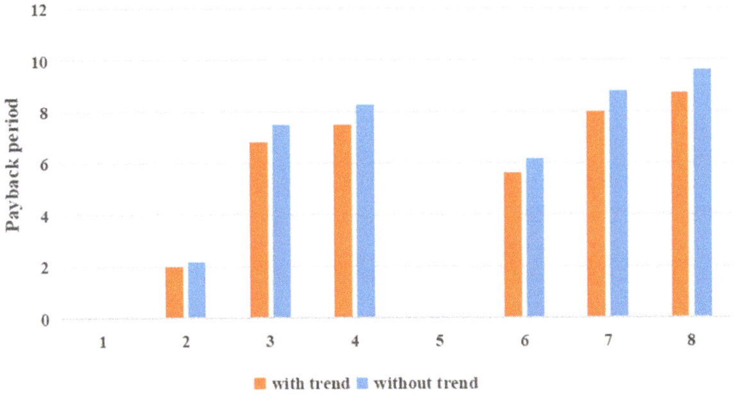

Figure 9. Results of payback period with and without linear carbon penalty trend.

This is consistent with the previous analysis. The same result also applies to the parameter of electricity prices.

5. Conclusions

In the context of electricity price reforms, power grid companies are urgently in need of tapping into potential loss reduction and optimizing loss mitigation measures to reduce grid losses through technical means. This necessity becomes even more critical in the backdrop of a low-carbon agenda. This paper proposes an optimization method for selecting loss reduction strategies in distribution networks that considers the benefits of low-carbon initiatives. It provides a theoretical foundation for the optimization of loss reduction strategies in distribution networks, addressing the challenge of quantitatively selecting loss reduction strategies. The model was validated through illustrative examples, demonstrating its good applicability and operability. We have also applied this method to the optimization of loss reduction in provincial power grids, and the results demonstrate that the method possesses excellent adaptability.

This paper contributes to proposing an optimization model for selecting loss reduction strategies in distribution networks, considering the low-carbon benefits. The model is designed to minimize the total cost, which includes the comprehensive investment in loss reduction measures (such as equipment purchase, removal, installation, or reconstruction), direct loss costs due to energy loss, and indirect loss costs. Voltage deviation and branch power serve as constraints in the model. Additionally, depending on the actual situation, the model can optionally include constraints on the number of loss reduction measures and total investment. This flexibility makes the model more adaptable to various scenarios.

While this study has demonstrated the potential of the proposed optimization model, several avenues for future research remain open. One promising direction is the integration of renewable energy sources into the grid loss reduction strategies. As renewable energy becomes more prevalent, the dynamic nature of these energy sources introduces new challenges and opportunities for grid management. Future work could explore how the proposed model can be adapted to incorporate renewable energy integration, thereby enhancing the grid's resilience and sustainability.

Another area for further investigation is the development of more advanced optimization algorithms to improve computational efficiency. As the complexity of power grids increases, the need for faster and more accurate optimization methods becomes critical. Research into hybrid algorithms or machine learning techniques could potentially enhance the model's performance and scalability.

Author Contributions: Conceptualization, W.L., Q.X., X.W., Z.L., T.L. and D.Z.; methodology, W.L., Q.X., X.W., Z.L., T.L. and D.Z.; software, W.L., Q.X., X.W., Z.L., T.L. and D.Z.; writing—original draft preparation, W.L., Q.X., X.W., Z.L., T.L. and D.Z. All authors have read and agreed to the published version of the manuscript.

Funding: This research was funded by the Technical Service Project of State Grid Jiangsu Electric Power Co., Ltd. (Grant No.: W24FZ2730030).

Data Availability Statement: The data are available from the corresponding author on reasonable request.

Conflicts of Interest: Authors Weiwu Li, Qing Xu, Xinying Wang, Zhengying Liu and Tianshou Li were employed by the State Grid Gansu Electric Power Co., Ltd. The remaining authors declare that the research was conducted in the absence of any commercial or financial relationships that could be construed as a potential conflict of interest.

References

1. Pang, B.; Zhang, L.H.; Ten, Y.F.; Chang, Z.W.; Tang, C.; Hu, C.Q.; Liu, Z.W.; Wang, B.L. An incentive scheme of peak-valley price based on differential privacy. *J. Chongqing Univ.* **2023**, *46*, 56–68.
2. Yu, X.; Dong, Z.; Zheng, D.; Deng, S. Analysis of critical peak electricity price optimization model considering coal consumption rate of power generation side. *Environ. Sci. Pollut. Res.* **2024**, *31*, 41514–41528. [CrossRef]
3. Tian, X.; Niu, X.; Zhu, X.; Zhao, G. Optimal strategy and assessment method for minimizing power loss of Shandong power network under lowcarbon background. *Autom. Electr. Power Syst.* **2014**, *38*, 67–72.
4. Bouakkaz, A.; Gil Mena, A.J.; Haddad, S.; Ferrari, M.L. Efficient energy scheduling considering cost reduction and energy saving in hybrid energy system with energy storage. *J. Energy Storage* **2021**, *33*, 101887. [CrossRef]
5. Eid, A. Cost-based analysis and optimization of distributed generations and shunt capacitors incorporated into distribution systems with nonlinear demand modeling. *Expert Syst. Appl.* **2022**, *198*, 116844. [CrossRef]
6. Li, C.; Xu, D. Family of enhanced ZCS single-stage single-phase isolated AC–DC converter for high-power high-voltage DC supply. *IEEE Trans. Ind. Electron.* **2017**, *64*, 3629–3639. [CrossRef]
7. Long, Y.; Wang, X.Y.; Zhou, Q.; An, H.Y.; Chen, Z. Distribution network reconfiguration for loss reduction considering maximum power supply capability index. *Proc. CSU-EPSA* **2017**, *29*, 131–134.
8. Sadeghian, O.; Moradzadeh, A.; Mohammadi-Ivatloo, B.; Abapour, M.; Anvari-Moghaddam, A.; Lim, J.S.; Marquez, F.P.G. A comprehensive review on energy saving options and saving potential in low voltage electricity distribution networks: Building and public lighting. *Sustain. Cities Soc.* **2021**, *72*, 103064. [CrossRef]
9. Sambaiah, K.S.; Jayabarathi, T. Loss minimization techniques for optimal operation and planning of distribution systems: A review of different methodologies. *Int. Trans. Electr. Energy Syst.* **2020**, *30*, e12230. [CrossRef]
10. Sun, K.; Xiao, H.; Liu, S.; You, S.; Yang, F.; Dong, Y.; Wang, W.; Liu, Y. A Review of Clean Electricity Policies—From Countries to Utilities. *Sustainability* **2020**, *12*, 7946. [CrossRef]
11. Kyriakopoulos, G.L. Energy Communities Overview: Managerial Policies, Economic Aspects, Technologies, and Models. *J. Risk Financ. Manag.* **2022**, *15*, 521. [CrossRef]
12. Lazzeroni, P.; Repetto, M. Optimal planning of battery systems for power losses reduction in distribution grids. *Electr. Power Syst. Res.* **2019**, *167*, 94–112. [CrossRef]
13. Xie, J.; Chen, C.; Long, H. A loss reduction optimization method for distribution network based on combined power loss reduction strategy. *Complexity* **2021**, *2021*, 9475754. [CrossRef]
14. Wang, W.; Cai, D.; Liu, H.; Zhou, C.; Cao, K.; Zhou, K.; Liu, C. Theoretical Analysis of Power Loss Reduction for Typical Power Grid. In Proceedings of the ICASIT 2020: 2020 International Conference on Aviation Safety and Information Technology, Weihai, China, 14–16 October 2020; pp. 383–387.
15. Suresh, M.; Edward, J.B. A hybrid algorithm based optimal placement of DG units for loss reduction in the distribution system. *Appl. Soft Comput.* **2020**, *91*, 106191. [CrossRef]
16. Pokhrel, B.R.; Karthikeyan, N.; Bak-Jensen, B.; Pillai, J.R. Loss optimization in distribution networks with distributed generation. In Proceedings of the 2017 52nd International Universities Power Engineering Conference (UPEC 2017), Heraklion, Greece, 28–31 August 2017; pp. 1–6.
17. Ahmadi, I.; Ahmadigorji, M.; Tohidifar, E. A novel approach for power loss reduction in distribution networks considering budget constraint. *Int. Trans. Electr. Energy Syst.* **2018**, *28*, e2635. [CrossRef]
18. Iweh, C.D.; Gyamfi, S.; Tanyi, E.; Effah-Donyina, E. Distributed Generation and Renewable Energy Integration into the Grid: Prerequisites, Push Factors, Practical Options, Issues and Merits. *Energies* **2021**, *14*, 5375. [CrossRef]

19. Kataray, T.; Nitesh, B.; Yarram, B.; Sinha, S.; Cuce, E.; Shaik, S.; Vigneshwaran, P.; Roy, A. Integration of smart grid with renewable energy sources: Opportunities and challenges–A comprehensive review. *Sustain. Energy Technol. Assess.* **2023**, *58*, 103363. [CrossRef]
20. Alotaibi, I.; Abido, M.A.; Khalid, M.; Savkin, A.V. A comprehensive review of recent advances in smart grids: A sustainable future with renewable energy resources. *Energies* **2020**, *13*, 6269. [CrossRef]
21. Mu, H.; Pei, Z.; Wang, H.; Li, N.; Duan, Y. Optimal Strategy for Low-Carbon Development of Power Industry in Northeast China Considering the 'Dual Carbon' Goal. *Energies* **2022**, *15*, 6455. [CrossRef]
22. Zhang, Y.; Wang, X.; Wang, J.; Zhang, Y. Deep Reinforcement Learning Based Volt-VAR Optimization in Smart Distribution Systems. *IEEE Trans. Smart Grid* **2020**, *12*, 361–371. [CrossRef]
23. Zhang, L.; Zhang, K.; Zhang, G. Power distribution system reconfiguration based on genetic algorithm. In Proceedings of the 2016 IEEE Advanced Information Management, Communicates, Electronic and Automation Control Conference (IMCEC), Xi'an, China, 3–5 October 2016; pp. 80–84.
24. Sehgal, S.; Swarnkar, A.; Gupta, N.; Niazi, K.R. Reconfiguration of distribution network for loss reduction at different load schemes. In Proceedings of the 2012 IEEE Students' Conference on Electrical, Electronics and Computer Science (SCEECS), Bhopal, India, 1–2 March 2012; pp. 1–4.
25. Gautam, M.; Bhusal, N.; Benidris, M.; Louis, S. A Spanning Tree-based Genetic Algorithm for Distribution Network Reconfiguration. In Proceedings of the 2020 IEEE Industry Applications Society Annual Meeting, Detroit, MI, USA, 10–16 October 2020; pp. 1–6.
26. Kamel, S.; Hamour, H.; Nasrat, L.; Yu, J.; Xie, K.; Khasanov, M. Radial distribution system reconfiguration for real power losses reduction by using salp swarm optimization algorithm. In Proceedings of the 2019 IEEE Innovative Smart Grid Technologies—Asia (ISGT Asia), Chengdu, China, 21–24 May 2019; pp. 720–725.
27. Wu, H.; Li, T.; Chen, H.; Li, B.; Liu, T. Distribution network line loss optimization method based on war strategy algorithm. In Proceedings of the 2023 8th Asia Conference on Power and Electrical Engineering (ACPEE), Tianjin, China, 14–16 April 2023; pp. 395–402.
28. Zhang, L.; Yan, Y.; Xu, W.; Sun, J.; Zhang, Y. Carbon emission calculation and influencing factor analysis based on industrial big data in the "double carbon" era. *Comput. Intell. Neurosci.* **2022**, *2022*, 2815940. [CrossRef] [PubMed]
29. Wan, Y.; Hou, H.; Ge, X.D.; Wang, X.H.; Lu, Y.C.; Wang, Y.X. Two-stage mixed-game win-win strategy for integrated energy system considering multi-stage utilization of carbon emission. *Autom. Electr. Power Syst.* **2025**, *49*, 69–79.
30. Zheng, W.; Wu, W.; Zhang, B.; Sun, H.; Liu, Y. A fully distributed reactive power optimization and control method for active distribution networks. *IEEE Trans. Smart Grid* **2015**, *7*, 1021–1033. [CrossRef]
31. Srinivas, M.; Patnaik, L. Adaptive probabilities of crossover and mutation in genetic algorithms. *IEEE Trans. Syst. Man. Cybern.* **1994**, *24*, 656–667. [CrossRef]
32. Zhai, L.; Feng, S. A novel evacuation path planning method based on improved genetic algorithm. *J. Intell. Fuzzy Syst.* **2022**, *42*, 1813–1823. [CrossRef]
33. Eberhart, R.; Kennedy, J. A new optimizer using particle swarm theory. In Proceedings of the Sixth International Symposium on Micro Machine and Human Science, Nagoya, Japan, 4–6 October 1995; pp. 39–43.
34. Zhan, Z.H.; Zhang, J.; Li, Y.; Chung, H.S.H. Adaptive Particle Swarm Optimization. *IEEE Trans. Syst. Man Cybern. Part B Cybern.* **2009**, *39*, 1362–1381. [CrossRef]

processes

MDPI

Review

Towards Reliable Prediction of Performance for Polymer Electrolyte Membrane Fuel Cells via Machine Learning-Integrated Hybrid Numerical Simulations

Rashed Kaiser [1], Chi-Yeong Ahn [2,3], Yun-Ho Kim [2,3] and Jong-Chun Park [1,*]

1 Department of Naval Architecture & Ocean Engineering, Pusan National University, Busan 46241, Republic of Korea; rashed.kaiser@pusan.ac.kr
2 Department of Green Mobility, Korea National University of Science and Technology (UST), Daejeon 34113, Republic of Korea; cyahn@kriso.re.kr (C.-Y.A.); yunhokim@kriso.re.kr (Y.-H.K.)
3 Alternative Fuels and Power System Research Division, Korea Research Institute of Ships and Ocean Engineering (KRISO), Daejeon 34103, Republic of Korea
* Correspondence: jcpark@pnu.edu

Abstract: For mitigating global warming, polymer electrolyte membrane fuel cells have become promising, clean, and sustainable alternatives to existing energy sources. To increase the energy density and efficiency of polymer electrolyte membrane fuel cells (PEMFC), a comprehensive numerical modeling approach that can adequately predict the multiphysics and performance relative to the actual test such as an acceptable depiction of the electrochemistry, mass/species transfer, thermal management, and water generation/transportation is required. However, existing models suffer from reliability issues due to their dependency on several assumptions made for the sake of modeling simplification, as well as poor choices and approximations in material characterization and electrochemical parameters. In this regard, data-driven machine learning models could provide the missing and more appropriate parameters in conventional computational fluid dynamics models. The purpose of the present overview is to explore the state of the art in computational fluid dynamics of individual components of the modeling of PEMFC, their issues and limitations, and how they can be significantly improved by hybrid modeling techniques integrating with machine learning approaches. Furthermore, a detailed future direction of the proposed solution related to PEMFC and its impact on the transportation sector is discussed.

Keywords: fuel cell; numerical modeling; mass transfer; electrochemical; progress; limitations; machine learning; PEMFC

1. Introduction

Fossil fuels, such as coal, oil, and natural gas, have been the primary sources of energy over the past 50 years, despite their significant impact on increasing CO_2 emissions and contributing to global warming [1,2]. This trend is unlikely to stop owing to the growing population and economic development unless a reliable and sustainable energy source replaces fossil fuels [3]. According to the modified Klass model [4], the depletion of oil, coal, and natural gas will take approximately 35, 107, and 37 years, respectively. The fuel infrastructure cannot be constructed overnight, even in light of the dire situation, because the process is costly and slow.

In this regard, hydrogen is seen as a suitable alternative to conventional fuels as both an energy source and carrier that could help alleviate the detrimental impacts of fossil fuel combustion on the environment [5,6]. Hydrogen utilizes electrochemical reactions to convert it into fuel and directly oxidize it for generating electricity [5]. Based on the electrolyte and working temperature, hydrogen fuel cells are classified as low temperature (20–100 °C), medium temperature (200–300 °C), and high temperature (600–1500 °C) fuel cells [7]. Low-temperature fuel cells have made notable advancements in transportation applications because of their rapid startup, compact size, and lighter weight in comparison to high-temperature fuel cells [8]. Proton exchange membrane fuel cells (PEMFCs) and alkaline fuel cells (AFCs) are among the most common types of low-temperature fuel cells. Both PEMFCs and AFCs are relatively low cost, though there is a lack of waste heat recovery. PEMFCs require costly platinum as an electrode and are prone to carbon monoxide (CO) and sulfur (S) poisoning, whereas AFCs require pure hydrogen (H_2) and oxygen (O_2) in addition to a potential risk of carbon dioxide (CO_2) poisoning [9]. However, based on their power densities, efficiency, and sensitivity to fuel impurities, PEMFCs are the most prominent options for low-temperature fuel cells [9–11]. PEMFCs have the potential to satisfy the power demands of various sectors, such as transportation (e.g., hybrid electric vehicles (HEV) [12], ships [13,14], and stationary and portable power generation [15]. However, PEMFCs have several shortcomings, including limited power density (<3 $kW{\cdot}L^{-1}$), low durability (<5000 h) [16], and high material cost [17], which hinder their market competitiveness [18]. Hence, further research is required to address the following issues to achieve the potential of PEMFCs:

Water management: Maintaining an appropriate amount of water inside the cell during PEMFC system operation is challenging because good proton conductivity must be achieved without compromising the performance by flooding the reaction site [19].

Thermal issue: In most cases, PEMFCs have an energy efficiency of approximately 50% because they produce waste heat almost equal to their electric power output [20]. Moreover, to maintain good proton conductivity, the temperature needs to be kept below 80 °C to prevent dehydration, and temperatures below 60 °C can lead to water condensation and cell flooding [21].

Cost and durability: The high Pt catalyst loading and membrane account for almost 80% of the cost of the PEMFC stack [22]. In addition, the materials used must have adequate temperature and humidity tolerance as well as suitable mechanical strength to withstand compression [23]. The degradation of materials and their impact on cell performance must also be considered [24].

To address the aforementioned issues, various experimental studies [25–30] have been conducted. However, experiments are expensive, sequential, time-consuming [31], limited to laboratory-scale models, and can be employed to measure only one quantity at a time, which is not convenient for improving the performance over existing designs [32]. By contrast, computational fluid dynamics (CFD) can significantly reduce the time and cost of experimentation by predicting real-life scenarios, and it can explore conditions that are not possible to observe experimentally [33]. Furthermore, advances in computational capabilities have enabled the implementation of more sophisticated and complicated three-dimensional (3D) PEMFC models, which are far more accurate than one-dimensional (1D) and two-dimensional (2D) models [33,34]. However, realistic modeling of PEMFCs still suffers from obstacles because material properties, environmental conditions, and material degradation are difficult to predict. Reasonable assumptions are still required, or a sensitivity analysis must be performed to obtain a model that reflects real-world scenarios [35].

Because of the accelerated growth of artificial intelligence (AI)-based methods such as machine and deep learning, a whole new dimension is opened for improving the current 3D CFD model of PEMFC. Machine learning is not only used for the prediction of performance [36–42], optimization of boundary conditions [43–47] for maximum output, water management [48–51], thermal condition [52,53], and degradation of PEMFC and various components [54–56], but also to provide accurate experiment parameters (linear and non-linear) [36,47,57,58] for reliable CFD modeling.

The purpose of this study is to provide an overview of the current state of the art in PEMFC simulations, identify their limitations, and potential for machine learning (ML) in PEMFC modeling, and discuss future directions for integrating ML in conventional CFD. In Section 2, the applications of PEMFC and their corresponding advantages and limitations are discussed, followed by the evaluation of the models required for each PEMFC component. Their limitations, various ML approaches, and ML-integrated modeling approaches are described in Section 3. In Sections 4 and 5, the final outlook of these integrated models and their potentials for solving PEMFC issues for future applications are reviewed, respectively.

2. Applications, Potentials, and Hurdles of PEMFC System Integration

PEMFCs utilize various integration styles to accommodate variations in operating environments and power demands across different application fields. Table 1 shows the present status of PEMFC system integration.

Table 1. Current status of PEMFC system integration across diverse sector.

Sector	Application (Power)	Potential	Advantages	Limitations	Reference
Portable	Laptops, cell phones, and military radio/communication devices. (5 to 50 W)	-Provide continuous power as long as hydrogen fuel is available. -Can be fabricated in small sizes without efficiency loss.	-Low acoustic and thermal signatures, high reliability, quick recharging, and high energy density.	-Complex system with water and heat management issues. -Hydrogen storage system. -Costly.	[59]
Stationary	Backup power system, off grid power supply, combined heat and power unit (CHP) (to 300 kW)	-Can be co-located other renewable power sources. -Significant cost benefits compared to battery-generator systems for shorter durations.	High energy and power densities, high modularity, longer operation times, compact size, and ability to operate under unkind ambient conditions.	-Coolant leakage for longer run. -Coolant and bipolar plate compatibility. -Reliability of components.	[60,61]
Vehicle	Passenger car, utility vehicles, and bus. (20 to 250 kW)	-Can be used in hybrid power system in addition to battery and supercapacitor.	-Efficiency is higher than the vehicles powered by internal combustion engines. -Low refueling time (<5 min) -No noise. -Zero emissions.	-Cost of the components (catalyst). -High operation cost. -Low durability (2500–3000 h).	[59,62]
Marine	Container, demonstrator, yachts, ferries, submarine (12–300 kW)	-Can be used both as main and auxiliary power system.	-High power to weight ratio. -Low maintenance cost. -Low noise. -Good modularity/part load performance	-Low power capacity. -Safety and reliability. -Durability. -High cost. -Storage issue.	[63]
Aviation	Small scaled manned/unmanned aircraft, drones. (100 W–33 kW)	Main power source of unmanned aerial vehicle (UAV), auxiliary power unit (APU) for large aircrafts.	-High power output.-Light weight. -Simple design.	-Additional space requirements for hydrogen storage. -Heat and water management.	[59,64,65]

3. PEMFC Modeling Approaches

PEMFC models are divided into three distinctive categories: (a) Black box, (b) grey box, and (c) white box [66,67] as can be seen in Table 2. Black box models are obtained from the experimental data and based on a statistical data-driven approach [67,68]. The experimental data is divided into two distinct sets: One specifically for training to identify the complex non-linear correlations between input and output, and another for validating the model. Therefore, the black box models do not consider the features and characteristics of physical phenomena inside the PEMFC. In addition to that, black box models require low computational costs and are mostly used to investigate the health of PEMFC and predict performance. Artificial neural network (ANN), fuzzy modeling, support vector machines (SVM), and gradient boosting are some common black box models [69]. Nevertheless, these models suffer from uncertainty as soon as the trained models encounter new operational parameters. On the other hand, white box models, known as mechanistic or theoretical models, are utilized to evaluate internal physical interactions in PEMFC, such as the polarization curve, water and heat management, by implementing a series of complicated algebraic and differential equations [67,70,71]. This equation includes Nernst–Planck, Butler–Volmer, and Fick's laws. Hence, the CFD simulations of PEMFC are white box models [72]. In contrast to these two models, grey box models utilize semi-empirical equations derived from the experimental data, balancing between complex and simple solutions [73,74]. Machine learning-assisted CFD simulation, which is a combination of black box and white box models, is referred to as grey box models [75].

Table 2. PEMFC model categories and their distinctive features, adapted from [68,70].

	Physical Insight →		
Category / Features	**Black Box (ML)**	**Grey Box (Semi-Empirical)**	**White Box (Mechanistic)**
Basis	Based on input and output data from the experiment.	Based on empirical equations validated by the experimental data.	Based on complicated algebraic and differential equations.
Dependency of test data	High	Medium	Low
Computational cost	Low	Medium	High
Accuracy of findings	Reasonable	Reasonable	High
Applications	Stack level	Stack level	Cell level
Online simulation	Possible	Possible	Not possible
	← Granularity		

4. PEMFC Fundamentals and Modeling Consideration of Components with ML

PEMFCs are devices that convert chemical energy into electrical energy. They consist of a membrane electrode assembly (MEA) and bipolar plates (BPs) on both sides, as shown in Figure 1.

The MEA consists of a membrane sandwiched between an anode (negatively charged) and a cathode (positively charged) catalyst layer (CL), followed by a gas diffusion layer (GDL). On the anode side, hydrogen enters as a fuel and is oxidized into a proton (H^+) after hydrogen oxidation reaction (HOR) occurs (Equation (1)), whereas on the cathode side, the oxygen reduction reaction (ORR) happens (Equation (2)). A potential difference is formed between the sides owing to the flow of electrons, and water is generated as a byproduct on the cathode side (Equation (3)). The electrochemical reactions are as follows [8,28]:

$$\text{Anode side: } H_2 \rightarrow 2H^+ + 2e^- \quad (1)$$

$$\text{Cathode side}: \ 2H^+ + 2e^- - \frac{1}{2}O_2 \rightarrow H_2O \quad (2)$$

$$\text{Overall reaction}: \ H_2 + \frac{1}{2}O_2 \rightarrow H_2O + \text{Electricity} + \text{Waste heat} \quad (3)$$

Table 3 shows the functions of each component, state of the art, and future trends.

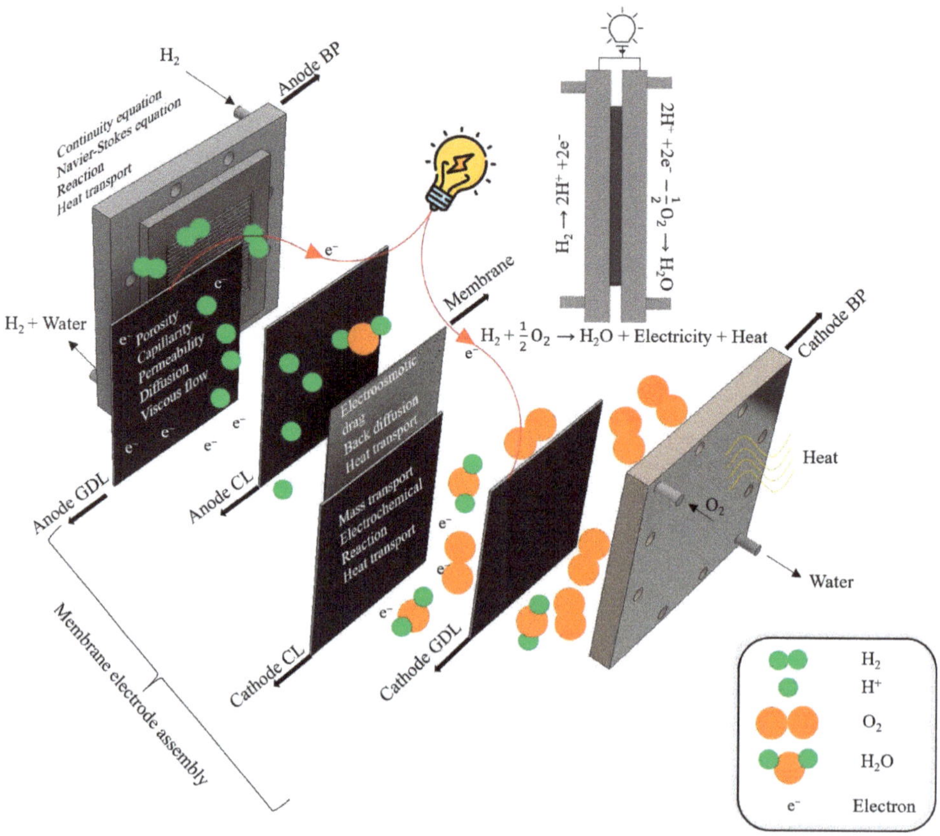

Figure 1. Schematic and operation of a conventional PEMFC.

Table 3. Functions and materials used for PEMFC components.

Components	Functions	Materials (Present)	Materials (Future)	Reference
Polymer electrolyte Membrane	-Conduction of protons -Separating barriers of reactants. -Blockade of electrons.	-Perfluorosulfonic acid (PFSA) membranes (~90 °C operations).	-Composite membranes such as Polytetrafluoroethylene (PTFE) and Polyvinylidene difluoride (PDDF) membranes (>100 °C operations). -PEM with ionogels	[23,76,77]
Catalyst layer	Site of electrochemical reactions.	-Platinum (Pt) loading (0.1–0.2 g kW^{-1}). -Heterogenous coverage of ionomer.	-Platinum (Pt) loading (<0.1 g kW^{-1}). -Lower and better distribution of ionomer. -Pt monolayer catalysts, nanometer film catalysts, controlled crystal shape catalysts, and non-PGM catalysts.	[23,76]

Table 3. Cont.

Components	Functions	Materials (Present)	Materials (Future)	Reference
Gas diffusion layer	-Distribution of reactants. -Water management. -Electricity and heat distribution through conduction.	-Carbon paper with fiber structure -Carbon black and hydrophobic agents	-Carbon paper with modified fiber configurations. -Carbon black with super hydrophobic additives. -Metal foam.	[23,76]
Bi-Polar plate	-Transport of reactants. -Water management. -Dissipation of heat. -Current collector.	-Graphite (~1 mm thick). -Metal (~0.6 mm thick).	-Metal foam. -Metal (~0.5 mm thick).	[23,76]

4.1. Modeling of Polymer Electrolyte Membrane

The PEM is one of the key components of the PEMFC. It enables protons (H^+) to move from the anode to the cathode to react with O_2 and produce water. An ideal membrane should possess high proton conductivity at low humidity, good chemical and mechanical stability, low electronic conductivity, low water permeability, and impermeability to reactants [78,79]. Moreover, manufacturing it should be easy and economical. The most commonly used membrane is Nafion, which is composed of perfluorosulfonic acid (PFSA). Membrane modeling is generally classified into two approaches: microscopic and macroscopic.

4.1.1. Microscopic

To understand the effect of ionic moieties [80], which are responsible for making the membrane more soluble in water despite being made of hydrophobic polymers, and the backbone dynamics [81,82] of Nafion as well as the oxygen permeation [83], conduction, and mobility of protons, various approaches such as statistical [84] and molecular dynamic (MD) simulations [85] have been used. Harvey et al. [84] proposed a one-dimensional (1D) MEA performance model, which includes liquid water transportation, an agglomerate catalyst structure, and various statistical MEA characteristic parameters. This model provided insights into the effect of Pt loading near the interface between the CL and membrane on the reaction distribution. Figure 2a shows the experimental validation of the statistical simulation model, and Figure 2b depicts the partial effect of high liquid water on a low oxygen partial pressure when the Pt loading is low. Kwon et al. [85] performed MD simulations to measure the solubility and permeability of O_2 in saturated PFSA ionomers on a Pt surface using two types of PFSA ionomers by varying the water content.

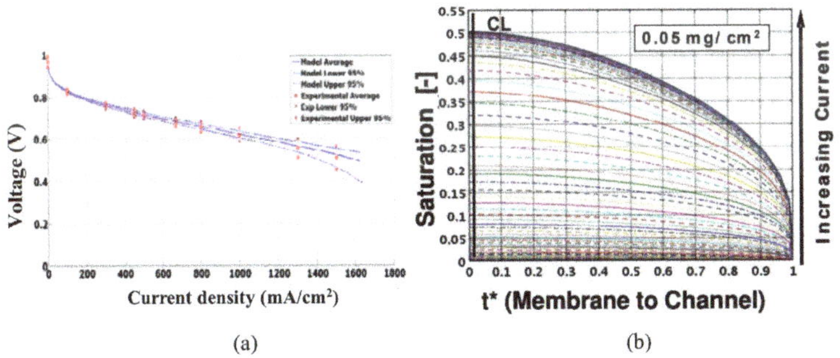

Figure 2. Cont.

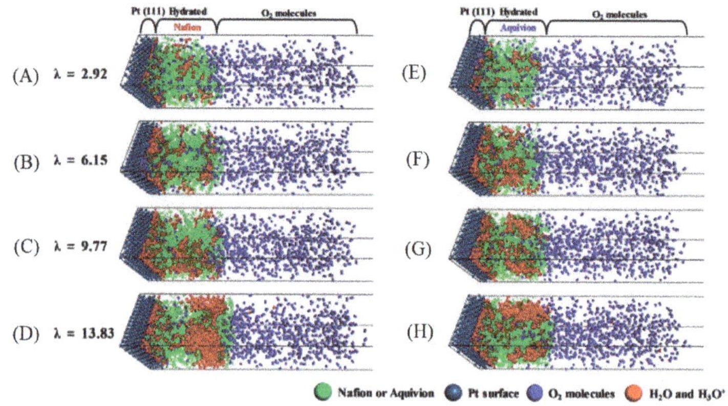

(c)

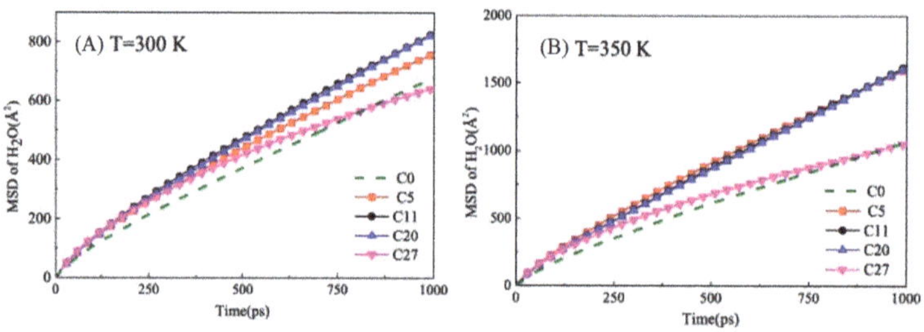

(d)

Figure 2. (**a**) Polarization curve of the statistical simulation model of the baseline MEA with experimental validation. (**b**) Liquid water distribution across the nondimensional distance (t*) of MEA where colored line indicates different saturation level; reproduced with permission [84]; copyright 2013, IOP publishers. (**c**) Snapshots of equilibrated systems comprising the Pt (111) surface, hydrated Nafion ionomers, and O_2 molecules with different λ (water molecules per sulfonic acid group). Illustrations show the hydrated Nafion thin film (**A–D**) and hydrated Aquivion thin film (**E–H**) on a Pt surface; reproduced with permission [85]; copyright 2021, Springer Nature. (**d**) Mean square displacement of water at 300 and 350 K measured from MD simulation; reproduced with permission [83]; copyright 2018, Elsevier Ltd.

Their simulation results provided guidance for designing better PFSA ionomers for PMFCs. Figure 2c presents an equilibrated structure where Nafion ionomers with high water content and water molecules are denser in the region of saturated Nafion–Pt interfaces because of their strong attraction. Additionally, dissolved O_2 molecules have the highest density, whereas water molecules (hydronium ions) have the lowest inside the hydrated Nafion. Table 4 lists some microscopic-level research methods and their results.

Despite the importance of the results of microscopic models, integrating them into a complete fuel cell model is difficult, which leads to the separation of the microscopic model research. However, one particular membrane property, that is, the water content (the number of H_2O moles per mole of sulfonic acid group), can be utilized in macroscopic models because it influences the membrane structure.

Table 4. Microscopic research of PEMFC membrane.

Research Method	Software/Code	Findings	Reference
MD simulation	Materials studio (MS) 4.3 COMPASS	Instead of Nafion, most widely used membrane material, a natural polymer Chitosan (a derivative of chitin a natural polysaccharide found in insects, fungi, and yeast) was used. The highest conductivity of 7.14×10^{-2} S/cm was observed when the system contained 40% of water.	[86]
MD simulation	Materials studio (MS) 4.4 COMPASS	The diffusion of water molecules and hydronium ions in Nafion 117 are measured using MD simulation. They found that the diffusion coefficient of both water molecules and hydrogen ions increase with the water content when the temperature in constant. Similar to this, both the diffusion coefficient increases with the temperature at constant water content.	[87]
MD simulation	COMPASS	Proton conductivity of three different types of polymer membrane; Dow, Nafion and Aciplex was compared. The Aciplex membrane has the optimum proton mobility and best water molecules and hydronium mobility at 350 K.	[88]
MD simulation	Materials Studio COMPASS II	Sulfonated polynorbornene-based (SPNB) membranes swell upon after getting hydrated and divided into hydrophobic and hydrophilic regions having sulfonic acid groups at their interfaces. Also, more water molecules can absorb the sulfur atoms and hydronium due to the rise of water at increasing temperature.	[89]
MD simulation	LAMPPS	Pt/C models were investigated with various PTFE-binder content in the presence of H_3PO_4 at 298.15 K and 433.15 K. It was found that the coverage of H_3PO_4 is higher at high temperature in comparison to lower temperature as the H_3PO_4 could still manage to contact the Pt surface through PTFE-binder due to favorable interactions.	[90]

4.1.2. Macroscopic

When modeling a membrane from a macroscopic perspective, two models are commonly used: sorption and transport.

Sorption Model

In PEMFCs, the membrane material adsorbs water when the environment is humid. The water content, λ, is defined as the number of water molecules per mole of sulfonic acid and is expressed as follows [91]:

$$\lambda \equiv \frac{n(H_2O)}{n(SO_3)} \tag{4}$$

The water content could be equivalently expressed as a function of total mass of water (m_w) within the same reference volume (V) [92]:

$$\lambda = \frac{m_w}{M_w c_f V} \tag{5}$$

Again, sorption of water is the measurement of the amount of water adsorbed by the membrane when the membrane is in equilibrium with water vapor/liquid at a given temperature. This term is known as the sorption isotherm, and it relates to the equilibrium water content λ_{eq}, which is the function of water activity a_w in the membrane phase [92]:

$$\lambda_{eq} = \lambda_{eq}(a_w) \tag{6}$$

Water could be in a position where it sorbed from or desorbed to a nearby phase. These conditions are referred to as vapor-equilibrated (VE), where a gas phase reacts with water vapor at a certain water activity, and liquid-equilibrated (LE), a liquid phase. Both saturated water vapor and pure liquid water are treated as same, as they equally processed the same $a_w = 1$, generates a thermodynamically unexpected scenario often known as "Schröder's paradox" [93]. Though some of the experiments did not observe this [94,95], research suggests VE water contents are closer to a typical LE value, suggesting the presence of thin water film.

In the membrane, the activity of water can be expressed as the activity of water vapor in the equilibrating gas phase under VE conditions [91]:

$$a_w = a_{w,vap}(equilibrium) \tag{7}$$

However, the a_w in the vapor phase can be approximated from the ration between the partial pressure of water, $p_{w,vap}$ and the saturated partial pressure of water, $p_{sat}(T)$ as a function of temperature (T) [92]:

$$a_{w,vap} \approx \frac{p_{w,vap}}{p_{sat}(T)} \equiv \frac{x_w p}{p_{sat}} \tag{8}$$

where partial pressure of water is the function of mole fraction of water vapor, x_w and p is the pressure of the gas mixture in the pore region.

The above equation ignores the fugacity correction considering the operating condition of PEMFC is close to the absolute pressure [92]. The saturated vapor pressure P_{sat} can be expressed as the following empirical equation [96]:

$$\log_{10}\left(\frac{p_{sat}}{p_0}\right) = a_0 + a_1(T - T_0) + a_2(T - T_0)^2 + a_3(T - T_0)^3 \tag{9}$$

Table 5 contains the water contain measurement equation at various conditions.

Table 5. Variations of water content, λ.

Equation	Remarks	Ref.
$\lambda = \begin{cases} 1.41 + 11.3a_w - 18.8a_w^2 + 16.2a_w^3 & 0 \le a_w \le 1 \\ 10.1 + 2.94(a_w - 1) & 0 < a_w \le 3 \\ 16.0 & 3 < a_w \end{cases}$	This equation is obtained from the fitting experimental data of water uptake in Nafion membrane at 80 °C	[97]
$\lambda = \begin{cases} 0.3 + 6a_w[1 - \tanh(a_w - 0.5)] + 3.9\sqrt{a_w}[1 + \tanh(\frac{a_w - 0.89}{0.23})] & s \le 0 \\ 16.8s + \lambda_{\mid(a=1)}(1 - s) & s > 0 \end{cases}$		[98,99]
$\lambda = \begin{cases} 0.043 + 17.18a_w - 39.85a_w^2 + 36a_w^3 & 0 \le a_w \le 1 \\ 14 + 1.4(a_w - 1) & 1 \le a_w \le 3 \end{cases}$	Although the relation in $a_w > 1$ obtained from 80 °C, the polynomial relation comes at 30 °C	[96]
$\lambda = 1.4089 + 11.26a_w - 18.768a_w^2 + 16.209a_w^3, 0 \le a_w \le 1$	A fit data obtained from the experiment under VE conditions at 80 °C	[100]

Transport Models

The macroscopic transport models are mainly distinguished into three approaches depending on the driving forces: (1) Chemical potential, (2) diffusion, and (3) hydraulic models.

Diffusive Model

This model considers the membrane as a system of single and homogeneous phases where protons and water move and dissolve by the diffusion process. Though both water and protons can be moved in a stationary membrane system, water movement is neglected or treated as constant for the sake of simplicity. Thus, the differential form of Ohms law is implemented for defining proton movement [97]:

$$i_2 = -\kappa \nabla \Phi_2 \tag{10}$$

where κ is the ionic conductivity of the membrane, Φ_2 is the chemical potential of the membrane. This model can be described using the dilute solution theory proposed by Newman [101]. This theory is evaluated only by the dilute concentration theory (DCT), which treats ionomers as solvents and water and protons as solute species. Assuming the negligible effect between the water and proton (most solute species are dilute in water), DCT implements the following Nernst–Planck equation to measure the interaction between the solute species and ionomer [101]:

$$\underbrace{N_i}_{\text{Flux}} = \underbrace{-z_i u_i F c_i \nabla \Phi_2}_{\text{migration}} - \underbrace{D_i \nabla c_i}_{\text{diffusion}} + \underbrace{c_i v_2}_{\text{convection}} \tag{11}$$

The migration term (1st part of the equation), containing the information of motion of charged species, related to potential gradient ($-\nabla\Phi_2$) and charged number, z_i, Faraday's constant, F, (9.6487×10^7 C/kmol) concentration, c_i and mobility, u_i. The midterm (diffusion) is composed of a concentration gradient, ∇c_i and a diffusion coefficient, D_i which is the function of mobility, u_i and can be expressed by the Nernst–Einstein equation [101–103]:

$$D_i = RTu_i \tag{12}$$

The convective term (final part of the equation) is the function of the concentration and motion of the solvent. However, this term will become null for the analysis of a single-phase system. The migration term of Equation (11) becomes zero as the water has a zero valence and is converted to Fick's law [104].

$$N_w = n_d \frac{i_2}{F} - D_w \nabla c_w \tag{13}$$

where ξ is the electroosmotic drag coefficient.

Chemical Potential Model

In this model, though the main force of transport is chemical potential, diffusion (gradients of species concentration) and convection (gradient of pressure) are also included. Based on the concentrated solute theory proposed by Bennion [105,106], three independent transport properties should be considered:

a. Chemical potential of proton, Φ_2
b. Transport coefficient of water, α_w
c. Chemical potential of water, μ_w

The final equation containing these transport systems is as follows [105]:

$$i_2 = -\frac{\kappa n_d}{F} \nabla \mu_w - \kappa \nabla \Phi_2 \tag{14}$$

$$N_w = n_d \frac{i_2}{F} - \alpha_w \nabla \mu_w \tag{15}$$

In addition to the above equations, the irreversible thermodynamics approach [107,108], generalized Stefan–Maxwell equations [109–112], and dusty fluid model [113–115] are also used. Unlike the diffusion model, proton–water interaction is considered in this model.

Though this interaction is very rare, it should be considered in situations where the gradient of water is very high, such as low humidity or high current density conditions. It should be noted that Equations (12) and (14) are similar except that the concentration and diffusion coefficient of water have been replaced by the chemical potential and transport coefficient of water, respectively. However, the chemical potential model is not well recognized for modeling PEMFC because of the difficulties in obtaining the transport parameters.

Hydraulic Model

This model treats the membrane as a two-phase system, unlike the diffusive and chemical potential models. Also, the addition of a second phase facilitates the hydraulic model to consider the pressure gradient responsible for the convective in the water. However, the water content of the membrane is assumed to remain constant ($\lambda = 22$) [116] as long as the water contains the pores to assist convective transportation. Bernardi et al. [117,118] proposed this model based on previous works [119–121], where a dilute solution approach with Equation (10) (Nernst–Planck equation) is used to characterize the movement of proton, and instead of zero, the velocity of the water is given by Schlogl's equation [122]:

$$v_w = -\left(\frac{k}{\mu}\right)\nabla p_L - \left(\frac{k_\Phi}{\mu}\right) z_f c_f F \nabla \Phi_2 \tag{16}$$

where k, k_Φ, p_L, μ, and $z_f c_f$ are the effective hydraulic permeability, effective electrokinetic permeability, liquid pressure, water viscosity, charge and concentration of fixed ionic sites, respectively.

This model also assumes a constant gas volume fraction in the membrane, though it does not represent the real experiment. This assumption (constant gas volume fraction) allows the gas to crossover through the membrane. However, the anode side near the membrane quickly dried out most of the time due to the vigorous EOD effect during PEMFC operation, which contradicts the assumption of a fully hydrated membrane. Thus, this model suffers from an unrealistic scenario by neglecting diffusive water transport [123].

Also, the concentration solution approach could be utilized in this model as previously conducted by Weber and Newman [124], where the same Equations (14) and (15) are used except the chemical potential is replaced by liquid pressure and the transport coefficient is changed to permeability through comparison of Darcy's law.

Hence, Equation (15) transforms [124]:

$$N_w = n_d \frac{i_2}{F} - \frac{k}{\mu \overline{V}_w} \nabla p_L \tag{17}$$

Here, $\overline{V}_w$ is the molar volume of water.

Combination Model

In order to portray the real experimental condition, diffusive and hydraulic models should be utilized together to cancel out the limitations of each model. A modeling of differentially pressurized PEMFC should consider both diffusive and convective water transport by adding diffusive, convective, and EOD terms, which result in a combination of diffusive and hydraulic models [114,125–128]. Hence, the new equation becomes [114]:

$$N_w = n_d \frac{i_2}{F} - \frac{k}{\mu} \nabla p_G - D_w \nabla c_w \tag{18}$$

4.1.3. Combined Model (Microscopic and Macroscopic)

Only one component of the microscopic model, namely the water content λ, can be successfully integrated into the macroscopic model [100]. Water is absorbed by the dry membrane to solvate the acid groups, and as the water content increases, the water droplets agglomerate and form interconnecting channels. Figure 3 displays the changes in the membrane structure depending on the water content, where the gray, black, and light gray areas represent the fluorocarbon matrix, polymer side chain, and liquid water, respectively, whereas the dotted line denotes the collapsed channel. When the amount of water exceeds the percolation threshold (λ = 2), a complete cluster channel is formed, as illustrated in Figure 3c. The formation of water channels indicates whether the water at the boundary is in the vapor (Figure 3c) or liquid (Figure 3d) state, termed VE or LE membranes, respectively. In the VE case, the channels are not well connected, whereas in the LE case, they are filled with water, expanded, and connected. These two distinctive structures form the foundation of the two macroscopic models: diffusive and hydraulic.

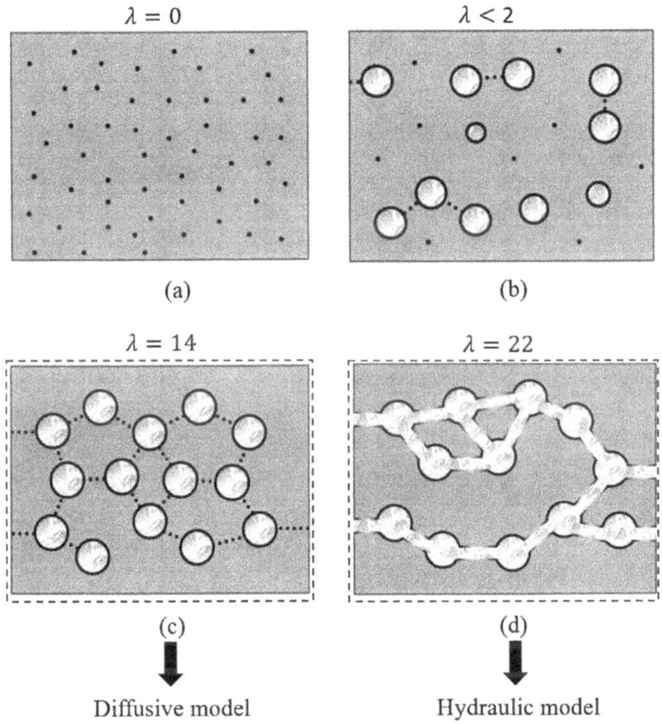

Figure 3. Transformation of membrane structure due to change in water content λ (**a–d**) reproduced with permission [100]; 2003 The Electrochemical Society, Inc.

4.1.4. Issues Related to State-of-Art Membrane Modeling

When it comes to water transport, the chemical potential model remains the most comprehensive solution among the membrane modeling approaches. However, a combined diffusive and hydraulic model with the addition of EOD remains a better choice because it considers most of the aspects of experimental findings [116]. Regardless of various options, current membrane modeling has the following issues:

Despite the technological advancement of membrane materials, the majority of research papers on the numerical investigation of PEMFC still rely on the springer model [96] and the Weber–Newmann model [124], which are based on experimental work [96,129–131], dating back to 1998 [92].

Electroosmotic drag (EOD) could have a different value depending on the electrodes (anode and cathode) [132]. As a result, the standard practice is to consider EOD as a source term on the cathode side and a sink term on the anode side, according to previous research [132,133].

Even though the EOD increases with the water content, λ based on a number of correlations observed from the experiment [96,134–136] as mentioned in Table 3, the differences among them introduce certain uncertainties during modeling [35].

Due to small pressure imbalances and low membrane permeability, the effect of cross-diffusion of reactants through the membrane is usually neglected [35], which is not true in some cases. The results indicate that this effect is not always insignificant, particularly when calculating cell efficiency. In some instances, the cross-over effect has been observed to impact cell efficiency by approximately 2% [137].

The recent introduction of reinforced membrane (typical thickness of 25–35 μm) in PEMFC helps to mitigate membrane degradation by reducing resistance and improving water management without compromising mechanical strength [138]. However, the composite nature of this type of membrane and the effect of its conductivity, diffusivity, water intake, and mechanical coupling in MEA to transport phenomena are not well understood [92].

Previous research [139,140] found that the convection type of transport, by considering the microporous membrane and its percolation phenomena, could give the membrane resistance close to the experimental value in comparison to the model that utilizes diffusion transport. However, convection models are not widely used due to their underestimation of dehydration.

Modeling approaches are still not considering the effect of free radical scavengers [141] or different additives [142] to alleviate the chemical degradation of modern membranes.

The effect of mechanical and chemical degradation [143] of membranes on the performance of PEMFC was not considered in the modeling.

4.1.5. ML in the Field of Membrane

Previous researchers are trying to use ML to design and optimize membranes [44,144–151], predict membrane properties [36,43,152], diagnose membrane conditions [153,154], and prevent membrane degradation [155–157]. Cho et al. [43] collected data from a 1.2 kW PEMFC in a MATLAB/Simulink environment and used that data to train a nonlinear autoregressive network (NARX) with Bayesian optimization to predict the voltage, temperature, and membrane water content (Max. error of MSE 2.14×10^{-4}). Figure 4a shows the architecture of the proposed NARX network. Huo et al. [154] utilized a genetic algorithm-based back propagation (GA-BP) neural network to predict membrane hydration in PEMFC. A dynamic model of PEMFC was used, which contains cathode mass flow, anode mass flow, membrane water content, and a stack voltage sub-model. Figure 4b depicts the GA-BP neural network, showing a better prediction of hydration in comparison to a least-squares support vector machine (LS-SVM) regarding mean square error (MSE), mean absolute error (MAE), and root mean square error (RMSE). In order to reduce the extensive work, resources, and time, Huo et al. [145] proposed a random forest feature selection process to identify the important features as input parameters, as verified by the previous studies. Subsequently, a convolutional neural network (CNN) with batch normalization and dropout methods is implemented to predict the performance of PEMFC, such as the I–V curve, as can be seen in Figure 4c. It is important to note that, among the important factors, the membrane contains hot press time (HPT), hot press pressure (HPP), hot press temperature (HPT), which is required for manufacturing, MEA, and the thickness of the membrane. Gu et al. [155] established a long short-term memory (LSTM) network model to diagnose the flooding fault of a bench test of a 92 kW vehicle with a fuel cell system, which is essential for the identification of membrane hydration conditions. Once the LSTM diagnosis network was built, it could effectively predict the condition of fuel cells instead of relying on a large number of sensors, which can effectively reduce the cost of an efficient system. Figure 4d shows the input parameters of the LSTM network and the results of the network in comparison to the diagnosis from actual THDA (total harmonic distortion analysis) diagnostic equipment.

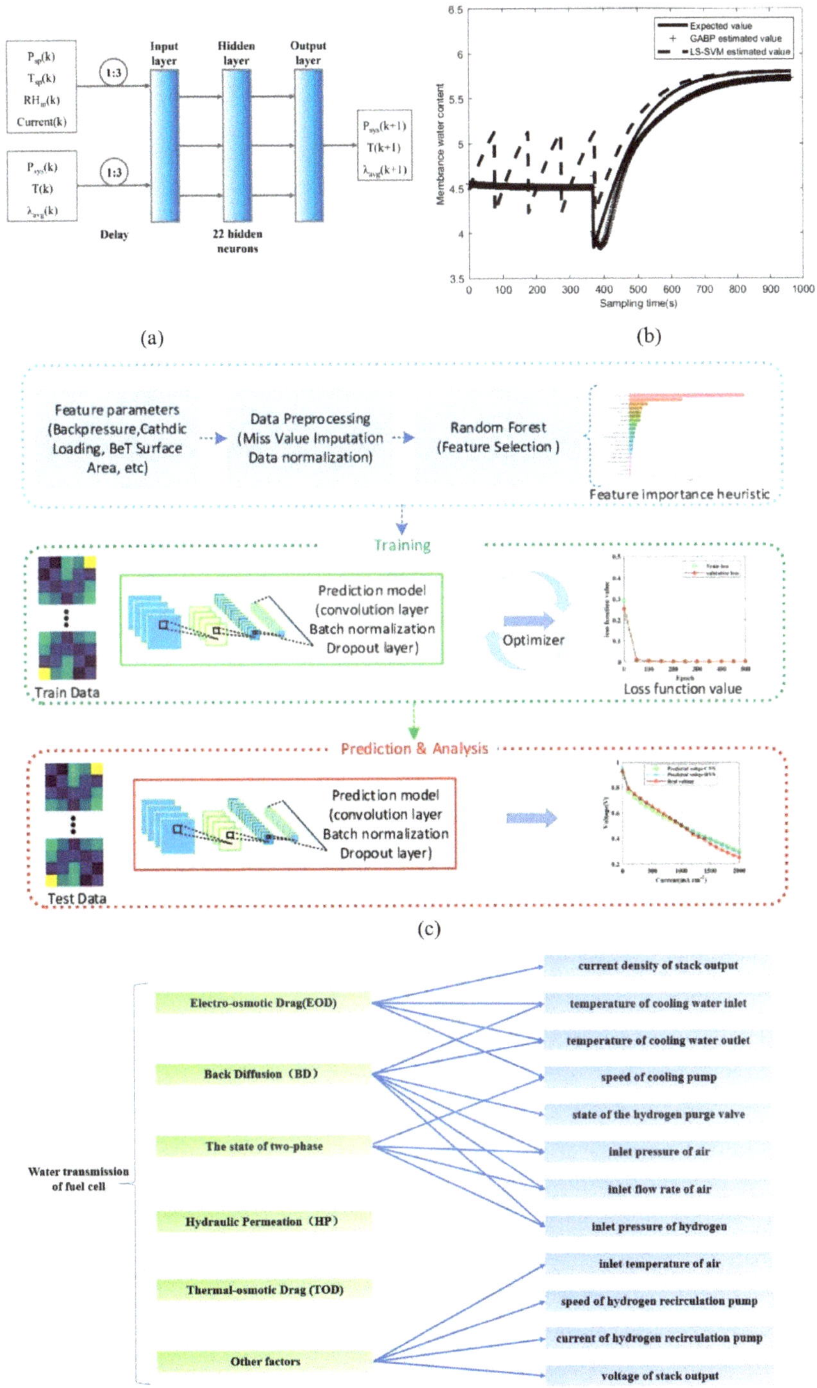

Figure 4. *Cont.*

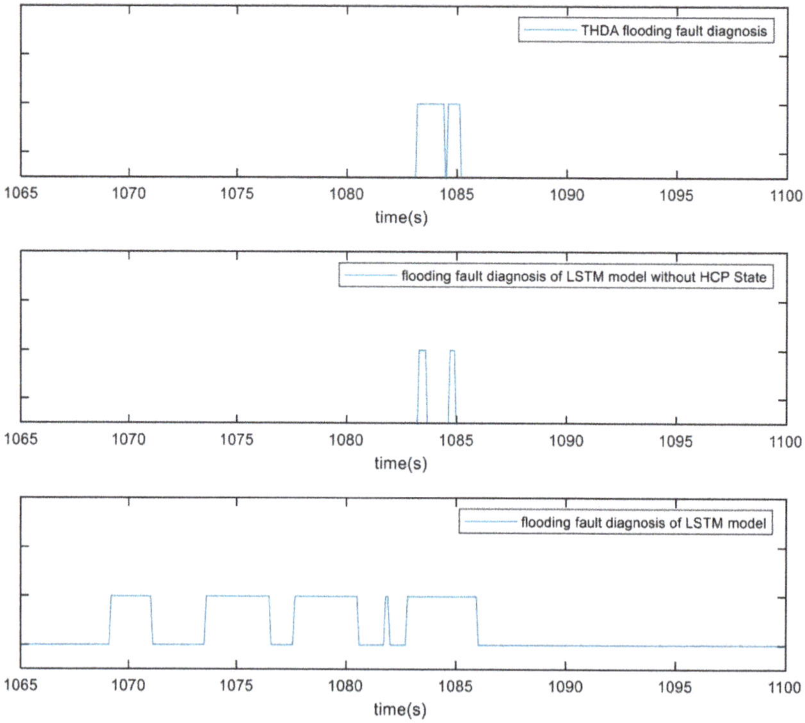

(e)

Figure 4. (**a**) The architecture of trained NARX network for membrane water content, λ prediction; reproduced with permission [43]; 2012 Elsevier Ltd. (**b**) Simulink dynamic model with GA-BP to estimate real-time water content of membrane; reproduced with permission [154]; 2023 MDPI (**c**) performance prediction model by CNN, where membrane thickness, hot press time, and pressure for making catalyst coated membrane (CCM) are some of the input parameters; reproduced with permission [145]; 2021 Elsevier Ltd.; (**d**) input vectors of LSTM that are responsible for PEMFC water transport across the membrane; and (**e**) comparison of flooding fault diagnosis by experiment and LSTM network; reproduced with permission [155]; 2021 Elsevier Ltd.

4.1.6. Integration of ML in Membrane Modeling

A potential integration of ML and CFD modeling is shown in Figure 5. For membrane modeling, a large number of datasets could be obtained from experiments and MD simulations. That includes boundary conditions from the real test, such as temperature, pressure, RH, etc.; water content from the MD simulation; and variables such as temperature, pressure, and time to manufacture MEA by the hot pressing method [158] as input. In addition to that, the ML method was used to optimize the input parameters and predict the corresponding output parameters by considering the degradation of membrane properties. Finally, the governing equations are modified depending on the change in properties due to more precise input parameters, the change in output due to material degradation, and membrane modeling.

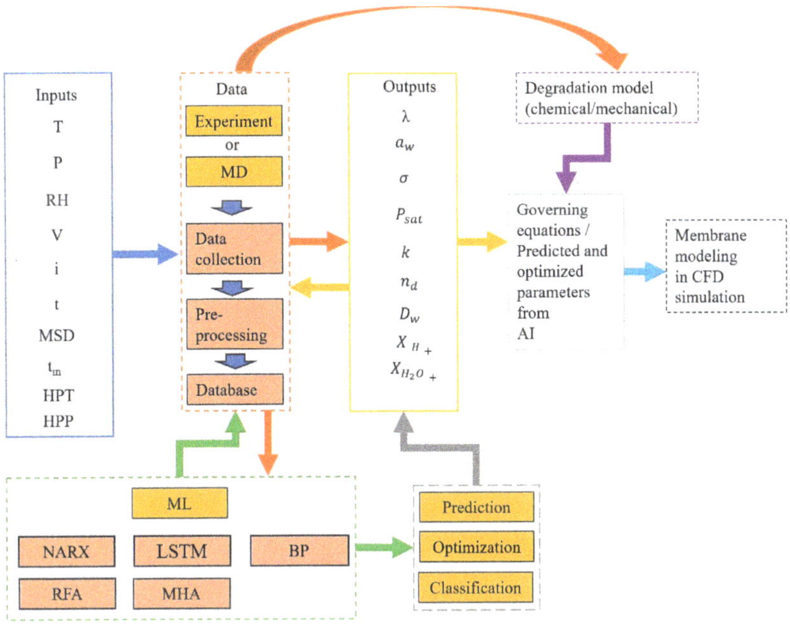

Figure 5. Framework of the PEMFC membrane model integrated with ML. (MSD—mean square displacement, HPT—hot press time (min), HPP—hot press pressure, MHA—meta-heuristic algorithms).

4.2. Modeling of Catalyst Layer (CL)

The CL is the location where the HOR and ORR occur (Equations (1) and (2)). It is often the thinnest component [159] (5 to 30 μm) in the PEMFC because of its high material cost, yet it handles the most complex transport phenomena [160]. Usually, fine particles of Pt and its alloys are sprayed on high-surface-area carbon in the active porous layer of the electrode to minimize Pt loading (0.1–0.4 mg·cm^{-2}) [161]. For effective usage, the CL must be positioned to form a triple-phase reaction interface at the intersection of e^-, H^+, and reactive gas transfer channels. The different CL models are as follows: interface, microscopic, single pore, and simple macrohomogeneous.

4.2.1. Interface Model

Among the CL models, the simplest model treats the CL as an interface between the GDL and membrane. For this purpose, the CL is assumed to be an infinitely thin layer, and its physical structure is completely ignored. Although computationally economical and efficient, this model is not appropriate when the main focus is on the CL [116]. Therefore, the CL can be considered the location where H_2 and O_2 are consumed and water is produced. The generation and consumption terms are expressed using Faraday's law [116]:

$$N_{i,k} = \sum_{h} S_{i,k,h} \frac{i_{h,1-k}}{n_h F} \tag{19}$$

Here, $S_{i,k,h}$ is the stoichiometric coefficient of species i in phase k; $i_{h,1-k}$ is the normal interfacial current transferred per unit interfacial area across the interface between the electronically conducting phase and phase k owing to the electron-transfer reaction h, and it is positive in the anodic direction.

For the aforementioned model, detailed information on the potential is not essential. In previous research [99,162–164] on water management, a similar technique was used by quantifying it with the current density. Both the resultant water on the cathode side and the electroosmotically dragged water are proportional to the current density [162]. In addition to Equation (19), researchers [125,165–168] have utilized a slightly more sophisticated approach by fitting a polarization curve to a single expression as follows [125]:

$$E = U' - R'i - \frac{RT}{\alpha_c F}\ln(i) + \frac{RT}{\alpha_c F}\ln\left(1 - \frac{i}{i_{lim}}\right) \quad (20)$$

where E is the cell potential and i is the superficial current density through the MEA, whereas α_c is the cathodic transfer coefficient.

To obtain a more sophisticated interface model, reaction kinetics such as the Tafel equation have been incorporated [169,170] with the Butler–Volmer equation [171–173], which can be expressed as:

$$i_{h,k-p} = i_{0_h}\left[\prod_{i}^{a}\frac{p_i}{p_i^{ref}}exp^{\left(\frac{\alpha_a F}{RT}(\Phi_k-\Phi_p-U_h^{ref})\right)} - \prod_{i}^{c}\frac{p_i}{p_i^{ref}}exp^{\left(\frac{-\alpha_c F}{RT}(\Phi_k-\Phi_p-U_h^{ref})\right)}\right] \quad (21)$$

New models have utilized a kinetic expression that considers multidimensional aspects, which is an improvement over the previous model (Equation (10)). However, this model (Equation (21)) is based on the assumption that the surface overpotential and concentration of reactant gases are uniform in the CL, and it uses the overpotential of the kinetic expression as a fitting parameter. Despite its simplicity and efficient calculation, this model is still limited because it does not consider multiple aspects of the CL that are not actually uniform. This model is not suitable for optimizing CL fabrication in terms of the effects of Pt loading, compressive load, and pore size.

4.2.2. Microscopic and Single Pore Models

Most of the earliest CL models were microscopic and single-pore models that were easily solvable. These models were utilized for the CL of phosphoric acid fuel cells with Teflon-coated pores for gas diffusion, whereas the rest of the area was flooded with a liquid electrolyte. Although a detailed microstructure is required for this model, effective values such as diffusivity and conductivity are not required because they are assumed to be homogeneous throughout the microstructure. Single-pore models are divided into two types according to their nature.

In the first model, known as the gas pore model, the pores are considered straight and cylindrical with a specific radius [174–177]. In addition, these pores extend to the length of the CL, and reactions occur on the surface.

Among the single-pore models, the flooded agglomerate model shows potential for real experimental data. This model still uses gas pores, although some of them are filled with electrolyte and catalyst [178–183]. In those filled pores, reactions, diffusion, and migration occur. Consequently, this model works better than the gas pore model because it covers a larger area.

However, pores with different porosities and tortuosities exist in real cases, and the electrolyte in PEMFCs is solid, unlike in phosphoric acid fuel cells, which are not supposed to penetrate the pores. Although the single-pore model is adequate, it is not suitable for simulating the CL in detail.

In addition to the two aforementioned models, a spherical agglomerate model, which considers special agglomerate structures in three-dimensional (3D) hexagonal arrays, was introduced by Antoine et al. [184]. Among the agglomerates, either gas pores or regions are flooded with electrolyte [185,186]. This model examines the interactions that depend on the placement of these agglomerates. In addition to Ohm's and Fick's laws, the reaction kinetic equation can be solved by employing this model.

The concentration contours around the electrocatalyst particles and can be validated by experimental current density trends. In contrast to the interface model, this model can depict the occurrence of HOR near the membrane and the effect of the packing and structure of agglomerate particles on the overall efficiency of the CL.

4.2.3. Simple Macrohomogeneous Model

Introduced initially by Tiedemann and Newman [187], the macrohomogeneous model neglects exact geometric details by treating the CL structure as a randomly arranged porous domain. Only a few variables, such as porosity, volume fraction, and surface area per volume, are used to represent the structure of the CL. Moreover, the transport properties should be averaged over the CL volume. Although this model does not include the detailed structure of the domain, it shares many similar theoretical expressions with the single-pore model. Two of the main length scales of this model are explained here [116].

Porous Electrode Model

Although it is based on the assumption of a single-pore model, the porous electrode model calculates the overall reaction without considering the detailed structure of the computational domain. In this model, the concentration and potential are assumed to be uniformly distributed in the agglomerates, indicating that the main interaction does not affect the agglomerates. The equations governing the simple porous electrode models are listed in Table 6. Although the thickness of the CLs serves as the characteristic length scale, there are some variations in the treatment of the simple porous electrode model.

Table 6. Important variables and governing equations for the CL [116].

Variable	Porous Electrode Model	Agglomerate Model
Overall liquid water flux, NL	$\frac{\partial \epsilon_k c_{i,k}}{\partial t} = -\nabla \cdot N_{i,k} - \sum_h a_{l,k} s_{i,k,h} \frac{i_{h,1-k}}{n_h F} + \sum_I S_{i,k,l} \sum_{p \neq k} a_{k,p} r_{i,k-p} + \sum_g s_{i,k,g} \epsilon_k R_{g,k}$	
Overall membrane water flux, NW	Equation above.	
Gas-phase component flux, NG,i	Equation above.	
Gas-phase component partial pressure, pG,i	$\nabla x_i = -\frac{N_i}{c_T D_{K_i}^{eff}} + \sum_{j \neq i} \frac{x_i N_j - x_j N_i}{c_T D_{i,j}^{eff}}$	
Liquid pressure, pL	$N_{w,L} = -\frac{k}{V_w \mu} \nabla p_L = -\frac{k}{V_w \mu} (\nabla p_c + \nabla p_G) = -\frac{k}{V_w \mu} \nabla p_c$	
Membrane water chemical potential, μw	Equation (14)/Equation (16)	
Electronic-phase current density, i1	$i_1 = -\sigma_0 \epsilon_1^{1.5} \nabla \Phi_1$	
Membrane current density, i2	Equation (9)/Equation (13)	
Electronic-phase potential, Φ1	$\sum_k \nabla \cdot i_k = 0$	
Membrane potential, ΦL	$i_s = -\kappa_s^{eff} \nabla \phi_s$	$\nabla \cdot i_s = 4F \frac{P_{O_2}}{H} \left(\frac{1}{E_r k_c (1-\varepsilon_{cat})} + \frac{(r_{agg}+\delta)\delta}{a_{agg} r_{agg} D} \right)^{-1}$ Where $E_r = \frac{1}{\Phi_L} \left(\frac{1}{\tanh(3\Phi_L)} - \frac{1}{3\Phi_L} \right)$

Table 6. *Cont.*

Variable	Porous Electrode Model	Agglomerate Model
Temperature, T	$\sum_k \rho_k \hat{C}_{P_k}\left(\frac{\partial T}{\partial t} + v_k \cdot \nabla T\right) = \sum_k h_{k,ext} a_{k,ext}(T_{ext} - T) + \nabla \cdot \left(k_T^{eff} \nabla T\right) - \sum_k \sum_i J_{i,k} \cdot \nabla \overline{H}_{i,k} + \sum_{1-k} \sum_h a_{1,k} i_{h,1-k}(\eta_{s_{ORR,1-k}} + \Pi_h) + \Delta H_{evap} r_{evap}$ Or $Q = \sum_{k-p} \sum_h a_{k,p} i_{h,k-p}(U_{H_h} - V)$	
Total gas pressure, p_G	$v_G = -\frac{k_G}{\mu_G} \nabla p_G$	
Liquid saturation, S	$\varepsilon_G = \varepsilon_0(1 - S)$	

In the first variation, the CLs are integrated and included as boundary conditions when calculating the cell potential or current density [128,188,189]. This treatment is similar to that of the interface models, except that it considers potential drops for the matrix and solution phases. The rationale for the integration is that it does not result in a loss of numerical accuracy and provides numerical stability and easier convergence, particularly when the reaction distribution is mainly uniform.

The second variation uses a simple porous electrode modeling approach, considering a CL of finite thickness and using the equation in Table 6. Although most of these models allow the diffusion mass transfer of reactant gases [112,118,190–192], in some previous studies, the concentration of the reactant gases has been assumed to be uniform throughout the CL [193–195]. In the final variation of the porous electrode model, rather than gas diffusion in the CL, the reactant gases are assumed to be dissolved in the electrolyte and transported by diffusion and reaction [196–199]. This model is closer to the thin film model, which does not consider gas pores [200,201]. The governing equations in Table 6 are well suited for solving this model, except that a concentration term should be added to the kinetic expressions instead of the partial pressure. Accordingly, a mass transport equation for the reactant or product gas should be added to account for diffusion effects on the membrane or water.

Agglomerate Model

This model considers the radius of the agglomerate as a characteristic length scale in addition to the CL thickness. Most of these agglomerates are assumed to be either large spheres, a combination of Nafion, carbon, and Pt particles [202–204], or small spheres, a void consisting of carbon and Pt particles filled with liquid water [205,206]. In the simple homogeneous agglomerate model, the main effects are assumed to occur at the agglomerate length scale. Hence, the reaction rate distribution, that is, gas concentration and surface overpotential, is uniform throughout the thickness of the CL. It is assumed that oxygen diffuses through the gas pores, dissolves into agglomerates through the electrolyte/water, and then reaches the reaction site by diffusion again. The mathematical expressions are similar to those of the microhomogeneous model indicated in Table 4, except that spherical and cylindrical coordinates are added for the gradients. Another agglomerate model, known as the embedded macrohomogeneous model, considers both the aforementioned macrohomogeneous model and reaction and overpotential distribution in the agglomerates that are disregarded in the simple homogeneous agglomerate model. These effects are essential when CL hydration and dehydration are considered. Figure 6 shows the distinctive features of the three catalyst models.

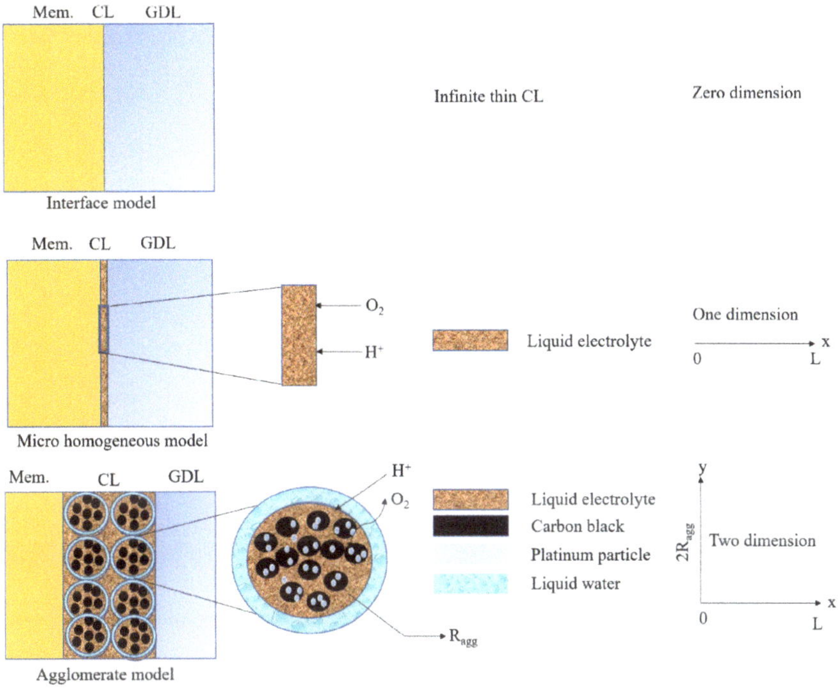

Figure 6. Distinctive features of different CL models.

In addition to the equations presented in Table 6, in this model, either a mass transfer term [207–209] (1st approach) or an effectiveness factor incorporating the agglomerate equation [210–212] (2nd approach) is used. Previous studies have reported that the agglomerate model is better than the macrohomogeneous model because it uses more empirically controlled parameters [204,213]. In particular, at higher current densities, the agglomerate model can predict the voltage drop, unlike the macrohomogeneous model. The differences between the equations of the macrohomogeneous and agglomerate models are displayed in Table 6.

Electrochemical Kinetics Equation of CL

The CL is the location where the HOR and ORR occur, consuming hydrogen and oxygen and producing water. Therefore, Faraday's law (Equation (19)), in combination with the molecular weight, is used to determine the rate of H_2, O_2, and H_2O consumption/production. Table 7 presents the source and sink terms related to the electrochemical reactions that are proportional to the exchange current density. The molecular weights of hydrogen M_{H_2} and oxygen M_{O_2} are 2 g·mol^{-1} and 32 g·mol^{-1}, respectively. Although some simulation studies do not consider the back diffusion of water from the cathode to the anode side [214,215], an additional electroosmotic drag coefficient $n_d = \frac{2.5\lambda}{22}$ is included in both the anode and cathode CL water species source terms for simplicity. In more sophisticated approaches, additional mass transfer rates of the vapor-to-liquid water phase change S_{v-l} and membrane absorption/desorption rates S_{n-v} are considered, as listed in Table 7. The Butler–Volmer equation is used to determine the electrochemical reaction rates for these source terms, as summarized in Table 8.

Table 7. Various source terms for species consumption and production from the literature.

	Anode Side			Cathode Side			Ref.
Source Term	S_{H2}	$S_{H2O,liq}$	$S_{H2O,vap}$	S_{O2}	$S_{H2O,liq}$	$S_{H2O,vap}$	
1	$-\left(\frac{M_{H_2}}{2F}\right)I_a$	0	0	$-\left(\frac{M_{O_2}}{4F}\right)I_c$	$\left(\frac{M_{H_2O}}{2F}\right)I_c$	0	[214,215]
2	$-\left(\frac{M_{H_2}}{2F}\right)I_a$	$-\left(\frac{M_{H_2O}}{2F}\right)n_d I_a$	0	$-\left(\frac{M_{O_2}}{4F}\right)I_c$	$\left(\frac{M_{H_2O}}{2F}\right)I_c + \left(\frac{M_{H_2O}}{F}\right)n_d I_c$	0	[123,216]
3	$-\left(\frac{M_{H_2}}{2F}\right)I_a$	S_{v-l}	$-S_{v-l} + S_{n-v}M_{H_2O}$	$-\left(\frac{M_{O_2}}{4F}\right)I_c$	S_{v-l}	$-S_{v-l} + S_{n-v}M_{H_2O}$	[123,216]
4	$-\left(\frac{M_{H_2}}{2F}\right)I_a - S_{v-l} + S_{n-v}$	S_{v-l}	0	$-\left(\frac{M_{O_2}}{4F}\right)I_c - S_{v-l} + S_{n-v}$	$S_{v-l} + S_{n-v}M_{H_2O}$	0	[217]

Table 8. Exchange current densities from the literature.

Anode Exchange Current Density, I_a	Cathode Exchange Current Density, I_c	Ref.
$-a_v I_{0,a}^{ref}\left[\frac{(\alpha_a+\alpha_c)F}{RT}\eta_{act}^{a}\right]$	$a_v I_{0,c}^{ref}\left(\frac{C_{O_2}}{C_{O_2}^{ref}}\right)e^{\left(-\frac{\alpha_c F}{RT}\eta_{act}^{c}\right)}$	[218]
$(1-s)I_{0.a}^{ref}\left(\frac{(1-s)\varepsilon C_{H_2}}{C_{H_2}^{ref}}\right)^{0.5}\left(e^{\left(\frac{2F\alpha_a}{RT}\eta_{act}^{a}\right)} - e^{\left(\frac{2F\alpha_c}{RT}\eta_{act}^{c}\right)}\right)$	$(1-s)I_{0.a}^{ref}\left(\frac{(1-s)\varepsilon C_{O_2}}{C_{O_2}^{ref}}\right)^{3.0} - \left(e^{\left(\frac{4F\alpha_a}{RT}\eta_{act}^{a}\right)} + e^{\left(\frac{4F\alpha_c}{RT}\eta_{act}^{c}\right)}\right)$	[123,219]
$\zeta_a I_{0.a}^{ref}\left(\frac{C_{H_2}}{C_{H_2}^{ref}}\right)^{0.5}\left(e^{\left(\frac{F\alpha_a}{RT}\eta_{act}^{a}\right)} - e^{\left(\frac{F\alpha_c}{RT}\eta_{act}^{c}\right)}\right)$	$\zeta_c I_{0.c}^{ref}\left(\frac{C_{O_2}}{C_{O_2}^{ref}}\right)^{1.0}\left(-e^{\left(\frac{F\alpha_a}{RT}\eta_{act}^{a}\right)} + e^{\left(\frac{F\alpha_c}{RT}\eta_{act}^{c}\right)}\right)$	[35]
$I_{0.a}^{ref}e^{\left(-1400\left(\frac{1}{T}-\frac{1}{298.15}\right)\right)}$	$J_{0.c}^{ref}e^{\left(-7900\left(\frac{1}{T}-\frac{1}{298.15}\right)\right)}$	[214,220]
$I_{0.a}^{ref}\left(\frac{P_{H_2}}{P_{H_2}^{ref}}\right)^{0.5}\left(e^{\left(\frac{F\alpha_a}{RT}\eta_{act}^{a}\right)} - e^{\left(\frac{F\alpha_c}{RT}\eta_{act}^{c}\right)}\right)$	$4F\frac{P_{O_2}}{H_{O_2}^{naf}}\frac{1}{\frac{\delta_{Naf}}{a_{ratio}D_{O_2}^{Naf}}+\frac{\delta_w}{a_{ratio}D_{O_2}^{Water}}\frac{H_{O_2}^{Water}}{H_{O_2}^{Naf}}+\frac{1}{\xi\|\eta}}$	[221]

4.2.4. Heat Transfer in the CL

Heat transfer in the CL is a critical aspect for accurately predicting the CL model because it affects the electrochemical reaction rate and species transport. Compared to other components of the PEMFC, the CL has the most complex thermal process, which involves the conduction of solid material parts and convection heat transfer during species transport. Additionally, the porous structure of the CL complicates the heat transfer between the solid and gas phases. The following two expressions are applied, depending on the absence or presence of additional latent heat of condensation and absorption/desorption [216,219]:

$$CL: S_{T,a} = \underbrace{I_a|\eta_{act}^{a}| + I_a\frac{\Delta S_a T}{2F}}_{Electrochemical\ reaction} + \underbrace{\|\nabla\varphi_e\|^2\kappa_e^{eff}}_{Ohmic}; \tag{22}$$

$$ACL: S_{T,an} = \underbrace{I_{an}|\eta_{act}^{an}| + I_{an}\frac{\Delta S_{an} T}{2F}}_{Electrochemical\ reaction} + \underbrace{\|\nabla\varphi_e\|^2\kappa_e^{eff} + \|\nabla\varphi_{ion}\|^2\kappa_{ion}^{eff}}_{Ohmic} + \underbrace{(S_{v-l}+S_{n-v})h_{evap}}_{Latent\ heat} \tag{23}$$

$$CCL: S_{T,cat} = I_{cat}|\eta_{act}^{cat}| + \|\nabla\varphi_e\|^2\kappa_e^{eff} + \|\nabla\varphi_{ion}\|^2\kappa_{ion}^{eff} + I_{cat}\frac{\Delta S_{cat} T}{2F} + (S_{v-l}+S_{n-v})h_{evap} \tag{24}$$

4.2.5. Issues Related to State-of-Art CL Modeling

When it comes to the modeling of CL of PEMFC, there is no proper experimental validation to clarify whether the governing equations are working or the parameters that are used are appropriate [222]. Regardless of experimental validation, CL modeling has the following issues:

Various CL modeling approaches were successfully utilized by the researchers with experimental validations, even though they had certain limitations. Interface modeling approaches suffer from overestimation of the current density, whereas macro-homogeneous modeling approaches are not suitable for the CL of complex structures with different materials [35]. Till now, the agglomerate model seems promising as it combines both the structural distribution and composition of CL material [35], including more parameters to fit experimental data [116]. Sui et al. [160] found that the agglomerate model had a better prediction of the polarization curve in comparison to other catalyst models. However, the agglomerated model still considers only one length scale and fails to fully consider the reaction distribution and proton migration across the CL [116].

Most of the numerical simulations related to PEMFC validated their simulation results with the polarization curve obtained from the experiment [223,224] though the majority of the important parameters such as reaction kinetics, ohmic resistance, and voltage drop off due to concentration loss depend on CL. Hence, the simulation results are not satisfactory, even though the validation was performed with an experimental polarization curve [223]. In order to solve this problem, validation and characterization should be performed on the same length scale, considering the real microstructure of the CL [160].

Contemporary simulation researchers heavily depend on the Butler–Volmer (BM) equation [171,172] for electrochemical reactions in CL (Table 8). In most simulation cases, the value of the reference exchange current density was measured from experiments [225], from previous solutions [215,217,226], and assumed [220] to match the polarization curve. Previous research [227,228] collected 10 papers and found 6 different values of reference exchange current densities and 9 different values of transfer coefficients. Based on their study, they found that the only polarization curves were not sufficient for the validation, as the two groups of parameters resulted in identical polarization curves. Furthermore, a curve-fitted exchange current density [229–231] (function of temperature) and variable exchange current density [232] (function of RH) were also proposed for CL. Dickinson et al. [233] heavily criticized the BM equation for its excessive parameterization, which makes PEMFC modeling complex and hampers experimental validation.

In most cases, the porosity, tortuosity, and contact angles (hydrophobic) of the pore walls of CL are considered homogenous, which is not real. A heterogeneous porous structure of CL could give a better prediction of PEMFC regarding mass transport and chemical reactions [234–236]. As a result, the contact angle between pore walls also changes depending on the size, Pt loading, C (support), and ionomer (Nafion/other perforluorosulfonic acid) [237].

Degradation of CL due to degeneration of the Pt-based catalyst, C support, and Nafion ionomer results in limiting the electrochemically active area, subsequently lowering the performance of fuel cells [238]. During the unsteady or lifetime of fuel cell modeling, modeling degradation is one challenge that researchers need to overcome. Franco et al. [239,240] proposed a mechanistic transient model, considering cathodic potential sensitivity to the boundary conditions due to catalyst layer aging. Later, this model is used to investigate the effect of CO contamination on CL and cell degradation [241,242]. Nevertheless, the majority of previous degradation models do not take into account the change in geometry and influence of material degradation on the local operating conditions of PEMFC [24] and provide an adequate model to predict mass transport in CL [243].

4.2.6. ML in the Field of CL

Machine learning approaches have been used in CL for feature extraction [244], optimization [58,147,245,246], predicting performance [247], and degradation of CL on PEMFC [156,248–252]. Wang et al. [244] implemented deep learning super-resolution and multi-label segmentation to process the images from X-ray micro-computed tomography, followed by LBM with multi-relaxation time (MRT) for water management modeling. Figure 7a shows the multilabel segmented image from super-resolution images for feature extraction such as void space, MP layer, membrane, CL, perpendicular, and parallel layers. To determine key parameters, Ding et al. [147] collected 64 high-quality journals related to the PEFMC experiment from 2010–2020, containing 10,000 datapoints with 140 IV curves. An ANN model was used where catalyst physical, chemical, and test operating conditions were used as inputs and voltages at different current densities were used as outputs (Figure 7b). The final algorithm can successfully predict the performance of the real test with a great accuracy of $R^2 = 0.99$. In addition to that, the trained model shows the best performance for predicting maximum power density on unexplored polarization curves with 26 inputs. Lou et al. [253] utilized an ML-assisted model that has two functions: (1) Quantitative sensitive analysis and (2) multi-objective optimization, in alignment with the main benefits of interpretability and prompt prediction, respectively, as can be seen in Figure 7c. Among the datasets, various structural parameters of cathode CL, including Pt loading, ratio of Pt to carbon-supported Pt, ionomer to carbon-supported Pt, agglomerate radius, C particle radius, Pt nanoparticle radius, pore diameters, thickness, and surface tension, were used. Four critical features were identified, and peak power density and limiting current density were increased to 9.96% and 10.47%, respectively, by optimizing the catalyst ratio and agglomeration. Elçiçek et al. [247] utilize a multilayer perceptron ANN model (Figure 7d), where they used reaction temperature, pH, and reaction duration as inputs to predict electrochemical active surface area (EASA) and reduction of Pt. The MLP-ANN model exhibited superior performance, standing out as the best among various machine learning algorithms when considering accuracy, overall performance, and generalization capabilities in comparison to SVR and RF. Moreover, the suggested model proves to be effective for optimizing electrocatalyst performance and prediction modeling, with an impressive R^2 of 99.99%. Figure 7e shows a framework for quantitative analysis and accurate prediction, proposed by Yao et al. [246], to improve the design efficiency of CL. A combination of the response surface method (RSM) and ANN is utilized to investigate the effect of CL composition on the performance of PEMFC regarding current density, thermal, and water management. Among the compositions, the volume fraction of dry ionomers has ben proven to be the most sensitive parameter. Data-driven ML has also been employed to predict performance reductions resulting from PEMFC degradation [248,251,252,254], although it cannot be clearly pinpointed whether it is due to membrane, CL, or GDL. Pt loss and reorganization are critical factors resulting from the high temperature, humidity, and load cycling [255]. Considering this, Ma et al. [256] proposed a grid LSTM-based recurrent neural network (RNN) (Figure 7f, top) to avoid vanishing gradients or investigating problems during training and effectively predict both short-term (Figure 7f, down) and long-term voltage degradation of PEMFC.

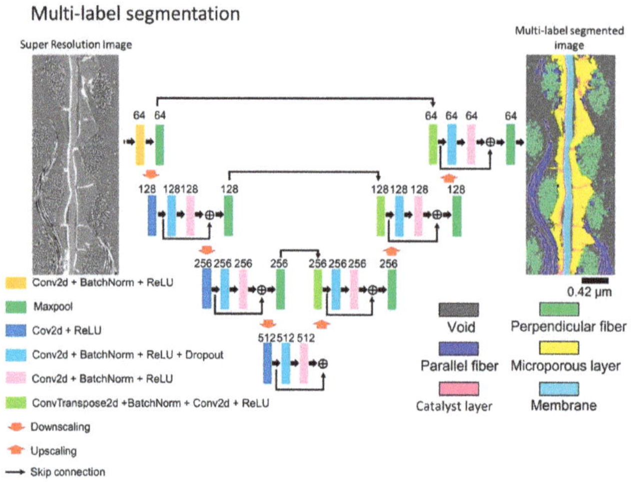

(a)

Figure 7. *Cont.*

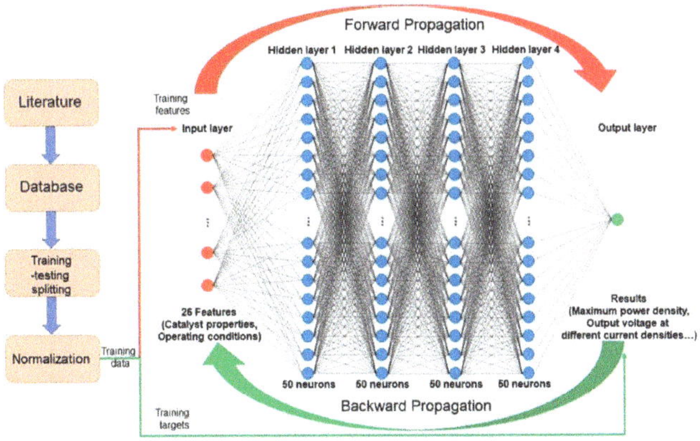

(b)

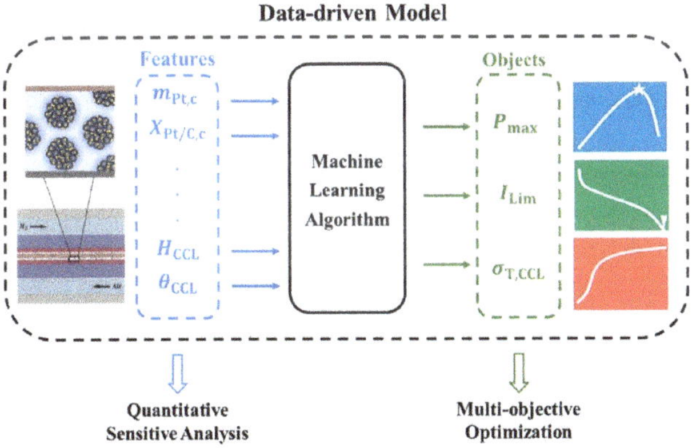

(c)

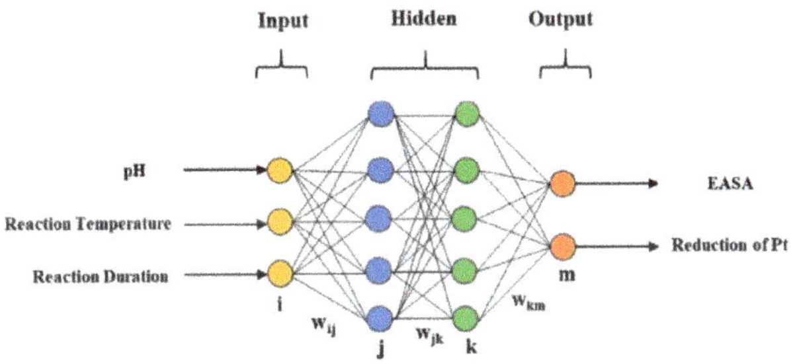

(d)

Figure 7. *Cont.*

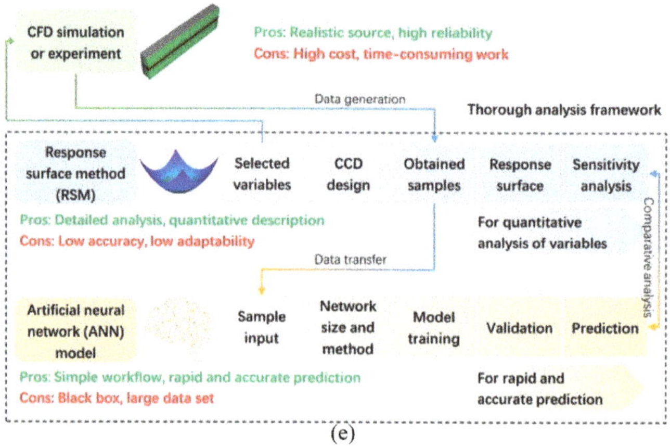

(e)

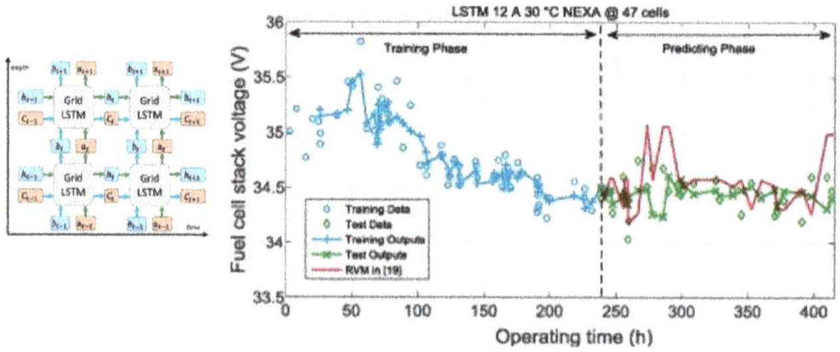

(f)

Figure 7. (**a**) Network architecture of the U-ResNet (bottom), for image feature extraction and useful feature decoding for various components of PEMFC, including CL; reproduced with permission [244]; 2023 Springer Nature, (**b**) ANN architecture for PEMFC performance prediction where CL properties are some of the input parameters; reproduced with permission [147]; 2020 John Wiley and Sons, (**c**) prediction of the CCL performance of a PEMFC by data-driven ML model where various structure parameters of CCL are utilized; reproduced with permission [253]; 2022 Elsevier Ltd., (**d**) architecture for ANN model for predicting electrocatalyst performance regarding EASA and reduction of Pt in CL surface where columns of circular nodes represents the layers for inputs, hidden and outputs; reproduced with permission [247]; 2022 John Wiley and Sons; (**e**) framework for CL parameter optimization, combining RSM and ANN model; reproduced with permission [245]; 2023 Elsevier Ltd. and (**f**) the structure of the G-LSTM with RNN for predicting voltage degradation due to carbon corrosion, Pt loss of CL, and membrane degradation.; reproduced with permission [256]; 2018 Elsevier Ltd.

4.2.7. Integration of ML in CL Modeling

Likewise, for membrane modeling, Figure 8 shows the framework of ML/CFD hybridization for modeling of CL. In the case of CL, image segmentation through CNN has proven to be a useful tool to extract features of complex porous CL structures such as porosity, tortuosity, and permeability. Instead of homogeneous approximations of material properties, CFD simulation can use more real-time heterogeneous properties of CL. Various ML methods, such as SVM, LSTM, and BP, can further optimize the properties based on experimental validation and improve the database. In addition to that, changes in current densities, conductivity, water, and thermal management due to mechanical and chemical degradation of CL can be modified in the governing equations of transport mechanisms of CFD simulation for more effective prediction.

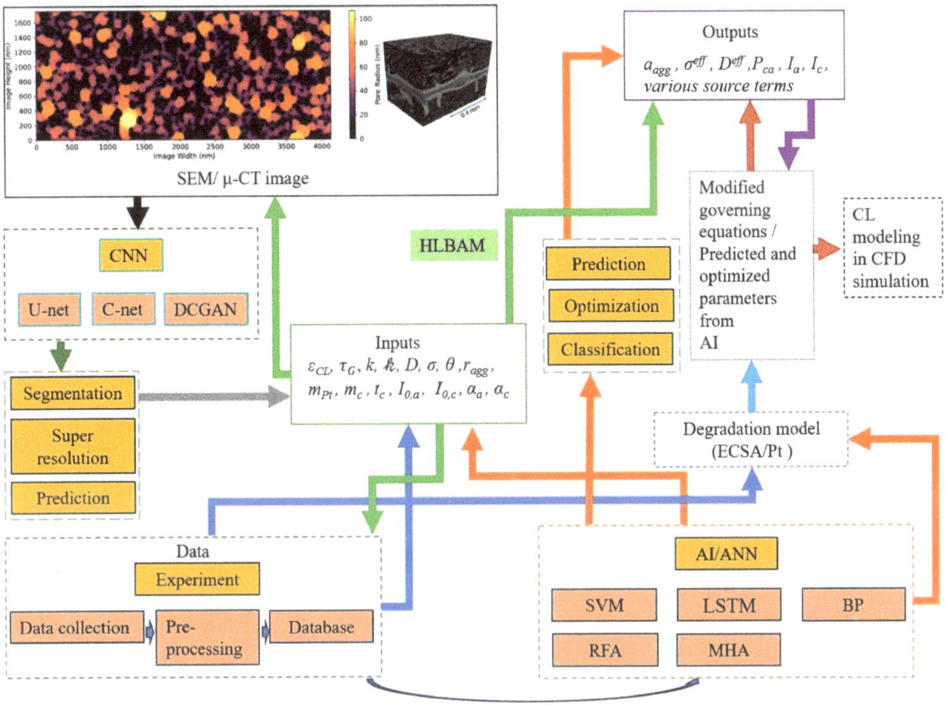

Figure 8. Framework of the PEMFC CL model integrated with ML.

4.3. Modeling of Gas Diffusion Layer (GDL)

The GDL is situated between the CL and flow channel, surrounding the two sides of the MEA. The purpose of the GDL is to create an electronic and thermal bridge between the flow channels and MEA, providing a transport path for the reactant gas flow and excess water for removal. Hence, an ideal GDL should be cost-effective and possess high thermal and electronic conductivities, appropriate wetting characteristics (hydrophobic/hydrophilic), and high chemical and mechanical durability. Carbon-based papers and woven/non-woven carbon fibers [257] are mainly used as GDLs, with pore sizes ranging from 1 to 100 μm, which are larger than those of the CL. The thickness of the GDL is generally between 200 and 400 μm, with a fiber diameter of 7–10 μm [258]. The modeling approach for the GDL is described in the following subsections.

4.3.1. Modeling Porous Structure

The GDL is primarily a carbon-based material and a wet-proofed, nonuniform porous medium. Hence, its transport properties suffer from heterogeneous porosity distributions [259]. To model such a structure, the following properties must be considered: porosity, wettability, permeability, capillary effect, and electrical conductivity.

Porosity

As mentioned previously, the porosity and pore size distribution affect the transport properties of the reactants and products [260], ohmic resistance [261], and variation in the liquid saturation profile [262]. Based on the pore size, GDLs can be categorized into three types. GDLs with micropores (pore sizes < 0.01 μm) mostly utilize Knudsen diffusion, whereas mesopores (pore sizes of 0.01–5 μm) employ bulk diffusion. By contrast, GDLs with large pores, known as macropores (pore sizes > 5 μm), contain gas molecules that mainly diffuse owing to molecular collisions. As mesopores remain in the transition between micro- and macropores, they are preferred because they have balanced transportation. The transport properties are discussed in Section 4.3.2.

Wettability, Permeability, and Capillary Effect

In the GDL, liquid water management strongly depends on the pore size and resulting wall adhesion effect. Based on the surface wettability (hydrophilic or hydrophobic), the water movement changes completely in the PEMFC. Polytetrafluoroethylene (PTFE) is added to the GDL to make it hydrophobic, which helps to remove water effectively. Different surface wettabilities of the pores produce a pressure difference, and subsequently, the water moves owing to capillary pressure. This capillary pressure is responsible for the liquid water transport in PEMFCs and is a function of the liquid water surface tension $\left(\sigma_{lq}\right)$, contact angle (θ), porosity (ε), permeability (K_0), and liquid water volume fraction (s_l) and can be expressed as follows [263,264]:

$$p_c = \begin{cases} \sigma_{lq}\cos\theta\left(\frac{\varepsilon}{K_0}\right)^{0.5}\left[1.42(1-s) - 2.12(1-s)^2 + 1.26(1-s)^3\right]\theta < 90^\circ \ (Hydrophilic) \\ \sigma_{lq}\cos\theta\left(\frac{\varepsilon}{K_0}\right)^{0.5}\left[1.42s - 2.12s^2 + 1.26s^3\right]\theta > 90^\circ \ (Hydrophobic) \end{cases} \quad (25)$$

Furthermore, during water transportation, a capillary pressure gradient is required to overcome the negative gas pressure gradient. Therefore, low permeability is required [257].

Electric Conductivity

Modeling the electrical conductivity of the GDL is vital because carbon has a high electrical conductivity. However, despite its high conductivity, the GDL causes ohmic losses owing to the periodic interfaces between the flow channel and the GDL [265]. In most CFD models, the electrical conductivity is assumed to be isotropic [266,267], which is not accurate in real-world situations [268]. Ohm's law is used to explain this [116]:

$$i_1 = -\sigma_0 \epsilon_1^{1.5} \nabla \Phi_1 \quad (26)$$

where ϵ_1 and σ_0 are the volume fraction and electrical conductivity of the electronically conducting phase, respectively. Subsequently, the aforementioned equation is corrected for additional porosity and tortuosity using the Bruggeman correction [269,270].

4.3.2. Transport Properties in GDL

There are two modes of transport in the GDL: gas phase and liquid phase.

Transport of Gas Phase

Almost every model uses the Stefan–Maxwell equation to treat gas transport, which is a function of the total concentration of gas species, c_T, mole fraction of species i, x_i, and effective binary interaction parameter between i and j, $D_{i,j}^{eff}$ [116]:

$$\nabla x_i = \sum_{j \neq i} \frac{x_i N_j - x_j N_i}{c_T D_{i,j}^{eff}} \quad (27)$$

where

$$D_{i,j}^{eff} = \frac{\epsilon_G}{\tau_G} D_{i,j} \quad (28)$$

Here, s_g and τ_G represent the volume fraction and tortuosity of the gas phase, respectively.

In the absence of liquid water, the gas-phase porosity (ε_G) is equivalent to the bulk porosity (ε_0) of the medium. However, in the presence of liquid water, a different approach is required, as explained in the following section. In these cases, a commonly used method is to determine the tortuosity value using the Bruggeman expression [269,270].

$$\tau_G = \epsilon_G^{-0.5} \tag{29}$$

For a GDL with microscopic pores, the effective Knudson diffusion coefficient, $D_{K_i}^{eff}$, is added to Equation (27) [104,271].

$$\nabla x_i = -\frac{N_i}{c_T D_{K_i}^{eff}} + \sum_{j \neq i} \frac{x_i N_j - x_j N_i}{c_T D_{i,j}^{eff}} \tag{30}$$

Although the diffusion mode has been emphasized in most previous models, convection has been considered in a few [111,169,189]. This can be accomplished by including an additional Darcy's law term [116]:

$$v_G = -\frac{k_G}{\mu_G} \nabla p_G \tag{31}$$

As per the dusty gas model [113–115], Equation (31) can be integrated into Equation (30) to account for the pressure-driven flow [116]:

$$\nabla x_i = -\frac{N_i}{c_T D_{K_i}^{eff}} + \sum_{j \neq i} \frac{x_i N_j - x_j N_i}{c_T D_{i,j}^{eff}} - \frac{x_i k_G}{D_{K_i}^{eff} \mu_G} \nabla p_G \tag{32}$$

Transport of Liquid Phase

For single-phase modeling, the liquid phase can be treated as a solid phase with a specific volume fraction [96,110,260,272] or as droplets carried by the gas stream [190,273,274]. The former approach provides an understanding of how flooding occurs and propagates, whereas the latter highlights the location of droplets and, in some cases, changes in water pressure and concentration.

Although these single-phase approaches are adequate to some extent, two-phase models are necessary for precise calculations because porous media such as GDLs have significant gas–liquid interactions. The two-phase model can be expressed as follows [263,275,276]:

$$p_c = p_L - p_G = -\frac{2\sigma_{lq} \cos\theta}{r} \tag{33}$$

where σ_{lq}, r, and θ are the surface tension of water, pore radius, and internal contact angle that a drop of water forms with a solid, respectively.

This two-phase model can also predict liquid saturation at different positions. The liquid saturation is defined as the amount of pore volume occupied by the liquid water and is expressed as [116]:

$$\epsilon_G = \epsilon_G(1 - s) \tag{34}$$

As the effective gas-phase diffusion coefficient is a function of liquid saturation, it is characterized by a water-flooding phenomenon. To calculate the saturation of the two-phase model, the gas and liquid capillary pressures should be measured using Darcy's law for each phase [116]:

$$N_{w,L} = -\frac{k}{\overline{V}_w \mu} \nabla p_L = -\frac{k}{\overline{V}_w \mu} (\nabla p_C + \nabla p_G) = -\frac{k}{\overline{V}_w \mu} \nabla p_C \tag{35}$$

The last part of the equation is derived under the assumption of uniform gas pressure in the PEMFC. The effective permeability, k, can be obtained by applying the following equation [116]:

$$k = k_r k_{sat} \tag{36}$$

Here, k_r and k_{sat} are the relative permeability and saturated permeability (or permeability at complete saturation), respectively.

4.3.3. Lattice Boltzmann Method (LBM)

The LBM is a mesoscopic CFD method based on the MD theory, which is used to simulate two-phase flow in porous GDL in PEMFCs [277–279]. This method enables a more straightforward simulation of the twin-flow model compared to conventional CFD methods based on the Navier–Stokes equations. In addition, the LBM is more appropriate for microscale simulations because it is formulated based on the kinetic theory. This method is mainly used to predict liquid water dynamics in a GDL under different nonuniform compression cases with nonuniform pore size distribution [279], oxygen diffusion [278], and water removal at the GDL wall with mixed wettability [280].

4.3.4. Issues Related to State-of-Art GDL Modeling

An accurate prediction of the effective transport properties of GDL is dependent on understanding and then utilizing the precise physical and electrochemical parameters in CFD simulation. State-of-the-art GDL modeling has the following issues:

The heterogeneity of GDL is not considered well during CFD simulation for the sake of simplicity. However, porosity, permeability, bulk resistance, and interfacial contact resistance (ICR) of GDL greatly affect the performance of PEMFC [281–283]. Nitta et al. [218] considered heterogeneous GDL in their 2D modeling and obtained better cell performance by optimizing porosity, effective thermal conductivity, and effective gas diffusion coefficient. Shinde et al. [284] implemented heterogeneous GDL with 8 different porosities with their corresponding permeability and ICR and compared their numerical model with a homogeneous one. In comparison to homogeneous GDL, heterogeneous models underestimate cell performance. Also, depending on the pore size, different diffusion kinetics or transport mechanisms also need to be considered.

The wettability of GDL greatly influences the water removal from the PEMFC. Hence, it is common practice to make GDL surfaces hydrophobic [285,286]. However, GDL surface wettability changes due to the loss of polytetrafluoroethylene (PTFE) during mechanical compression cycles [287,288] and various degradation mechanisms [289,290], which are responsible for altering PTFE distribution. Even though GDL has a heterogeneous wettability [291,292] and its corresponding capillary pressure, P_c [293], most of the CFD modeling involves implementing a single contact angle for the whole GDL. In addition to that, the effect of the surface roughness of GDL is important for effective water management [294], which is always neglected during numerical investigation.

For a long period of time, GDL corrosion, such as carbon corrosion and loss of PTFE, is responsible for limiting mass transfer, water management, and electrical conductivity [290]. Mostly, two types of degradation can be seen in GDL: (1) During compression, fiber breakage, and carbon corrosion, and (2) changes in wettability because of oxidation and PEFE loss. Though there is individual research related to the change in GDL wettability [295] and porosity [284], to the best of the authors' knowledge, the integration of GDL surfaces with heterogeneous characteristics into PEMFC modeling has yet to be conducted.

4.3.5. ML in the Field of GDL

Likewise, CL and ML have also been implemented in the field of GDL for feature extraction [296–298], optimization [290–301], and degradation [302,303], though they particularly do not consider the degradation of GDL. Mahdaviara et al. [297] implemented 2D and 3D U-net deep learning models for multiphase segmentation of images from high-resolution X-ray tomography (micro-CT). These images include GDL with different percentages of hydrophobic polytetrafluoroethylene (PTFE). Figure 9a shows the workflow of the deep learning model for segmentation of wet GDL images (top) with distinctive features such as water, air, and fiber colored as blue, red, and green, respectively (bottom). Finally, they were able to measure the absolute permeability, which is close to the findings from the experiment. Froning et al. [298] utilize a CNN-based prediction model based on the LB simulation of GDL with stochastic arrangement of fibers, four types of binder distribution, and up to 50% compression, as can be seen in Figure 9b. The model could successfully predict GDL characteristics such as permeability and tortuosity without the need for high-performance computing (HPC) simulations. To optimize the properties of GDL, Lobato et al. [299] collected 110 experimental data, where the inputs were PTFE content (%), porosity from Hg-porosity (%), mean pore diameter (mm), permeability (m^2), hydrophobicity level (%), and current density (mA/cm^2), and the outputs were cell voltage (mV), tortuosity, and trained direct neural modeling. Furthermore, inverse neural modeling was utilized to inversely produce the GDL properties, making the neural network model an effective optimization tool (Figure 9c). Although there is a reduction in the performance of PEMFC due to various components, it is not possible to pinpoint the reason specifically to the CL, GDL, or other parts. However, it is a well-known fact that the hydrophobic deterioration of the GDL stands out as a primary factor leading to both ohmic and mass transfer degradation, ultimately causing a decrease in the power output of PEMFC [304]. By keeping that in mind, Nagulapati et al. [302] utilize three data-driven models, including Gaussian process regression (GPR), support vector machine (SVM) and artificial neural network (ANN), with raw data obtained from a dynamic load durability test of 10,000 h for a single PEMFC. In their investigations, the GPR model demonstrates the highest predictive accuracy as the training data size increases to 50% and beyond, achieving the lowest RMSE value of 0.0071 using 70% training data, making it more suitable for machine learning-based fault detection and state-of-health SOH estimation (Figure 9d).

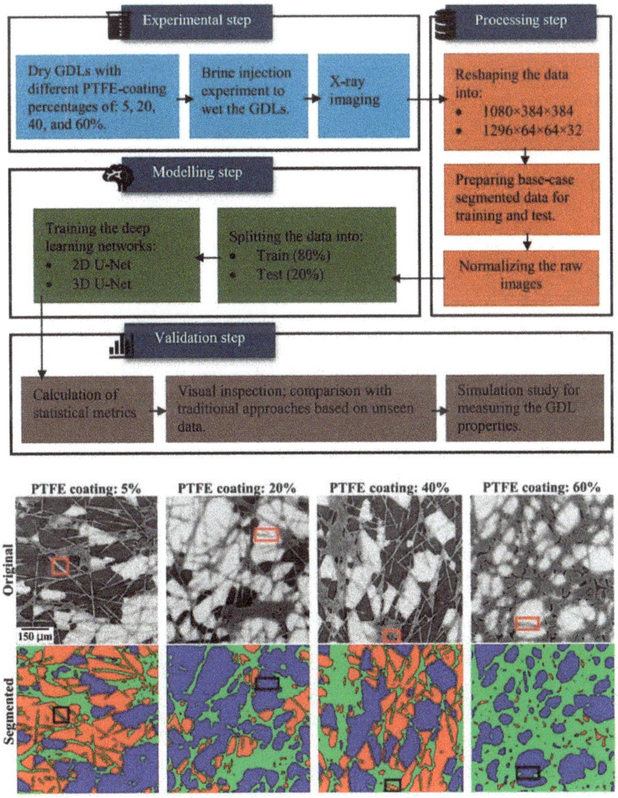

(a)

Figure 9. *Cont.*

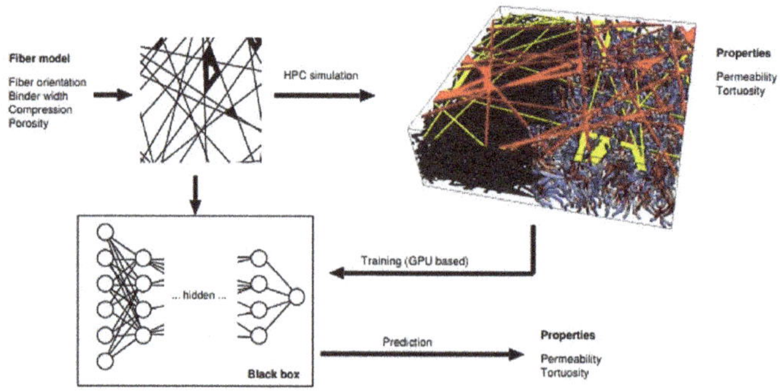

(b)

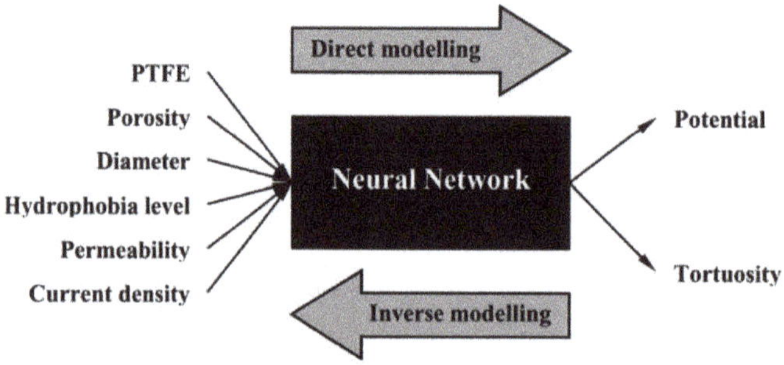

(c)

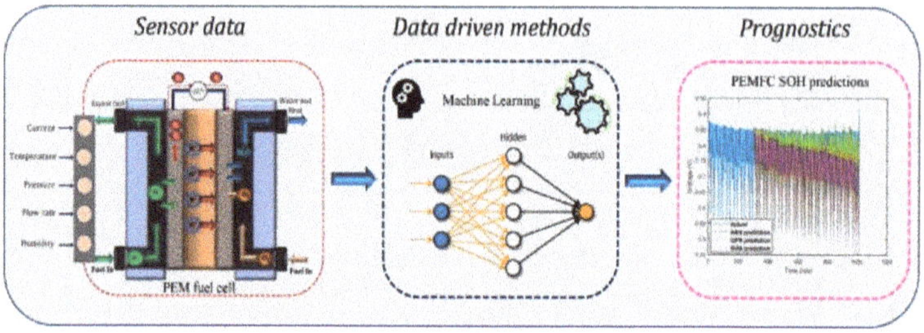

(d)

Figure 9. (**a**) Workflow for developing deep learning-based model for segmentation from wet images of GDL (top), original GDL images at a resolution of 384 × 384 pixels, along with the segmented images produced using trainable 3D Weka segmentation (down). The aqueous phase is shown in blue, air in red, and fibers in green in the segmented images whereas rectangles highlight examples of potential segmentation errors; reproduced with permission [297]; 2023 Elsevier Ltd., (**b**) ML approach for predicting the permeability of GDL from image data from previous lattice Boltzmann (LB) simulations; reproduced with permission [298]; 2022 MDPI, (**c**) architecture of neural network to investigate effect of GDL properties on PEMFC; reproduced with permission [299]; 2010 Elsevier Ltd., and (**d**) schematic of ML-based prognostics of PEMFC where carbon corrosion and hydrophobic loss of GDL are among the reasons for degradation; reproduced with permission [302]; 2023 Elsevier Ltd.

4.3.6. Integration of ML in GDL Modeling

Figure 10 shows the framework for ML integration of GDL modeling, which is almost close to the CL framework. Unlike CL modeling, GDL does not need to consider additional electromechanical parameters. The degradation of GDL, which affects the wettability of its surface, should be a vital parameter for GDL modeling.

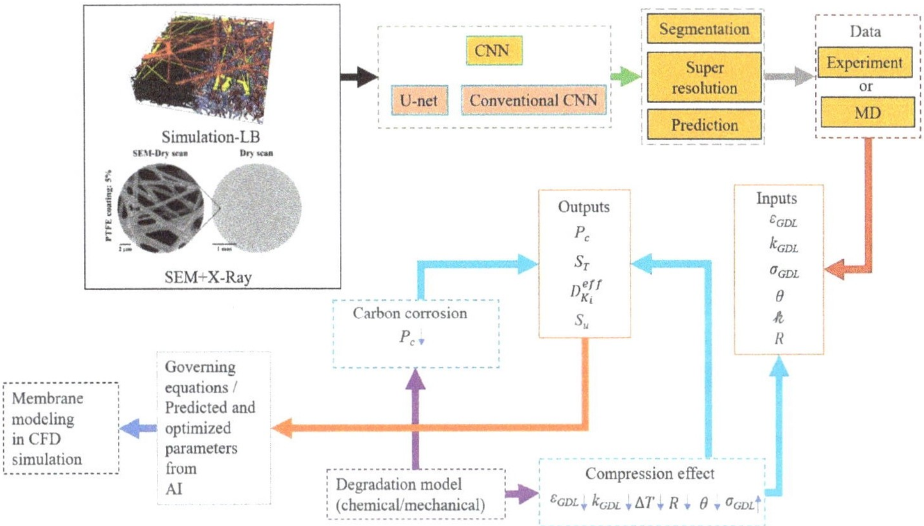

Figure 10. Framework of the PEMFC GDL model integrated with ML. Inset picture represents reproduced with permission [297]; 2023 Elsevier Ltd images from LB simulation (top left); reproduced with permission [298].

4.4. Modeling of Bi-Polar Plate (BP)

Among the components of PEMFCs, BPs are the most important because they provide an adequate number of reactants to the GDL and CL, effectively remove excess water (which is responsible for flooding), broaden the contact area between the flow channel and GDL to maximize proton transport, and minimize the pressure drop between the inlet and outlet [305,306]. In addition, BPs account for 80% of the total weight and 45% of the cost of the fuel cell stacks [307,308]. Graphite plates are mainly used for BPs because of their high electrical conductivity [309–311].

4.4.1. Flow Inside the Channel

It is well established that the flow inside the channel is primarily laminar [60,312]. The steady-state continuity equation for the gas phase is expressed by the following equation [60]:

$$\frac{\partial \rho_{mix} \cdot u_{mix,i}}{\partial x_i} = S_m \tag{37}$$

where ρ_{mix} and $u_{mix,i}$ are the mixture multiphase density and velocity in direction i, respectively, and S_m is the mass consumption/production in the flow channels. Although S_m remains null in most cases, gas-to-liquid and liquid-to-dissolution transitions are considered in flow channels [35].

4.4.2. Modeling of Pressure Drop

The fluid flow inside the channel depends on the pressure difference between the inlet and outlet of the channel and is proportional to the flow rate in the channel. Thus, the pressure drop can be approximated by using the following expression for incompressible flow in pipes [159]:

$$\Delta P = f\frac{L_{chan}}{D_H}\rho\frac{\overline{V}^2}{2} + \sum K_L\rho\frac{\overline{V}^2}{2} \tag{38}$$

where $f, L_{chan}, D_H, \rho, \overline{V},$ and K_L are the friction factor, channel length, hydraulic diameter, fluid density, average velocity, and local resistance, respectively.

The hydraulic diameter of a rectangular flow field can be expressed as [159]

$$D_H = \frac{2w_c d_c}{(w_c + d_c)} \tag{39}$$

where w_c and d_c are the width and depth of the channel, respectively.

However, for porous flow fields, Equation (38) can be replaced by [159]

$$\Delta P = \mu\frac{Q_c}{kA_c}L_{chan} \tag{40}$$

where $\mu, Q_c, k,$ and A_c are the fluid viscosity, geometric flow rate through the cell, permeability, and cross-sectional area of the flow field, respectively.

4.4.3. Issues Related to State-of-Art BP Modeling

Though the flow and transport of the reactants are not complicated in the flow channel of BP, the following issues remain during the 3D modeling of the PEMFC:

The reactant flow rate, velocity, or stoichiometry from each inlet of the anode and cathode can be either implemented as a constant value from experiments [215,313,314], a function of pressure, temperature, RH, etc. [217,315,316], or an exact solution considering various cross-sectional areas with laminar profiles [214,317]. However, despite the procedure to choose this flow condition, researchers are able to validate their simulation results with experiments. Hence, there should be some clear-cut guidance for flow conditions.

The degradation of the bi-polar plate is clearly neglected during the PEMFC modeling. However, bi-polar plates can significantly induce the degradation of MEA, especially in high-temperature PEMEC [318] and in metallic bi-polar plates [319].

The recent PEMFC research trend introduces new types of porous flow channels instead of conventional flow channels, which show great potential regarding high mass transport, heat removal, electrical conductivity, and enhanced performance [320,321]. However, numerical modeling is yet to be conducted for the optimization of this type of flow field. It is important to note that, unlike the conventional hollow flow channel, the transport phenomena and flow field-GDL interactions would be completely changed.

4.4.4. ML in the Field of BP

In order to optimize the design parameters of the bi-polar plate [40,322,323], ML has proven to be an efficient tool. In addition to that, data-driven ML also utilizes PEMFC optimization of operating conditions and performance predictions [324–328]. Seyhan et al. [40] implement an ANN to optimize a wavy serpentine flow channel. Experimental parameters such as hydrogen, air flow rate, cell temperature, and the amplitude of the channel were used to train the ANN model (Figure 11a). Furthermore, the train model was able to optimize the operating conditions and design parameters of the flow channel. Figure 11b shows the work schematic of a single PEMFC with flow field configuration and the dimensions of the divided unit (top) [322]. A CFD, combined with ANN and an intelligent optimization algorithm, was used to extract the key geometry of the divided part based on the power density and oxygen uniformity index (Figure 11c). Moreover, in comparison to the base model and design of single-objective optimization (SOO), multi-objective optimization (MOO) is superior according to the oxygen concentration, the oxygen uniformity index, and the water removal capacity. Li et al. [327] used a ML base bagging neural network (BNN) (Figure 11d) to predict the PEMFC performance from the parameters of different block structures. The database was created using the optimized height and width of an imitated water block, and the prediction performance of BNN was compared to that of the BP neural network (Figure 11e). The performance prediction model by BNN produces high precision predictions with less data and has proven to be an effective optimization tool for flow channels.

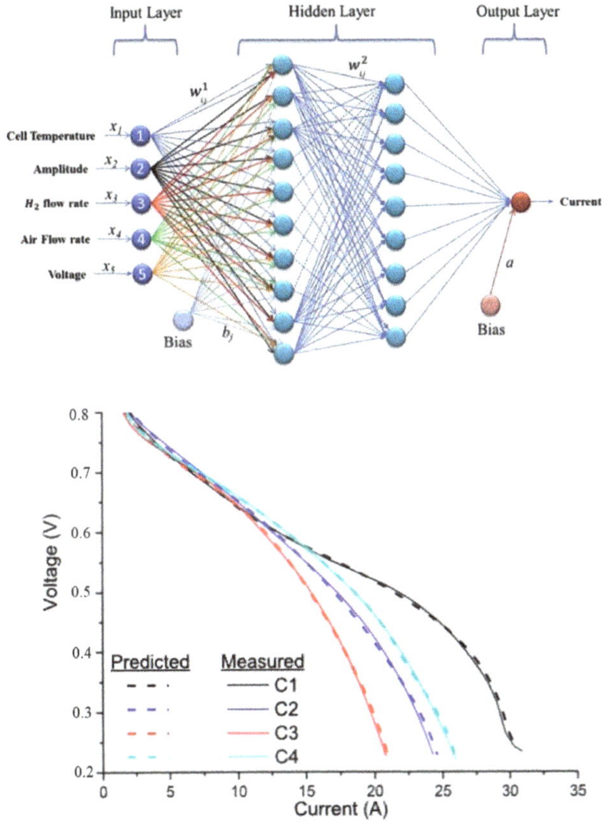

(a)

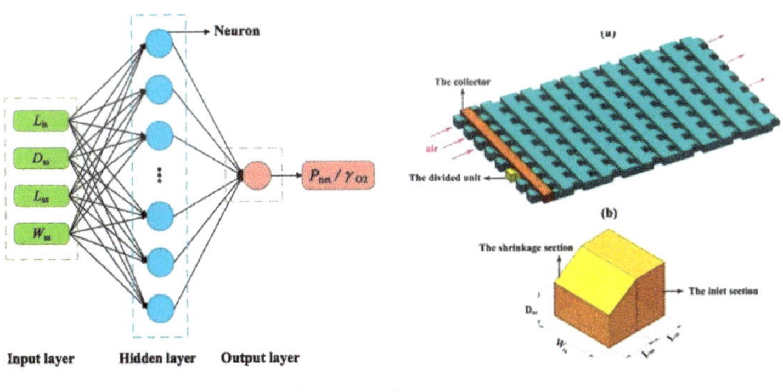

(b)

Figure 11. *Cont.*

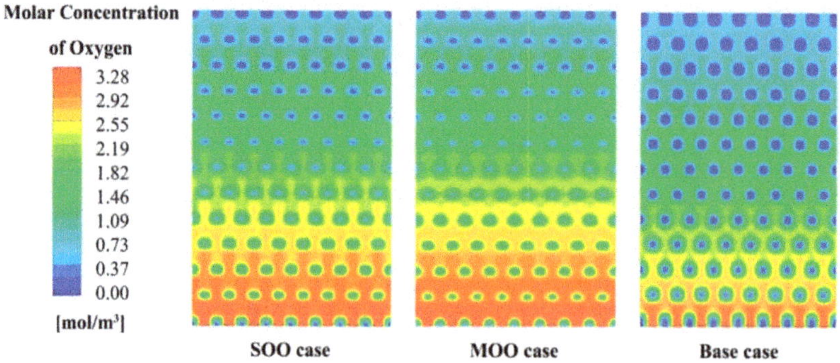

(c)

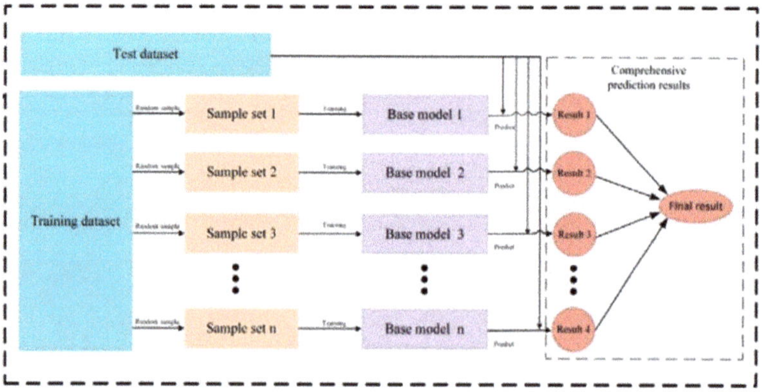

(d)

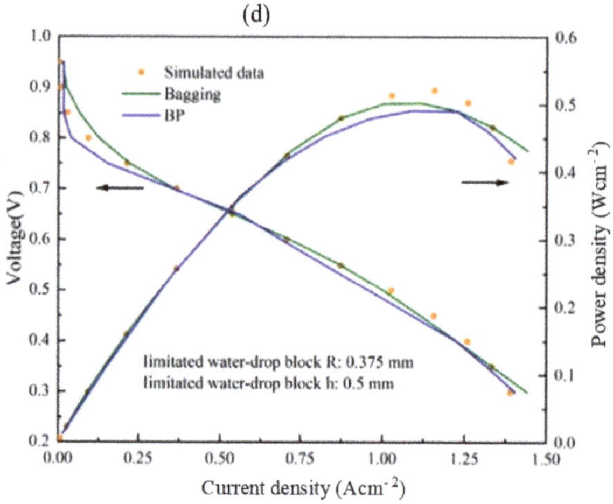

(e)

Figure 11. (**a**) Schematic of ANN model (top) for predicting current densities (bottom) of different types of flow channels where w_{ij}^1 and w_{ij}^2 represent the weights between the input and the first hidden layer, and the weights between the first and second hidden layers, respectively; reproduced with permission [40]; 2017 Elsevier Ltd.; (**b**) ANN structure to improve the 3D fine-mesh flow field; (**c**) comparison of performance for 2 optimized flow configuration with the base model regarding molar concentration of O_2; reproduced with permission [322]; 2022 Elsevier Ltd.; (**d**) diagram of bagging ensemble algorithm to predicting the performance of PEMFC using different block arrangements in the flow channel; and (**e**) corresponding polarization in comparison to BP and simulation; reproduced with permission [327]; 2022 Elsevier Ltd.

4.4.5. Integration of ML in BP Modeling

The input parameters for back propagation (BP) are easily obtainable, such as flow rates, pressure, and temperature, from external sensors. However, more optimized parameters and the nonlinear relationship between the given input and output can be accurately predicted using ML methods, as illustrated in Figure 12. Though ML approaches are also employed for design optimization of the flow channel and the selection of better materials for BP, these aspects are not included in the proposed framework as they are not essential for improving CFD modeling.

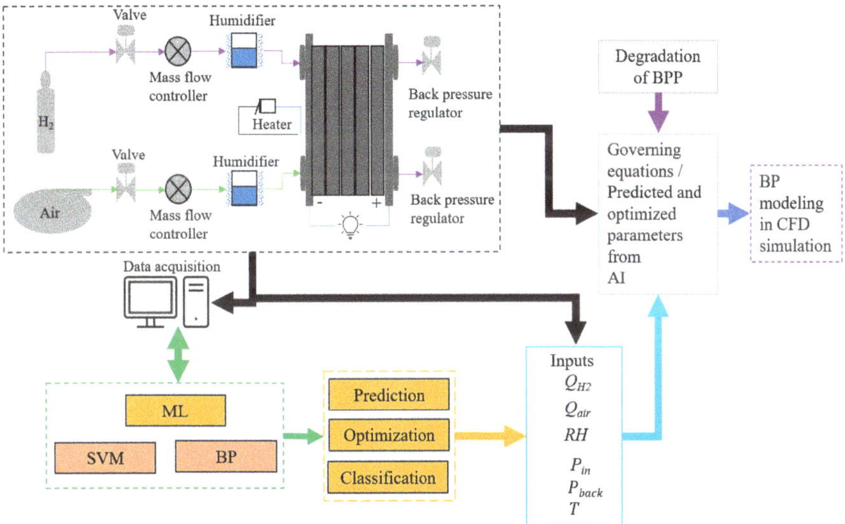

Figure 12. Framework of the PEMFC BP model integrated with ML.

5. Final Overview

Despite ongoing research to enhance the efficiency and commercialization of large-scale PEMFC, achieving this goal through standalone experiments is unlikely. In this regard, computational fluid dynamics (CFD) modeling emerges as an effective tool, circumventing the time-consuming, costly, and labor-intensive nature of experimental processes. However, as we deviate from assumptions and dynamic parameters for more accurate modeling, the model describing physical phenomena becomes increasingly complex.

Fortunately, computational costs are no longer a significant issue. However, the challenge lies in the adequacy of governing equations to solve the flow and its corresponding effects. Additionally, the intricate pattern of PEMFC generates copious data with a non-linear relationship between parameters, necessitating more processing time and complex flow equations. ML algorithms, with their superior accuracy in pattern recognition and optimization techniques, hold the potential to address certain aspects of CFD modeling. As illustrated in Figure 13, the PEMFC development cycle involves a combination of experiments, CFD modeling, and ML approaches. Experimental results inform operational parameters, establish databases for future development, and subsequently validate simulations and ML algorithms. CFD modeling optimizes component design, contributes new data in conjunction with experiments for ML, and, based on data from experiments and CFD, ML provides more generalized yet accurate equations. This enhances the heterogeneity of components in CFD simulations, ultimately leading to optimized designs and material properties for maximizing PEMFC efficiency. Therefore, the ML-assisted CFD model will have the following impacts:

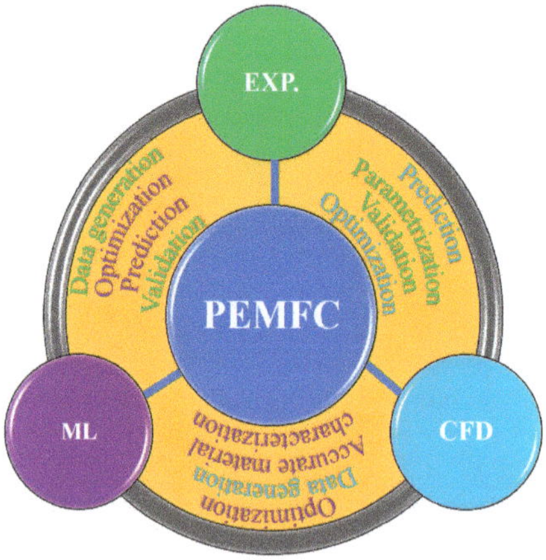

Figure 13. Cycle of PEMFC development through experiment, CFD and ML.

Development of materials: In terms of cost, weight, and durability, state-of-the-art PEMFC materials suffer from heavy bi-polar plates, costly and degradable catalyst layers, and PEM. Innovative materials with enhanced chemical stability, mechanical strength, and degradation resistance hold the promise of elevating PEMFC components to unprecedented levels of longevity and performance. Therefore, ML-assisted approaches could provide more diverse sets of material options that will be more durable, efficient, and cost-effective, depending on the operation condition.

Optimum design: After developing the material for each component, integration into one single cell needs further research. These include optimum wettability of GDL for better water drainage, efficient flow channel design for uniform reactant distribution, thermal and water management, and adequate pressure drop. In this regard, ML-integrated modeling approaches will replace the lengthy and costly experimental setup.

Future transport sector: Based on Table 1, future transportation will require more durable, cost-effective, and efficient PEMFC or stacks. Thus, new materials and designs are vital for replacing conventional fuel with PEMFC.

6. Conclusions and Outlooks

The purpose of this overview is to provide a comprehensive modeling guideline for conducting PEMFC simulations to optimize the design and enhance performance with the assistance of machine learning (ML). This study covers individual modeling methods for each component of PEMFCs, detailing the advantages and disadvantages of each model, addressing state-of-the-art modeling issues, reviewing previous ML studies, and proposing a framework to combine computational fluid dynamics (CFD) with ML. ML can either use experimental data to establish correlations between the parameter of interest and the expected output or utilize CFD data as a surrogate model to find optimal parameters, thus saving computational costs. Despite the potential for higher accuracy in ML-integrated CFD, the limited additional data from experiments and CFD could sometimes lead to overfitting and over/underestimation of generalization. Moreover, understanding the selection of input parameters to train the model for specific problems and the relationships obtained between the data for prediction is an aspect that requires further exploration.

Hence, the future integration of data-driven ML methods and CFD methods should involve a trade-off between model accuracy and computational cost. For CFD-based modeling, accuracy can be enhanced by obtaining curve-fitting transport properties from the experiment at various conditions, generalizing the relation, reducing the dimensionality, and avoiding the less sensitive parameters with the data-driven ML method to improve the governing equations of numerical models. However, this is not work that can be accomplished overnight, and further research is required to achieve high model accuracy with low computational costs.

Author Contributions: Conceptualization, R.K. and J.-C.P.; methodology, R.K.; formal analysis, R.K.; investigation, R.K.; data curation, R.K.; writing—original draft preparation, R.K.; writing—review and editing, J.-C.P.; supervision, J.-C.P.; project administration, J.-C.P., C.-Y.A. and Y.-H.K.; funding acquisition, J.-C.P., C.-Y.A. and Y.-H.K. All authors have read and agreed to the published version of the manuscript.

Funding: This research was supported by a grant (PMS5470) from "Development on operation and reliability verification technology of 1 mw class eco-friendly ship fuel and power system under ocean environment, 1525013774" program funded by ministry of oceans and fisheries of Korean government.

Data Availability Statement: Data are not available.

Conflicts of Interest: The authors declare no conflicts of interest.

Abbreviations

CFD	computational fluid dynamics
CL	catalyst layer
GDL	gas diffusion layer
HOR	hydrogen oxidation reaction
HPT	hot press time
HPP	hot press pressure
ORR	oxygen reduction reaction
MD	molecular dynamics
MEA	membrane electrode assembly
MSD	mean square displacement
MHA	Meta-heuristic algorithms
PFSA	perfluorosulfonic acid
Pt	platinum
PTFE	polytetrafluorethylene
LE	liquid equilibrated
SPNB	sulfonated polynorbornene
PEMFC	proton electrolyte membrane fuel cell
VE	vapor equilibrated
Symbols	
a	water activity (1/m)
A	specific external surface area, cm^2
$a_{k,p}$	interfacial surface area between phases k and p per unit volume, 1/cm
c	molar concentration, $kmol/m^3$
c_b	solubility of oxygen, mol/cm^3
$c_{i,k}$	interstitial concentration of species *i* in phase *k*, mol/cm^3
c_T	total solution concentration or molar density, mol/cm^3
$\hat{C}_{P_k}$	heat capacity of phase *k*, J/(g·K)
D	mass diffusivity (m^2/s)
D_H	hydraulic diameter, cm
D_{K_i}	Knudsen diffusion coefficient of species i, cm^2/s
$D_{i,j}$	diffusion coefficient of *i* in *j*, cm^2/s
E_r	effectiveness factor
E	cell potential, V
F_s	interaction force between phases, N/cm^3
f	friction factor
F	Faraday's constant
$J_{i,k}$	flux density of species *i* in phase *k* relative to the mass-averaged velocity of phase *k*, $mol/(cm^2{\cdot}s)$

J_{diff}	oxygen flux per unit volume, mol/(cm^2·s)
h	heat transfer coefficient, J/(cm^2·s·K)
h_{evap}	latent heat of evaporation, J/kg
$\overline{H}_{i,k}$	partial molar enthalpy of species *i* in phase *k*, J/mol
ΔH_{evap}	heat or enthalpy of evaporation, J/mol
$i_{h,1-k}$	normal interfacial current transferred per unit interfacial area across the interface between the electronically conducting phase and phase *k* due to electron-transfer reaction *h*, A/cm
$i_{h,k-p}$	transfer current density of reaction *h* per unit interfacial area between phases *k* and *p*, A/cm
i_{0_h}	exchange current density for reaction *h*, A/cm^2
L	length, cm
k	permeability, m^2
$\acute{k}$	ORR rate constant
m	mass
M	molecular weight, g/mol
n_d	electroosmotic drag coefficient
$N_{i,k}$	superficial flux density of species *i* in phase *k*, mol/(cm^2·s)
p	partial pressure, atm
Q_c	geometric flow rate, cm^3/s
$r_{i,k-p}$	rate of reaction per unit of interfacial area between phases *k* and *p*, mol/(cm^2·s)
R'	total ohmic resistance, Ω/cm^2
R	ideal gas constant, 8.3143 J/(mol·K)
$R_{g,k}$	rate of homogeneous reaction g in phase *k*, mol/(cm^3·s)
r_{evap}	rate of evaporation, mol/(cm^3·s)
s	water volume fraction
s_g	gas volume fraction
S	source term
$S_{i,k,h}$	stoichiometric coefficient of species *i* residing in phase *k* and participating in electron transfer reaction *h*
$s_{i,k,g}$	stoichiometric coefficient of species *i* residing in phase *k* and participating in electron transfer reaction *g*
T	temperature, K
u	mobility, (m^2·kmol)/(J·s)
$\vec{u}_m$	velocity vector, cm/s
U'	potential intercept for a polarization equation, V
U_h	reversible cell potential of reaction *h*, V
U_{H_h}	enthalpy potential, V
V	volume, m^3
$\overline{V}$	molar volume, cm^3/mol
x_i	mole fraction of species *i*
z	charge number of valence
Greek letters	
α	transfer coefficient, water transport coefficient kmol^2/(J·m·s)
γ	roughness factor
δ_{film}	electrolyte thickness, cm
ε	porosity
η	overpotential, V
θ	contact angle
κ	ionic conductivity of the membrane, S/cm
λ	water content
μ	dynamic viscosity, kg/(m·s)
μ_w	membrane water chemical potential
v	velocity, cm/s
$\overline{v}_{avg}$	fluid density average velocity, cm/s
ρ	density, g/cm^3
σ	standard conductivity in the electronically conducting phase, S/cm
σ_{lq}	liquid–water surface tension, N/m

τ_G	tortuosity
Φ	chemical potential
ϕ	Thiele modulus
$\|\|$	thermal conductivity, W/(m·K)
Subscripts and superscripts	
agg	agglomerate
act	activation
an	anode
cat	cathode
chan	channel
eq	equilibrium
ext	external to the control volume
f	sulfonic acid group
G	gas phase
H_2	hydrogen
H_2O	water
i, j	ith and jth components
ion	ionic
lim	limiting
L, l	liquid phase
k	phase
O_2	oxygen
ref	reference
sat	saturated
sol	solvent (ionomer)
$n-v$	non-frozen membrane water to vapor
Naf	Nafion
w	water phase
vap	vapor
$v-l$	water to liquid (and vice versa)

References

1. Caetano, N.S.; Mata, T.M.; Martins, A.A.; Felgueiras, M.C. New Trends in Energy Production and Utilization. *Energy Procedia* **2017**, *107*, 7–14. [CrossRef]
2. Martins, F.; Felgueiras, C.; Smitkova, M.; Caetano, N. Analysis of fossil fuel energy consumption and environmental impacts in European countries. *Energies* **2019**, *12*, 964. [CrossRef]
3. Sugiawan, Y.; Managi, S. New evidence of energy-growth nexus from inclusive wealth. *Renew. Sustain. Energy Rev.* **2019**, *103*, 40–48. [CrossRef]
4. Shafiee, S.; Topal, E. When will fossil fuel reserves be diminished? *Energy Policy* **2009**, *37*, 181–189. [CrossRef]
5. Singla, M.K.; Nijhawan, P.; Oberoi, A.S. Hydrogen fuel and fuel cell technology for cleaner future: A review. *Environ. Sci. Pollut. Res.* **2021**, *28*, 15607–15626.
6. Nicoletti, G.; Arcuri, N.; Nicoletti, G.; Bruno, R. A technical and environmental comparison between hydrogen and some fossil fuels. *Energy Convers. Manag.* **2015**, *89*, 205–213. [CrossRef]
7. Felseghi, R.-A.; Carcadea, E.; Raboaca, M.S.; Trufin, C.N.; Filote, C. Hydrogen fuel cell technology for the sustainable future of stationary applications. *Energies* **2019**, *12*, 4593. [CrossRef]
8. Dincer, I. Hydrogen and fuel cell technologies for sustainable future. *Jordan J. Mech. Ind. Eng.* **2008**, *2*, 1–14.
9. Xing, H.; Stuart, C.; Spence, S.; Chen, H. Fuel cell power systems for maritime applications: Progress and perspectives. *Sustainability* **2021**, *13*, 1213. [CrossRef]
10. Mallouppas, G.; Yfantis, E.A. Decarbonization in shipping industry: A review of research, technology development, and innovation proposals. *J. Mar. Sci. Eng.* **2021**, *9*, 415. [CrossRef]
11. Mekhilef, S.; Saidur, R.; Safari, A. Comparative study of different fuel cell technologies. *Renew. Sustain. Energy Rev.* **2012**, *16*, 981–989. [CrossRef]
12. Sulaiman, N.; Hannan, M.A.; Mohamed, A.; Ker, P.J.; Majlan, E.H.; Daud, W.R.W. Optimization of energy management system for fuel-cell hybrid electric vehicles: Issues and recommendations. *Appl. Energy* **2018**, *228*, 2061–2079. [CrossRef]
13. Dall'Armi, C.; Pivetta, D.; Taccani, R. Health-Conscious Optimization of Long-Term Operation for Hybrid PEMFC Ship Propulsion Systems. *Energies* **2021**, *14*, 3813. [CrossRef]
14. Lee, H.; Ryu, B.; Anh, D.P.; Roh, G.; Lee, S.; Kang, H. Thermodynamic analysis and assessment of novel ORC-DEC integrated PEMFC system for liquid hydrogen fueled ship application. *Int. J. Hydrogen Energy* **2023**, *48*, 3135–3153. [CrossRef]

15. Wu, D.; Peng, C.; Yin, C.; Tang, H. Review of system integration and control of proton exchange membrane fuel cells. *Electrochem. Energy Rev.* **2020**, *3*, 466–505. [CrossRef]
16. Xie, B.; Zhang, G.; Jiang, Y.; Wang, R.; Sheng, X.; Xi, F.; Zhao, Z.; Chen, W.; Zhu, Y.; Wang, Y.; et al. "3D+ 1D" modeling approach toward large-scale PEM fuel cell simulation and partitioned optimization study on flow field. *eTransportation* **2020**, *6*, 100090. [CrossRef]
17. He, C.; Desai, S.; Brown, G.; Bolleplli, S. PEM fuel cell catalysts: Cost, performance, and durability. *Electrochem. Soc. Interface* **2005**, *14*, 41. [CrossRef]
18. Raj, A.; Shamim, T. Investigation of the effect of multidimensionality in PEM fuel cells. *Energy Convers. Manag.* **2014**, *86*, 443–452. [CrossRef]
19. Ji, M.; Wei, Z. A review of water management in polymer electrolyte membrane fuel cells. *Energies* **2009**, *2*, 1057–1106. [CrossRef]
20. Goebel, S.G. Evaporative Cooled Fuel Cell. US Patent 6960404 B2; General Motors Corporation, 1 November 2005.
21. Kandlikar, S.G.; Lu, Z. Thermal management issues in a PEMFC stack–A brief review of current status. *Appl. Therm. Eng.* **2009**, *29*, 1276–1280. [CrossRef]
22. Moreno, N.G.; Molina, M.C.; Gervasio, D.; Robles, J.F.P. Approaches to polymer electrolyte membrane fuel cells (PEMFCs) and their cost. *Renew. Sustain. Energy Rev.* **2015**, *52*, 897–906. [CrossRef]
23. Jiao, K.; Xuan, J.; Du, Q.; Bao, Z.; Xie, B.; Wang, B.; Zhao, Y.; Fan, L.; Wang, H.; Hou, Z. Designing the next generation of proton-exchange membrane fuel cells. *Nature* **2021**, *595*, 361–369. [CrossRef] [PubMed]
24. Fink, C.; Karpenko-Jereb, L.; Ashton, S. Advanced CFD Analysis of an Air-cooled PEM Fuel Cell Stack Predicting the Loss of Performance with Time. *Fuel Cells* **2016**, *16*, 490–503. [CrossRef]
25. Wang, L.; Quan, Z.; Zhao, Y.; Yang, M.; Zhang, J. Experimental investigation on thermal management of proton exchange membrane fuel cell stack using micro heat pipe array. *Appl. Therm. Eng.* **2022**, *214*, 118831. [CrossRef]
26. Huang, Z.; Jian, Q.; Zhao, J. Thermal management of open-cathode proton exchange membrane fuel cell stack with thin vapor chambers. *J. Power Sources* **2021**, *485*, 229314. [CrossRef]
27. Kandlikar, S.G.; Lu, Z.; Domigan, W.E.; White, A.D.; Benedict, M.W. Measurement of flow maldistribution in parallel channels and its application to ex-situ and in-situ experiments in PEMFC water management studies. *Int. J. Heat Mass Transf.* **2009**, *52*, 1741–1752. [CrossRef]
28. Werner, C.; Busemeyer, L.; Kallo, J. The impact of operating parameters and system architecture on the water management of a multifunctional PEMFC system. *Int. J. Hydrogen Energy* **2015**, *40*, 11595–11603. [CrossRef]
29. Li, W.; Jing, S.; Wang, S.; Wang, C.; Xie, X. Experimental investigation of expanded graphite/phenolic resin composite bipolar plate. *Int. J. Hydrogen Energy* **2016**, *41*, 16240–16246. [CrossRef]
30. Kahveci, E.E.; Taymaz, I. Experimental study on performance evaluation of PEM fuel cell by coating bipolar plate with materials having different contact angle. *Fuel* **2019**, *253*, 1274–1281. [CrossRef]
31. Haraldsson, K.; Wipke, K. Evaluating PEM fuel cell system models. *J. Power Sources* **2004**, *126*, 88–97. [CrossRef]
32. Arvay, A.; French, J.; Wang, J.-C.; Peng, X.-H.; Kannan, A. Nature inspired flow field designs for proton exchange membrane fuel cell. *Int. J. Hydrogen Energy* **2013**, *38*, 3717–3726. [CrossRef]
33. Hamdollahi, S.; Jun, L. A Review on Modeling of Proton Exchange Membrane Fuel Cell. *Chem. Ind. Chem. Eng. Q.* **2023**, *29*, 61–74. [CrossRef]
34. Arif, M.; Cheung, S.C.; Andrews, J. Different approaches used for modeling and simulation of polymer electrolyte membrane fuel cells: A review. *Energy Fuels* **2020**, *34*, 11897–11915. [CrossRef]
35. D'adamo, A.; Haslinger, M.; Corda, G.; Höflinger, J.; Fontanesi, S.; Lauer, T. Modelling Methods and Validation Techniques for CFD Simulations of PEM Fuel Cells. *Process* **2021**, *9*, 688. [CrossRef]
36. Legala, A.; Zhao, J.; Li, X. Machine learning modeling for proton exchange membrane fuel cell performance. *Energy AI* **2022**, *10*, 100183. [CrossRef]
37. Zhang, Z.; Bai, F.; Quan, H.-B.; Yin, R.-J.; Tao, W.-Q. PEMFC Output Voltage Prediction Based on Different Machine Learning Regression Models. In Proceedings of the 2022 5th International Conference on Energy, Electrical and Power Engineering (CEEPE), Chongqing, China, 22–24 April 2022.
38. Kheirandish, A.; Shafiabady, N.; Dahari, M.; Kazemi, M.S.; Isa, D. Modeling of commercial proton exchange membrane fuel cell using support vector machine. *Int. J. Hydrogen Energy* **2016**, *41*, 11351–11358. [CrossRef]
39. Ghosh, S.; Routh, A.; Rahaman, M.; Ghosh, A. Modeling and control of a PEM fuel cell performance using Artificial Neural Networks to maximize the real time efficiency. In Proceedings of the 2019 International Conference on Energy Management for Green Environment (UEMGREEN), Kolkata, India, 25–27 September 2019.
40. Seyhan, M.; Akansu, Y.E.; Murat, M.; Korkmaz, Y.; Akansu, S.O. Performance prediction of PEM fuel cell with wavy serpentine flow channel by using artificial neural network. *Int. J. Hydrogen Energy* **2017**, *42*, 25619–25629. [CrossRef]
41. Jemeï, S.; Hissel, D.; Péra, M.; Kauffmann, J. On-board fuel cell power supply modeling on the basis of neural network methodology. *J. Power Sources* **2003**, *124*, 479–486. [CrossRef]
42. Jemeïjemei, S.; Hissel, D.; Pérapera, M.-C.; Kauffmann, J.M. A new modeling approach of embedded fuel-cell power generators based on artificial neural network. *IEEE Trans. Ind. Electron.* **2008**, *55*, 437–447. [CrossRef]
43. Cho, Y.; Hwang, G.; Gbadago, D.Q.; Hwang, S. Artificial neural network-based model predictive control for optimal operating conditions in proton exchange membrane fuel cells. *J. Clean. Prod.* **2022**, *380*, 135049. [CrossRef]

44. Nanadegani, F.S.; Lay, E.N.; Iranzo, A.; Salva, J.A.; Sunden, B. On neural network modeling to maximize the power output of PEMFCs. *Electrochim. Acta* **2020**, *348*, 136345. [CrossRef]
45. Salva, J.A.; Iranzo, A.; Rosa, F.; Tapia, E.; Lopez, E.; Isorna, F. Optimization of a PEM fuel cell operating conditions: Obtaining the maximum performance polarization curve. *Int. J. Hydrogen Energy* **2016**, *41*, 19713–19723. [CrossRef]
46. Mawardi, A.; Yang, F.; Pitchumani, R. Optimization of the operating parameters of a proton exchange membrane fuel cell for maximum power density. *J. Fuel Cell Sci. Technol.* **2005**, *2*, 121–135. [CrossRef]
47. Wang, J.; Jiang, H.; Chen, G.; Wang, H.; Lu, L.; Liu, J.; Xing, L. Integration of multi-physics and machine learning-based surrogate modelling approaches for multi-objective optimization of deformed GDL of PEM fuel cells. *Energy AI* **2023**, *14*, 100261. [CrossRef]
48. Wan, W.; Yang, Y.; Li, Y.; Xie, C.; Song, J.; Deng, Z.; Tan, J.; Zhang, R. Operating Conditions Combination Analysis Method of Optimal Water Management State for PEM Fuel Cell. *Green Energy Intell. Transp.* **2023**, *2*, 100105. [CrossRef]
49. Zhang, L.; Pan, M.; Quan, S. Model predictive control of water management in PEMFC. *J. Power Sources* **2008**, *180*, 322–329. [CrossRef]
50. Zhou, S.; Wang, K.; Zhou, S.; Shan, J.; Bao, D. Features selection and substitution in PEM fuel cell water management failures diagnosis. *Fuel Cells* **2021**, *21*, 512–522. [CrossRef]
51. Laribi, S.; Mammar, K.; Sahli, Y.; Koussa, K. Analysis and diagnosis of PEM fuel cell failure modes (flooding & drying) across the physical parameters of electrochemical impedance model: Using neural networks method. *Sustain. Energy Technol. Assess.* **2019**, *34*, 35–42.
52. Laribi, S.; Mammar, K.; Sahli, Y.; Koussa, K. Fault diagnosis of thermal management system in a polymer electrolyte membrane fuel cell. *Energy* **2021**, *214*, 119062.
53. Pourrahmani, H.; Siavashi, M.; Moghimi, M. Design optimization and thermal management of the PEMFC using artificial neural networks. *Energy* **2019**, *182*, 443–459. [CrossRef]
54. Maleki, E.; Maleki, N. Artificial neural network modeling of Pt/C cathode degradation in PEM fuel cells. *J. Electron. Mater.* **2016**, *45*, 3822–3834. [CrossRef]
55. Kui, C.; Laghrouche, S.; Djerdir, A. Proton exchange membrane fuel cell degradation and remaining useful life prediction based on artificial neural network. In Proceedings of the 2018 7th International Conference on Renewable Energy Research and Applications (ICRERA), Paris, France, 14–17 October 2018.
56. Matsuura, T.; Chen, J.; Siegel, J.B.; Stefanopoulou, A.G. Degradation phenomena in PEM fuel cell with dead-ended anode. *Int. J. Hydrogen Energy* **2013**, *38*, 11346–11356. [CrossRef]
57. Han, I.-S.; Chung, C.-B. A hybrid model combining a support vector machine with an empirical equation for predicting polarization curves of PEM fuel cells. *Int. J. Hydrogen Energy* **2017**, *42*, 7023–7028. [CrossRef]
58. Khajeh-Hosseini-Dalasm, N.; Ahadian, S.; Fushinobu, K.; Okazaki, K.; Kawazoe, Y. Prediction and analysis of the cathode catalyst layer performance of proton exchange membrane fuel cells using artificial neural network and statistical methods. *J. Power Sources* **2011**, *196*, 3750–3756. [CrossRef]
59. Wang, Y.; Diaz, D.F.R.; Chen, K.S.; Wang, Z.; Adroher, X.C. Materials, technological status, and fundamentals of PEM fuel cells–a review. *Mater. Today* **2020**, *32*, 178–203. [CrossRef]
60. Wilberforce, T.; Alaswad, A.; Palumbo, A.; Dassisti, M.; Olabi, A.G. Advances in stationary and portable fuel cell applications. *Int. J. Hydrogen Energy* **2016**, *41*, 16509–16522. [CrossRef]
61. Du, B.; Guo, Q.; Pollard, R.; Rodriguez, D.; Smith, C.; Elter, J. PEM fuel cells: Status and challenges for commercial stationary power applications. *JOM* **2006**, *58*, 45–49. [CrossRef]
62. Mancino, A.N.; Menale, C.; Vellucci, F.; Pasquali, M.; Bubbico, R. PEM Fuel Cell Applications in Road Transport. *Energies* **2023**, *16*, 6129. [CrossRef]
63. Sürer, M.G.; Arat, H.T. Advancements and current technologies on hydrogen fuel cell applications for marine vehicles. *Int. J. Hydrogen Energy* **2022**, *47*, 19865–19875. [CrossRef]
64. Dyantyi, N.; Parsons, A.; Sita, C.; Pasupathi, S. PEMFC for aeronautic applications: A review on the durability aspects. *Open Eng.* **2017**, *7*, 287–302. [CrossRef]
65. Karlsson, V.; Ahlmark, D. An Environmental Perspective on the Feasibility of Using Existing PEMFC Technology in General Aviation. Bachelor Thesis, Royal Institute of Technology, Stockholm, Sweden, 26 May 2013. Available online: https://urn.kb.se/resolve?urn=urn:nbn:se:kth:diva-141020 (accessed on 22 May 2024).
66. Jahnke, T.; Futter, G.; Latz, A.; Malkow, T.; Papakonstantinou, G.; Tsotridis, G.; Schott, P.; Gérard, M.; Quinaud, M.; Quiroga, M.; et al. Performance and degradation of Proton Exchange Membrane Fuel Cells: State of the art in modeling from atomistic to system scale. *J. Power Sources* **2016**, *304*, 207–233. [CrossRef]
67. Petrone, R.; Zheng, Z.; Hissel, D.; Péra, M.; Pianese, C.; Sorrentino, M.; Becherif, M.; Yousfi-Steiner, N. A review on model-based diagnosis methodologies for PEMFCs. *Int. J. Hydrogen Energy* **2013**, *38*, 7077–7091. [CrossRef]
68. Kandidayeni, M.; Macias, A.; Amamou, A.; Boulon, L.; Kelouwani, S.; Chaoui, H. Overview and benchmark analysis of fuel cell parameters estimation for energy management purposes. *J. Power Sources* **2018**, *380*, 92–104. [CrossRef]
69. Costa, M.A.; Wullt, B.; Norrlöf, M.; Gunnarsson, S. Failure detection in robotic arms using statistical modeling, machine learning and hybrid gradient boosting. *Measurement* **2019**, *146*, 425–436. [CrossRef]
70. Ashraf, H.; Abdellatif, S.O.; Elkholy, M.M.; El-Fergany, A.A. Computational techniques based on artificial intelligence for extracting optimal parameters of PEMFCs: Survey and insights. *Arch. Comput. Methods Eng.* **2022**, *29*, 3943–3972. [CrossRef]

71. Shah, A.; Luo, K.; Ralph, T.; Walsh, F. Recent trends and developments in polymer electrolyte membrane fuel cell modelling. *Electrochim. Acta* **2011**, *56*, 3731–3757. [CrossRef]
72. Shah, A.; Luo, K.; Ralph, T.; Walsh, F. CFD assisted modeling for control system design: A case study. *Simul. Model. Pract. Theory* **2009**, *17*, 730–742.
73. Al-Baghdadi, M.A.S. Modelling of proton exchange membrane fuel cell performance based on semi-empirical equations. *Renew. Energy* **2005**, *30*, 1587–1599. [CrossRef]
74. Pisani, L.; Murgia, G.; Valentini, M.; D'aguanno, B. A new semi-empirical approach to performance curves of polymer electrolyte fuel cells. *J. Power Sources* **2002**, *108*, 192–203. [CrossRef]
75. Menesklou, P.; Sinn, T.; Nirschl, H.; Gleiss, M. Grey box modelling of decanter centrifuges by coupling a numerical process model with a neural network. *Minerals* **2021**, *11*, 755. [CrossRef]
76. Wang, C.; Wang, S.; Peng, L.; Zhang, J.; Shao, Z.; Huang, J.; Sun, C.; Ouyang, M.; He, X. Recent progress on the key materials and components for proton exchange membrane fuel cells in vehicle applications. *Energies* **2016**, *9*, 603. [CrossRef]
77. Zhou, Y.; Wang, B.; Ling, Z.; Liu, Q.; Fu, X.; Zhang, Y.; Zhang, R.; Hu, S.; Zhao, F.; Li, X.; et al. Advances in ionogels for proton-exchange membranes. *Sci. Total Environ.* **2024**, *921*, 171099. [CrossRef] [PubMed]
78. Kraytsberg, A.; Ein-Eli, Y. Review of advanced materials for proton exchange membrane fuel cells. *Energy Fuels* **2014**, *28*, 7303–7330. [CrossRef]
79. Beuscher, U.; Cleghorn, S.J.; Johnson, W.B. Challenges for PEM fuel cell membranes. *Int. J. Energy Res.* **2005**, *29*, 1103–1112. [CrossRef]
80. Strathmann, H. *Ion-Exchange Membrane Separation Processes*; Elsevier: Amsterdam, The Netherlands, 2004.
81. Valenzuela, E.; Gamboa, S.; Sebastian, P.; Moreira, J.; Pantoja, J.; Ibañez, G.; Reyes, A.; Campillo, B.; Serna, S. Proton charge transport in nafion nanochannels. *J. Nano Res.* **2009**, *5*, 31–36. [CrossRef]
82. Chen, Q.; Schmidt-Rohr, K. Backbone Dynamics of the Nafion Ionomer Studied by 19F-13C Solid-State NMR. *Macromol. Chem. Phys.* **2007**, *208*, 2189–2203. [CrossRef]
83. Rao, Z.; Zheng, C.; Geng, F. Proton conduction of fuel cell polymer membranes: Molecular dynamics simulation. *Comput. Mater. Sci.* **2018**, *142*, 122–128. [CrossRef]
84. Harvey, D.A.; Bellemare-Davis, A.; Karan, K.; Jayansankar, B.; Pharoah, J.G.; Colbow, V.; Young, A.; Wessel, S. Statistical simulation of the performance and degradation of a PEMFC membrane electrode assembly. *ECS Trans.* **2013**, *50*, 147. [CrossRef]
85. Kwon, S.H.; Kang, H.; Sohn, Y.-J.; Lee, J.; Shim, S.; Lee, S.G. Molecular dynamics simulation study on the effect of perfluorosulfonic acid side chains on oxygen permeation in hydrated ionomers of PEMFCs. *Sci. Rep.* **2021**, *11*, 8702. [CrossRef]
86. Srinophakun, T.; Martkumchan, S. Ionic conductivity in a chitosan membrane for a PEM fuel cell using molecular dynamics simulation. *Carbohydr. Polym.* **2012**, *88*, 194–200. [CrossRef]
87. Chen, L.; He, Y.-L.; Tao, W.-Q. The temperature effect on the diffusion processes of water and proton in the proton exchange membrane using molecular dynamics simulation. *Numer. Heat Transf. Part A Appl.* **2014**, *65*, 216–228. [CrossRef]
88. Zheng, C.; Geng, F.; Rao, Z. Proton mobility and thermal conductivities of fuel cell polymer membranes: Molecular dynamics simulation. *Comput. Mater. Sci.* **2017**, *132*, 55–61. [CrossRef]
89. Park, C.H.; Kim, T.-H.; Kim, D.J.; Nam, S.Y. Molecular dynamics simulation of the functional group effect in hydrocarbon anionic exchange membranes. *Int. J. Hydrogen Energy* **2017**, *42*, 20895–20903. [CrossRef]
90. Kwon, S.H.; Lee, S.Y.; Kim, H.-J.; Jang, S.S.; Lee, S.G. Distribution characteristics of phosphoric acid and PTFE binder on Pt/C surfaces in high-temperature polymer electrolyte membrane fuel cells: Molecular dynamics simulation approach. *Int. J. Hydrogen Energy* **2021**, *46*, 17295–17305. [CrossRef]
91. Kusoglu, A.; Weber, A.Z. New insights into perfluorinated sulfonic-acid ionomers. *Chem. Rev.* **2017**, *117*, 987–1104. [CrossRef] [PubMed]
92. Dickinson, E.J.; Smith, G. Modelling the proton-conductive membrane in practical polymer electrolyte membrane fuel cell (PEMFC) simulation: A review. *Membranes* **2020**, *10*, 310. [CrossRef] [PubMed]
93. von Schröder, P. Über Erstarrungs-und Quellugserscheinungen von Gelatine. *Z. Für Phys. Chem.* **1903**, *45*, 75–117. [CrossRef]
94. Onishi, L.M.; Prausnitz, J.M.; Newman, J. Water– Nafion equilibria. Absence of schroeder's paradox. *J. Phys. Chem. B* **2007**, *111*, 10166–10173. [CrossRef]
95. Jeck, S.; Scharfer, P.; Kind, M. Absence of Schroeder's paradox: Experimental evidence for water-swollen Nafion®membranes. *J. Membr. Sci.* **2011**, *373*, 74–79. [CrossRef]
96. Springer, T.E.; Zawodzinski, T.; Gottesfeld, S. Polymer electrolyte fuel cell model. *J. Electrochem. Soc.* **1991**, *138*, 2334. [CrossRef]
97. Newman, J.; Balsara, N.P. *Electrochemical Systems*; John Wiley & Sons: Hoboken, NJ, USA, 2021.
98. Ye, Q.; Van Nguyen, T. Three-dimensional simulation of liquid water distribution in a PEMFC with experimentally measured capillary functions. *J. Electrochem. Soc.* **2007**, *154*, B1242. [CrossRef]
99. Kulikovsky, A. Quasi-3D modeling of water transport in polymer electrolyte fuel cells. *J. Electrochem. Soc.* **2003**, *150*, A1432. [CrossRef]
100. Weber, A.Z.; Newman, J. Transport in polymer-electrolyte membranes: I. Physical model. *J. Electrochem. Soc.* **2003**, *150*, A1008. [CrossRef]
101. Pasaogullari, U.; Wang, C.-Y.; Chen, K.S. Two-phase transport in polymer electrolyte fuel cells with bilayer cathode gas diffusion media. *J. Electrochem. Soc.* **2005**, *152*, A1574. [CrossRef]

102. Einstein, A. Über die von der molekularkinetischen Theorie der Wärme geforderte Bewegung von in ruhenden Flüssigkeiten suspendierten Teilchen [AdP 17, 549 (1905)]. *Ann. Der Phys.* **2005**, *14* (Suppl. S1), 182–193. [CrossRef]
103. Nernst, W. Zur kinetik der in lösung befindlichen körper. *Z. Für Phys. Chem.* **1888**, *2*, 613–637. [CrossRef]
104. Bird, R.B.; Stewart, W.E.; Lightfoot, E.N. *Transport Phenomena*; John Wiley & Sons: New York, NY, USA, 1960; Volume 413, p. 362.
105. Bennion, D.N. *Mass Transport of Binary Electrolyte Solutions in Membranes*; Department of Engineering, University of California: Berkeley, CA, USA, 1966.
106. Pintauro, P.N.; Bennion, D.N. Mass transport of electrolytes in membranes. 1. Development of mathematical transport model. *Ind. Eng. Chem. Fundam.* **1984**, *23*, 230–234. [CrossRef]
107. Koter, S.; Hamann, C. Characteristics of ion-exchange membranes for electrodialysis on the basis of irreversible thermodynamics. *J. Non-Equilibrium Thermodyn.* **1990**, *15*, 315–334. [CrossRef]
108. Auclair, B.; Nikonenko, V.; Larchet, C.; Métayer, M.; Dammak, L. Correlation between transport parameters of ion-exchange membranes. *J. Membr. Sci.* **2002**, *195*, 89–102. [CrossRef]
109. Baschuk, J.; Li, X. A general formulation for a mathematical PEM fuel cell model. *J. Power Sources* **2005**, *142*, 134–153. [CrossRef]
110. Baschuk, J.; Li, X. A comprehensive, consistent and systematic mathematical model of PEM fuel cells. *Appl. Energy* **2009**, *86*, 181–193. [CrossRef]
111. Fimrite, J.; Struchtrup, H.; Djilali, N. Transport phenomena in polymer electrolyte membranes: I. modeling framework. *J. Electrochem. Soc.* **2005**, *152*, A1804. [CrossRef]
112. Wöhr, M.; Bolwin, K.; Schnurnberger, W.; Fischer, M.; Neubrand, W.; Eigenberger, G. Dynamic modelling and simulation of a polymer membrane fuel cell including mass transport limitation. *Int. J. Hydrogen Energy* **1998**, *23*, 213–218. [CrossRef]
113. Berg, P.; Promislow, K.; Pierre, J.S.; Stumper, J.; Wetton, B. Water management in PEM fuel cells. *J. Electrochem. Soc.* **2004**, *151*, A341. [CrossRef]
114. Thampan, T.; Malhotra, S.; Tang, H.; Datta, R. Modeling of conductive transport in proton-exchange membranes for fuel cells. *J. Electrochem. Soc.* **2000**, *147*, 3242. [CrossRef]
115. Mason, E.A.; Malinauskas, A.P. *Gas Transportin Porous Media: The Dusty-Gas Model*; Elsevier: New York, NY, USA, 1983.
116. Weber, A.Z.; Newman, J. Modeling transport in polymer-electrolyte fuel cells. *Chem. Rev.* **2004**, *104*, 4679–4726. [CrossRef]
117. Bernardi, D.M.; Verbrugge, M.W. Mathematical model of a gas diffusion electrode bonded to a polymer electrolyte. *AIChE J.* **1991**, *37*, 1151–1163. [CrossRef]
118. Bernardi, D.M.; Verbrugge, M.W. A mathematical model of the solid-polymer-electrolyte fuel cell. *J. Electrochem. Soc.* **1992**, *139*, 2477. [CrossRef]
119. Verbrugge, M.W.; Hill, R.F. Ion and Solvent Transport in Ion-Exchange Membranes: I. A Macrohomogeneous Mathematical Model. *J. Electrochem. Soc.* **1990**, *137*, 886. [CrossRef]
120. Verbrugge, M.W.; Hill, R.F. Analysis of Promising Perfluorosulfonic Acid Membranes for Fuel-Cell Electrolytes. *J. Electrochem. Soc.* **1990**, *137*, 3770. [CrossRef]
121. Verbrugge, M.W.; Hill, R.F. Transport phenomena in perfluorosulfonic acid membranes during the passage of current. *J. Electrochem. Soc.* **1990**, *137*, 1131. [CrossRef]
122. Schlögl, R. Zur Theorie der anomalen Osmose. *Z. Für Phys. Chem.* **1955**, *3*, 73–102. [CrossRef]
123. Jiao, K.; Li, X. Water transport in polymer electrolyte membrane fuel cells. *Prog. Energy Combust. Sci.* **2011**, *37*, 221–291. [CrossRef]
124. Weber, A.Z.; Newman, J. Transport in polymer-electrolyte membranes: II. Mathematical model. *J. Electrochem. Soc.* **2004**, *151*, A311. [CrossRef]
125. Costamagna, P. Transport phenomena in polymeric membrane fuel cells. *Chem. Eng. Sci.* **2001**, *56*, 323–332. [CrossRef]
126. Jung, S.Y.; Nguyen, T.V. An along-the-channel model for proton exchange membrane fuel cells. *J. Electrochem. Soc.* **1998**, *145*, 1149.
127. You, L.; Liu, H. A two-phase flow and transport model for the cathode of PEM fuel cells. *Int. J. Heat Mass Transf.* **2002**, *45*, 2277–2287. [CrossRef]
128. Hertwig, K.; Martens, L.; Karwoth, R. Mathematical Modelling and Simulation of Polymer Electrolyte Membrane Fuel Cells. Part I: Model Structures and Solving an Isothermal One-Cell Model. *Fuel Cells* **2002**, *2*, 61–77. [CrossRef]
129. Zawodzinski, T.A.; Derouin, C.; Radzinski, S.; Sherman, R.J.; Smith, V.T.; Springer, T.E.; Gottesfeld, S. Water uptake by and transport through Nafion®117 membranes. *J. Electrochem. Soc.* **1993**, *140*, 1041. [CrossRef]
130. Zawodzinski, T.A., Jr.; Springer, T.E.; Uribe, F.; Gottesfeld, S., Jr. Characterization of polymer electrolytes for fuel cell applications. *Solid State Ion.* **1993**, *60*, 199–211. [CrossRef]
131. van Bussel, H.P.; Koene, F.G.; Mallant, R.K. Dynamic model of solid polymer fuel cell water management. *J. Power Sources* **1998**, *71*, 218–222. [CrossRef]
132. Berning, T. On the Nature of Electro-Osmotic Drag. *Energies* **2020**, *13*, 4726. [CrossRef]
133. Berning, T. On water transport in polymer electrolyte membranes during the passage of current. *Int. J. Hydrogen Energy* **2011**, *36*, 9341–9344. [CrossRef]
134. Fuller, T.F. *Solid-Polymer-Electrolyte Fuel Cells*; University of California: Berkeley, CA, USA, 1992.
135. Carcadea, E.; Ingham, D.B.; Stefanescu, I.; Ionete, R.; Ene, H. The influence of permeability changes for a 7-serpentine channel pem fuel cell performance. *Int. J. Hydrogen Energy* **2011**, *36*, 10376–10383. [CrossRef]
136. Zawodzinski, T.A.; Davey, J.; Valerio, J.; Gottesfeld, S. The water content dependence of electro-osmotic drag in proton-conducting polymer electrolytes. *Electrochim. Acta* **1995**, *40*, 297–302. [CrossRef]

137. Seddiq, M.; Khaleghi, H.; Mirzaei, M. Numerical analysis of gas cross-over through the membrane in a proton exchange membrane fuel cell. *J. Power Sources* **2006**, *161*, 371–379. [CrossRef]
138. de Bruijn, F.A.; Makkus, R.C.; Mallant, R.K.A.; Janssen, G.J.M. Materials for state-of-the-art PEM fuel cells, and their suitability for operation above 100 C. *Adv. Fuel Cells* **2007**, *1*, 235–336.
139. Eikerling, M.; Kharkats, Y.I.; Kornyshev, A.A.; Volfkovich, Y.M. Phenomenological theory of electro-osmotic effect and water management in polymer electrolyte proton-conducting membranes. *J. Electrochem. Soc.* **1998**, *145*, 2684. [CrossRef]
140. Eikerling, M.; Kornyshev, A.; Stimming, U. Electrophysical properties of polymer electrolyte membranes: A random network model. *J. Phys. Chem. B* **1997**, *101*, 10807–10820. [CrossRef]
141. Rui, Z.; Liu, J. Understanding of free radical scavengers used in highly durable proton exchange membranes. *Prog. Nat. Sci. Mater. Int.* **2020**, *30*, 732–742. [CrossRef]
142. Park, S.; Lee, H.; Shin, S.-H.; Kim, N.; Shin, D.; Bae, B. Increasing the durability of polymer electrolyte membranes using organic additives. *ACS Omega* **2018**, *3*, 11262–11269. [CrossRef] [PubMed]
143. Prabhakaran, V.; Arges, C.G.; Ramani, V. Investigation of polymer electrolyte membrane chemical degradation and degradation mitigation using in situ fluorescence spectroscopy. *Proc. Natl. Acad. Sci. USA* **2012**, *109*, 1029–1034. [CrossRef] [PubMed]
144. Zhu, G.; Chen, W.; Lu, S.; Chen, X. Parameter study of high-temperature proton exchange membrane fuel cell using data-driven models. *Int. J. Hydrogen Energy* **2019**, *44*, 28958–28967. [CrossRef]
145. Huo, W.; Li, W.; Zhang, Z.; Sun, C.; Zhou, F.; Gong, G. Performance prediction of proton-exchange membrane fuel cell based on convolutional neural network and random forest feature selection. *Energy Convers. Manag.* **2021**, *243*, 114367. [CrossRef]
146. Briceno-Mena, L.A.; Venugopalan, G.; Romagnoli, J.A.; Arges, C.G. Machine learning for guiding high-temperature PEM fuel cells with greater power density. *Patterns* **2021**, *2*, 100187. [CrossRef] [PubMed]
147. Ding, R.; Wang, R.; Ding, Y.; Yin, W.; Liu, Y.; Li, J.; Liu, J. Designing AI-Aided Analysis and Prediction Models for Nonprecious Metal Electrocatalyst-Based Proton-Exchange Membrane Fuel Cells. *Angew. Chem.* **2020**, *132*, 19337–19345. [CrossRef]
148. Amirinejad, M.; Tavajohi-Hasankiadeh, N.; Madaeni, S.S.; Navarra, M.A.; Rafiee, E.; Scrosati, B. Adaptive neuro-fuzzy inference system and artificial neural network modeling of proton exchange membrane fuel cells based on nanocomposite and recast Nafion membranes. *Int. J. Energy Res.* **2013**, *37*, 347–357. [CrossRef]
149. Musharavati, F. Four dimensional bio-inspired optimization approach with artificial intelligence for proton exchange membrane fuel cell. *Int. J. Energy Res.* **2022**, *46*, 21424–21440. [CrossRef]
150. Briceno-Mena, L.A.; Romagnoli, J.A.; Arges, C.G. PemNet: A Transfer Learning-Based Modeling Approach of High-Temperature Polymer Electrolyte Membrane Electrochemical Systems. *Ind. Eng. Chem. Res.* **2022**, *61*, 3350–3357. [CrossRef]
151. Foughali, Y.; Mankour, M.; Sekour, M.; Azzeddine, H.A.; Larbaoui, A.; Chaouch, D.-E.; Berka, M. A RBF artificial neural network to predict a fuel cell maximum power point. *Prz. Elektrotechniczny* **2022**, *1*, 102–106. [CrossRef]
152. Li, G.; Zhu, Y.; Guo, Y.; Mabuchi, T.; Li, D.; Huang, S.; Wang, S.; Sun, H.; Tokumasu, T. Deep Learning to Reveal the Distribution and Diffusion of Water Molecules in Fuel Cell Catalyst Layers. *ACS Appl. Mater. Interfaces* **2023**, *15*, 5099–5108. [CrossRef] [PubMed]
153. Zhang, H.; Liu, Z.; Liu, W.; Mao, L. Diagnosing Improper Membrane Water Content in Proton Exchange Membrane Fuel Cell Using Two-Dimensional Convolutional Neural Network. *Energies* **2022**, *15*, 4247. [CrossRef]
154. Huo, H.; Chen, J.; Wang, K.; Wang, F.; Jin, G.; Chen, F. State Estimation of Membrane Water Content of PEMFC Based on GA-BP Neural Network. *Sustainability* **2023**, *15*, 9094. [CrossRef]
155. Gu, X.; Hou, Z.; Cai, J. Data-based flooding fault diagnosis of proton exchange membrane fuel cell systems using LSTM networks. *Energy AI* **2021**, *4*, 100056. [CrossRef]
156. Nourizadeh, A. *Machine Learning of PEM Fuel Cell Degradation: Artificial Neural Network and Long Short-Term Memory Recurrent Neural Network*; University of California: Irvine, CA, USA, 2023.
157. Nguyen, H.-L.; Lee, S.-M.; Yu, S. A Comprehensive Review of Degradation Prediction Methods for an Automotive Proton Exchange Membrane Fuel Cell. *Energies* **2023**, *16*, 4772. [CrossRef]
158. Gatto, I.; Saccà, A.; Baglio, V.; Aricò, A.S.; Oldani, C.; Merlo, L.; Carbone, A. Evaluation of hot pressing parameters on the electrochemical performance of MEAs based on Aquivion®PFSA membranes. *J. Energy Chem.* **2019**, *35*, 168–173. [CrossRef]
159. Spiegel, C. *PEM Fuel Cell Modeling and Simulation Using MATLAB*; Elsevier: Amsterdam, The Netherlands, 2011.
160. Sui, P.-C.; Zhu, X.; Djilali, N. Modeling of PEM fuel cell catalyst layers: Status and outlook. *Electrochem. Energy Rev.* **2019**, *2*, 428–466. [CrossRef]
161. Wilson, M.S.; Gottesfeld, S. Thin-film catalyst layers for polymer electrolyte fuel cell electrodes. *J. Appl. Electrochem.* **1992**, *22*, 1–7. [CrossRef]
162. Janssen, G.; Overvelde, M. Water transport in the proton-exchange-membrane fuel cell: Measurements of the effective drag coefficient. *J. Power Sources* **2001**, *101*, 117–125. [CrossRef]
163. Weber, A.Z.; Newman, J. Transport in polymer-electrolyte membranes: III. model validation in a simple fuel-cell model. *J. Electrochem. Soc.* **2004**, *151*, A326. [CrossRef]
164. Bradean, R.; Promislow, K.; Wetton, B. Transport phenomena in the porous cathode of a proton exchange membrane fuel cell. *Numer. Heat Transf. Part A Appl.* **2002**, *42*, 121–138. [CrossRef]
165. Weber, A.Z.; Darling, R.M.; Newman, J. Modeling two-phase behavior in PEFCs. *J. Electrochem. Soc.* **2004**, *151*, A1715. [CrossRef]

166. Nam, J.H.; Kaviany, M. Effective diffusivity and water-saturation distribution in single-and two-layer PEMFC diffusion medium. *Int. J. Heat Mass Transf.* **2003**, *46*, 4595–4611. [CrossRef]
167. Newman, J. Optimization of potential and hydrogen utilization in an acid fuel cell. *Electrochim. Acta* **1979**, *24*, 223–229. [CrossRef]
168. Mann, R.F.; Amphlett, J.C.; Hooper, M.A.I.; Jensen, H.M.; Peppley, B.A.; Roberge, P.R. Development and application of a generalised steady-state electrochemical model for a PEM fuel cell. *J. Power Sources* **2000**, *86*, 173–180. [CrossRef]
169. West, A.; Fuller, T. Influence of rib spacing in proton-exchange membrane electrode assemblies. *J. Appl. Electrochem.* **1996**, *26*, 557–565. [CrossRef]
170. Yi, J.S.; Van Nguyen, T. Multicomponent transport in porous electrodes of proton exchange membrane fuel cells using the interdigitated gas distributors. *J. Electrochem. Soc.* **1999**, *146*, 38. [CrossRef]
171. Berning, T.; Lu, D.M.; Djilali, N. Three-dimensional computational analysis of transport phenomena in a PEM fuel cell. *J. Power Sources* **2002**, *106*, 284–294. [CrossRef]
172. Kazim, A.; Liu, H.; Forges, P. Modelling of performance of PEM fuel cells with conventional and interdigitated flow fields. *J. Appl. Electrochem.* **1999**, *29*, 1409–1416. [CrossRef]
173. Singh, D.; Lu, D.; Djilali, N. A two-dimensional analysis of mass transport in proton exchange membrane fuel cells. *Int. J. Eng. Sci.* **1999**, *37*, 431–452. [CrossRef]
174. Giner, J.; Hunter, C. The Mechanism of Operation of the Teflon-Bonded Gas Diffusion Electrode: A Mathematical Model. *J. Electrochem. Soc.* **1969**, *116*, 1124. [CrossRef]
175. Grens, E.A.; Turner, R.M.; Katan, T. A model for analysis of porous gas electrodes. *Adv. Energy Convers.* **1964**, *4*, 109–119. [CrossRef]
176. Chirkov, Y.G. Difference between hydrophobized and hydrophilic electrodes. III. Cylindrical gas pore model. *Sov. Electrochem.* **1975**, *11*, 36.
177. Chirkov, Y.G. Mechanism of filling wetproofed electrodes with gas. *Elektrokhimiya* **1972**, *8*, 11.
178. Cutlip, M. An approximate model for mass transfer with reaction inporous gas diffusion electrodes. *Electrochim. Acta* **1975**, *20*, 767–773. [CrossRef]
179. Cutlip, M.; Yang, S.; Stonehart, P. Model development for porous gas-diffusion electrodes and application to the phosphoric-acid fuel-cell cathode. *J. Electrochem. Soc.* **1986**, *86*, 12.
180. Iczkowski, R.P.; Cutlip, M.B. Voltage losses in fuel cell cathodes. *J. Electrochem. Soc.* **1980**, *127*, 1433. [CrossRef]
181. Cutlip, M.; Yang, S.; Stonehart, P. Simulation and optimization of porous gas-diffusion electrodes used in hydrogen oxygen phosphoric acid fuel cells—II development of a detailed anode model. *Electrochim. Acta* **1991**, *36*, 547–553. [CrossRef]
182. Vogel, W.; Lundquist, J.; Bradford, A. Reduction of oxygen on Teflon-backed platinum black electrodes. *Electrochim. Acta* **1972**, *17*, 1735–1744. [CrossRef]
183. Björnbom, P. Modelling of a double-layered PTFE-bonded oxygen electrode. *Electrochim. Acta* **1987**, *32*, 115–119. [CrossRef]
184. Antoine, O.; Bultel, Y.; Durand, R.; Ozil, P. Electrocatalysis, diffusion and ohmic drop in PEMFC: Particle size and spatial discrete distribution effects. *Electrochim. Acta* **1998**, *43*, 3681–3691. [CrossRef]
185. Bultel, Y.; Ozil, P.; Durand, R. Modelling the mode of operation of PEMFC electrodes at the particle level: Influence of ohmic drop within the active layer on electrode performance. *J. Appl. Electrochem.* **1998**, *28*, 269–276. [CrossRef]
186. Bultel, Y.; Ozil, P.; Durand, R. Modelling of mass transfer within the PEM fuel cell active layer: Limitations at the particle level. *J. Appl. Electrochem.* **1999**, *29*, 1025–1033. [CrossRef]
187. Tiedemann, W.; Newman, J. Maximum effective capacity in an ohmically limited porous electrode. *J. Electrochem. Soc.* **1975**, *122*, 1482. [CrossRef]
188. Kulikovsky, A.; Divisek, J.; Kornyshev, A. Modeling the cathode compartment of polymer electrolyte fuel cells: Dead and active reaction zones. *J. Electrochem. Soc.* **1999**, *146*, 3981. [CrossRef]
189. Murgia, G.; Pisani, L.; Valentini, M.; D'aguanno, B. Electrochemistry and mass transport in polymer electrolyte membrane fuel cells I. Model. *J. Electrochem. Soc.* **2001**, *149*, A31. [CrossRef]
190. Gurau, V.; Liu, H.; Kakac, S. Two-dimensional model for proton exchange membrane fuel cells. *AIChE J.* **1998**, *44*, 2410–2422. [CrossRef]
191. Bevers, D.; WO¨HR, M.; Yasuda, K.; Oguro, A.K. Simulation of a polymer electrolyte fuel cell electrode. *J. Appl. Electrochem.* **1997**, *27*, 1254–1264. [CrossRef]
192. Eikerling, M.; Kornyshev, A. Modelling the performance of the cathode catalyst layer of polymer electrolyte fuel cells. *J. Electroanal. Chem.* **1998**, *453*, 89–106. [CrossRef]
193. Fuller, T.F.; Newman, J. Water and thermal management in solid-polymer-electrolyte fuel cells. *J. Electrochem. Soc.* **1993**, *140*, 1218. [CrossRef]
194. Kornyshev, A.; Kulikovsky, A. Characteristic length of fuel and oxygen consumption in feed channels of polymer electrolyte fuel cells. *Electrochim. Acta* **2001**, *46*, 4389–4395. [CrossRef]
195. Kulikovsky, A.A. Quasi Three–Dimensional Modelling of the PEM Fuel Cell: Comparison of the Catalyst Layers Performance. *Fuel Cells* **2001**, *1*, 162–169. [CrossRef]
196. Wang, J.-T.; Savinell, R.F. Simulation studies on the fuel electrode of a H_2O_2 polymer electrolyte fuel cell. *Electrochim. Acta* **1992**, *37*, 2737–2745. [CrossRef]

197. Perry, M.L.; Newman, J.; Cairns, E.J. Mass transport in gas-diffusion electrodes: A diagnostic tool for fuel-cell cathodes. *J. Electrochem. Soc.* **1998**, *145*, 5. [CrossRef]
198. Um, S.; Wang, C.Y.; Chen, K. Computational fluid dynamics modeling of proton exchange membrane fuel cells. *J. Electrochem. Soc.* **2000**, *147*, 4485. [CrossRef]
199. Marr, C.; Li, X. Composition and performance modelling of catalyst layer in a proton exchange membrane fuel cell. *J. Power Sources* **1999**, *77*, 17–27. [CrossRef]
200. Gloaguen, F.; Convert, P.; Gamburzev, S.; Velev, O.; Srinivasan, S. An evaluation of the macro-homogeneous and agglomerate model for oxygen reduction in PEMFCs. *Electrochim. Acta* **1998**, *43*, 3767–3772. [CrossRef]
201. Gloaguen, F.; Durand, R. Simulations of PEFC cathodes: An effectiveness factor approach. *J. Appl. Electrochem.* **1997**, *27*, 1029–1035. [CrossRef]
202. Jaouen, F.; Lindbergh, G.; Sundholm, G. Investigation of Mass-Transport Limitations in the Solid Polymer Fuel Cell Cathode: I. Mathematical Model. *J. Electrochem. Soc.* **2002**, *149*, A437. [CrossRef]
203. Siegel, N.P.; Ellis, M.W.; Nelson, D.J.; Von Spakovsky, M.R. Single Domain Pemfc Model Based on Agglomerate Catalyst Geometry. *J. Power Sources* **2003**, *115*, 81–89. [CrossRef]
204. Broka, K.; Ekdunge, P. Modelling the PEM fuel cell cathode. *J. Appl. Electrochem.* **1997**, *27*, 281–289. [CrossRef]
205. Mehta, V.; Cooper, J.S. Review and analysis of PEM fuel cell design and manufacturing. *J. Power Sources* **2003**, *114*, 32–53. [CrossRef]
206. Antolini, E. Review in applied electrochemistry. Number 54 recent developments in polymer electrolyte fuel cell electrodes. *J. Appl. Electrochem.* **2004**, *34*, 563–576. [CrossRef]
207. Mazumder, S.; Cole, J.V. Rigorous 3-D mathematical modeling of PEM fuel cells: I. Model predictions without liquid water transport. *J. Electrochem. Soc.* **2003**, *150*, A1503. [CrossRef]
208. Mazumder, S.; Cole, J.V. Rigorous 3-D mathematical modeling of PEM fuel cells: II. Model predictions with liquid water transport. *J. Electrochem. Soc.* **2003**, *150*, A1510. [CrossRef]
209. Rho, Y.W.; Srinivasan, S.; Kho, Y.T. Mass Transport Phenomena in Proton Exchange Membrane Fuel Cells Using O_2/He, O_2/Ar, and O_2/N_2 Mixtures: II. Theoretical Analysis. *J. Electrochem. Soc.* **1994**, *141*, 2089. [CrossRef]
210. Maja, M.; Tosco, P.; Vanni, M. A One-Dimensional Model of Gas-Diffusion Electrodes for O_2 Reduction. *J. Electrochem. Soc.* **2001**, *148*, A1368. [CrossRef]
211. Pisani, L.; Murgia, G.; Valentini, M.; D'aguanno, B. A working model of polymer electrolyte fuel cells: Comparisons between theory and experiments. *J. Electrochem. Soc.* **2002**, *149*, A898. [CrossRef]
212. Dannenberg, K.; Ekdunge, P.; Lindbergh, G. Mathematical model of the PEMFC. *J. Appl. Electrochem.* **2000**, *30*, 1377–1387. [CrossRef]
213. Sui, P.-C.; Chen, L.-D.; Seaba, J.P.; Wariishi, Y. *Modeling and Optimization of a PEMFC Catalyst Layer*; SAE Transactions: Warrendale, PA, USA, 1999; pp. 729–737.
214. Sivertsen, B.R.; Djilali, N. CFD-based modelling of proton exchange membrane fuel cells. *J. Power Sources* **2005**, *141*, 65–78. [CrossRef]
215. d'Adamo, A.; Riccardi, M.; Borghi, M.; Fontanesi, S. CFD Modelling of a Hydrogen/Air PEM Fuel Cell with a Serpentine Gas Distributor. *Processes* **2021**, *9*, 564. [CrossRef]
216. Wang, C.-Y. Fundamental Models for Fuel Cell Engineering. *Chem. Rev.* **2004**, *104*, 4727–4766. [CrossRef] [PubMed]
217. Zhang, G.; Fan, L.; Sun, J.; Jiao, K. A 3D model of PEMFC considering detailed multiphase flow and anisotropic transport properties. *Int. J. Heat Mass Transf.* **2017**, *115*, 714–724. [CrossRef]
218. Nitta, I.; Karvonen, S.; Himanen, O.; Mikkola, M. Modelling the effect of inhomogeneous compression of GDL on local transport phenomena in a PEM fuel cell. *Fuel Cells* **2008**, *8*, 410–421. [CrossRef]
219. Zhang, G.; Jiao, K. Three-dimensional multi-phase simulation of PEMFC at high current density utilizing Eulerian-Eulerian model and two-fluid model. *Energy Convers. Manag.* **2018**, *176*, 409–421. [CrossRef]
220. Wang, B.; Wu, K.; Xi, F.; Xuan, J.; Xie, X.; Wang, X.; Jiao, K. Numerical analysis of operating conditions effects on PEMFC with anode recirculation. *Energy* **2019**, *173*, 844–856. [CrossRef]
221. Yang, X.-G.; Ye, Q.; Cheng, P. Matching of water and temperature fields in proton exchange membrane fuel cells with non-uniform distributions. *Int. J. Hydrogen Energy* **2011**, *36*, 12524–12537. [CrossRef]
222. De Lile, J.R.; Zhou, S. Theoretical modeling of the PEMFC catalyst layer: A review of atomistic methods. *Electrochim. Acta* **2015**, *177*, 4–20. [CrossRef]
223. Ju, H.; Wang, C.-Y. Experimental validation of a PEM fuel cell model by current distribution data. *J. Electrochem. Soc.* **2004**, *151*, A1954. [CrossRef]
224. Kone, J.-P.; Zhang, X.; Yan, Y.; Hu, G.; Ahmadi, G. Three-dimensional multiphase flow computational fluid dynamics models for proton exchange membrane fuel cell: A theoretical development. *J. Comput. Multiph. Flows* **2017**, *9*, 3–25. [CrossRef]
225. Ferreira, R.B.; Falcão, D.; Oliveira, V.; Pinto, A. 1D+ 3D two-phase flow numerical model of a proton exchange membrane fuel cell. *Appl. Energy* **2017**, *203*, 474–495. [CrossRef]
226. Chiu, H.-C.; Jang, J.-H.; Yan, W.-M.; Li, H.-Y.; Liao, C.-C. A three-dimensional modeling of transport phenomena of proton exchange membrane fuel cells with various flow fields. *Appl. Energy* **2012**, *96*, 359–370. [CrossRef]

227. Min, C.; He, Y.; Liu, X.; Yin, B.; Jiang, W.; Tao, W. Parameter sensitivity examination and discussion of PEM fuel cell simulation model validation: Part II: Results of sensitivity analysis and validation of the model. *J. Power Sources* **2006**, *160*, 374–385. [CrossRef]
228. Tao, W.; Min, C.; Liu, X.; He, Y.; Yin, B.; Jiang, W. Parameter sensitivity examination and discussion of PEM fuel cell simulation model validation: Part I. Current status of modeling research and model development. *J. Power Sources* **2006**, *160*, 359–373. [CrossRef]
229. Parthasarathy, A.; Srinivasan, S.; Appleby, A.J.; Martin, C.R. Temperature dependence of the electrode kinetics of oxygen reduction at the platinum/Nafion®interface—A microelectrode investigation. *J. Electrochem. Soc.* **1992**, *139*, 2530. [CrossRef]
230. Heidary, H.; Kermani, M.J.; Khajeh-Hosseini-Dalasm, N. Performance analysis of PEM fuel cells cathode catalyst layer at various operating conditions. *Int. J. Hydrogen Energy* **2016**, *41*, 22274–22284. [CrossRef]
231. Harvey, D.; Pharoah, J.; Karan, K. A comparison of different approaches to modelling the PEMFC catalyst layer. *J. Power Sources* **2008**, *179*, 209–219. [CrossRef]
232. Zhang, J.; Tang, Y.; Song, C.; Xia, Z.; Li, H.; Wang, H.; Zhang, J. PEM fuel cell relative humidity (RH) and its effect on performance at high temperatures. *Electrochim. Acta* **2008**, *53*, 5315–5321. [CrossRef]
233. Dickinson, E.J.; Hinds, G. The Butler-Volmer equation for polymer electrolyte membrane fuel cell (PEMFC) electrode kinetics: A critical discussion. *J. Electrochem. Soc.* **2019**, *166*, F221. [CrossRef]
234. Tomizawa, M.; Inoue, G.; Nagato, K.; Tanaka, A.; Park, K.; Nakao, M. Heterogeneous pore-scale model analysis of micro-patterned PEMFC cathodes. *J. Power Sources* **2023**, *556*, 232507. [CrossRef]
235. So, M.; Park, K.; Ohnishi, T.; Ono, M.; Tsuge, Y.; Inoue, G. A discrete particle packing model for the formation of a catalyst layer in polymer electrolyte fuel cells. *Int. J. Hydrogen Energy* **2019**, *44*, 32170–32183. [CrossRef]
236. Inoue, G.; Park, K.; So, M.; Kimura, N.; Tsuge, Y. Microscale simulations of reaction and mass transport in cathode catalyst layer of polymer electrolyte fuel cell. *Int. J. Hydrogen Energy* **2022**, *47*, 12665–12683. [CrossRef]
237. Olbrich, W.; Kadyk, T.; Sauter, U.; Eikerling, M. Wetting phenomena in catalyst layers of PEM fuel cells: Novel approaches for modeling and materials research. *J. Electrochem. Soc.* **2022**, *169*, 054521. [CrossRef]
238. Xu, K.; Zhao, X.; Hu, X.; Guo, Z.; Ye, Q.; Li, L.; Song, J.; Song, P. The review of the degradation mechanism of the catalyst layer of membrane electrode assembly in the proton exchange membrane fuel cell. In *IOP Conference Series: Earth and Environmental Science*; IOP Publishing: Bristol, UK, 2020.
239. Franco, A.A.; Schott, P.; Jallut, C.; Maschke, B. A dynamic mechanistic model of an electrochemical interface. *J. Electrochem. Soc.* **2006**, *153*, A1053. [CrossRef]
240. Franco, A.A.; Tembely, M. Transient multiscale modeling of aging mechanisms in a PEFC cathode. *J. Electrochem. Soc.* **2007**, *154*, B712. [CrossRef]
241. Franco, A.A.; Guinard, M.; Barthe, B.; Lemaire, O. Impact of carbon monoxide on PEFC catalyst carbon support degradation under current-cycled operating conditions. *Electrochim. Acta* **2009**, *54*, 5267–5279. [CrossRef]
242. Malek, K.; Franco, A.A. Microstructure-based modeling of aging mechanisms in catalyst layers of polymer electrolyte fuel cells. *J. Phys. Chem. B* **2011**, *115*, 8088–8101. [CrossRef] [PubMed]
243. Weber, A.Z.; Borup, R.L.; Darling, R.M.; Das, P.K.; Dursch, T.J.; Gu, W.; Harvey, D.; Kusoglu, A.; Litster, S.; Mench, M.M.; et al. A critical review of modeling transport phenomena in polymer-electrolyte fuel cells. *J. Electrochem. Soc.* **2014**, *161*, F1254. [CrossRef]
244. Da Wang, Y.; Meyer, Q.; Tang, K.; McClure, J.E.; White, R.T.; Kelly, S.T.; Crawford, M.M.; Iacoviello, F.; Brett, D.J.L.; Shearing, P.R.; et al. Large-scale physically accurate modelling of real proton exchange membrane fuel cell with deep learning. *Nat. Commun.* **2023**, *14*, 745. [CrossRef]
245. Yao, J.; Yang, Y.; Hou, X.; Yang, Y.; Yang, F.; Wu, Z.; Zhang, Z. Fast design of catalyst layer with optimal electrical-thermal-water performance for proton exchange membrane fuel cells. *J. Energy Chem.* **2023**, *81*, 642–655. [CrossRef]
246. Park, Y.; Hwang, C.-K.; Bang, K.; Hong, D.; Nam, H.; Kwon, S.; Yeo, B.C.; Go, D.; An, J.; Ju, B.-K.; et al. Machine learning filters out efficient electrocatalysts in the massive ternary alloy space for fuel cells. *Appl. Catal. B Environ.* **2023**, *339*, 123128. [CrossRef]
247. Elçiçek, H.; Özdemir, O.K. Prediction of electrocatalyst performance of Pt/C using response surface optimization algorithm-based machine learning approaches. *Int. J. Energy Res.* **2022**, *46*, 21353–21372. [CrossRef]
248. Zheng, L.; Hou, Y.; Zhang, T.; Pan, X. Performance prediction of fuel cells using long short-term memory recurrent neural network. *Int. J. Energy Res.* **2021**, *45*, 9141–9161. [CrossRef]
249. Xia, Z.; Wang, Y.; Ma, L.; Zhu, Y.; Li, Y.; Tao, J.; Tian, G. A hybrid prognostic method for proton-exchange-membrane fuel cell with decomposition forecasting framework based on AEKF and LSTM. *Sensors* **2022**, *23*, 166. [CrossRef] [PubMed]
250. Wang, Y.; Wu, K.; Zhao, H.; Li, J.; Sheng, X.; Yin, Y.; Du, Q.; Zu, B.; Han, L.; Jiao, K. Degradation prediction of proton exchange membrane fuel cell stack using semi-empirical and data-driven methods. *Energy AI* **2023**, *11*, 100205. [CrossRef]
251. Sun, B.; Liu, X.; Wang, J.; Wei, X.; Yuan, H.; Dai, H. Short-term performance degradation prediction of a commercial vehicle fuel cell system based on CNN and LSTM hybrid neural network. *Int. J. Hydrogen Energy* **2023**, *48*, 8613–8628. [CrossRef]
252. Sun, B.; Liu, X.; Wang, J.; Wei, X.; Yuan, H.; Dai, H. An improved neural network model for predicting the remaining useful life of proton exchange membrane fuel cells. *Int. J. Hydrogen Energy* **2023**, *48*, 25499–25511. [CrossRef]
253. Lou, Y.; Hao, M.; Li, Y. Machine-learning-assisted insight into the cathode catalyst layer in proton exchange membrane fuel cells. *J. Power Sources* **2022**, *543*, 231827. [CrossRef]
254. Yang, Z.; Wang, B.; Sheng, X.; Wang, Y.; Ren, Q.; He, S.; Xuan, J.; Jiao, K. An artificial intelligence solution for predicting short-term degradation behaviors of proton exchange membrane fuel cell. *Appl. Sci.* **2021**, *11*, 6348. [CrossRef]

255. Bi, W.; Gray, G.E.; Fuller, T.F. PEM fuel cell Pt/C dissolution and deposition in nafion electrolyte. *Electrochem. Solid-State Lett.* **2007**, *10*, B101. [CrossRef]
256. Ma, R.; Yang, T.; Breaz, E.; Li, Z.; Briois, P.; Gao, F. Data-driven proton exchange membrane fuel cell degradation predication through deep learning method. *Appl. Energy* **2018**, *231*, 102–115. [CrossRef]
257. Cindrella, L.; Kannan, A.; Lin, J.; Saminathan, K.; Ho, Y.; Lin, C.; Wertz, J. Gas diffusion layer for proton exchange membrane fuel cells—A review. *J. Power Sources* **2009**, *194*, 146–160. [CrossRef]
258. Nabovati, A.; Hinebaugh, J.; Bazylak, A.; Amon, C.H. Effect of porosity heterogeneity on the permeability and tortuosity of gas diffusion layers in polymer electrolyte membrane fuel cells. *J. Power Sources* **2014**, *248*, 83–90. [CrossRef]
259. Fadzillah, D.; Rosli, M.; Talib, M.; Kamarudin, S.; Daud, W. Review on microstructure modelling of a gas diffusion layer for proton exchange membrane fuel cells. *Renew. Sustain. Energy Rev.* **2017**, *77*, 1001–1009. [CrossRef]
260. Chu, H.-S.; Yeh, C.; Chen, F. Effects of porosity change of gas diffuser on performance of proton exchange membrane fuel cell. *J. Power Sources* **2003**, *123*, 1–9. [CrossRef]
261. Xia, L.; Ni, M.; He, Q.; Xu, Q.; Cheng, C. Optimization of gas diffusion layer in high temperature PEMFC with the focuses on thickness and porosity. *Appl. Energy* **2021**, *300*, 117357. [CrossRef]
262. Pasaogullari, U.; Wang, C.-Y. Two-phase transport and the role of micro-porous layer in polymer electrolyte fuel cells. *Electrochim. Acta* **2004**, *49*, 4359–4369. [CrossRef]
263. Dullien, F.A. *Porous Media: Fluid Transport and Pore Structure*; Academic Press: Cambridge, MA, USA, 2012.
264. Udell, K.S. Heat transfer in porous media considering phase change and capillarity—The heat pipe effect. *Int. J. Heat Mass Transf.* **1985**, *28*, 485–495. [CrossRef]
265. Ozden, A.; Alaefour, I.E.; Shahgaldi, S.; Li, X.; Colpan, C.O.; Hamdullahpur, F. Gas diffusion layers for PEM fuel cells: Ex-and in-situ characterization. In *Exergetic, Energetic and Environmental Dimensions*; Elsevier: Amsterdam, The Netherlands, 2018; pp. 695–727.
266. Nam, J.; Chippar, P.; Kim, W.; Ju, H. Numerical analysis of gas crossover effects in polymer electrolyte fuel cells (PEFCs). *Appl. Energy* **2010**, *87*, 3699–3709. [CrossRef]
267. Perng, S.-W.; Wu, H.-W. Non-isothermal transport phenomenon and cell performance of a cathodic PEM fuel cell with a baffle plate in a tapered channel. *Appl. Energy* **2011**, *88*, 52–67. [CrossRef]
268. Ismail, M.S.; Hughes, K.J.; Ingham, D.B.; Ma, L.; Pourkashanian, M. Effects of anisotropic permeability and electrical conductivity of gas diffusion layers on the performance of proton exchange membrane fuel cells. *Appl. Energy* **2012**, *95*, 50–63. [CrossRef]
269. Bruggeman, D. Dielectric constant and conductivity of mixtures of isotropic materials. *Ann. Phys.* **1935**, *24*, 636–679. [CrossRef]
270. Li, G.; Pickup, P.G. Ionic conductivity of PEMFC electrodes: Effect of Nafion loading. *J. Electrochem. Soc.* **2003**, *150*, C745. [CrossRef]
271. Rothfeld, L.B. Gaseous counterdiffusion in catalyst pellets. *AIChE J.* **1963**, *9*, 19–24. [CrossRef]
272. Thampan, T.; Malhotra, S.; Zhang, J.; Datta, R. PEM fuel cell as a membrane reactor. *Catal. Today* **2001**, *67*, 15–32. [CrossRef]
273. Nguyen, T.V.; White, R.E. A water and heat management model for proton-exchange-membrane fuel cells. *J. Electrochem. Soc.* **1993**, *140*, 2178. [CrossRef]
274. Dohle, H.; Kornyshev, A.; Kulikovsky, A.; Mergel, J.; Stolten, D. The current voltage plot of PEM fuel cell with long feed channels. *Electrochem. Commun.* **2001**, *3*, 73–80. [CrossRef]
275. Bear, J. *Dynamics of Fluids in Porous Media Dover Publications*; INC: New York, NY, USA, 1988.
276. Leverett, M. Capillary behavior in porous solids. *Trans. AIME* **1941**, *142*, 152–169. [CrossRef]
277. Kim, K.N.; Kang, J.H.; Lee, S.G.; Nam, J.H.; Kim, C.-J. Lattice Boltzmann simulation of liquid water transport in microporous and gas diffusion layers of polymer electrolyte membrane fuel cells. *J. Power Sources* **2015**, *278*, 703–717. [CrossRef]
278. Zhang, D.; Cai, Q.; Gu, S. Three-dimensional lattice-Boltzmann model for liquid water transport and oxygen diffusion in cathode of polymer electrolyte membrane fuel cell with electrochemical reaction. *Electrochim. Acta* **2018**, *262*, 282–296. [CrossRef]
279. Yang, M.; Jiang, Y.; Liu, J.; Xu, S.; Du, A. Lattice Boltzmann method modeling and experimental study on liquid water characteristics in the gas diffusion layer of proton exchange membrane fuel cells. *Int. J. Hydrogen Energy* **2022**, *47*, 10366–10380. [CrossRef]
280. Salah, Y.B.; Tabe, Y.; Chikahisa, T. Gas channel optimisation for PEM fuel cell using the lattice Boltzmann method. *Energy Procedia* **2012**, *28*, 125–133. [CrossRef]
281. Koorata, P.K. Deformation Mechanics of Fuel Cell Gas Diffusion Layer: Cyclic Response and Constitutive Model. *J. Electrochem. Soc.* **2022**, *169*, 104505. [CrossRef]
282. Wang, Y.; Guan, C.; Zhang, P.; Zhu, T.; Wang, S.; Zhu, Y.; Wang, X. Optimal design of a cathode flow field with a new arrangement of baffle plates for a high clean power generation of a polymer electrolyte membrane fuel cell. *J. Clean. Prod.* **2022**, *375*, 134187. [CrossRef]
283. Poornesh, K.; Cho, C. Stability of polymer electrolyte membranes in fuel cells: Initial attempts to bridge physical and chemical degradation modes. *Fuel Cells* **2015**, *15*, 196–203. [CrossRef]
284. Shinde, U.; Koorata, P.K.; Padavu, P. Electrical/flow heterogeneity of gas diffusion layer and inlet humidity induced performance variation in polymer electrolyte fuel cells. *Int. J. Hydrogen Energy* **2023**, *48*, 12877–12892. [CrossRef]
285. Mortazavi, M.; Tajiri, K. Effect of the PTFE content in the gas diffusion layer on water transport in polymer electrolyte fuel cells (PEFCs). *J. Power Sources* **2014**, *245*, 236–244. [CrossRef]

286. Mathias, M.; Roth, J.; Fleming, J.; Lehnert, W.; Vielstich, W. Handbook of fuel cells—Fundamentals, technology and applications. *Fuel Cell Technol. Appl.* **2003**, *3*, 517–537.
287. Bazylak, A.; Sinton, D.; Liu, Z.-S.; Djilali, N. Effect of compression on liquid water transport and microstructure of PEMFC gas diffusion layers. *J. Power Sources* **2007**, *163*, 784–792. [CrossRef]
288. Radhakrishnan, V.; Haridoss, P. Effect of cyclic compression on structure and properties of a Gas Diffusion Layer used in PEM fuel cells. *Int. J. Hydrogen Energy* **2010**, *35*, 11107–11118. [CrossRef]
289. Wu, J.; Martin, J.J.; Orfino, F.P.; Wang, H.; Legzdins, C.; Yuan, X.-Z.; Sun, C. In situ accelerated degradation of gas diffusion layer in proton exchange membrane fuel cell: Part I: Effect of elevated temperature and flow rate. *J. Power Sources* **2010**, *195*, 1888–1894. [CrossRef]
290. Pan, Y.; Wang, H.; Brandon, N.P. Gas diffusion layer degradation in proton exchange membrane fuel cells: Mechanisms, characterization techniques and modelling approaches. *J. Power Sources* **2021**, *513*, 230560. [CrossRef]
291. El-Kharouf, A.; Mason, T.J.; Brett, D.J.; Pollet, B.G. Ex-situ characterisation of gas diffusion layers for proton exchange membrane fuel cells. *J. Power Sources* **2012**, *218*, 393–404. [CrossRef]
292. Qiu, D.; Janssen, H.; Peng, L.; Irmscher, P.; Lai, X.; Lehnert, W. Electrical resistance and microstructure of typical gas diffusion layers for proton exchange membrane fuel cell under compression. *Appl. Energy* **2018**, *231*, 127–137. [CrossRef]
293. Pourrahmani, H.; Van Herle, J. The impacts of the gas diffusion layer contact angle on the water management of the proton exchange membrane fuel cells: Three-dimensional simulation and optimization. *Int. J. Energy Res.* **2022**, *46*, 16027–16040. [CrossRef]
294. Bao, Y.; Gan, Y. Roughness effects of gas diffusion layers on droplet dynamics in PEMFC flow channels. *Int. J. Hydrogen Energy* **2020**, *45*, 17869–17881. [CrossRef]
295. Ira, Y.; Bakhshan, Y.; Khorshidimalahmadi, J. Effect of wettability heterogeneity and compression on liquid water transport in gas diffusion layer coated with microporous layer of PEMFC. *Int. J. Hydrogen Energy* **2021**, *46*, 17397–17413. [CrossRef]
296. Shum, A.D.; Liu, C.P.; Lim, W.H.; Parkinson, D.Y.; Zenyuk, I.V. Using Machine Learning Algorithms for Water Segmentation in Gas Diffusion Layers of Polymer Electrolyte Fuel Cells. *Transp. Porous Media* **2022**, *144*, 715–737. [CrossRef]
297. Mahdaviara, M.; Shojaei, M.J.; Siavashi, J.; Sharifi, M.; Blunt, M.J. Deep learning for multiphase segmentation of X-ray images of gas diffusion layers. *Fuel* **2023**, *345*, 128180. [CrossRef]
298. Froning, D.; Wirtz, J.; Hoppe, E.; Lehnert, W. Flow Characteristics of Fibrous Gas Diffusion Layers Using Machine Learning Methods. *Appl. Sci.* **2022**, *12*, 12193. [CrossRef]
299. Lobato, J.; Cañizares, P.; Rodrigo, M.A.; Piuleac, C.-G.; Curteanu, S.; Linares, J.J. Direct and inverse neural networks modelling applied to study the influence of the gas diffusion layer properties on PBI-based PEM fuel cells. *Int. J. Hydrogen Energy* **2010**, *35*, 7889–7897. [CrossRef]
300. Pourrahmani, H. Water management of the proton exchange membrane fuel cells: Optimizing the effect of microstructural properties on the gas diffusion layer liquid removal. *Energy* **2022**, *256*, 124712. [CrossRef]
301. Lei, H.; Xing, L.; Jiang, H.; Wang, Y.; Bin Xu, B.; Xuan, J.; Liu, T.X. Designing graded fuel cell electrodes for proton exchange membrane (PEM) fuel cells with recurrent neural network (RNN) approaches. *Chem. Eng. Sci.* **2023**, *267*, 118350. [CrossRef]
302. Nagulapati, V.M.; Kumar, S.S.; Annadurai, V.; Lim, H. Machine learning based fault detection and state of health estimation of proton exchange membrane fuel cells. *Energy AI* **2023**, *12*, 100237. [CrossRef]
303. Wang, B.; Yang, Z.; Ji, M.; Shan, J.; Ni, M.; Hou, Z.; Cai, J.; Gu, X.; Yuan, X.; Gong, Z.; et al. Long short-term memory deep learning model for predicting the dynamic performance of automotive PEMFC system. *Energy AI* **2023**, *14*, 100278. [CrossRef]
304. Yang, Y.; Li, X.; Tang, F.; Ming, P.; Li, B.; Zhang, C. Power evolution of fuel cell stack driven by anode gas diffusion layer degradation. *Appl. Energy* **2022**, *313*, 118858. [CrossRef]
305. Marappan, M.; Palaniswamy, K.; Velumani, T.; Chul, K.B.; Velayutham, R.; Shivakumar, P.; Sundaram, S. Performance Studies of Proton Exchange Membrane Fuel Cells with Different Flow Field Designs—Review. *Chem. Rec.* **2021**, *21*, 663–714. [CrossRef]
306. Li, X.; Sabir, I. Review of bipolar plates in PEM fuel cells: Flow-field designs. *Int. J. Hydrogen Energy* **2005**, *30*, 359–371. [CrossRef]
307. Hermann, A.; Chaudhuri, T.; Spagnol, P. Bipolar plates for PEM fuel cells: A review. *Int. J. Hydrogen Energy* **2005**, *30*, 1297–1302. [CrossRef]
308. Cunningham, B.; Baird, D.G. The development of economical bipolar plates for fuel cells. *J. Mater. Chem.* **2006**, *16*, 4385–4388. [CrossRef]
309. Tawfik, H.; Hung, Y.; Mahajan, D. Metal bipolar plates for PEM fuel cell—A review. *J. Power Sources* **2007**, *163*, 755–767. [CrossRef]
310. Antunes, R.A.; Oliveira, M.C.L.; Ett, G.; Ett, V. Corrosion of metal bipolar plates for PEM fuel cells: A review. *Int. J. Hydrogen Energy* **2010**, *35*, 3632–3647. [CrossRef]
311. Antunes, R.A.; Oliveira, M.C.L.; Ett, G.; Ett, V. Contact resistance prediction of proton exchange membrane fuel cell considering fabrication characteristics of metallic bipolar plates. *Energy Convers. Manag.* **2018**, *169*, 334–344.
312. Ramos-Alvarado, B.; Hernandez-Guerrero, A.; Elizalde-Blancas, F.; Ellis, M.W. Constructal flow distributor as a bipolar plate for proton exchange membrane fuel cells. *Int. J. Hydrogen Energy* **2011**, *36*, 12965–12976. [CrossRef]
313. Atyabi, S.A.; Afshari, E. A numerical multiphase CFD simulation for PEMFC with parallel sinusoidal flow fields. *J. Therm. Anal. Calorim.* **2019**, *135*, 1823–1833. [CrossRef]
314. Choi, K.-S.; Kim, H.-M.; Moon, S.-M. Numerical studies on the geometrical characterization of serpentine flow-field for efficient PEMFC. *Int. J. Hydrogen Energy* **2011**, *36*, 1613–1627. [CrossRef]

315. Perng, S.-W.; Wu, H.-W. A three-dimensional numerical investigation of trapezoid baffles effect on non-isothermal reactant transport and cell net power in a PEMFC. *Appl. Energy* **2015**, *143*, 81–95. [CrossRef]
316. Wu, H.; Li, X.; Berg, P. On the modeling of water transport in polymer electrolyte membrane fuel cells. *Electrochim. Acta* **2009**, *54*, 6913–6927. [CrossRef]
317. Kakaç, S.; Shah, R.K.; Aung, W. *Handbook of Single-Phase Convective Heat Transfer*; Wiley-Interscience: Hoboken, NJ, USA, 1987.
318. Hartnig, C.; Schmidt, T.J. On a new degradation mode for high-temperature polymer electrolyte fuel cells: How bipolar plate degradation affects cell performance. *Electrochim. Acta* **2011**, *56*, 4237–4242. [CrossRef]
319. Eom, K.; Cho, E.; Nam, S.-W.; Lim, T.-H.; Jang, J.H.; Kim, H.-J.; Hong, B.K.; Yang, Y.C. Degradation behavior of a polymer electrolyte membrane fuel cell employing metallic bipolar plates under reverse current condition. *Electrochim. Acta* **2012**, *78*, 324–330. [CrossRef]
320. Zhang, G.; Qu, Z.; Tao, W.-Q.; Wang, X.; Wu, L.; Wu, S.; Xie, X.; Tongsh, C.; Huo, W.; Bao, Z.; et al. Porous flow field for next-generation proton exchange membrane fuel cells: Materials, characterization, design, and challenges. *Chem. Rev.* **2022**, *123*, 989–1039. [CrossRef]
321. Ahn, C.; Lim, M.S.; Hwang, W.; Kim, S.; Park, J.E.; Lim, J.; Choi, I.; Cho, Y.; Sung, Y. Effect of porous metal flow field in polymer electrolyte membrane fuel cell under pressurized condition. *Fuel Cells* **2017**, *17*, 652–661. [CrossRef]
322. Yu, Z.; Xia, L.; Xu, G.; Wang, C.; Wang, D. Improvement of the three-dimensional fine-mesh flow field of proton exchange membrane fuel cell (PEMFC) using CFD modeling, artificial neural network and genetic algorithm. *Int. J. Hydrogen Energy* **2022**, *47*, 35038–35054. [CrossRef]
323. Zheng, J.; Qin, Y.; Guo, Q.; Dong, Z.; Zhu, C.; Wang, Y. Block structure optimization in PEMFC flow channels using a data-driven surrogate model based on random forest. *Int. J. Green Energy* **2023**, *20*, 816–822. [CrossRef]
324. Wilberforce, T.; Biswas, M. A study into proton exchange membrane fuel cell power and voltage prediction using artificial neural network. *Energy Rep.* **2022**, *8*, 12843–12852. [CrossRef]
325. Yang, J.; Wu, Y.; Liu, X. Proton Exchange Membrane Fuel Cell Power Prediction Based on Ridge Regression and Convolutional Neural Network Data-Driven Model. *Sustainability* **2023**, *15*, 11010. [CrossRef]
326. Cao, J.; Yin, C.; Feng, Y.; Su, Y.; Lu, P.; Tang, H. A Dimension-Reduced Artificial Neural Network Model for the Cell Voltage Consistency Prediction of a Proton Exchange Membrane Fuel Cell Stack. *Appl. Sci.* **2022**, *12*, 11602. [CrossRef]
327. Li, H.-W.; Liu, J.-N.; Yang, Y.; Lu, G.-L.; Qiao, B.-X. Coupling flow channel optimization and Bagging neural network to achieve performance prediction for proton exchange membrane fuel cells with varying imitated water-drop block channel. *Int. J. Hydrogen Energy* **2022**, *47*, 39987–40007. [CrossRef]
328. Li, H.-W.; Liu, J.-N.; Yang, Y.; Lu, G.-L.; Qiao, B.-X. A coupled and interactive influence of operational parameters for optimizing power output of cleaner energy production systems under uncertain conditions. *Int. J. Energy Res.* **2019**, *43*, 1294–1302.

processes

MDPI

Article

Primary-Side Indirect Control of the Battery Charging Current in a Wireless Power Transfer Charger Using Adaptive Hill-Climbing Control Technique

Abdellah Lassioui [1,*], Marouane El Ancary [1], Zakariae El Idrissi [1], Hassan El Fadil [1], Kamal Rachid [1] and Aziz Rachid [2]

1 ISA Laboratory, National School of Applied Sciences (ENSA), Ibn Tofail University, Kenitra 14000, Morocco; marouane.elancary@gmail.com (M.E.A.); zakariae.elidrissi@gmail.com (Z.E.I.); elfadilhassan@yahoo.fr (H.E.F.); kamal.rachid01@gmail.com (K.R.)

2 LSIB Laboratory, Faculty of Sciences and Techniques (FST), Hassan II University of Casablanca, Mohammedia 28806, Morocco; rachidaziz03@gmail.com

* Correspondence: abdellah.lassioui@uit.ac.ma

Abstract: This paper addresses the control task of a wireless power transfer (WPT) charger designed for electric vehicles (EVs). The challenge is to maintain a constant battery charging current when the WPT is controlled on the ground side. Indeed, the intermittent latency involved in the wireless data communication between the ground and vehicle sides leads to system instability. To overcome this issue, a new control approach has been proposed in this paper. The proposed technique ensures indirect control of the battery charging current through control of the current on the ground side. The control technique relies on an adaptive hill-climbing algorithm in conjunction with a PI-based controller. The adaptive parameter is adjusted online, during the operation of the charger, only when a new measure of the battery charging current is received on the primary side. This makes it possible to avoid the need for real-time wireless data communication. It should be noted that this aspect is crucial in ensuring the controller's robustness and stability of the system regardless of potential delays in wireless communication and large misalignments between the coils. The validity of the proposed control technique has been confirmed through simulation. In addition, experimental validation, using a laboratory test bed, demonstrated satisfactory results.

Keywords: electric vehicles; wireless power transfer; battery charging; hill-climbing; indirect control

1. Introduction

Wireless power transfer (WPT) refers to a technique enabling power transfer without physical contact between the electric vehicles (EV) and the charging station [1,2]. It is worth noting that the integration of WPT technology in EVs represents a significant advancement in automotive technology. Indeed, WPT eliminates the need for physical connectors or cables, offering an enhanced and user-centric charging experience for owners of EVs. This can lead to increased adoption of EVs by addressing concerns about charging infrastructure availability and accessibility [3–5].

According to the literature, the WPT technology could be segmented into three categories: static charging mode (for parked vehicles), stationary charging (brief charging stops in urban areas such as bus stops and traffic lights), and dynamic charging (charging while in motion) [6,7]. Static charging, typically using low-power chargers, occurs when the vehicle is stationary in either a private garage or public parking area. Stationary charging briefly boosts EV range without turning off the engine. Dynamic WPT charging occurs while the vehicle is in motion, enabling continuous charging and eliminating the need for frequent stops to recharge [3,4].

The SAE J2954 standard outlines a schematic overview of the WPT charging system which could be divided into two subsystems as shown in Figure 1: a primary side, also known as ground assembly (GA), and a secondary side known as vehicle assembly (VA). The GA begins with rectifying the AC grid voltage using an AC/DC converter which, guarantees a power factor of unity [8–10]. The rectified DC voltage is supplied to the GA inverter, which operates typically at 85 kHz and is controlled using phase shift techniques [11–13]. A reactive power compensation block follows, eliminating current harmonics and compensating for reactive power [14]. The transmitting coil, powered by the inverter, generates a time-variable magnetic field (MF), transferred wirelessly and guided to the receiving coil via ferrite cores [15,16]. The receiving coil induces an AC voltage, rectified by either controlled or diode-based rectifiers, depending on the system type (bidirectional WPT or unidirectional WPT) [17]. The rectified voltage is used to supply the battery with the recommended charging current and voltage [18–22].

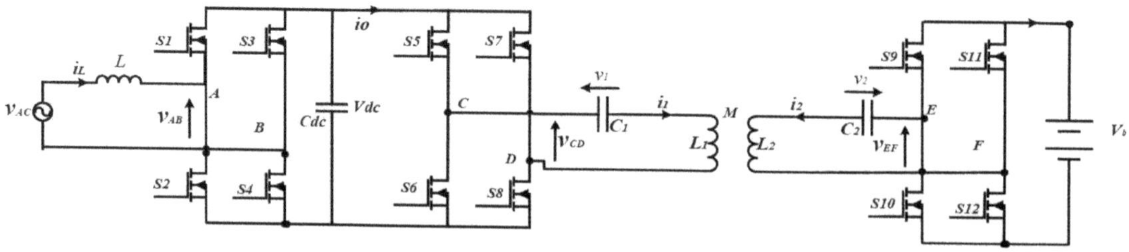

Figure 1. Overall block diagram of a WPT charger.

The SAEJ2954 also covers the interoperability issue in WPT chargers as a crucial aspect for user convenience, safety, and market growth. Indeed, it ensures seamless charging across different brands and models. In addition, interoperability standards also promote safety and protect users from potential hazards. Overall, the interoperability evaluation of WPT chargers can be categorized into three categories: (i) interoperability based on the coil's geometry and parameters, (ii) interoperability based on compensation networks, and (iii) interoperability based on efficiency test [23].

- Interoperability based on the coil's geometry and parameters

In this case, the interoperability is evaluated based on magnetic flux distribution and the coupling coefficient or mutual inductance between the ground side coil and vehicle side coil. Research indicates that circular-circular and circular-rectangular coil pairs generally offer medium coupling coefficients, while DD-DD, Circular-DD and DD-DDQP coil pairs have high mutual coupling and transmission efficiency demonstrating good interoperability [23]. Still, the complex design of these coils could be the main drawback limiting their large adoption in WPT chargers [16].

- Interoperability based on compensation networks

The compensation network in WPT chargers ensures impedance matching and reactive power compensation. Even with interoperable coils, an inappropriate compensation network can cause impedance mismatch, reducing the active power received by the load. Common compensation networks in EV wireless charging include series (S), parallel (P), and LCC. Research shows that only the S-S topology maintains frequency stability despite load changes, indicating why the S-S topology is widely adopted for WPT charging [24].

- Interoperability evaluation based on efficiency test

Evaluating interoperability based on power and efficiency is the most common and intuitive method and is widely used in industries. According to SAE J2954, efficiency is defined as the power ratio from the grid to the battery, which must be at least 85% of the rated working power when the ground and vehicle side coils are perfectly aligned. In addition, the efficiency of the WPT charger must be more than 80% of the maximum misalignment between the coils [25].

For the WPT charger to operate safely and reliably, it is essential to have efficient data communication between the ground assembly and the vehicle assembly. Indeed, in feedback control systems, delayed or inconsistent communication can result in oscillations, overshoot, or even instability of the control loop.

From the control viewpoint, several techniques have been proposed in the literature [26,27]. A ground-side approach is proposed in [28]. The chosen approach prioritizes primary side power regulation to reduce the complexity, size, and cost of vehicle onboard components.

A ground-side control technique of a WPT charger has been proposed in [29]. In this technique, the power transferred to the battery is controlled through the control of DC-DC power converters. The authors claim the universal characteristic of the technique; however, it still suffers from complex power architecture as additional DC-DC converters are used on both sides of the charger. In order to reduce the complexity and size of the WPT charger, a primary-side current and voltage control technique has been proposed in [30]. The technique enables key benefits such as rapid mutual inductance determination, elimination of receiver-side DC-DC converters, and automatic charging mode selection. However, the proposed technique eliminates the use of wireless communication, which is mandatory in a WPT charger as it allows battery state monitoring for safe charging.

Another ground-side technique based on two-level control is proposed in [31]. The proposed technique proves its effectiveness in enhancing the start-up transient time of a WPT charger. Apart from its complexity, the proposed technique is validated using a resistive load which does not model the real behaviour of a real battery.

In [32], the authors propose a charging technique that maintains constant current and voltage using a primary-side controller. This approach utilizes a load identification technique to estimate the battery's equivalent resistance and eliminates the need for real-time wireless communication.

Similar to [32], a primary side control with an improved load identification approach is proposed in [33].

A dual-side communication-free control technique is introduced in [34]. The battery charging current and voltage are controlled using a model predictive control (MPC) technique implemented on the primary side of the WPT charger. However, the complexity and computational burden could be the main limitations of this technique.

The authors in [35] propose a method to simplify the control and eliminate wireless communication. Indeed, the primary side is controlled to limit the output voltage while the secondary side is controlled to achieve CC/CV charging through a DC-DC converter. Besides its simplicity and effectiveness in ensuring good charging performances under large voltage and misalignment variations, this technique suffers from additional power converters and the need for dual-side controllers, which increases its cost and implementation complexity.

The authors in [36,37] propose a simultaneous data and power transfer for a WPT charger. In this technique, both data and power are transmitted through the magnetic coupler using the phase shift keying technique. The proposed method eliminates the need for wireless communication boards; however, additional decoding circuits are mandatory to decode and process data received on the ground side.

From the literature review provided, several common points emerge regarding control techniques for WPT chargers. Indeed, many techniques prioritize primary side power regulation to reduce the complexity, size, and cost of vehicle onboard components by eliminating receiver-side DC-DC converters. While some techniques eliminate the need for wireless communication, it is acknowledged as essential for battery state monitoring for safe charging.

In this paper, the novelty of the proposed technique is that it uses an indirect control of the battery current through the control of the primary side current while maintaining wireless communication between the GA and VA for monitoring and safe operation of the battery charging parameters. The technique is based on an adaptive hill-climbing algorithm in conjunction with a PI-based controller. The adaptive parameter is updated online, during the operation of the charger, only when a new measure of the battery charging current is available. This makes it possible to avoid the need for real-time wireless data communication. Additionally, it maintains the continuous operation of the controller during the latency intervals.

The structure of this paper is as follows: Section 2 focuses on the modelling of the WPT charger, followed by the presentation of the proposed control technique in Section 3. Section 4 delves into the simulation results, while Section 5 covers the experimental test bed setup and the corresponding results. Section 6 addresses the main results and the supremacy of the proposed control technique over existing methods. Finally, the paper concludes with Section 7.

2. Modelling of the Wireless Power Transfer Charger

This section is devoted to deriving an accurate mathematical model for the WPT charger. This model will be used later for testing the proposed control technique. Note that the modelling task of the AC/DC PFC converter is omitted in this section as the proposed technique is applied only to the DC-DC part of the WPT charger. In addition, the battery has been modelled by an equivalent nonlinear circuit consisting of an open circuit voltage v_{OC} in series with a resistance r_b. Applying Kirchhoff's laws to the circuit in Figure 1 yields (1)–(5):

$$v_{CD} = v_{C1} + L_1 \frac{di_1}{dt} - M \frac{di_2}{dt} \tag{1}$$

$$M \frac{di_1}{dt} = L_2 \frac{di_2}{dt} + v_{C2} + v_{EF} \tag{2}$$

$$i_1 = C_1 \frac{dv_{C1}}{dt} \tag{3}$$

$$i_2 = C_2 \frac{dv_{C2}}{dt} \tag{4}$$

$$i_0 = |i_2| = \frac{v_0 - v_{OC}}{r_b} + C_f \frac{dv_0}{dt} \tag{5}$$

where v_{cd} is the output voltage of the ground side inverter, v_{c1} and v_{c2}, are, respectively, the voltage across the ground and vehicle side compensation capacitors, L_{GA} and L_{VA} are the inductances of the ground and vehicle side coils, i_1 and i_2 are current in the ground side and vehicle side coils, M is the mutual inductance between the coils, v_{ef} and v_0 are, respectively, the voltage at the input and output of the vehicle side rectifier, C_f is the filtering capacitor at the output of the vehicle side rectifier.

Let us select the state variable vector as depicted in (6). Thus, rearranging (1)–(5), one obtains the state space representation of the WPT charger as shown in (7)–(11).

$$x = [v_{C1}, v_{C2}, i_1, i_2, v_0]^T = [x_1, x_2, x_3, x_4, x_5]^T \tag{6}$$

$$\dot{x}_1 = \frac{1}{C_1} x_3 \tag{7}$$

$$\dot{x}_2 = \frac{1}{C_2} x_4 \tag{8}$$

$$\dot{x}_3 = \frac{1}{\rho_1}(v_{CD} - x_1) - \frac{1}{\rho_2}(x_2 - sgn(x_4)x_5) \tag{9}$$

$$\dot{x}_4 = \frac{1}{\rho_2}(v_{CD} - x_1) - \frac{1}{\rho_3}(x_2 - sgn(x_4)x_5) \tag{10}$$

$$\dot{x}_5 = \frac{1}{C_f} sgn(x_4) x_4 - \frac{x_5 - v_{OC}}{C_f r_b} \tag{11}$$

where $\rho_1 = \frac{L_{GA}L_{VA}-M^2}{L_{VA}}$, $\rho_2 = \frac{L_{GA}L_{VA}-M^2}{M}$, $\rho_3 = \frac{L_{GA}L_{VA}-M^2}{L_{GA}}$ and $sgn(x)$ takes 1 when x is >0, -1 when x is negative and 0 when x is null.

The inverter output voltage is a square wave function that could be approximated, as expressed in (12), using the first harmonic approximation. Thus, the expression of the power transferred from the ground side is given in (13).

$$v_{CD} = \frac{4V_{dc}}{\pi} \sin\left(\frac{\phi}{2}\right) \sin(\omega t) \tag{12}$$

where ϕ is the phase shift control signal and V_{dc} is the ground side DC voltage.

Equations (12) and (13) indicate that the amount of power transferred from the primary side to the secondary side could be controlled on the ground side by adjusting the phase shift control signal.

$$P_{GA} = \frac{8}{\pi^2} \frac{V_{dc}V_b}{\omega M} \sin\left(\frac{\phi}{2}\right) \tag{13}$$

It is worth noting that the phase shift control technique is often used in WPT chargers instead of the conventional PWM control technique. The main reason is that the generated output voltage of the ground-side inverter needs to vary at a frequency of 85 kHz. In this case, using the PWM control technique would require a higher carrier frequency (>> 85 kHz), which involves considerable stress and power switching loss in the ground-side inverter. This is why phase shift control is preferred over PWM in WPT chargers. Indeed, the inverter switching frequency is the same as that of the desired output voltage. A typical driver scheme is illustrated in Figure 2.

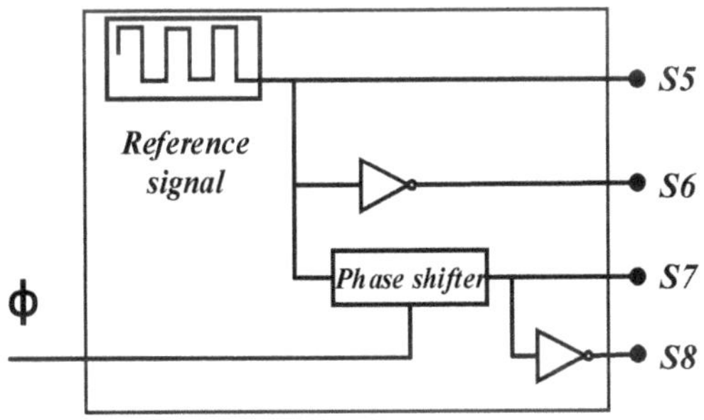

Figure 2. Phase shift driver.

3. Controller Design

This subsection is devoted to designing a controller for the WPT charger. Note that the proposed control approach deals only with the CC charging mode of the battery as it is the most widely adopted in real-life EVs [38]. The main control objective is to ensure a well-regulated battery charging current regardless of the misalignments between the coils and the intermittent communication latencies between the ground side and the vehicle side. It should be noted that the proposed control technique will be implemented on the ground side. This is due to the significant modifications required for implementing these controllers on the vehicle side, as well as the necessity for authorization from the car manufacturer to access the CAN (Controller Area Network).

3.1. Problem Statement

When trying to implement a controller for the WPT charger based on real-time data communication between the ground side and the vehicle side, we noticed an unstable behaviour of the system even if the parameters of the controller were properly tuned. This instability is due to the intermittent latency introduced by the communication boards. Indeed, the ground-side controller performs a periodic check of the received value of the battery charging current with a period that is mainly fixed by the sampling time of the controller board. As some incontrollable latencies are introduced by wireless communication (based on Wi-Fi as recommended by SAEJ2954 standard [39]), the measured battery charging current value may not be available at the time the controller performs the check. In this case, a maximum of the control signal will be generated leading to an overshoot of the battery current. The opposite scenario occurs when the controller checks the received battery charging current and finds that it is superior to the desired current. In this case, the control signal will be set to its minimum. The repetition of these two scenarios leads to an unstable and oscillating behaviour of the WPT charger. To overcome these issues, a novel control strategy has been proposed in the sequel.

3.2. PI-Based Adaptive Hill-Climbing Controller Design

The general concept of the proposed control approach is illustrated in Figure 3. In this control approach, the battery charging current is controlled indirectly through the control of the average DC on the ground side. Thus, the reference of the average DC $I_{DC,avg,reference}$ is calculated based on the desired battery charging current $I_{Bat,reference}$ as shown in (14):

$$I_{DC,avg,reference} = \frac{I_{Bat,reference} * V_{Bat}}{\eta * V_{DC}} = I_{Bat,reference} * \theta \tag{14}$$

with $\theta = \frac{V_{Bat}}{\eta * V_{DC}}$.

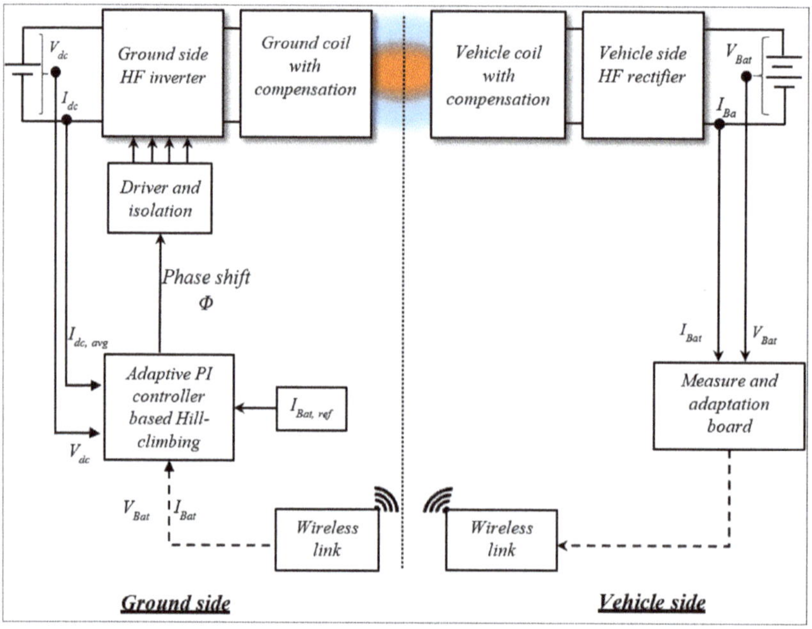

Figure 3. General concept of the proposed control technique.

It is worth noting that the adaptive parameter θ depends on the efficiency η of the charger and the battery voltage V_{Bat} which are subject to change during the charging operation. This parameter will be updated online, during the operation of the charger, only when a new measure of the battery charging current is available. This makes it possible to avoid the need for real-time wireless data communication. Additionally, to maintain the continuous operation of the controller during the latency intervals, introduced by the Wi-Fi communication, the latest calculated value of θ will be used by the controller.

It should be noted that the online adaptation of θ is performed using an adaptive hill-climbing algorithm. Indeed, depending on the sign of the error of the charging current shown in (15), the parameter θ will be either incremented or decremented by a variable step $\Delta\theta$ as in (16). The latter varies as a function of the error as shown in (17). The main advantage of choosing a variable step instead of a fixed one is that it ensures a tight regulation of the battery charging current. Indeed, $\Delta\theta$ decreases as far as the error vanishes, which avoids an overshoot in the battery charging current.

$$e(k) = I_{Bat,ref} - I_{Bat}(k) \tag{15}$$

$$\theta(k) = \theta(k-1) \pm \Delta\theta(k) \tag{16}$$

$$\Delta\theta(k) = \alpha * |e(k)| \tag{17}$$

With $k = 0, 1 \ldots N$ is a positive integer and α is a real parameter that will be chosen depending on the desired control performances (rapidity and precision).

Note that the initial value of the adaptive parameter $\theta(0)$ is calculated based on the nominal values of the battery voltage, the efficiency of the WPT charger, and the ground DC voltage.

The flowchart of the proposed adaptive hill-climbing algorithm combined with the PI controller is illustrated in Figure 4.

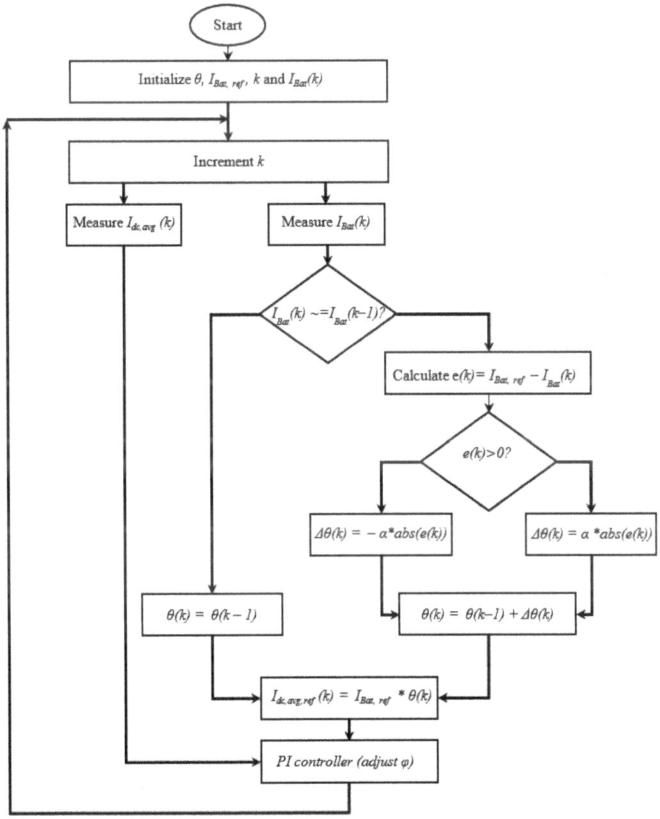

Figure 4. Flowchart of the proposed adaptive hill-climbing algorithm combined with the PI controller.

4. Simulation Results of the Proposed Control Approach

The performances of the control approach have been tested, first using simulation tools. The simulation parameters are listed in Table 1. It should be noted that these parameters are chosen based on those measured in the experimental setup. It is worth noting that The PI controller in our system has been tuned using a trial-and-error approach. This method involved iterative adjustments of the proportional and integral gains to achieve the desired performance in terms of stability, response time, and robustness.

Table 1. Summary of the WPT parameters used during simulation tests.

Parameter	Value
Primary side DC voltage (V_{dc})	120 V
Voltage of the battery (V_{bat})	48 V
Operating frequency (f_{sw})	82.986 kHz
Mutual inductance (M)	44 µH
Ground side coil's inductance L_{GA}	116 µH
Vehicle side coil's inductance L_{VA}	416 µH
Ground side compensation capacitance C_1	8.9 nF
Vehicle side compensation capacitance C_2	31.5 nF
Proportional gain	10
Integral gain	100

The adaptive parameter θ, generated by the hill-climbing algorithm, is plotted in Figure 5. It was initialized by a value of 0.42, which is calculated based on nominal values, namely, an efficiency of 95%, a DC voltage of 120 V, and a battery voltage of 48 V. It is noticed, in the zoomed part of the figure, that θ varies with a variable step ($\Delta\theta$), which takes higher values when the tracking error is larger and small values as far as the tracking error vanishes.

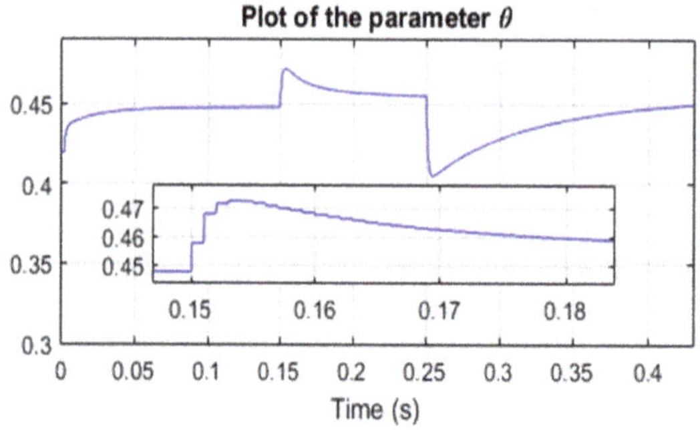

Figure 5. Evolution of the adaptive parameter as a function of time.

Based on the generated θ and the desired battery reference current, an average DC reference current is generated. The latter and the actual average DC are plotted in Figure 6 where it can be seen that the average current reference is dynamically updated based on the desired battery charging current. The figure clearly shows a perfect and smooth tracking of the reference. It follows, according to the indirect control technique, that the battery charging current is well-regulated to the desired value. This could be verified as shown in Figure 7, where we plotted the battery charging current and its reference. The latter was stepped up from 2 A to 3 A at 0.15 s and stepped down again to 1 A at 0.25 s. One can see that all the desired battery charging currents have been reached with a convergence time of less than 0.1 s.

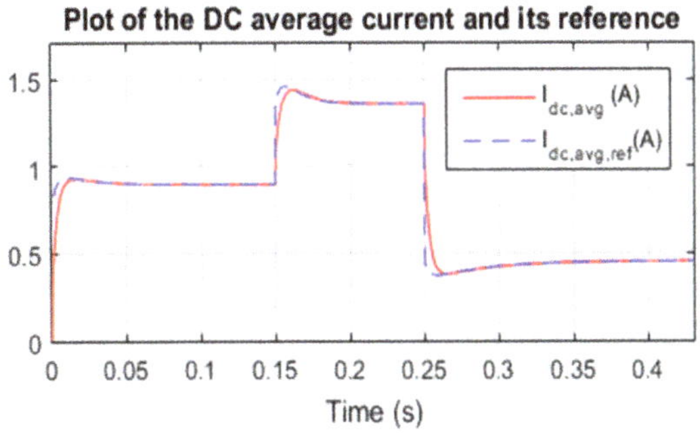

Figure 6. Evolution of the average DC and its reference.

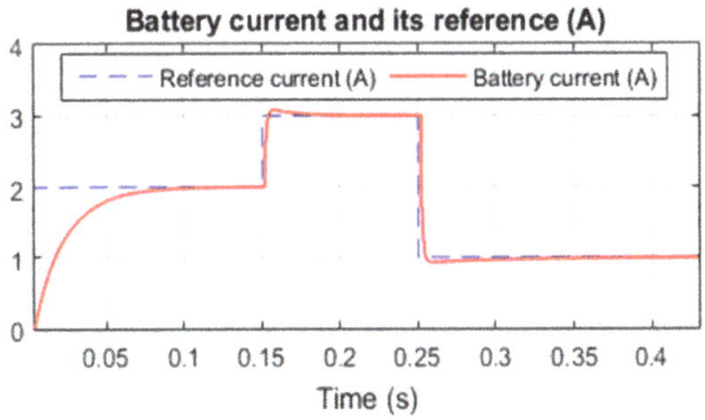

Figure 7. Plot of the battery charging current and its reference during simulation test.

The phase shift control signal, generated by the PI controller, is plotted in Figure 8. We can see that the latter evolves smoothly without any saturation. This helps reduce the stress and ensures optimal working of the power converters.

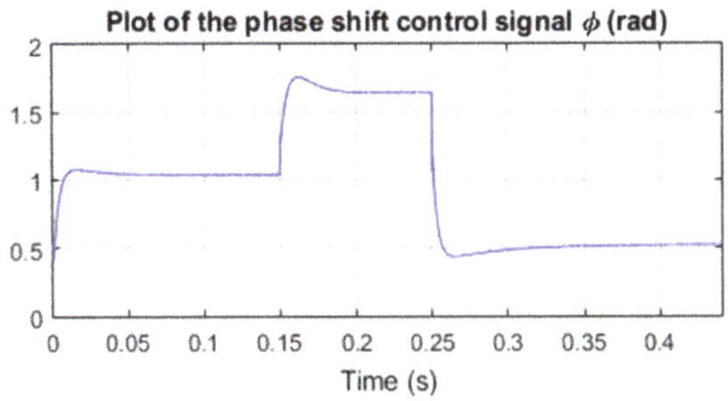

Figure 8. Evolution of the phase shift control signal as a function of time.

Figure 9 shows the evolution of the battery voltage as a function of the charging current. We notice a small increase in the voltage around the nominal value of 48 V when the battery is being charged.

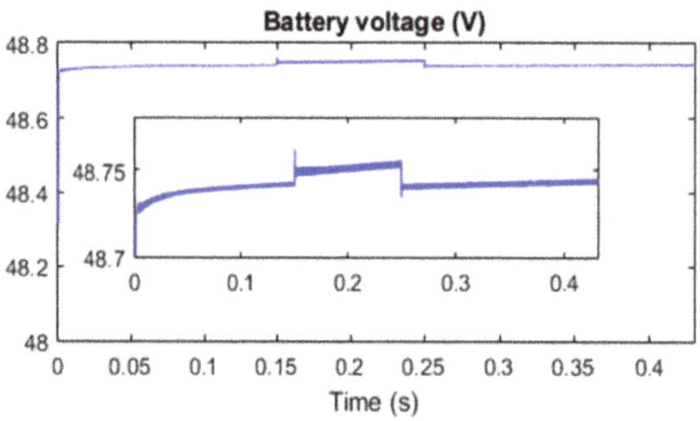

Figure 9. Evolution of the battery voltage.

The inverter output voltage and current are plotted in Figure 10. The zoomed parts of the figure (Figure 10a,b) show that the power transferred from the ground side to the vehicle side is adjusted by changing the RMS value of the voltage through the phase shift control signal. It is also evident that the shape of the primary coil's current looks like a sine wave despite the voltage being a square wave with infinite harmonics. This is due to the filtering behaviour of the resonant LC network (ground side coil with its associated compensation series capacitor) tuned at the operating frequency.

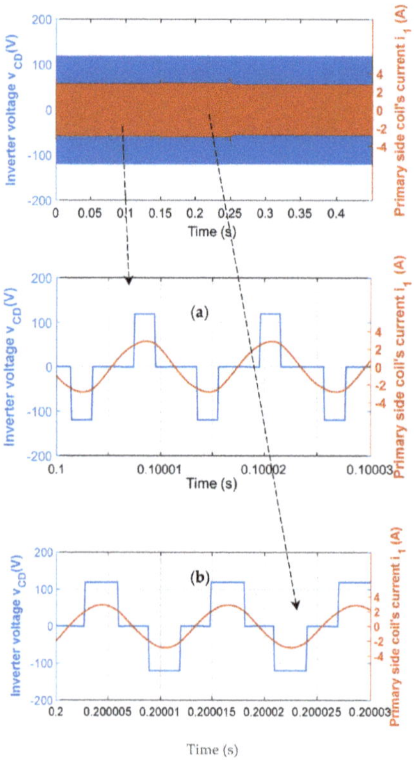

Figure 10. Inverter output voltage and primary side coil's current, (**a**) zoom at time 0.1 s and (**b**) zoom at 0.2 s.

The vehicle side rectifier input voltage and the coil's current are plotted in Figure 11. The voltage is a square wave with a maximum value that is fixed by the actual battery voltage. However, due to the filtering behaviour of the vehicle side LC network, the current is a sine wave with a frequency of 85 kHz. We notice that both of these signals are in phase with variations in the coil's current as a function of the desired battery charging current.

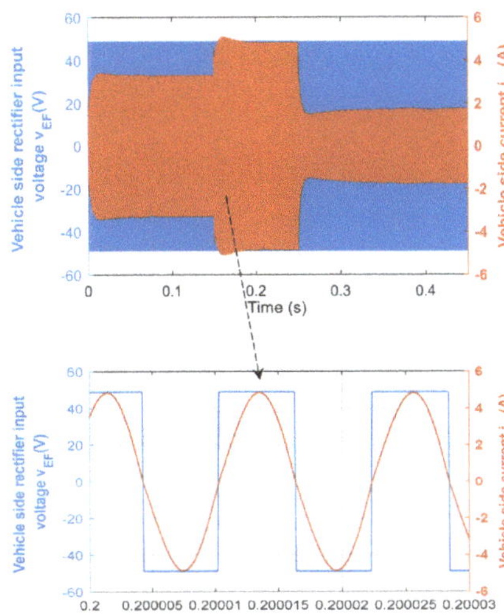

Figure 11. Voltage at the rectifier input and vehicle side coil's current with a zoom at time 0.2 s.

The obtained simulation results show that all the desired control objectives have been realized, which verifies the efficacy of the PI-based adaptive hill-climbing control technique.

5. Experimental Results of the Proposed Control Approach

5.1. Description of the Laboratory Test Bed

To validate the efficacy of the proposed control technique, a test setup was constructed. It is worth noting that Silicon Carbide (SiC) and Galium Nitride (GaN) MOSFET-based power converters are emerging technologies best suited for WPT chargers. Indeed, their performances remain higher even when operating at high frequency and high power. SiC MOSFET technology is adopted in the present paper (reference CRD 8FF1217P-1 from Wolfspeed, Inc., Durham, NC, USA).

To minimize the proximity and skin effect, the GA and VA coils are made of 1050-strand twisted Litz wire. In addition, Mn–Zn ferrite cores (from TDK Corporation, Japan) are also added to guide the flux between the coils.

From a design viewpoint, key parameters for a circular planar coil include the inner radius, the outer radius, the spacing between the turns, and the number of turns. These must be carefully selected to achieve the desired performance. It is worth noting that increasing the spacing between turns can help achieve the desired inductance with fewer turns, reducing the coil's weight and cost. However, it also decreases mutual coupling. To ensure better coupling, turns are brought closer together to eliminate spacing, leaving the number of turns and the inner and outer radii as the primary adjustable parameters for desired coil performance.

The ground-side coil has an outer diameter of 48 cm. It is worth noting that the outer diameter is kept constant to ensure proper coupling at a nominal air gap of 12 cm, as the height of the created magnetic flux is approximately one-quarter of the outer diameter of the coil. The inner diameter has been adjusted to 5.5 cm to achieve the desired coil inductance in conjunction with the compensation capacitor to ensure resonance operation. In addition, the number of spires is 40, leading to an inductance of 416 μH with six ferrite bars.

The geometry for the vehicle-side coil has an outer diameter of 30 cm and an inner diameter of 8.5 cm. The smaller outer diameter, compared to that of the ground-side coil, ensures that the vehicle-side coil remains within the coverage area of the ground-side coil, even in the presence of large misalignments, in order to ensure good coupling and keep the power transfer efficiency as high as possible. Once again, the inner diameter is adjusted to ensure the desired coil inductance to ensure the resonance operation in conjunction with the compensation capacitors. Note that, with this configuration, 19 spires, and 4 ferrites bars, the obtained inductance value of the ground side coil is 116 μH.

Figure 12 illustrates the realized coils.

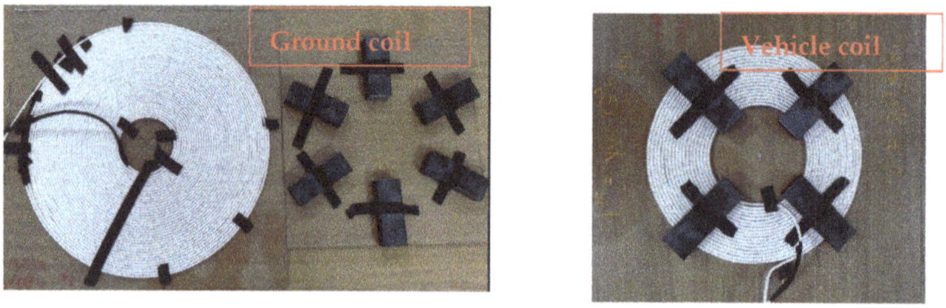

Figure 12. Image of the ground-side coil and the vehicle-side coil.

The series compensation topology, which consists of adding capacitors in series with the coils, is used in our case. The main advantage of series compensation is that the calculated capacitor values are independent of the mutual inductance between the coils. This is generally desired in WPT chargers as misalignments lead to variable mutual coupling. Still, one issue with this compensation topology is the high voltage involved across the capacitors, which complicates the design task. Indeed, several capacitors are connected in series, forming a branch, to withstand the involved high voltage. Then, several branches could be connected in parallel to achieve the desired capacitance value. The realized ground side (8.9 nF) and vehicle side (31.5 nF) compensation boards are illustrated in Figure 13.

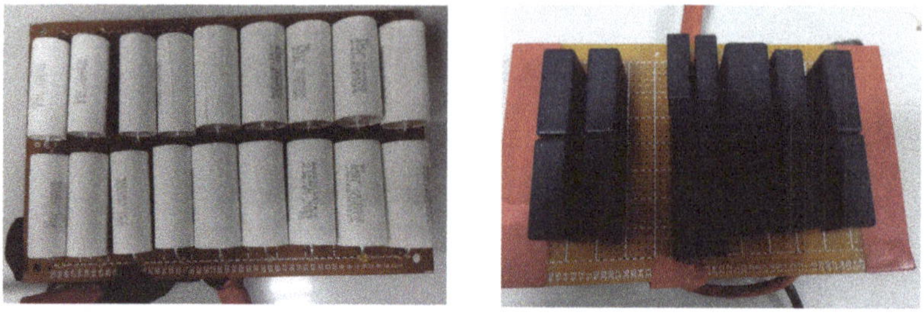

Figure 13. Compensation boards. Ground side (**left**) and ground side (**right**).

It is worth noting that the operating frequency is calculated based on the realized compensation capacitors and coils. Indeed, the capacitance values of the compensation boards on the ground and vehicle sides are C1 = 8.9 nF and C2 = 31.5 nF, respectively. In addition, the inductances of the coils on the ground and vehicle sides are L1 = 416 μH and L2 = 116 μH, respectively. These values result in two resonance frequencies: f1 = 82,713 Hz and f2 = 83,259 Hz. Therefore, the resonance frequency is chosen as the mean value of f1 and f2, which is f = (f1 + f2)/2 = 82,986 Hz. Note that this frequency falls within the recommended range of the SAE J2954 standard (81–91 kHz).

The air-gap distance between the coils is set to 125 mm (nominal distance of the Z1 class of the SAEJ2954 standard). This air-gap distance gives rise to a mutual coupling of 44 μH between the coils. As for the control board, TI C2000 f28335 (from Texas Instrument, Texas, USA) is used. Raspberry Pi4 is used to ensure Wi-Fi communication between the ground and vehicle parts of the WPT charger. In addition, MicroLabBox (from dSPACE

GmbH, Paderborn Germany) and measurement boards are used for data acquisition and logging. Moreover, a 48 V battery pack is used to emulate the vehicle's battery. The detailed realization of the prototype can be found in [40]. Figure 14 illustrates the realized prototype.

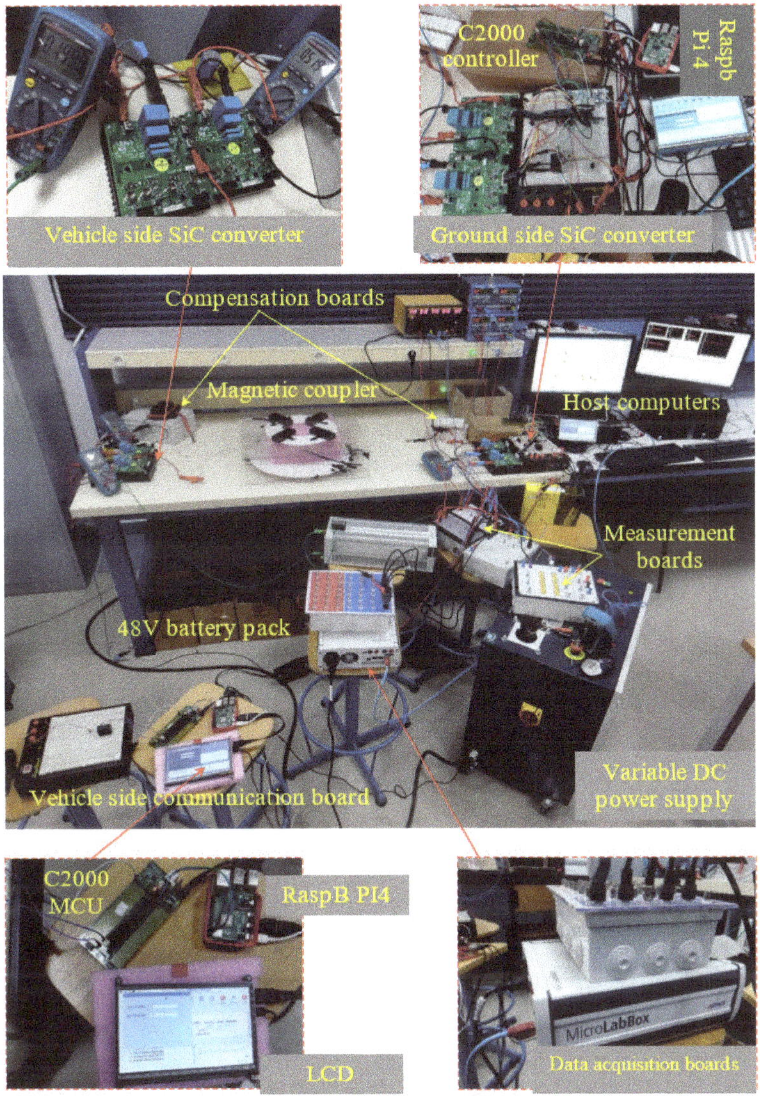

Figure 14. The experimental testbed used for testing the control approach.

5.2. Experimental Results

The performances of the proposed control technique are tested experimentally using two coil positions. In the first positioning, the two coils are laterally misaligned ($\Delta x = \Delta y = 5$ cm) and spaced at a distance of 12 cm. In the second positioning, the air gap was increased to 15 cm in the presence of large misalignments ($\Delta x = \Delta y = 10$ cm), which are above the testing limits of the SAEJ2954 standard. The aim is to test the controller under extreme conditions. The results obtained for each scenario are presented in the sequel.

Starting with the first scenario, the battery charging current and its reference are plotted in Figure 15. The desired current was set first to 1 A then stepped up to 3 A and down to 2 A, 1 A, and 0 A. The point is to check the tracking performances of the controller for different values of the charging current. As illustrated, all the desired reference currents are reached without any overshoot, which proves the effectiveness of the proposed control technique. The average DC is plotted in Figure 16 where it can be seen that it steps up and down as a function of the variations in the battery current.

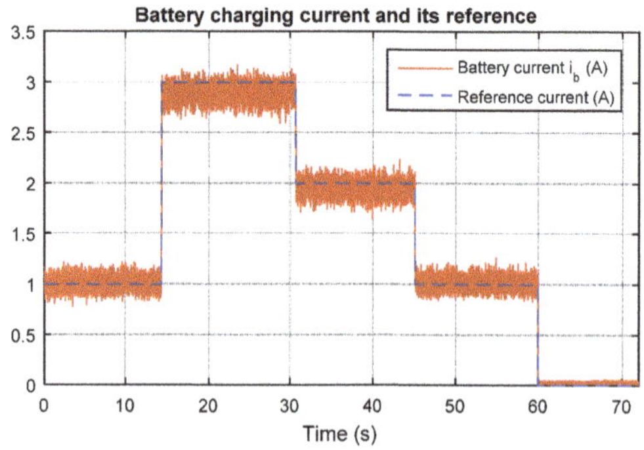

Figure 15. Evolution of the battery charging current and its reference.

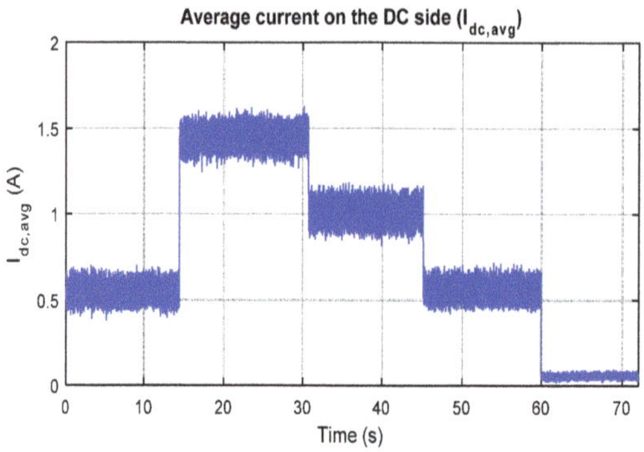

Figure 16. Evolution of the average DC on the primary side.

The DC side and battery voltages are plotted in Figures 17 and 18. Thus, the calculation, using the values of the currents and voltages on both sides, shows that the efficiency of the WPT charger varies according to the power transferred to the battery. Indeed, when the battery charging current is set to 1 A, the corresponding battery voltage, DC side voltage and DC side current are, respectively, 51 V, 120 V, and 0.5 A. These values give rise to an efficiency of approximately 85%. Similarly, the power transfer efficiencies corresponding to the case where the battery charging currents are set to 3 A and 2 A are, respectively, 90% and 87%. These results show that the efficiency varies according to the power transferred to the battery. Indeed, the efficiency increases as the power transferred increases. In addition, these results prove the effectiveness of the adaptive control technique in maintaining the battery charging currents at the desired values despite the power transfer efficiency variations.

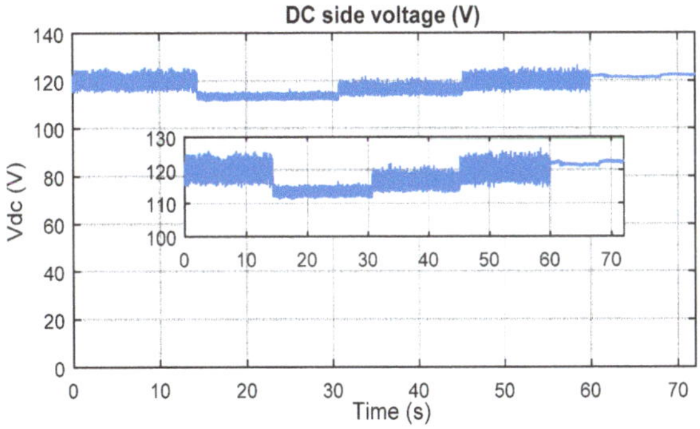

Figure 17. Evolution of the DC voltage on the primary side.

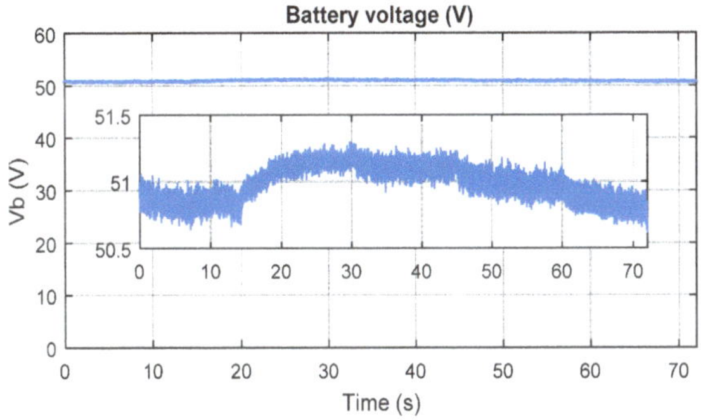

Figure 18. Plot of the battery voltage.

Moving to the second scenario, the efficiency of the WPT charger is expected to be lower than that of the first scenario as the coils are largely misaligned and the air gap is increased. The positioning of the primary and secondary coils is illustrated in Figure 19.

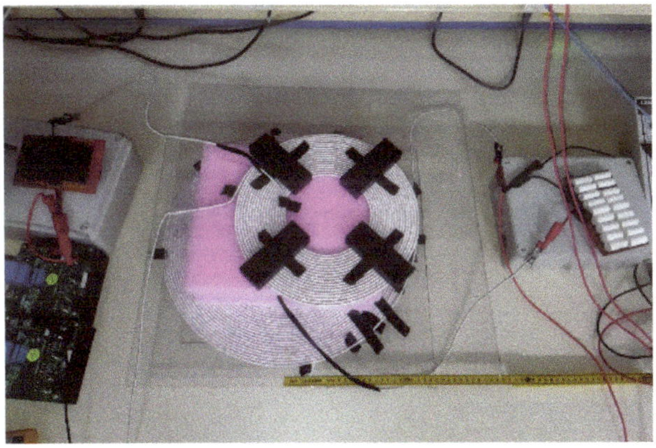

Figure 19. Image of the primary (**bottom**) and secondary (**top**) coils under large misalignments and air gaps.

During this test, as shown in Figure 20, the battery reference current is first set to 2 A, stepped up to 3 A, and then stepped down to 1 A and 0 A. In the same figure, one can see that the actual battery charging current tracks these references well, despite the fact that the test is performed considering the worst positioning scenario of the coils.

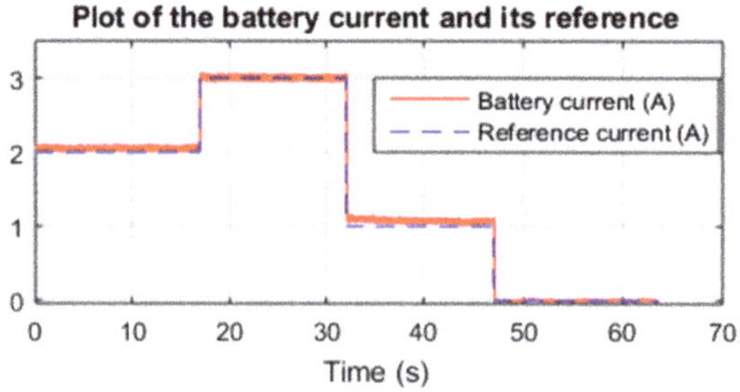

Figure 20. Plot of the battery current and its reference during the second testing scenario.

The corresponding average DC is plotted in Figure 21, where it can be seen that it varies according to the changes in the battery current.

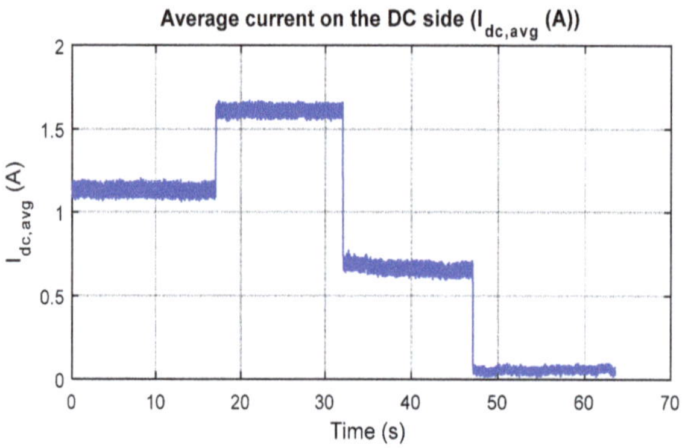

Figure 21. Plot of the average current on the primary side during the second testing scenario.

In addition, considering the voltages on the battery side and the DC side, plotted in Figures 22 and 23, and the currents in the DC side and battery side show that the efficiency in this scenario drops to approximately 75%. This test reveals that the proposed control technique has good performance even in the presence of large air gaps and misalignments between the coils. In addition, the system exhibits good stability against communication latency.

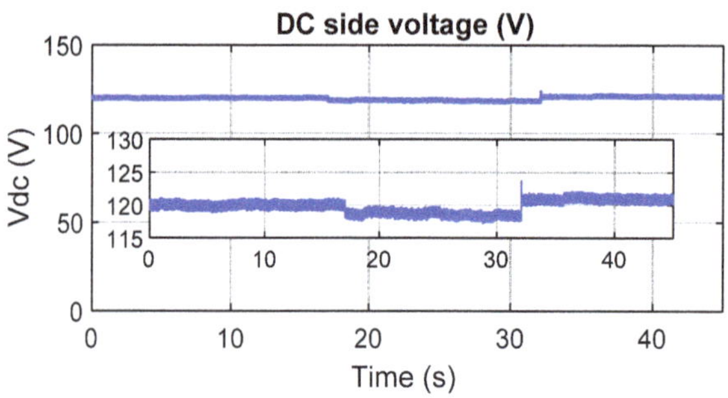

Figure 22. Plot of the DC voltage on the primary side during the second testing scenario.

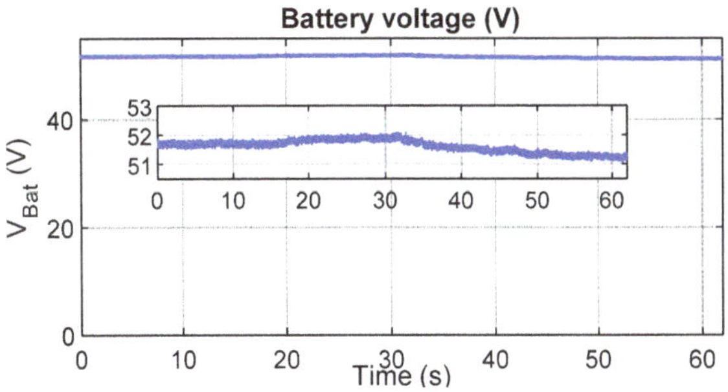

Figure 23. Plot of the battery voltage during the second testing scenario.

6. Results, Discussion, and Performance Evaluation of the Proposed Control Technique

The novelty of our proposed technique lies in its indirect control of the battery current through the regulation of the primary side current while maintaining wireless communication between the ground assembly (GA) and vehicle assembly (VA) for monitoring and safe operation of the battery charging current. This technique utilizes an adaptive hill-climbing algorithm in conjunction with a PI-based controller. The adaptive parameter is updated online during charger operation, only when a new measurement of the battery charging current is available. This approach avoids the need for real-time wireless data communication and ensures continuous controller operation during latency intervals.

The obtained simulation and experimental results illustrate the successful application of the proposed control technique. Thus, the experimental validation confirms the control technique's robustness under real-world conditions, including significant misalignments and varied air gaps. The main results are summarized as follows:

- Adaptive parameter (θ) behaviour

The adaptive parameter θ is dynamically adjusted based on the tracking error. Larger steps occur with significant errors and smaller steps as errors decrease, ensuring adaptive system stability and efficient performance.

- Accurate and robust current tracking

Both the average DC and the battery charging current closely follow their respective references, even during current steps, indicating the controller's precision in maintaining desired charging currents. In addition, in scenarios with both moderate ($\Delta x = \Delta y = 5$ cm, 12 cm air gap) and large misalignments ($\Delta x = \Delta y = 10$ cm, 15 cm air gap), the battery charging current accurately tracks the reference values without overshoot, highlighting the controller's robustness and stability.

- Smooth control signal

The phase shift control signal evolves smoothly without saturation, minimizing stress on power converters and ensuring optimal operation.

- Adaptive Power transfer

The average DC is adjusted effectively in response to changes in the battery current, validating the indirect control approach.

It is worth noting that, from the literature review given in the introduction section, it was found that several common points emerge regarding control techniques for WPT chargers. Many techniques prioritize primary-side power regulation to reduce the complexity, size, and cost of vehicle onboard components by eliminating receiver-side DC-DC converters. While some methods eliminate the need for wireless communication, it remains essential for monitoring battery current to ensure safe charging.

Compared to existing techniques, our approach not only maintains the essential wireless communication for battery state monitoring but also achieves stable and efficient charging control. This combination of indirect battery current control and adaptive algorithmic adjustments enhances the reliability and performance of WPT chargers, addressing the limitations identified in the literature.

7. Conclusions

This paper addresses the critical control task of wireless power transfer (WPT) chargers for electric vehicles (EVs), focusing on maintaining a stable battery charging current amidst challenges such as coil misalignment and intermittent communication latencies between the ground and vehicle sides. The paper introduces a novel control approach that employs a combination of a hill-climbing algorithm and a PI controller, enabling indirect control of the battery charging current through the ground-side current control.

The proposed technique ensures the continuous and stable operation of the controller, effectively mitigating the impact of wireless communication latencies. By leveraging an adaptive parameter updated online during charger operation, we circumvent the need for real-time wireless data communication, thus enhancing system robustness and stability regardless of potential delays and large coil misalignments.

Simulation and experimental validation on a laboratory test bed confirm the effectiveness and robustness of our proposed approach, even under worst-case testing scenarios. The obtained results validate the viability of the proposed technique for real-world applications.

For future work, the focus will be on further optimization and validation of the proposed technique under diverse operating conditions, including the use of different coil designs and compensation topologies. Additionally, we intend to incorporate battery parameter estimation to further enhance the reliability and performance of the proposed control technique.

Author Contributions: Conceptualization, A.L. and H.E.F.; methodology, A.L.; software, M.E.A.; validation, A.L. and Z.E.I.; formal analysis, A.L.; investigation, H.E.F.; resources, A.L.; data curation, H.E.F.; writing—original draft preparation, A.L.; writing—review and editing, A.R. and K.R.; visualization, H.E.F.; supervision, H.E.F. All authors have read and agreed to the published version of the manuscript.

Funding: This research received no external funding.

Data Availability Statement: The original contributions presented in the study are included in the article, further inquiries can be directed to the corresponding author.

Conflicts of Interest: The authors declare no conflicts of interest.

References

1. Lukic, S.M.; Pantic, Z. Cutting the Cord: Static and Dynamic Inductive Wireless Charging of Electric Vehicles. *Electrif. Mag. IEEE* **2013**, *1*, 57–64. [CrossRef]
2. Patil, D.; McDonough, M.K.; Miller, J.M.; Fahimi, B.; Balsara, P.T. Wireless Power Transfer for Vehicular Applications: Overview and Challenges. *IEEE Trans. Transp. Electrific.* **2018**, *4*, 3–37. [CrossRef]
3. Laha, A.; Kalathy, A.; Pahlevani, M.; Jain, P. A Comprehensive Review on Wireless Power Transfer Systems for Charging Portable Electronics. *Eng* **2023**, *4*, 1023–1057. [CrossRef]
4. Lassioui, A.; Fadil, H.E.; Rachid, A.; Belhaj, F.Z.; Tarkany, O.; Bajit, A. Characterestics Analysis of Wireless Power Transfer System for Electric Vehicle Charging Applications. In Proceedings of the 2018 International Symposium on Advanced Electrical and Communication Technologies (ISAECT), Rabat, Morocco, 21–23 November 2018; pp. 1–6.
5. Li, S.; Mi, C.C. Wireless Power Transfer for Electric Vehicle Applications. *IEEE J. Emerg. Sel. Top. Power Electron.* **2015**, *3*, 4–17. [CrossRef]
6. Machura, P.; De Santis, V.; Li, Q. Driving Range of Electric Vehicles Charged by Wireless Power Transfer. *IEEE Trans. Veh. Technol.* **2020**, *69*, 5968–5982. [CrossRef]
7. Zaid, S.A.; Albalawi, H.; Alatwi, A.M.; Elemary, A. Performance Improvement of an Electric Vehicle Charging Station Using Brain Emotional Learning-Based Intelligent Control. *Processes* **2024**, *12*, 1014. [CrossRef]
8. Huang, Q.; Huang, A.Q. Review of GaN Totem-Pole Bridgeless PFC. *CPSS Trans. Power Electron. Appl.* **2017**, *2*, 187–196. [CrossRef]
9. Bentalhik, I.; El Fadil, H.; Lassioui, A.; Koundi, M.; El Idrissi, Z.; Giri, F. Adaptive Observer Design for a Bidirectional AC/DC Totem Pole PFC Power Converter. *IFAC-PapersOnLine* **2022**, *55*, 438–443. [CrossRef]

10. Mounica, M.; Rajpathak, B.A.; Kolhe, M.L.; Naik, K.R.; Moparthi, J.R.; Kotha, S.K. Piece-Wise Droop Controller for Enhanced Stability in DC-Microgrid-Based Electric Vehicle Fast Charging Station. *Processes* **2024**, *12*, 892. [CrossRef]
11. Berger, A.; Agostinelli, M.; Vesti, S.; Oliver, J.A.; Cobos, J.A.; Huemer, M. A Wireless Charging System Applying Phase-Shift and Amplitude Control to Maximize Efficiency and Extractable Power. *IEEE Trans. Power Electron.* **2015**, *30*, 6338–6348. [CrossRef]
12. Li, Y.; Hu, J.; Chen, F.; Li, Z.; He, Z.; Mai, R. Dual-Phase-Shift Control Scheme With Current-Stress and Efficiency Optimization for Wireless Power Transfer Systems. *IEEE Trans. Circuits Syst. I Regul. Pap.* **2018**, *65*, 3110–3121. [CrossRef]
13. Lassioui, A.; Fadil, H.E.; Rachid, A.; El-Idrissi, Z.; Bouanou, T.; Belhaj, F.Z.; Giri, F. Modelling and Sliding Mode Control of a Wireless Power Transfer System for BEV Charger. *Int. J. Model. Identif. Control* **2020**, *34*, 171–186. [CrossRef]
14. Mohamed, A.A.S.; Berzoy, A.; de Almeida, F.G.N.; Mohammed, O. Modeling and Assessment Analysis of Various Compensation Topologies in Bidirectional IWPT System for EV Applications. *IEEE Trans. Ind. Appl.* **2017**, *53*, 4973–4984. [CrossRef]
15. Bouanou, T.; Fadil, H.E.; Lassioui, A. Analysis and Design of Circular Coil Transformer in a Wireless Power Transfer System for Electric Vehicle Charging Application. In Proceedings of the 2020 International Conference on Electrical and Information Technologies (ICEIT), Rabat, Morocco, 4–7 March 2020; pp. 1–6.
16. Bouanou, T.; El Fadil, H.; Lassioui, A.; Assaddiki, O.; Njili, S. Analysis of Coil Parameters and Comparison of Circular, Rectangular, and Hexagonal Coils Used in WPT System for Electric Vehicle Charging. *World Electr. Veh. J.* **2021**, *12*, 45. [CrossRef]
17. Rachid, A.; El Fadil, H.; Gaouzi, K.; Rachid, K.; Lassioui, A.; El Idrissi, Z.; Koundi, M. Electric Vehicle Charging Systems: Comprehensive Review. *Energies* **2023**, *16*, 255. [CrossRef]
18. Rachid, A.; Fadil, H.E.; Giri, F. Dual Stage CC-CV Charge Method for Controlling DC-DC Power Converter in BEV Charger. In Proceedings of the 2018 19th IEEE Mediterranean Electrotechnical Conference (MELECON), Marrakech, Morocco, 2–7 May 2018; pp. 74–79.
19. Lassioui, A.; EL Fadil, H.; Rachid, A.; Bouanou, T.; Giri, F. Adaptive Output Feedback Nonlinear Control of a Wireless Power Transfer Charger for Battery Electric Vehicle. *J. Control Autom. Electr. Syst.* **2021**, *32*, 492–506. [CrossRef]
20. Jia, Q.-S.; Long, T. A Review on Charging Behavior of Electric Vehicles: Data, Model, and Control. *Control Theory Technol.* **2020**, *18*, 217–230. [CrossRef]
21. Wu, W.; Zeng, J.; Jian, Q.; Tang, L.; Hou, J.; Han, C.; Song, Q.; Luo, Y. A Li-Ion Battery State of Charge Estimation Strategy Based on the Suboptimal Multiple Fading Factor Extended Kalman Filter Algorithm. *Processes* **2024**, *12*, 998. [CrossRef]
22. Zhang, R.; Li, X.; Sun, C.; Yang, S.; Tian, Y.; Tian, J. State of Charge and Temperature Joint Estimation Based on Ultrasonic Reflection Waves for Lithium-Ion Battery Applications. *Batteries* **2023**, *9*, 335. [CrossRef]
23. Song, K.; Lan, Y.; Zhang, X.; Jiang, J.; Sun, C.; Yang, G.; Yang, F.; Lan, H. A Review on Interoperability of Wireless Charging Systems for Electric Vehicles. *Energies* **2023**, *16*, 1653. [CrossRef]
24. Triviño, A.; González-González, J.M.; Aguado, J.A. Wireless Power Transfer Technologies Applied to Electric Vehicles: A Review. *Energies* **2021**, *14*, 1547. [CrossRef]
25. Venkatesan, R.; Savio, A.D.; Balaji, C.; Narayanamoorthi, R.; Kotb, H.; ELrashidi, A.; Nureldeen, W. A Comprehensive Review on Efficiency Enhancement of Wireless Charging System for the Electric Vehicles Applications. *IEEE Access* **2024**, *12*, 46967–46994. [CrossRef]
26. Nguyen, H.T.; Alsawalhi, J.Y.; Hosani, K.A.; Al-Sumaiti, A.S.; Jaafari, K.A.A.; Byon, Y.-J.; Moursi, M.S.E. Review Map of Comparative Designs for Wireless High-Power Transfer Systems in EV Applications: Maximum Efficiency, ZPA, and CC/CV Modes at Fixed Resonance Frequency Independent From Coupling Coefficient. *IEEE Trans. Power Electron.* **2022**, *37*, 4857–4876. [CrossRef]
27. Sagar, A.; Kashyap, A.; Nasab, M.A.; Padmanaban, S.; Bertoluzzo, M.; Kumar, A.; Blaabjerg, F. A Comprehensive Review of the Recent Development of Wireless Power Transfer Technologies for Electric Vehicle Charging Systems. *IEEE Access* **2023**, *11*, 83703–83751. [CrossRef]
28. Miller, J.M.; Onar, O.C.; Chinthavali, M. Primary-Side Power Flow Control of Wireless Power Transfer for Electric Vehicle Charging. *IEEE J. Emerg. Sel. Top. Power Electron.* **2015**, *3*, 147–162. [CrossRef]
29. Baros, D.; Rigogiannis, N.; Drougas, P.; Voglitsis, D.; Papanikolaou, N.P. Transmitter Side Control of a Wireless EV Charger Employing IoT. *IEEE Access* **2020**, *8*, 227834–227846. [CrossRef]
30. Yang, Y.; Zeng, J.; Hui, S.Y.R. A Fast Primary-Side Current and Voltage Control for Direct Wireless Battery Chargers. *IEEE J. Emerg. Sel. Top. Power Electron.* **2024**, *12*, 355–366. [CrossRef]
31. Shi, W.; Deng, J.; Wang, Z.; Cheng, X. The Start-up Dynamic Analysis and One Cycle Control-PD Control Combined Strategy for Primary-Side Controlled Wireless Power Transfer System. *IEEE Access* **2018**, *6*, 14439–14450. [CrossRef]
32. Song, K.; Li, Z.; Jiang, J.; Zhu, C. Constant Current/Voltage Charging Operation for Series–Series and Series–Parallel Compensated Wireless Power Transfer Systems Employing Primary-Side Controller. *IEEE Trans. Power Electron.* **2018**, *33*, 8065–8080. [CrossRef]
33. Liu, Z.; Wang, L.; Guo, Y.; Li, S. Primary-Side Linear Control for Constant Current/Voltage Charging of the Wireless Power Transfer System Based on the LCC-N Compensation Topology. *IEEE Trans. Ind. Electron.* **2022**, *69*, 8895–8904. [CrossRef]
34. Kalat, S.N.; Vaez-Zadeh, S.; Zakerian, A.; Babaki, A.; Ebel, T. A Communication-Free and Model-Free Predictive Control for a Dynamic IPT System With High Power Factor for Electric Vehicles. *IEEE Access* **2023**, *11*, 96773–96783. [CrossRef]
35. Kim, M.; Joo, D.-M.; Lee, B.K. Design and Control of Inductive Power Transfer System for Electric Vehicles Considering Wide Variation of Output Voltage and Coupling Coefficient. *IEEE Trans. Power Electron.* **2019**, *34*, 1197–1208. [CrossRef]

36. Huang, C.-C.; Lin, C.-L.; Wu, Y.-K. Simultaneous Wireless Power/Data Transfer for Electric Vehicle Charging. *IEEE Trans. Ind. Electron.* **2017**, *64*, 682–690. [CrossRef]
37. Chen, W.; Liu, Z.; Liu, J.; Ye, J.; Su, H. A New Simultaneous Wireless Power and Information Transfer System With Integrated Signal Magnetic Coupler. *IEEE Trans. Ind. Electron.* **2023**, *70*, 12708–12718. [CrossRef]
38. Onar, O.C.; Campbell, S.L.; Seiber, L.E.; White, C.P.; Chinthavali, M. Vehicular Integration of Wireless Power Transfer Systems and Hardware Interoperability Case Studies. In Proceedings of the 2016 IEEE Energy Conversion Congress and Exposition (ECCE), Milwaukee, WI, USA, 18–22 September 2016; pp. 1–8.
39. *J2954_202010*; Wireless Power Transfer for Light-Duty Plug-In/Electric Vehicles and Alignment Methodology. SAE International: Warrendale, PA, USA, 2020.
40. Bentalhik, I.; Lassioui, A.; EL Fadil, H.; Bouanou, T.; Rachid, A.; EL Idrissi, Z.; Hamed, A.M. Analysis, Design and Realization of a Wireless Power Transfer Charger for Electric Vehicles: Theoretical Approach and Experimental Results. *World Electr. Veh. J.* **2022**, *13*, 121. [CrossRef]

processes

MDPI

Article

Power Quality Disturbance Identification Method Based on Improved CEEMDAN-HT-ELM Model

Ke Liu [1], Jun Han [1], Song Chen [2,*], Liang Ruan [2], Yutong Liu [1] and Yang Wang [2]

1 Electric Power Research Institute of State Grid Qinghai Power Grid Corporation, Xining 810008, China; liuke19882024@163.com (K.L.); hanjun19942024@163.com (J.H.); qhkeji2024@163.com (Y.L.)

2 College of Electrical Engineering, Sichuan University, Chengdu 610065, China; ruan_liang@stu.scu.edu.cn (L.R.); fwang@scu.edu.cn (Y.W.)

* Correspondence: chen_song@stu.scu.edu.cn

Abstract: The issue of power quality disturbances in modern power systems has become increasingly complex and severe, with multiple disturbances occurring simultaneously, leading to a decrease in the recognition accuracy of traditional algorithms. This paper proposes a composite power quality disturbance identification method based on the integration of improved Complementary Ensemble Empirical Mode Decomposition (CEEMDAN), Hilbert Transform (HT), and Extreme Learning Machine (ELM). Addressing the limitations of traditional signal processing techniques in handling nonlinear and non-stationary signals, this study first preprocesses the collected initial power quality signals using the improved CEEMDAN method to reduce modal aliasing and spurious components, thereby enabling a more precise decomposition of noisy signals into multiple Intrinsic Mode Functions (IMFs). Subsequently, the HT is utilized to conduct a thorough analysis of the reconstructed signals, extracting their time-amplitude information and instantaneous frequency characteristics. This feature information provides a rich data foundation for subsequent classification and identification. On this basis, an improved ELM is introduced as the classifier, leveraging its powerful nonlinear mapping capabilities and fast learning speed to perform pattern recognition on the extracted features, achieving accurate identification of composite power quality disturbances. To validate the effectiveness and practicality of the proposed method, a simulation experiment is designed. Upon examination, the approach introduced in this study retains a fault diagnosis accuracy exceeding 95%, even amidst significant noise disturbances. In contrast to conventional techniques, such as Convolutional Neural Network (CNN) and Support Vector Machine (SVM), this method achieves an accuracy enhancement of up to 5%. Following optimization via the Particle Swarm Optimization (PSO) algorithm, the model's accuracy is boosted by 3.6%, showcasing its favorable adaptability.

Keywords: power quality disturbances; complementary ensemble empirical mode decomposition (CEEMDAN); Hilbert transform (HT); extreme learning machine (ELM)

1. Introduction

The stability and purity of power quality are crucial to the normal operation of power systems and the stability of electrical equipment. Power quality issues can manifest as voltage instability, frequency fluctuations, increased harmonic content, and the presence of current asymmetry [1]. Among these, voltage stability is directly related to the normal operation of electrical devices, frequency stability is vital for the operation of synchronous equipment, and harmonic content and current asymmetry can lead to equipment damage and reduced system energy efficiency [2]. The root causes of power quality issues include nonlinear loads, the integration of renewable energy sources, and equipment failures [3]. Harmonics and current asymmetry generated by nonlinear loads are common problems, while the instability of renewable energy sources also poses challenges to power quality. Equipment failures may cause voltage sags, voltage swells, and other issues, further affecting the stability of power systems [4]. Maintaining good power quality requires the integrated application of traditional and novel power quality improvement technologies, including power electronic devices, static reactive power compensation devices, intelligent dispatching and control systems, and more. Comprehensive analysis and monitoring can promptly identify and resolve power quality issues, safeguarding the normal operation of power systems, improving energy utilization efficiency, extending equipment lifespan, and ensuring the safe and stable operation of user equipment [5]. Therefore, maintaining power quality is not only a fundamental requirement for the operation of power systems but also a critical aspect of promoting sustainable energy development.

The detection of harmonics and inter-harmonics is the foundation and prerequisite for harmonic analysis and management. At present, domestic and foreign scholars have achieved remarkable results in the research of harmonic detection, while inter-harmonic detection remains a research difficulty and hotspot in power grid power quality due to its high demand for algorithm resolution [6]. Fourier Transform is a classic algorithm for spectrum analysis, known for its practicality, speed, and efficiency. Several widely used Fourier Transform algorithms include Fast Fourier Transform (FFT) [7], Discrete Fourier Transform (DFT) [8], and Short-Time Fourier Transform (STFT) [9]. However, Fourier Transform algorithms require synchronous sampling, which is practically impossible to achieve strictly. To approach synchronous sampling and reduce synchronization errors, two main methods are employed: hardware synchronization and software synchronization [10,11]. Hardware synchronization achieves synchronous sampling through the addition of phase-locked loop hardware circuits, which are widely used in practical applications. Scholars have also proposed improved phase-locked loops and digital phase-locked loops. Conventional software synchronization methods first measure the signal period, divide it by the number of sampling points within that period to obtain the sampling interval, and finally achieve synchronous sampling through timed interrupts [12]. Additionally, dual-rate sampling and optimized selection of sampling points have also been proposed to achieve synchronous sampling. Non-synchronous sampling can easily lead to spectral leakage and the fence effect with Fourier Transform, affecting the accuracy of spectrum detection [13]. To reduce errors caused by spectral leakage and the fence effect due to non-synchronous sampling, some scholars have proposed windowed interpolation algorithms and were the first to introduce the idea of windowed interpolation into harmonic detection [14]. Windowing can reduce spectral leakage, while interpolation can eliminate the fence effect, making this algorithm the most popular solution in practical engineering. A window function with a narrower main lobe width and lower sidelobe amplitude will have better time-domain characteristics. Furthermore, wavelet transform analysis is also widely used in harmonic detection. The principle involves decomposing the measured signal into different frequency bands using wavelet decomposition according to a certain scale, and then reconstructing the original signal based on these frequency bands [15]. Due to its excellent time-domain and frequency-domain local characteristics, wavelet transform achieves adaptive changes in the time–frequency window, overcoming the fixed-time–frequency window width and low-resolution drawbacks of Fourier Transform. Reference [16] proposes the Mallat harmonic detection method, which combines multi-resolution analysis with wavelet transform, offering good measurement accuracy, and dynamic performance. However, the more wavelet decomposition times, the greater the error in reconstructing the signal, and multi-resolution analysis has uneven frequency band decomposition, leading to wide high-frequency bands and narrow low-frequency bands, resulting in accurate low-frequency signal detection but significantly reduced high-frequency signal detection accuracy. Reference [17] utilizes Empirical Mode Decomposition (EMD) to extract features, but this method suffers from modal aliasing and mode overlap issues. Reference [18] adopts S transform to extract features, which can identify disturbances but is relatively complex.

The second phase in the identification process of power quality disturbance signals involves pattern recognition of the extracted feature information. Pattern recognition typically employs artificial intelligence techniques for decision-making and classification. Machine learning, as the fastest-growing branch of artificial intelligence, has become the most commonly used method for pattern recognition. Machine learning involves algorithms that enable machines to analyze and discover patterns from vast amounts of data, enabling intelligent recognition and prediction of new samples. Currently, machine learning has undergone two major stages of development: shallow learning and deep learning (DL). Common shallow machine learning recognition algorithms include Artificial Neural Networks (ANN) [19], Decision Trees (DT) [20], Random Forests (RF) [21], and SVM [22]. ANN is an algorithmic mathematical model that mimics the characteristics of human neural networks for distributed parallel processing. It can adjust the biases and weights between different neurons to achieve functions such as pattern recognition and function approximation. ANN excels in handling nonlinear data and is widely used in power quality disturbance identification due to its simple structure. Reference [23] proposes a composite disturbance identification method based on dual neural networks, which utilizes an adaptive linear network to obtain feature vectors for detecting and classifying signal disturbances such as surges, harmonics, and power outages. Reference [24] combines ANN with HT, using EMD to decompose signals into IMFs and extract instantaneous amplitude and spectral information to classify transient signals. In practical applications of ANN, a large number of samples are often required, and the training time can be lengthy, limiting its application in power quality disturbance identification. DT mimics human logical thinking to establish rules and achieve classification, characterized by a simple structure and ease of implementation. Reference [25] constructs a rule-based DT classification method based on features extracted using FFT transform, dynamic measure algorithm, and S transform. Reference [26] extracts classification features based on the curve differences obtained after HT transformation and adaptively determines feature thresholds through the construction of a classification DT model. However, DT suffers from issues such as slow convergence, high sensitivity to noise, and high dependence on training samples. RF is a popular learning algorithm based on DT in recent years. RF is a large ensemble classifier composed of multiple tree-structured DT classifiers, combining the random subspace algorithm with the Bagging algorithm [27]. Its basic principle is to improve prediction and classification accuracy through the combination of multiple DTs. Reference [28] achieves a high accuracy in power quality disturbance signal identification by combining an improved S transform with the RF algorithm. However, RF's drawback lies in the significant time and space required for training when there are too many DTs, limiting its development in the field of power quality disturbance identification. SVM is a supervised learning method for nonlinear system modeling and pattern recognition [29]. It overcomes ANN's shortcomings of slow convergence and local optimal solutions, offering advantages such as good global optimization, strong generalization ability, and short training time, making it suitable for small sample classification problems. Reference [30] combines an improved wavelet transform with SVM to construct an integrated model for power quality disturbance identification. Reference [31] combines the Genetic Algorithm with SVM to build an identification model, achieving promising experimental results.

Despite the progress made in the aforementioned research, several issues and challenges still persist:

(1) When faced with multiple power quality disturbances occurring simultaneously, the recognition accuracy of traditional algorithms drops significantly. Furthermore, existing signal processing techniques struggle to fully extract the time–frequency characteristics underlying complex, nonlinear, and non-stationary power signals, making it particularly challenging to accurately identify composite power quality disturbances.
(2) Difficulty in Processing Nonlinear and Non-Stationary Signals: Since power quality signals are typically nonlinear and non-stationary, this complicates the effective decomposition and feature extraction of the signals. Traditional signal processing methods may not be able to accurately decompose noisy signals into multiple IMFs, thereby affecting the accuracy of subsequent classification and identification.

To address the aforementioned issues, this paper proposes a power quality disturbance identification method based on an improved CEEMDAN-HT-ELM model. The key innovations of this paper can be summarized as follows:

(1) This paper proposes a hybrid power quality disturbance identification method based on improved CEEMDAN, HT, and ELM. This method can effectively handle nonlinear and non-stationary signals, enhancing the identification accuracy.
(2) This paper introduces an improved ELM as the classifier, leveraging its powerful nonlinear mapping capability and fast learning speed for pattern recognition of extracted features, achieving accurate identification of hybrid power quality disturbances.

The organization of the full paper is as follows: Sections 2 and 3 constitute the methodology, which introduces in detail the denoising decomposition algorithm and the power disturbance identification model. Section 4 presents a case study, demonstrating the feasibility and effectiveness of the method proposed in this paper. Finally, Section 5 summarizes the entire paper.

2. CEEMDAN Algorithm Principle and Mathematical Model

2.1. Basic Principles

First, the noisy signal containing power quality disturbances is decomposed into several IMFs using the CEEMDAN algorithm. The process is as follows:

According to Equation (1), the signal $X(t)$ is constructed as [5]:

$$X(t) = x(t) + \varepsilon_0 w_i(t) \tag{1}$$

where $x(t)$ represents the original signal; $X(t)$ represents the signal with white noise added; ε_0 is the amplitude coefficient of the white noise; $\omega_i(t)$ represents the white noise.

By applying EMD to the signal $X(t)$, the first-order IMF component and the corresponding residue component can be obtained [7], as shown in Equations (2) and (3):

$$IMF_1(n) = \frac{1}{N}\sum_{i=1}^{N} IMF_1^i(n) \tag{2}$$

$$r_1(n) = x(n) - IMF_1(n) \tag{3}$$

Here, N represents the number of modal components generated, and n is an index variable used to iterate through or index all data samples of the first intrinsic mode function. IMF_1^i represents the intrinsic mode components obtained through CEEMDAN decomposition, and Equation (3) is used to calculate the residual after removing the first modal component. The residue component $r_1(n)$ obtained from Equation (3) is then decomposed again to yield the second-order IMF component [11], as shown in Equation (4):

$$IMF_2(n) = \frac{1}{N}\sum_{i=1}^{N} E_1\{r_1(n) + \varepsilon_1 E_1[w_i(n)]\} \quad (4)$$

In the equation, $E_j(\cdot)$ represents the j-th order component. Following this process, the j-th residual signal component can be expressed as [5]:

$$r_j(n) = r_{j-1}(n) - IMF_n(n) \quad (5)$$

The $(j + 1)$-th order IMF component can be expressed as:

$$IMF_{j+1}(n) = \frac{1}{N}\sum_{i=1}^{N} E_1\{r_{j+1}(n) + \varepsilon_j E_j[w_i(n)]\} \quad (6)$$

By repeating the above steps until the signal can no longer be decomposed further, the algorithm stops, yielding a total of the j-th IMF components. Using an IMF component selection criterion, the decomposed IMF components are categorized into signal IMF components, aliased IMF components, and noise IMF components. The original signal after decomposition can be represented as:

$$x(n) = r(n) + \sum_{j=1}^{k} IMF_j + \sum_{j=k+1}^{m} IMF_j + \sum_{j=m+1}^{n} IMF_j \quad (7)$$

In the above equation, $r(n)$ represents the final residual component; $\sum_{j=1}^{k} IMF_j$ represents the signal IMF components; $\sum_{j=k+1}^{m} IMF_j$ represents the aliased IMF components; and $\sum_{j=m+1}^{n} IMF_j$ represents the noise IMF components.

Given the different frequency ranges associated with the aliased IMF components and noise IMF components, a combined denoising approach is utilized, leveraging the advantages of different denoising methods. The final denoised signal can be expressed as:

$$x(n)^{\Delta} = r(n) + \sum_{j=1}^{k} IMF_j + \sum_{j=k+1}^{m} IMF_j^{\Delta} + \sum_{j=m+1}^{n} IMF_j^{\Delta} \quad (8)$$

where $\sum_{j=k+1}^{m} IMF_j^{\Delta}$ represents the denoised aliased IMF components, and $\sum_{j=m+1}^{n} IMF_j^{\Delta}$ represents the denoised noise IMF components.

2.2. IMF Component Selection Criterion

In order to achieve better denoising results, this paper attempts to introduce the Renyi's entropy-based distance method to subdivide the IMF components obtained from CEEMDAN decomposition into three categories: signal IMF components, aliased IMF components, and noise IMF components. The Probability Density Function (PDF) contains complete characteristic information of a signal. This paper employs the kernel-smoothed PDF method to calculate the PDF of each IMF component and the original signal separately. An improved Renyi's distance is then used to evaluate the similarity between the PDF of each IMF component and the PDF of the original signal. For two sets $M_i = \{m_1, m_2, \ldots, m_k\}$ and $N_i = \{n_1, n_2, \ldots, n_k\}$, the improved Renyi's distance between these two sets is defined as:

$$D(M_i, N_i) = \frac{1}{k}\sum_{j=1}^{k} \frac{|m_i - n_i|}{|m_i|} \tag{9}$$

In the above equation, a smaller value of $D(M_i, N_i)$ indicates a higher similarity between the two sets. It is generally believed that noise is primarily distributed in the high-frequency portion of the signal represented by the low-order IMF components. After the minimum value point of the improved Renyi's distance is reached, as the order of the IMF components increases, the noise contained in the IMF components rapidly decreases. In all, the improved CEEMDAN method is used to preprocess the collected initial power quality signals. The entire process can be described through Equations (1)–(9). Through Equations (1)–(8), the initial noise signal is decomposed into multiple IMFs, and Equation (9) is further applied to reduce modal mixing and artifact components. In addition to the distance method based on Rényi entropy, there are various other methods that can be used for the subdivision of IMF components, such as those based on mutual information, correlation coefficient, and Euclidean distance. Rényi entropy is capable of quantifying the uncertainty or complexity of signals and exhibits high sensitivity to minute changes in signals. In the classification of power quality disturbances, IMF components of different disturbance types possess distinct complexity characteristics, and the distance method based on Rényi entropy can effectively capture these differences, thereby achieving more precise subdivision. Furthermore, this method possesses a certain degree of robustness against noise, capable of suppressing noise interference to a certain extent and enhancing the accuracy of classification. However, the distance method based on Rényi entropy involves relatively complex calculations, increasing the computational burden of the algorithm. Therefore, in situations where computational resources are limited, a balance needs to be struck between computational complexity and classification accuracy.

3. Power Quality Disturbance Identification Model Based on Hilbert Transform and Improved Extreme Learning Machine

3.1. Introduction to Hilbert Transform

Given a random signal $z(t)$ in the real domain, its HT expression [4] is:

$$y(t) = H[z(t)] = \frac{1}{\pi}\int_{-\infty}^{+\infty} \frac{z(\tau)}{t-\tau} d\tau \tag{10}$$

The inverse transform of the above Equation (10) can be expressed as:

$$z(t) = H^{-1}(y(t)) = -\frac{1}{\pi}\int_{-\infty}^{+\infty} \frac{y(\tau)}{t-\tau} d\tau \tag{11}$$

Rewriting Equations (10) and (11) in the form of convolution gives:

$$\begin{cases} y(t) = \frac{1}{\pi t} * z(t) \\ z(t) = -\frac{1}{\pi t} * y(t) \end{cases} \tag{12}$$

Using $z(t)$ and $y(t)$, we construct a complex-valued function $s(t)$ as:

$$s(t) = z(t) + jy(t) = z(t) + jH[z(t)] \tag{13}$$

Denoting $s(t) = A[z(t)]$, then the following equation holds:

$$A[z(t)] = z(t) + jH[z(t)] = a(t)e^{j\theta(t)} \tag{14}$$

$$a(t) = \sqrt{z^2(t) + y^2(t)} = \sqrt{z^2(t) + H^2[z(t)]} \tag{15}$$

$$\theta(t) = \arctan\frac{y(t)}{z(t)} = \arctan\frac{H[z(t)]}{z(t)} \tag{16}$$

In the Equations (14) and (16), $a(t)$ and $\theta(t)$ represent the instantaneous amplitude and instantaneous phase of the random signal function $z(t)$ using Equation (12), respectively. The frequency can be obtained by differentiating the phase angle, and the instantaneous frequency is given by:

$$\varpi(t) = \frac{d\theta(t)}{dt} = \frac{H'[z(t)]z(t) - H[z(t)]z'(t)}{z^2(t) + H^2[z(t)]} \tag{17}$$

Using Equations (14)–(17), the computation of the instantaneous amplitude, instantaneous phase, and instantaneous frequency of the signal $z(t)$ through HT can be completed.

3.2. Fault Diagnosis Model Based on Extreme Learning Machine

The ELM algorithm is a network algorithm proposed based on feedforward neural networks [12,16]. It is a typical single-hidden layer feedforward neural network, with a structure as shown in Figure 1. The initial power quality signals are preprocessed using the improved CEEMDAN method to reduce modal aliasing and spurious components, thereby enabling a more precise decomposition of the signals into multiple IMF. The HT is then utilized to conduct an in-depth analysis of the reconstructed IMF signals, extracting their time–amplitude information and instantaneous frequency characteristics. This extracted feature information, including time–amplitude information and instantaneous frequency characteristics, constitutes the input for the ELM. The output is the encoded type of power quality disturbance; for example, harmonic disturbances are encoded as 1, voltage sag disturbances as 2, and so on.

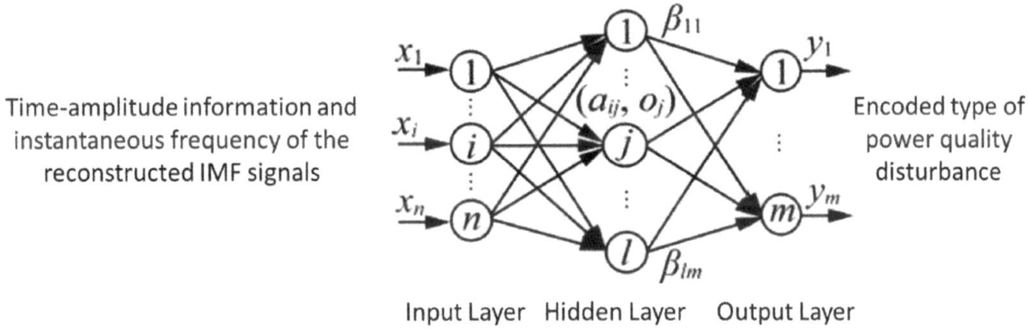

Figure 1. Schematic Diagram of ELM Structure Framework.

The ELM algorithm architecture consists of three layers of neurons: the input layer, the hidden layer, and the output layer. The neurons within each layer are fully connected. The input layer has n neurons, the hidden layer has l neurons, and the output layer has m neurons. a_{ij} represents the connection weight between the input layer and the hidden layer, o_j represents the threshold of each hidden layer neuron, and β_{ij} represents the connection weight between the hidden layer and the output layer.

Assuming $g(x)$ is the activation function of the hidden layer [13], the output is expressed as shown in Equation (18), where $i = 1, 2, \ldots, n; j = 1, 2, \ldots, l; k = 1, 2, \ldots, m$.

$$y_k = \sum_i^n \beta_{jk} g(a_{ij} x_i + o_j) \tag{18}$$

Let H be the output matrix of the hidden layer, as shown in Equation (19).

$$H(a_1, a_2, \ldots, a_l, o_1, o_2, \ldots, o_l, x_1, x_2, \ldots, x_n) = \begin{bmatrix} g(a_1 x_1 + o_1) & \cdots & g(a_l x_1 + o_l) \\ \cdots & \cdots & \cdots \\ g(a_1 x_n + o_1) & \cdots & g(a_l x_n + o_l) \end{bmatrix}_{n \times l} \tag{19}$$

Given the output matrix as Y, we have $H\beta = YT$, leading to Equation (8).

$$\beta = H^{-1} Y^T \tag{20}$$

In the ELM algorithm, the connection weights between the input layer and the hidden layer are randomly generated, and the thresholds of the hidden layer neurons are also randomly set. During training, after selecting the activation function, only the number of hidden layer neurons needs to be determined, and then the connection weight matrix β_{ij} between the hidden layer and the output layer can be calculated. Through the weight matrix β_{ij}, untrained data can be further tested, offering advantages such as fast learning speed and strong generalization performance. However, there are also certain issues. The random generation of connection weights between the input layer and the hidden layer, as well as the thresholds of the hidden layer neurons, can affect the training results of the ELM algorithm each time, which in turn affects its generalization performance.

$$\beta = \left(H^T H\right)^{-1} H^T Y^T \tag{21}$$

$$\beta = \left(H^T H + \lambda I\right)^{-1} H^T Y^T \tag{22}$$

The random generation of connection weights between the input layer and the hidden layer, as well as the thresholds of the hidden layer neurons, can lead to uncertainty in the calculation results, making it difficult to achieve the desired results through just one or a few calculations. To reduce the error caused by random data, the PSO algorithm is used to iteratively optimize the input connection weights and thresholds of the ELM algorithm with regularization parameters [32,33]. The specific steps are as follows:

1. Establish training and testing samples based on the extracted signal features.
2. Introduce regularization parameters and perform optimal parameter setting.
3. Set the input connection weights and hidden layer neuron thresholds of the Regularized ELM (RELM) algorithm as particles, and determine the particle length.
4. Configure the relevant parameters of the PSO algorithm, including speed update parameters, iteration count, population size, maximum and minimum speed values, etc.
5. Use the root mean square error (RMSE) value of the initialized particles in the training samples as the fitness value for each particle. Continuously iterate and select the larger fitness value as the new individual best and global best.
6. Stop the iteration when the RMSE value meets the specified condition or the maximum iteration count is reached. The corresponding particle at this point represents the optimal input connection weights and hidden layer neuron thresholds to be found.

7. Use the optimal parameters to assign values to the input layer and hidden layer of the RELM algorithm, thereby obtaining the optimal RELM algorithm parameter model. This model can then be used to classify and identify power quality disturbance signals.

To gain a better understanding, the flowchart of the proposed method is given in Figure 2. With the construction of digital power systems, the widely installed smart meters and fault recording devices can provide data for the ELM training set. The fault voltage waveforms recorded by the fault recording devices can be extracted and analyzed in real-time by the control center. These data are representative and authentic, reflecting various possible disturbance scenarios in the power system. The timing for retraining the ELM model depends on the following three factors: 1. Environmental changes: If the operating environment of the power system changes (such as the introduction of new loads, changes in grid structure, etc.), the ELM may need to be retrained to adapt to new disturbance characteristics. 2. Performance decline: If the performance of the ELM in practical applications declines to a certain extent (e.g., accuracy falls below a certain threshold), retraining is necessary to enhance its performance. 3. Data updates: As new data are continuously collected and processed, the training set can be periodically updated, and the ELM retrained to capture the latest disturbance characteristics and improve the model's generalization ability. Retraining of the model is required when any of the above three situations occur.

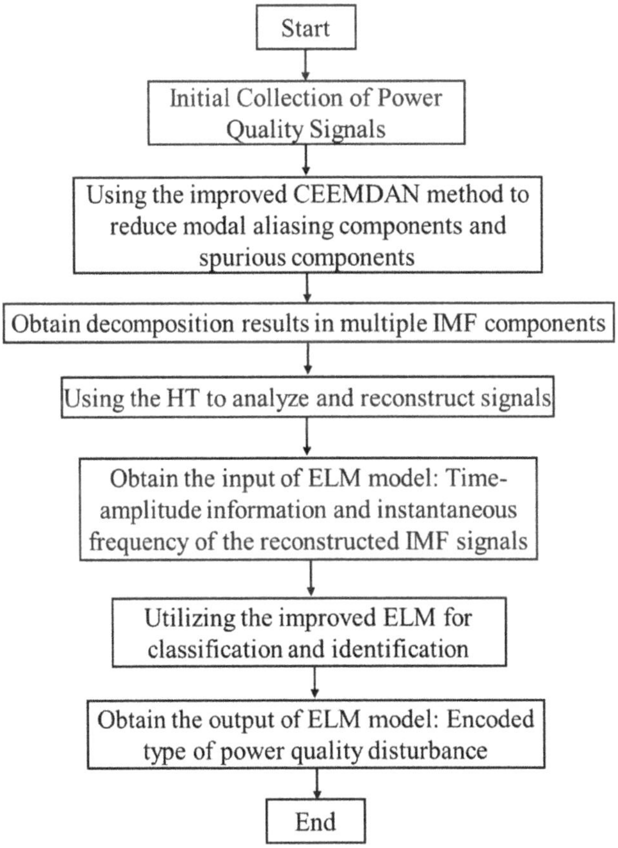

Figure 2. The flowchart of the proposed method.

4. Case Study with Discussions

In the process of identifying power quality disturbance signals, the collected voltage signals are selected for analysis and diagnosis. Issues related to power quality, such as voltage fluctuations, voltage sags, voltage swells, voltage interruptions, and harmonics, are directly reflected in the voltage signals. These disturbance types were chosen and emphasized based primarily on their prevalence in power systems and their impact on system stability and equipment operation. Voltage fluctuations refer to variations in voltage levels around its average value over a period of time. Such fluctuations can lead to decreased equipment performance or even damage. Harmonics cause distortion in voltage and current, affecting the normal operation of power equipment, shortening equipment lifespan, and reducing the stability and efficiency of power systems. Voltage sags and swells may cause sensitive equipment to shut down or fail to restart, significantly impacting industrial production, commercial operations, and residential life. Voltage interruptions can result in equipment shutdown, data loss, and other severe consequences, posing a threat to the reliability and stability of power systems. It should be noted that, within the context of this paper, the primary focus is on harmonics. Harmonics are ubiquitous in power systems and are becoming increasingly severe with the increase in nonlinear loads. The monitoring of harmonics requires high-precision measurement equipment and professional analysis techniques. The mitigation of harmonics necessitates comprehensive consideration of factors such as grid structure, load characteristics, and filtering devices, posing significant challenges. In other words, if the method proposed in this paper can accurately identify harmonics, it can also achieve efficient identification and classification of other types of disturbances. Therefore, by monitoring and analyzing these voltage signals, one can intuitively understand the status of power quality. In power systems, voltage signals are generally easy to collect using sensors or measurement equipment. Furthermore, the processing and analysis of voltage signals are relatively straightforward, allowing for feature extraction and pattern recognition using existing signal processing techniques and algorithms. Power quality disturbances often lead to equipment failures or performance degradation. There is a close relationship between changes in voltage signals and these failures. For instance, voltage sags may cause motors to shut down or fail to restart, while voltage interruptions may result in equipment shutdown or data loss. By analyzing voltage signals, these potential issues can be promptly identified, and corresponding measures can be taken for prevention or repair.

4.1. Effectiveness Validation of the CEEMDAN Algorithm

In the identification of power quality disturbances in electric power systems, harmonics are considered one of the most complex and challenging disturbances to extract features from. The harmonic data utilized in this Section were collected from smart meters of State Grid Jiangsu Electric Power Company in real-world scenarios, with a total sample size of 36,000. The relevant data have been published on the open-source data website CSDN, and researchers can download them via the link [34]. The model of the smart meter employed here is APM510 which is produced by China Jiangsu Ankui Electric Appliance Manufacturing Co., Ltd., as shown in Figure 3. It can accurately measure electrical parameters such as current, voltage, and power factor. Harmonics not only consist of frequency components that are integer multiples of the fundamental frequency but can also occur simultaneously in multiple forms, forming a complex harmonic spectrum. Furthermore, the amplitude, phase, and other parameters of harmonics may vary over time, adding to their complexity. In power systems, multiple harmonics may coexist and interact with each other, making it even more difficult to extract the features of a single harmonic. Additionally, there may be overlapping or similar frequencies and amplitudes among different harmonics, further complicating feature extraction. To address complex harmonic signals, the proposed method and the traditional Ensemble Empirical Mode decomposition (EEMD) algorithm are compared and illustrated in Figures 4 and 5 below, respectively.

<table>
<tr><td rowspan="3">

Product Appearance</td><td>Model</td><td>APM510</td></tr>
<tr><td>Performance</td><td>It can accurately measure electrical parameters such as current, voltage, and power factor, and supports multiple communication protocols like RS485 and Ethernet, which facilitates connection with remote systems and data transmission, enabling real-time data transmission and calculation.</td></tr>
<tr><td>Applicable Types of Disturbance Analysis</td><td>Harmonics, Inter-harmonics, Flicker, and Voltage Fluctuations</td></tr>
</table>

Figure 3. Structural Diagram and Related Introduction of the Smart Meter.

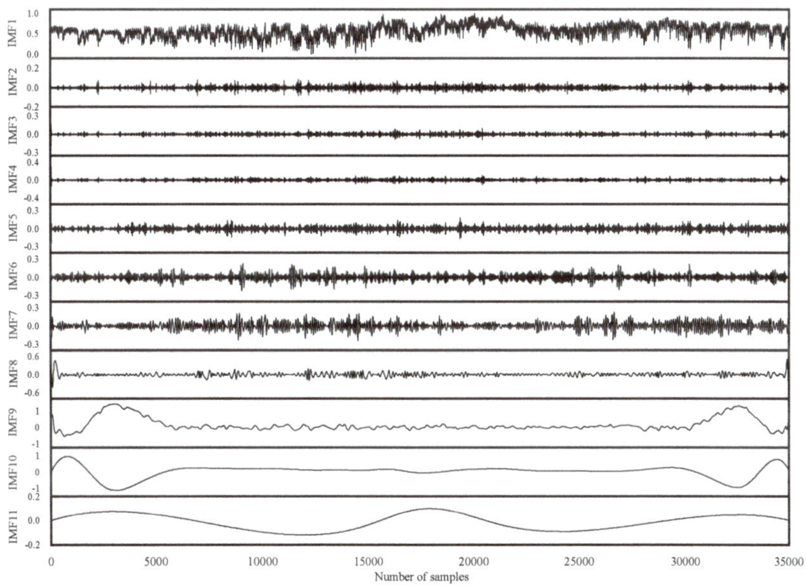

Figure 4. The decomposition results of harmonic signals obtained through the CEEMDAN algorithm.

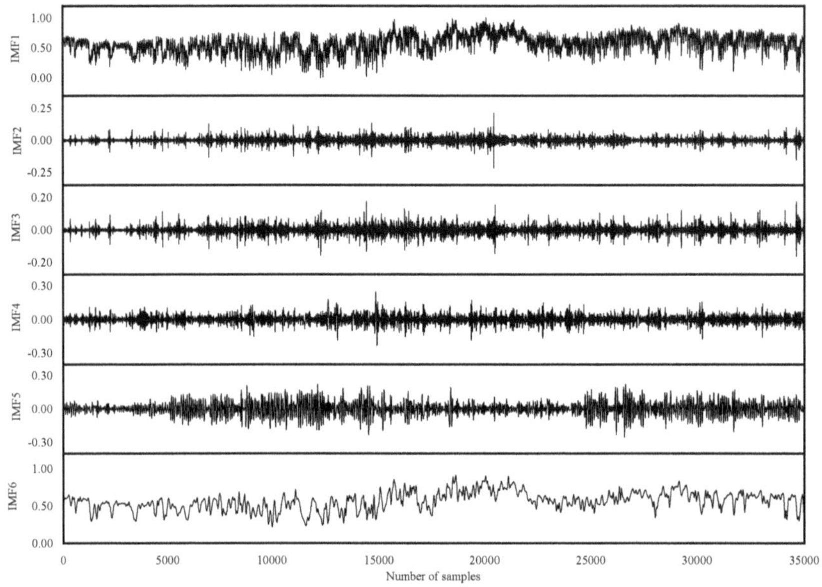

Figure 5. The decomposition results of harmonic signals obtained through the EEMD algorithm.

Upon observing Figures 4 and 5, it becomes evident that the CEEMDAN algorithm, during the decomposition process, not only adds Gaussian white noise to the original signal but also adaptively incorporates specific noise based on the outcome of the previous decomposition stage at each iteration. This adaptive noise addition approach facilitates a more detailed revelation of different frequency components within the signal, resulting in a greater number of decomposed components. While the EEMD algorithm addresses the mode mixing issue in EMD by introducing white noise, the noise added is global and remains constant throughout the entire decomposition process. In contrast, CEEMDAN further mitigates mode mixing by incorporating noise related to the previous stage at each step, leading to a finer decomposition outcome with potentially more components. When decomposing harmonic signals, the 11 IMF components obtained each represent the time evolution of different frequency components within the signal. Generally, the first few IMF components (such as IMF1, IMF2, etc.) represent high-frequency components or noise in the signal. These components capture the rapidly changing parts of the data, which may include noise, interference, or other high-frequency characteristics within the signal. As the sequence number of the IMF components increases, the frequencies gradually decrease. The intermediate IMF components (such as IMF3 to IMF6, etc.) typically represent medium-frequency components in the signal. These components reflect the fluctuation characteristics of the signal within the medium-frequency range. The later IMF components (such as IMF7 to IMF11, etc.) represent low-frequency components in the signal. These components capture the slower-changing parts of the signal, which may include trend components or very low-frequency fluctuations.

In this paper, two indicators—modal aliasing degree and reconstruction error are selected to quantitatively assess the effectiveness of the method. Through analysis, the modal aliasing degree and reconstruction error of the strategy proposed in this paper are 0.0028 and 1.78%, respectively, while those of the EEMD strategy are 0.0108 and 4.59%. This suggests that the IMF components decomposed using the strategy proposed in this paper can more accurately reflect the different frequency components in the signal and preserve more complete original signal information. Due to the use of adaptive noise addition and a more refined decomposition strategy, the CEEMDAN algorithm typically achieves higher decomposition accuracy. This enhances the precision in identifying and extracting characteristic information from power quality harmonic disturbance signals. Compared to EEMD, CEEMDAN's gradual addition and removal of noise during decomposition more effectively suppresses the influence of noise on the decomposition results, which is crucial for improving the accuracy of harmonic disturbance signal analysis.

Not only is the CEEMDAN algorithm suitable for decomposing power quality harmonic disturbance signals, but it is also widely applied in signal processing tasks across various fields. Its robust signal decomposition capabilities and broad applicability make it a significant player in the signal processing domain. The multiple IMFs obtained through CEEMDAN decomposition represent different frequency scales within the signal. These components can be further utilized for multi-scale analysis, providing a more comprehensive understanding of signal characteristics and variation patterns. The large number of relatively independent IMFs from CEEMDAN decomposition facilitates subsequent signal processing and analysis tasks, such as reconstruction, filtering, or feature extraction, tailored to meet diverse application requirements. The proposed method integrates advanced signal processing techniques and machine learning algorithms, enabling it to identify multiple concurrent power quality disturbance signals, including voltage fluctuations, voltage sags, voltage swells, voltage interruptions, harmonics, frequency deviations, and so on. The actual power grid structure is complex, with numerous branch lines and load variations, which complicates the propagation and distribution of harmonics. Various noise interferences exist in the power grid, such as background noise and measurement noise, which interfere with the detection and analysis of harmonics. Therefore, this paper focuses on the analysis of harmonic disturbances in power quality.

4.2. Verification of Power Quality Disturbance Identification Accuracy

Based on the RELM algorithm, five training and testing trials were conducted for each of the four different activation functions: sigmoid, sin, hardlim, and radbas. The test accuracy results are shown in Figure 6. It can be observed that among the three other activation functions, radbas achieves the highest average test accuracy, thus it was selected as the activation function. The radbas function possesses powerful nonlinear mapping capabilities. In power quality disturbance signals, which often contain complex nonlinear features, the radbas function is better able to capture these features, thereby enhancing the identification accuracy of the model.

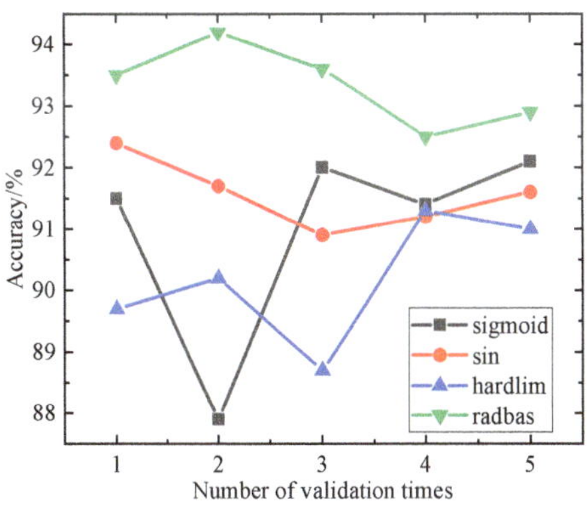

Figure 6. Verification of Accuracy with Different Activation Functions.

Compared to other activation functions, the radbas function exhibits more flexible local response characteristics. This means it can produce different output responses within different input ranges, enabling a more precise description of the local features of disturbance signals. During the training process, the radbas function may assist the ELM model in learning more generalized feature representations, enabling the model to maintain high identification accuracy even when confronted with unseen disturbance signals.

Selecting an appropriate activation function helps balance the model's complexity and generalization ability, avoiding overfitting. While traditional activation functions like sigmoid are widely used in deep learning, they may not be the optimal choice for certain specific tasks, such as power quality disturbance identification. In contrast, the radbas function may better enable the model to maintain high identification accuracy while avoiding overfitting to the training data. Through multiple training and testing trials, the advantage of the radbas function in terms of average test accuracy becomes evident. This further validates its effectiveness as an activation function for power quality disturbance identification tasks.

To further demonstrate the accuracy of the proposed method in this paper, a comparison is made with traditional ELM, SVM, and CNN models. The power quality disturbance identification accuracies are presented in Table 1 below. In this paper, the PSO algorithm is employed to iteratively refine the input connection weights, thresholds, and regularization parameters of the ELM algorithm compared to traditional methods. The total data sample consists of 200 records, obtained from fault recording devices of State Grid Jiangsu Electric Power Company in China Nanjing. The relevant data can be downloaded from an open-source data website [34]. The fault recording devices utilized in this paper are of the ZH-6 dynamic recording unit model which is produced by China Wuhan Zhongyuan Huadian Technology Co., Ltd. as illustrated in Figure 7.

In practical operation of power systems, complex situations often arise where multiple disturbances occur simultaneously. Table 2 presents the identification accuracy of the proposed method in cases where multiple disturbances occur simultaneously. The number of samples in the training set used is 150. The considered combinations of disturbance faults include the following categories:

Combination 1: Voltage sags, voltage swells, voltage interruptions, and harmonics.
Combination 2: Voltage fluctuations, voltage swells, voltage interruptions, and harmonics.
Combination 3: Voltage fluctuations, voltage sags, voltage interruptions, and harmonics.
Combination 4: Voltage fluctuations, voltage sags, voltage swells, and harmonics.

Table 1. Comparison of Power Quality Disturbance Identification Accuracies Across Different Models.

Noise/dB	Identification Accuracy/%				
	The Proposed Method	ELM Model	SVM	Transformer	CNN
5	98.7	94.3	95.6	96.8	94.9
10	97.6	93.2	95.0	95.4	93.8
15	97.2	92.4	94.8	94.2	93.6
18	95.2	90.6	93.4	93.7	91.2

Table 2. Comparison of Diagnostic Accuracy Among Different Models in Cases of Multiple Disturbances Occurring Simultaneously.

Combination	Identification Accuracy/%				
	The Proposed Method	ELM Model	SVM	Transformer	CNN
1	95.0	93.2	91.6	93.8	92.5
2	95.6	89.7	92.4	94.6	89.4
3	96.1	91.1	92.0	93.2	87.8
4	95.4	93.2	91.8	91.7	93.0

<table>
<tr><td rowspan="3">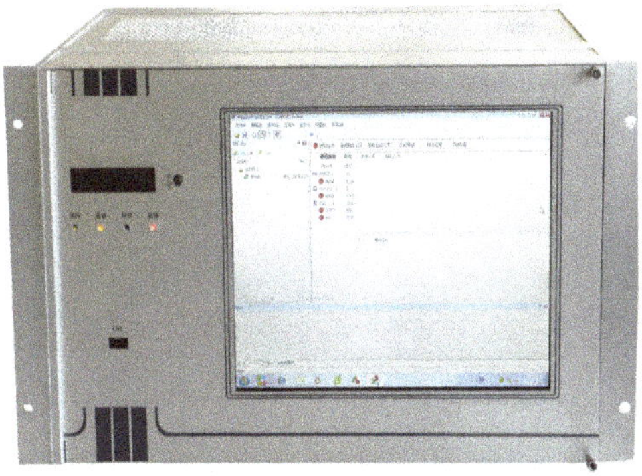
Product Appearance</td><td>Model</td><td>ZH-6 dynamic recording unit</td></tr>
<tr><td>Performance</td><td>It achieves synchronous dual storage of transient-triggered waveform data, making it suitable for substations with voltage levels of 35(66)kV and above. Field data acquisition is implemented under Linux using VXD technology.</td></tr>
<tr><td>Applicable Types of Disturbance Analysis</td><td>Short-circuit faults, system oscillations, frequency collapses, voltage collapses, harmonics</td></tr>
</table>

Figure 7. Structural Diagram and Related Introduction of fault recording devices.

Upon examining the table above, it is evident that the method proposed in this paper exhibits robustness and effectively copes with various noise interferences, maintaining a disturbance identification accuracy above 95%. Regularization, a common technique in machine learning, is used to prevent overfitting and enhance the model's generalization ability. In the context of ELM algorithm, incorporating a regularization parameter constrains the model's complexity, ensuring that during training, the model not only strives for high accuracy on the training set but also considers its performance on the test set or unseen data. This helps the model maintain stable performance even in the presence of noise. The ELM algorithm, based on single-hidden-layer feedforward neural networks, boasts rapid learning capabilities and strong nonlinear approximation abilities. During training, it randomly generates hidden layer node parameters and utilizes the least squares method to directly solve for the output layer weights, eliminating the need for iterative adjustment of hidden layer parameters, thereby significantly improving learning efficiency. This rapid learning capability renders ELM particularly advantageous for power quality disturbance identification and classification tasks with high real-time requirements. In this paper, the swarm size is set to 40, the number of iterations to 200, and the inertia weight is dynamically adjusted during the iteration process, gradually decreasing from 0.9 to 0.4. The cognitive learning factor and social learning factor are both set to 2. Both the PSO algorithm and the GA are widely used heuristic optimization algorithms currently. When optimizing the parameters of ELM and identifying power quality disturbance signals, the ELM model optimized by PSO exhibits higher fault diagnosis accuracy. When the number of test samples is 1000, the average diagnosis accuracies of the models optimized by PSO and GA are 98.2% and 94.6%, respectively. This is because the PSO algorithm simulates the foraging behavior of bird flocks, utilizing information sharing and collaboration among particles to search for the optimal solution. It is characterized by its ability to quickly converge to a local optimal solution and, in some cases, to escape from local optima, demonstrating good global search capabilities. When optimizing the connection weights and hidden layer neuron thresholds of the ELM model, PSO can quickly locate better parameter combinations, thereby improving the fault diagnosis accuracy of the model. In contrast, the GA simulates natural selection and genetic processes, such as crossover and mutation, to generate new individuals and optimize solutions. The GA possesses strong global search capabilities but may be time-consuming during the search process and, in some cases, prone to premature convergence, i.e., converging prematurely to a non-global optimal solution. Additionally, the parameters in the PSO algorithm (such as inertia weight and learning factors) can be adjusted according to the characteristics of specific problems to better adapt to the optimization process. This flexibility enables PSO to more precisely control the search direction and step size when optimizing ELM model parameters, thereby finding better solutions. Although the GA also has a certain parameter adjustment space, the impact of its parameter settings on the optimization results may not be as intuitive and significant as that of PSO. When multiple disturbances occur simultaneously, their characteristics can interfere with each other, making the features of individual disturbances less obvious or difficult to distinguish. This may pose difficulties for algorithms in feature extraction, thereby reducing the subsequent classification and identification accuracy. The occurrence of multiple disturbances simultaneously implies the need to process more complex signals, which increases computational complexity and processing time. Although the Transformer model has achieved remarkable results in many fields, when dealing with issues such as power quality disturbances that exhibit complexity and nonlinearity, it may require more training data and a more complex model structure to achieve high accuracy. In contrast, the method proposed in this paper achieves high accuracy with less training data and a simpler model structure through meticulous signal preprocessing, precise feature extraction, and efficient classifier selection. Compared to the Transformer model, another significant advantage of the strategy proposed in this paper is its higher computational efficiency. When performing identification of multiple types of mixed disturbances, the computation time for the strategy proposed in this paper is 127.4 s, while the computation time for the Transformer model is 557.4 s. This is due to the complexity and high number of parameters in the Transformer model, which requires substantial computational resources for both training and inference processes. ELM is a rapid single-layer feedforward neural network training method, whose basic idea is to randomly select the weights and biases from the input layer to the hidden layer, and then directly calculate the output layer weights. When dealing with large-scale datasets or real-time fault diagnosis tasks, the ELM model may have more advantages. Considering that the dispatching time scale in power systems is generally 5 min, the method proposed in this paper can meet the needs of real-time dispatching.

Incorporating regularization parameters into the ELM algorithm significantly boosts its efficacy. This regularization component serves as a stabilizer in the objective function, facilitating a trade-off between minimizing training errors and controlling model complexity. This equilibrium aids in sustaining the model's performance stability in noisy scenarios, thus enhancing its robustness. In the realm of power quality disturbance identification and classification, practical signals frequently encounter numerous noise sources stemming from equipment, transmission lines, and other power system elements. Thanks to its robust generalization capabilities and resilience, the regularized ELM algorithm can somewhat mitigate noise interference and precisely discern types of power quality disturbances. Furthermore, the diagnostic precision of our proposed method outstrips that of conventional ELM models, SVMs, and CNNs.

While ELM algorithms excel in swift learning and nonlinear approximation, they may fall prey to overfitting without regularization, thereby undermining their performance on test sets. By fine-tuning model complexity via regularization, the ELM algorithm attains superior classification accuracy. SVMs, despite their proficiency in managing small samples and nonlinear challenges, frequently entail more intricate training procedures and demonstrate lower processing efficiency for large-scale datasets. Conversely, ELM algorithms maintain high classification accuracy while offering quicker training times and superior processing efficiency. Meanwhile, CNNs, while formidable in image processing, may not exhibit the same advantages when tackling one-dimensional time-series data, such as power quality disturbance signals, and they often necessitate extensive data and computational resources for training. In contrast, ELM algorithms can effectively classify even with constrained resources.

4.3. Effectiveness Testing of Industrial Application

In order to further validate the effectiveness and feasibility of the method proposed in this paper within an industrial testing system, the IEEE-33-bus testing system is adopted, and its structural topology is illustrated in Figure 8 below. Four disturbance signal sources are installed at nodes 14, 20, 23, and 28 in the diagram. The noise sources can generate and superimpose different types of disturbance signals according to requirements. Figure 9 demonstrates the diagnostic accuracy under different numbers of disturbance signal sources. It should be noted that the four disturbance signal sources in Figure 6 consist of 100 sets of disturbance signals, which are derived from actual regional grid data of State Grid Jiangsu Electric Power Company. These signals can be downloaded from the open-source data website CSDN [34]. The types of faults considered include voltage sags, voltage swells, voltage interruptions, harmonics, and combinations of the aforementioned faults.

Observing Figure 9, it can be found that under the superposition of multiple types of disturbance signals, the method proposed in this paper can still maintain a diagnostic accuracy above 94.5%. Traditional algorithms have limitations when dealing with nonlinear and non-stationary signals. However, the proposed method preprocesses the initial power quality signals by introducing improved CEEMDAN, effectively reducing modal aliasing and spurious components, enabling noisy signals to be more accurately decomposed into multiple IMFs. This step significantly improves the accuracy and reliability of signal decomposition, laying a solid foundation for subsequent feature extraction and classification recognition. Secondly, the HT is utilized to conduct an in-depth analysis of the reconstructed signals, extracting their time–amplitude information and instantaneous frequency characteristics. This feature information not only contains the key characteristics of power quality disturbances but also provides abundant data support for subsequent classification and recognition. Through the transformation of HT, different types of power quality disturbance signals become more separable in the feature space. Finally, an improved ELM is introduced as the classifier, leveraging its powerful nonlinear mapping capabilities and fast learning speed to perform pattern recognition on the extracted features. The introduction of ELM not only improves the accuracy and efficiency of classification and recognition but also adapts well to the complexity of superposition of multiple types of power quality disturbance signals.

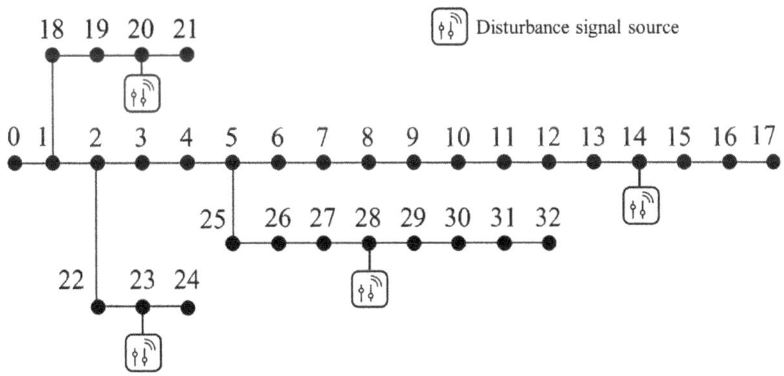

Figure 8. Structural topology of the IEEE-33-bus testing system.

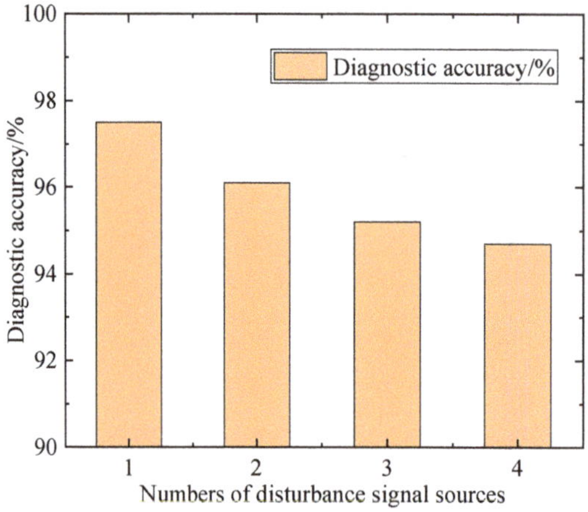

Figure 9. Diagnostic accuracy under different numbers of disturbance signal sources.

In summary, the proposed method, through improved signal preprocessing techniques, feature extraction methods, and classification recognition algorithms, effectively solves the problem of decreased recognition accuracy when traditional algorithms deal with the superposition of multiple types of power quality disturbance signals. Thus, it achieves the goal of maintaining high diagnostic accuracy even when multiple types of power quality disturbance signals are superimposed. In practical applications, this method requires high-precision sensors and data acquisition cards to capture power quality signals. Due to the involvement of complex signal processing and machine learning algorithms, high-performance computing devices (such as high-performance computers or dedicated servers) are needed to support real-time or near-real-time data processing and analysis. The physical conditions of the installation site, such as temperature, humidity, electromagnetic interference, etc., must be considered to ensure stable operation of the equipment. This adds a degree of complexity to the application and promotion. The performance of the method proposed in this paper will also be affected by parameter settings. For example, the number of decomposition layers in CEEMDAN, the choice of window function in HT, and the number of hidden layer neurons in ELM will all influence the final recognition results. Therefore, these parameters require careful tuning. Although the proposed method may achieve high recognition accuracy when dealing with complex nonlinear power quality disturbances, its computational complexity is also relatively high. Especially when processing large-scale datasets, it requires longer computation time and higher computational resources.

In the future, with the rapid development of big data and cloud computing technologies, it will become feasible to apply these technologies to the identification of power quality disturbance signals. By collecting and analyzing vast amounts of power quality data, potential patterns and characteristics of the disturbance signals can be revealed, providing data support for the optimization of identification algorithms. Meanwhile, cloud computing technology enables real-time processing and analysis of large-scale data, offering robust support for real-time monitoring and intelligent recognition in power systems.

5. Conclusions

This paper proposes an innovative hybrid power quality disturbance identification method aimed at addressing the increasingly complex and severe power quality disturbance issues in modern power systems. By integrating the improved CEEMDAN technique, HT, and an ELM classifier, the method effectively overcomes the limitations of traditional algorithms in processing nonlinear and non-stationary signals, significantly enhancing the identification accuracy and robustness of power quality disturbances. After analysis, the method proposed in this paper maintains fault diagnosis accuracy above 95% even under strong noise interference. Compared with traditional methods, including CNN and SVM, the accuracy can be improved by up to 5%. After optimization using the PSO algorithm, the model's accuracy increases by 3.6%, which demonstrates good applicability and robustness.

Specifically, the modified CEEMDAN approach successfully mitigates modal aliasing and spurious components, enabling a more precise decomposition of noisy signals into multiple IMFs, which lays a solid foundation for subsequent feature extraction. The introduction of HT further exploits the time–amplitude information and instantaneous frequency characteristics of the reconstructed signals, providing a rich and effective feature set for the classifier. Building upon this, the improved ELM is utilized as the classifier, leveraging its rapid learning speed and powerful nonlinear mapping capability to achieve precise identification of composite power quality disturbances.

Experimental results demonstrate that the proposed method not only excels in identifying various types of power quality disturbances but also maintains high accuracy even in noisy environments, fully validating its effectiveness and practicality in real-world applications. Furthermore, the method exhibits robustness and superiority in complex grid environments, offering potent technical support for power quality monitoring and management in modern power systems.

Author Contributions: Conceptualization, K.L., J.H., S.C., L.R., Y.L. and Y.W.; methodology, K.L., J.H., S.C., L.R., Y.L. and Y.W.; software, K.L., J.H., S.C., L.R., Y.L. and Y.W.; writing—original draft preparation, K.L., J.H., S.C., L.R., Y.L. and Y.W. All authors have read and agreed to the published version of the manuscript.

Funding: This research received no external funding.

Data Availability Statement: The data can be obtained by contacting the corresponding author.

Conflicts of Interest: Ke Liu, Jun Han and Yutong Liu were employed by Electric Power Research Institute of State Grid Qinghai Power Grid Corporation. The remaining authors declare that the research was conducted in the absence of any commercial or financial relationships that could be construed as a potential conflict of interest.

Nomenclature

Parameters and Variables	Meaning
$x(t)$	The original signal
$X(t)$	The signal with white noise added
ε_0	The amplitude coefficient of the white noise
$\omega_i(t)$	The white noise
$r_1(n)$	The residue component
$E_j(\cdot)$	The j-th order component
$r(n)$	The final residual component
N	The number of modal components generated
n	An index variable used to iterate through or index all data samples of the first intrinsic mode function
IMF_1^i	The intrinsic mode components obtained through CEEMDAN decomposition

$\sum_{j=1}^{k} IMF_j$	The signal IMF components
$\sum_{j=k+1}^{m} IMF_j$	The aliased IMF components
$\sum_{j=m+1}^{n} IMF_j$	The noise IMF components
$z(t)$	The random signal
$s(t)$	The complex-valued function
$a(t)$ and $\theta(t)$	The instantaneous amplitude and instantaneous phase
a_{ij}	The connection weight between the input layer and the hidden layer
o_j	The threshold of each hidden layer neuron
β_{ij}	The connection weight between the hidden layer and the output layer.
$g(x)$	The activation function of the hidden layer
H	The output matrix of the hidden layer
Y	The output matrix

References

1. Chen, J.; Zhao, Y.; Lin, H.; Wei, Y.; Liu, W.; Guo, Q.; Li, Y.R.; Mantooth, H.A. Analysis and Control of Cascaded Energy Storage System for Energy Efficiency and Power Quality Improvement in Electrified Railways. *IEEE Trans. Transp. Electrif.* **2024**, *10*, 1299–1313. [CrossRef]
2. Zhou, J.; Gao, Y.; Wu, C.; Gu, T.; Xu, C.; Lv, Q. Detection of Power Quality Disturbances Based on Improved Wavelet Threshold Function and CEEMD. *J. Electron. Meas. Instrum.* **2023**, *33*, 141–148.
3. Zu, L.; Liu, Z.; Sheng, N. Research on Denoising Method of Wavelet Transform under Non-stationary Acoustic Signals. *Mod. Electron. Technol.* **2022**, *45*, 35–40.
4. Zhang, Z.; Liu, M.; Yu, B.; Zhang, Z.; Wang, Q. Research on Denoising of Improved Wavelet Threshold New Algorithm in Power Grid Disturbance. *Mod. Electron. Technol.* **2024**, *44*, 53–57.
5. Boopathi, R.; Indragandhi, V. Experimental Investigations on Photovoltaic Interface Neutral Point Clamped Multilevel Inverter-Based Shunt Active Power Filter to Enhance Grid Power Quality. *IEEE Access* **2024**, *12*, 74482–74498. [CrossRef]
6. Monteiro, V.; Moreira, C.; Lopes, J.A.P.; Antunes, C.H.; Osório, G.J.; Catalão, J.P.S.; Afonso, J.L. A Novel Three-Phase Multiobjective Unified Power Quality Conditioner. *IEEE Trans. Ind. Electron.* **2024**, *71*, 59–70. [CrossRef]
7. Tang, F. Spectral Analysis of Interharmonics Based on Spatial Spectrum Estimation Algorithm. Master's Thesis, Lanzhou University of Technology, Lanzhou, China, 2022.
8. Grandke, T. Interpolation Algorithms for Discrete Fourier Transforms of Weighted Signals. *IEEE Trans. Instrum. Meas.* **2013**, *32*, 350–355. [CrossRef]
9. Harris, F.J. On the use of windows for harmonic analysis with the DFT. *Proc. IEEE* **1978**, *66*, 51–83. [CrossRef]
10. Ji, B.; Liang, Z.; Niu, S. FFT Harmonic Analysis Based on Nine-Term Cosine Combination Window with Minimum Sidelobes. *Electr. Power Sci. Eng.* **2022**, *28*, 16–21.
11. Wen, H.; Teng, Z.; Guo, S. Triangular Self-Convolution Window With Desirable Sidelobe Behaviors for Harmonic Analysis of Power System. *IEEE Trans. Instrum. Meas.* **2023**, *59*, 543–552. [CrossRef]
12. Luo, Y.; Li, K.; Li, L.; Huang, J.; Zhao, C.; Xiao, X. A Novel Recognition Method for Complex Power Quality Disturbances Based on Rotation Vector and Fuzzy Transfer Field. *IEEE Trans. Ind. Inform.* **2024**, *20*, 12502–12512. [CrossRef]
13. Shui, L.I.U.; Zhenxin, Z.H.O.N.G.; Ming, C.H.E.N.; Jiehuan, L.I.N.; Yang, Z.H.A.N.G.; Xianghua, P.E.N.G.; Jingyuan, Y.I.N. Power Quality Disturbance Detection Method Based on Improved Empirical Fourier Decomposition. *Mod. Electr. Power* **2023**, *40*, 465–473.
14. Sun, D.; Gao, Q.; Zhu, D.; Lu, Y. Adaptive Circular Power Harmonic Analysis Algorithm Based on Discrete Fourier Transform. *J. Anhui Univ. (Nat. Sci. Ed.)* **2023**, *37*, 57–64.
15. Hu, G.; Li, X.; Liang, D. Thin Cloud Removal from Remote Sensing Images Based on Dual-Tree Complex Wavelet Transform. *J. Anhui Univ. (Nat. Sci. Ed.)* **2024**, *40*, 50–58.
16. Li, B.; Cao, M.; Li, S.; Bao, W. Power Quality Disturbance Recognition Method Based on WT and GA-SVM. *Power Electron. Technol.* **2023**, *54*, 52–55.

17. Lopez-Ramirez, M.; Ledesma-Carrillo, L.; Cabal-Yepez, E.; Rodriguez-Donate, C.; Miranda-Vidales, H.; Garcia-Perez, A. EMD-based feature extraction for power quality disturbance classification using moments. *Energies* **2016**, *9*, 565. [CrossRef]
18. Cheng, Z.; Wu, J.; Liu, D.; Chen, L. Composite Power Quality Disturbance Detection Based on AMIST. *New Technol. Electr. Power Energy* **2020**, *39*, 59–65.
19. Yin, B.; He, Y.; Zhu, Y. A Method for Multi-Disturbance Detection and Recognition of Power Quality Based on Generalized S-Transform and Fuzzy SOM Network. *Proc. CSEE* **2023**, *35*, 866–872.
20. Fang, Y.; Pang, H.; Chen, Y. Detection Method of Voltage Sag Disturbance Based on Improved HHT. In *IOP Conference Series: Earth and Environmental Science, Volume 645, Proceedings of the 2020 International Conference on Smart Grid and Energy Engineering 13–15 November 2020, Guilin, China*; IOP Publishing Ltd.: Bristol, UK, 2021; p. 012069.
21. Tian, Z.; Fu, C. Localization and Classification of Power Quality Disturbances Based on HHT. *Power Syst. Prot. Control* **2023**, *43*, 36–42.
22. Xu, J.; Guo, T. Simulation Analysis of Power Quality Disturbance Detection Based on Improved HHT Combined with Improved S-Transform. *Autom. Technol. Appl.* **2024**, *37*, 98–102+109.
23. Sun, S.; Wang, Q.; Yan, H.; Lin, X.; Du, T. Application of EEMD in Power Harmonic Detection. *Proc. CSU-EPSA* **2024**, *28*, 25–31.
24. Xu, Y.C.; Gao, Y.K.; Li, Z.X.; Xi, L. Application of Improved LMD Algorithm in Power Quality Disturbance Signal Detection of Microgrid. *Power Syst. Technol.* **2019**, *43*, 332–341.
25. Dragomiretskiy, K.; Zosso, D. Variational Mode Decomposition. *IEEE Trans. Signal Process.* **2014**, *62*, 531–544. [CrossRef]
26. Xu, Y.; Gao, Y.; Li, Z.; Lu, M. Detection and Classification of Power Quality Disturbances in Distribution Network Based on VMD and DFA. *CSEE J. Power Energy Syst.* **2023**, *6*, 122–130.
27. Huang, C.; Zhou, T. A New Method for Power Quality Disturbance Detection Based on Variational Mode Decomposition. *Electr. Power Autom. Equip.* **2022**, *38*, 116–123.
28. Yu, K.; Jia, L.; Chen, Y.; Xu, W. The Past, Present and Future of Deep Learning. *J. Comput. Res. Dev.* **2013**, *50*, 1799–1804.
29. Valtierra-Rodriguez, M.; de Jesus Romero-Troncoso, R.; Osornio-Rios, R.A.; Garcia-Perez, A. Detection and Classification of Single and Combined Power Quality Disturbances Using Neural Networks. *IEEE Trans. Ind. Electron.* **2013**, *61*, 2473–2482. [CrossRef]
30. Dong, G.; Li, D.; Chen, Y.; Ma, X.; Fu, A.; Mu, G.; Xiao, B. Power Quality Disturbance Classification Method Based on Feature Fusion Parallel Optimization Model. *Proc. CSEE* **2023**, *43*, 1017–1027.
31. Ruan, J.; Jiang, H.; Li, X.; Shi, Y.; Chan, F.T.S.; Rao, W. A Granular GA-SVM Predictor for Big Data in Agricultural Cyber-Physical Systems. *IEEE Trans. Ind. Inform.* **2019**, *15*, 6510–6521. [CrossRef]
32. Wu, F.; Shi, G.; Wang, S. Inverse Kinematics Solution of Manipulator Based on PSO-ELM. In Proceedings of the 2019 IEEE International Conference on Cybernetics and Intelligent Systems (CIS) and IEEE Conference on Robotics, Automation and Mechatronics (RAM), Bangkok, Thailand, 18–20 November 2019; pp. 293–297.
33. Figueiredo, E.M.N.; Ludermir, T.B. Effect of the PSO Topologies on the Performance of the PSO-ELM. In Proceedings of the 2012 Brazilian Symposium on Neural Networks, Curitiba, Brazil, 20–25 October 2012; pp. 178–183.
34. Available online: https://download.csdn.net/download/ilovejohn/10338755?ops_request_misc=%257B%2522request%255Fid%2522%253A%25221b0063fc0b487f4f112ee63204460044%2522%252C%2522scm%2522%253A%252220140713.130102334.pc%255Fdownload.%2522%257D&request_id=1b0063fc0b487f4f112ee63204460044&biz_id=1&utm_medium=distribute.pc_search_result.none-task-download-2~download~first_rank_ecpm_v1~rank_v31_ecpm-3-10338755-null-null.269%5Ev2%5Econtrol&utm_term=IEEE33&spm=1018.2226.3001.4451.4 (accessed on 22 October 2024).

processes

MDPI

Article

An Advanced Power System Modeling Approach for Transformer Oil Temperature Prediction Integrating SOFTS and Enhanced Bayesian Optimization

Zhixiang Tong [1], Yan Xu [1], Xianyu Meng [2], Yongshun Zheng [2], Tian Peng [2] and Chu Zhang [2,*]

1 Wudongde Hydropower Plant, China Yangtze Power Co., Ltd., Kunming 651512, China; tong_zhixiang@ctg.com.cn (Z.T.); xu_yan4@ctg.com.cn (Y.X.)

2 NARI Engineering Technology Co., Ltd., Nanjing 211100, China; mengxy@nrec.com (X.M.); zys19352399371@163.com (Y.Z.); husthydropt@126.com (T.P.)

* Correspondence: zhangchuhust@foxmail.com; Tel.: +86-150-0523-9070

Abstract

Accurate prediction of transformer top-oil temperature is crucial for insulation ageing assessment and fault warning. This paper proposes a novel prediction method based on Variational Mode Decomposition (VMD), kernel principal component analysis (Kernel PCA), a Time-aware Shapley Additive Explanations–Multilayer Perceptron (TSHAP-MLP) feature selection method, enhanced Bayesian optimization, and a Self-organized Time Series Forecasting System (SOFTS). First, the top-oil temperature signal is decomposed using VMD to extract components of different frequency bands. Then, Kernel PCA is employed to perform non-linear dimensionality reduction on the resulting intrinsic mode functions (IMFs). Subsequently, a TSHAP-MLP approach—incorporating temporal weighting and a sliding window mechanism—is used to evaluate the dynamic contributions of historical monitoring data and IMF features over time. Features with SHAP values greater than 1 are selected to reduce input dimensionality. Finally, an enhanced hierarchical Bayesian optimization algorithm is used to fine-tune the SOFTS model parameters, thereby improving prediction accuracy. Experimental results demonstrate that the proposed model outperforms transformer, TimesNet, LSTM, and BP in terms of error metrics, confirming its effectiveness for accurate transformer top-oil temperature prediction.

Keywords: power transformer; oil temperature prediction; deep learning; feature selection; Bayesian optimization

1. Introduction

In high-voltage transformer condition monitoring, the top-oil temperature is a key indicator for evaluating insulation performance. Transformer oil functions as both an insulating medium and a coolant, and maintaining an optimal oil temperature is essential for ensuring the transformer's safe and stable operation. An abnormal increase in top-oil temperature often signals internal thermal accumulation, which can accelerate the aging of insulating oil and diminish its dielectric strength. In severe cases, this may result in insulation failure, leading to equipment malfunction or forced shutdown. With the increasing integration of clean and renewable energy sources into modern power systems, the operational profiles of transformers have become more dynamic and complex, making effective thermal management more critical than ever. Consequently, accurate and timely prediction of top-oil temperature trends is crucial not only for enhancing operational reliability but also for supporting maintenance planning, load scheduling, and informed decision-making [1,2].

In recent years, driven by rapid advances in intelligent technologies and big data analytics, the prediction of transformer top-oil temperature based on historical operational data has gained significant research interest. Current methods predominantly employ data-driven models, particularly neural network architectures such as multilayer perceptrons, convolutional neural networks (CNN) [3], and long short-term memory (LSTM) networks [4], alongside various other machine learning techniques. These approaches are increasingly replacing traditional physics-based prediction models [5].

Currently, neural network-based data-driven approaches for predicting transformer top-oil temperature (TOT) are widely used both domestically and internationally. Dong et al. [6] proposed a TOT prediction method based on data quality enhancement. This method first employs a Markov model to impute missing values, then applies Ensemble Empirical Mode Decomposition (EEMD) to decompose the time series into multiple subsequences, thereby reducing interference from information at different time scales. Finally, Extreme Learning Machine (ELM) models are used to predict each component separately, and the overall top-oil temperature is reconstructed from these predictions. Deng et al. [7] employed EEMD and an LSTM) network to predict transformer top-oil temperature. Cao et al. [8] used Complete Ensemble Empirical Mode Decomposition with Adaptive Noise (CEEMDAN) in conjunction with the Prophet algorithm. Hu et al. [9] applied a Multi-Stage Temporal Convolutional Network (MS-TCN) for transformer top-oil temperature forecasting. Chang et al. [10] utilized a CNN implemented on a Hadoop + Spark big data platform and its ecosystem components for temperature prediction. Wang et al. [11] proposed a prediction method based on VMD and Gated Recurrent Unit (GRU) networks. Wang et al. [12] extracted spatiotemporal features from load data using a CNN and a Bidirectional LSTM network, incorporating an attention mechanism to emphasize key information. Lei et al. [13] developed a GRU-based time series model enhanced by an improved attention mechanism for intelligent forecasting. Wang et al. [14] established a real-time optimal estimation model for top-oil temperature using the Kalman filtering algorithm. Dong et al. [15] proposed a transformer top-oil temperature prediction model based on an LSTM network.

Some studies have also adopted hybrid approaches that integrate optimization algorithms with predictive models to enhance the accuracy of transformer top-oil temperature forecasting. Yang et al. [16] proposed a prediction model based on an LSTM network, in which Particle Swarm Optimization (PSO) was employed to determine the optimal time step, thereby improving the prediction precision of top-oil temperature. Zou et al. [17] developed a transformer oil temperature prediction model by integrating an Improved Whale Optimization Algorithm (IWOA) with an LSTM neural network enhanced by a Self-Attention (SA) mechanism. Liu et al. [18] introduced an advanced hybrid neural network model, BWO-TCN-BiGRU-Attention, for accurate top-oil temperature prediction. This model combines Black Widow Optimization (BWO), Temporal Convolutional Network (TCN), Bidirectional Gated Recurrent Unit (BiGRU), and an attention mechanism, effectively balancing global feature extraction and temporal dependency modeling in its architecture. Hao et al. [19] proposed a transformer winding hot-spot temperature inversion method based on a Self-Attention Gate Recurrent Unit (SA-GRU) model. Zhang et al. [20] developed a transformer oil temperature prediction model using an improved PSO neural network. In this method, asymmetric learning factors and mutation operators are introduced into the traditional Backpropagation (BP) neural network to enhance the global search capability and convergence speed of the algorithm. Li et al. [21] presented a top-oil temperature prediction model that integrates PSO with a Hybrid Kernel Extreme Learning

Machine (HKELM). Li et al. [22] proposed an interval prediction model for transformer top-oil temperature by combining Kernel Extreme Learning Machine (KELM) with the bootstrap method. Yuan et al. [23] developed a similar prediction model that integrates PSO with HKELM to improve accuracy. Li et al. [24] constructed an improved Weighted Support Vector Regression (WSVR) model optimized by PSO for accurately estimating the top-oil temperature.

Inspired by advances in hybrid learning and interpretable feature selection techniques, this study proposes an intelligent prediction method for transformer top-oil temperature by integrating VMD, Kernel PCA, TSHAP-MLP, enhanced Bayesian optimization, and the SOFTS model. The main contributions of this work are as follows:

(1) The top-oil temperature signal is first decomposed using VMD to extract intrinsic mode functions (IMFs) across different frequency bands. These IMFs are then processed via Kernel PCA to perform non-linear dimensionality reduction, thereby mitigating data redundancy and improving computational efficiency.
(2) A TSHAP-MLP approach is introduced to dynamically evaluate the temporal contribution of each feature, incorporating both temporal weighting and a sliding window mechanism. Features with SHAP values exceeding one are retained to reduce the input dimensionality while preserving critical information.
(3) The SOFTS model is constructed as the core forecasting framework, and its parameters are fine-tuned using an enhanced hierarchical Bayesian optimization algorithm to boost prediction accuracy.

Unlike earlier hybrid frameworks that mainly combine signal decomposition with traditional neural models, our method uniquely integrates VMD, Kernel PCA, and time-aware SHAP with a self-organized forecasting backbone optimized by hierarchical Bayesian search. This synergy enables both higher accuracy and greater adaptability under varying operating conditions.

In this study, Section 2 outlines normalization, Variational Mode Decomposition (VMD), Kernel PCA, and feature extraction. Section 3 details the TSHAP-MLP feature selection and SHAP-based strategy. Section 4 develops the SOFTS model with hierarchical Bayesian optimization. Section 5 covers the case study with data sources and evaluation metrics. Section 6 compares methods for transformer top-oil temperature prediction. Section 7 concludes this study.

2. Methods

2.1. Min-Max Normalization

Normalize each sample by scaling its feature values to a standard range. Use Min-Max normalization to scale the feature values to the [0, 1] interval [25]. The normalized feature formula is as follows:

$$X' = \frac{X - X_{\min}}{X_{\max} - X_{\min}} \tag{1}$$

where X is the original data value, X' is the normalized value, and $X_{\max}$ and $X_{\min}$ are the minimum and maximum values of the feature values.

After model training is complete, the normalized results are mapped back to the original value range. The denormalization formula is

$$X = X' \times (X_{\max} - X_{\min}) + X_{\min} \tag{2}$$

2.2. Variational Mode Decomposition

By applying VMD, the original signal is adaptively decomposed into a finite number of intrinsic mode functions. (IMF) [26,27]. Each IMF corresponds to a different frequency component of the signal. VMD automatically extracts these components through a variational framework, which can effectively handle non-stationary and non-linear signals.

The goal of VMD is to solve the following optimization problem using variational methods:

$$\min_{\{u_k\},\{\omega_k\}} \left(\sum_{k=1}^{K} \partial_t \left[(f(t) - u_k(t)) e^{-j\omega_k t} \right]_2^2 \right) \tag{3}$$

where $f(t)$ is the original input signal, $u_k(t)$ is the kth modal function (IMF) to be solved, ω_k is the corresponding modal centre frequency, and K is the number of decomposed IMFs. By introducing Lagrange multipliers and the multiplier method, VMD continuously adjusts the IMFs and frequencies to ultimately obtain the optimal solution.

Variational Mode Decomposition (VMD) decomposes oil temperature signals into multiple intrinsic mode functions (IMFs), each corresponding to a specific frequency band, thereby enabling the extraction of latent time–frequency features. These IMFs facilitate a more detailed analysis of transformer temperature dynamics.

2.3. Kernel Principal Component Analysis (Kernel PCA)

Kernel PCA uses kernel functions to map data to a higher-dimensional feature space [28], where non-linear relationships become linear, and then applies PCA for dimension reduction.

Kernel PCA projects the original data b into a high-dimensional feature space using the kernel function $\phi(x)$:

$$\phi : x \rightarrow \phi(x) \tag{4}$$

Then, PCA is applied in this high-dimensional space. The core of Kernel PCA is to calculate the similarity between samples through kernel functions without explicitly constructing a high-dimensional space. Its optimization objective is

$$\max \sum_{i=1}^{N} \lambda_i \left(k(x_i, x_j) - \frac{1}{N} \sum_{j=1}^{N} k(x_i, x_j) \right) \tag{5}$$

where λ_i is the eigenvalue of PCA, and $k(x_i, x_j)$ is the kernel function used to calculate the similarity between samples x_i and x_j. The kernel function used here is the Gaussian kernel (RBF kernel):

$$k(x_i, x_j) = \left(x_i^{\top} x_j + c \right)^d \tag{6}$$

2.4. Feature Extraction Steps

When analyzing transformer oil temperature signals, the following steps are performed:

Step 1: VMD decomposes the transformer oil temperature signal into five IMF components, each representing a characteristic signal of a different frequency.

Step 2: Kernel PCA is used to perform non-linear dimensionality reduction on these IMF components to extract the main non-linear features.

3. TSHAP-MLP Feature Selection

3.1. The Principle Behind TSHAP

SHAP is a game theory-based interpretation method that aims to evaluate the contribution of each feature to the model prediction [29]. Its mathematical formula is

$$\begin{gathered} \phi_i(f, x) = \sum_{S \subseteq N \setminus \{i\}} \frac{|S|!(|N|-|S|-1)!}{|N|!} \Delta f(S, i) \\ \Delta f(S, i) = f(S \cup \{i\}) - f(S) \end{gathered} \tag{7}$$

TSHAP (time-aware SHAP) is an improved method that combines game theory interpretation with time series structure to evaluate the marginal contribution of each time slice feature to the model prediction. Based on the original SHAP, it introduces time series weights and a sliding window mechanism to more reasonably measure the dynamic impact of each time step feature on the prediction target. Its improved mathematical expression is

$$\phi_i^{(t)}(f,x) = \sum_{S \subseteq N \setminus \{i\}} \frac{|S|!(|N|-|S|-1)!}{|N|!} \cdot w(t) \cdot \Delta f(S,i) \tag{8}$$

where ϕ_i is the SHAP value of the ith feature, $\phi_i^{(t)}$ is the temporal weight of the ith feature at time step t, and $w(t)$ is the temporal weight of the current feature at time step t. N is the set of all features, S is the subset of features excluding the ith feature, $f(S)$ denotes the model's prediction output on the subset S, and $f(S \cup \{i\})$ denotes the model's output after including the ith feature. This method fully considers the temporal position and historical influence of features, enabling more accurate identification of critical moments and high-contribution features in sequence prediction, thereby effectively enhancing the model's interpretability.

3.2. Multilayer Perceptron

In a multilayer perceptron (MLP), the input, hidden, and output layers are sequentially arranged, with full connectivity between neurons of adjacent layers. MLPs are mainly used for classification, regression, and other tasks and have powerful non-linear modeling capabilities [30].

The input layer of an MLP receives feature vectors of raw data, with each feature corresponding to an input node. The hidden layer is located between the input layer and the output layer and consists of several neurons. The hidden layer introduces a non-linear activation function, enabling the MLP to handle non-linear problems.

Assuming that the input feature is $X = [x_1, x_2, \ldots, x_n]$, the weight matrix is W, the bias vector is b, and the activation function is f, the forward propagation process of MLP can be represented by the following steps and formulas:

Perform linear transformation on the input vector:

$$z^{(1)} = W^{(1)}X + b^{(1)} \tag{9}$$

where $z^{(1)}$ is the input of the hidden layer, $W^{(1)}$ is the weight matrix, and $b^{(1)}$ is the bias term.

Introduce non-linearity by applying activation function f to the results of the linear transformation:

$$a^{(1)} = f(z^{(1)}) \tag{10}$$

where $a^{(1)}$ is the output of the hidden layer, and commonly used activation functions include ReLU (rectified linear unit), Sigmoid, Tanh, etc.

The output of the hidden layer $a^{(1)}$ is used as the input to the output layer, which then undergoes another linear transformation and activation function:

$$\begin{aligned} z^{(2)} &= W^{(2)}a^{(1)} + b^{(2)} \\ a^{(2)} &= f(z^{(2)}) \end{aligned} \tag{11}$$

where $z^{(2)}$ is the input of the output layer, $W^{(2)}$ is the weight matrix of the output layer, $b^{(2)}$ is the bias of the output layer, and $a^{(2)}$ is the final output.

3.3. SHAP-MLP Feature Selection

By integrating the multilayer perceptron (MLP) with SHAP analysis, we utilize SHAP values to quantify the contribution of each feature to the MLP's predictions, facilitating the selection of key features. Initially, the MLP model is trained on the full feature set, enabling it to automatically optimize feature weights to achieve the best input–output mapping. Subsequently, the SHAP method calculates the contribution of each feature, allowing evaluation of its influence on the prediction outcomes. Finally, features are ranked according to their SHAP values, with the most significant ones retained and those with minimal contributions excluded.

4. SOFTS Model—Hierarchical Bayesian Optimization Algorithm

4.1. SOFTS Model

SOFTS is a deep learning architecture built upon the multilayer perceptron (MLP) framework that employs a centralized approach to capture interactions among multiple sequences, achieving state-of-the-art results in multivariate prediction tasks. A key innovation of the SOFTS model is the introduction of the STAR aggregation–redistribution (STAR) module, designed to enhance both the efficiency and accuracy of multivariate time series forecasting [31]. The self-organized mechanism of SOFTS, combined with the STAR random pooling operation, enhances robustness to unusual operating conditions, allowing the model to adapt to input patterns not explicitly observed during training. Figure 1 shows the overall framework of SOFTS, with the specific steps as follows:

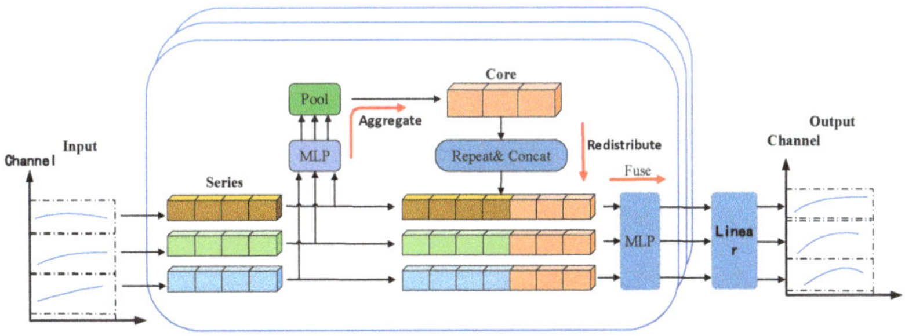

Figure 1. The overall framework diagram of SOFTS [31].

(1) Normalize the input multivariate time series, adjusting the series to a state where the mean is zero and the variance is unity, and restore the statistical characteristics prior to normalization after prediction.

(2) Sequence embedding involves linearly projecting the time series data of each channel and embedding it into a d-dimensional vector space.

The input is time series data $X \in R^{(C \times L)}$, where C is the number of channels and L is the time window length. The output is the sequence embedding representation $S_0 \in R^{(C \times d)}$.

(3) Multilayer STAR modules interact with each other, with each layer of STAR modules responsible for capturing the interrelationships between different channels.

The input is the sequence embedding representation S_{i-1} from the previous layer. The embedding representation is projected through a multilayer perceptron (MLP) to generate a core representation $o \in R^d$, which summarizes the global information of all channels. The core representation o is then concatenated with the sequence embedding representations of each channel to obtain a new fusion representation F_i.

The fused representation is processed using an MLP, and the embedded representation S_i is updated via residual connections. The output is the new sequence embedded representation, S_i.

(4) Linear projection prediction

After passing through N layers of STAR modules, linear projection is used to convert the final embedded representation SN into the prediction result $\hat{Y}$.

The input is the output S_N of the STAR module.

The output is the prediction value $\hat{Y} \in R^{(C \times H)}$, where H is the prediction time step length.

(5) Normalization Restoration: After obtaining the prediction results, the statistical characteristics removed during the normalization process are restored to the prediction values.

STAR Aggregation–Redistribution Module: The STAR module is the core innovation of the SOFTS model [31], designed to efficiently capture dependencies between channels in multivariate time series. The module reduces computational complexity through a centralized structure and improves the model's robustness to abnormal channels. Although the training data did not cover all possible abnormal oil temperature fluctuations, the design of SOFTS allows the model to generalize to unseen scenarios.

Figure 2 shows the STAR module. The specific working steps of the STAR module are as follows:

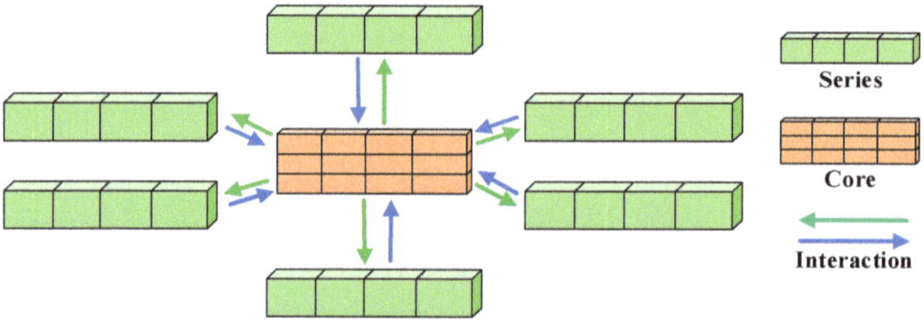

Figure 2. The structure of the STAR module [31].

Unlike conventional max/average pooling, the STAR module applies a random pooling operation. This introduces controlled stochasticity during feature aggregation, which enhances generalization by preventing the model from overfitting to fixed temporal patterns. It also improves robustness by exposing the model to diverse feature views, similar to an implicit ensemble effect.

(1) Core representation generation: In each layer of the STAR module, a multilayer perceptron (MLP) is first used to generate a global core representation. The core representation reflects the comprehensive information of the multi-channel time series. The calculation formula is as follows:

$$o = \text{Stoch\_Pool}(\text{MLP1}(S_{i-1})) \tag{12}$$

where S_{i-1} is the sequence embedding representation from the previous layer, MLP1 is a mapping function that projects the hidden layer dimension from d to d′, and Stoch_Pool is a random pooling operation.

(2) Core and sequence fusion: Concatenate the core representation o with the sequence embedding representation S_{i-1} of each channel to obtain a new representation F_i:

$$F_i = \text{Repeat\_Concat}(S_{i-1}, o) \tag{13}$$

The operation copies the core representation and connects it to the sequence representation of each channel.

(3) Non-linear transformation: Process the connected representation through another MLP to obtain the updated embedded representation S_i:

$$S_i = \mathrm{MLP2}(F_i) + S_{i-1} \tag{14}$$

where MLP2 is another mapping function that projects the connected representation from d + d′ dimensions back to d dimensions and uses residual connections to enhance the feature representation.

4.2. Hierarchical Bayesian Optimization Algorithm

Bayesian optimization is a method that uses probabilistic models to predict the objective function and explores the most valuable optimization methods through function sampling [32]. This paper improves the Gaussian process (GP) optimizer (GP-Tuner) within the Neural Network Intelligence (NNI) framework to enhance model performance and reduce search costs. Hierarchical Bayesian optimization (HBO) divides parameters into control and base layers, reducing coupling and dimensionality issues while improving convergence. In our experiments, HBO converged faster than standard Bayesian optimization and achieved lower final errors. The specific steps are as follows:

(1) Define the parameter space to be optimized. The parameters to be optimized are learning rate (lr), dropout rate (d_ff), training epochs (train_epochs), batch size (batch_size), and model dimension (d_model). The parameter ranges are defined as follows:

$$\theta = [\mathrm{lr}, \mathrm{top\_k}, \mathrm{d\_model}, \mathrm{d\_ff}, \mathrm{dropout}, \mathrm{train\_epochs}, \mathrm{batch\_size}]$$

(2) Define the objective function to minimize the mean squared error (MSE) on the validation set. Given the parameter combination, the objective function is

$$f(\theta) = \frac{1}{N}\sum_{i=1}^{N}(\hat{y}_i - y_i)^2 \tag{15}$$

where $\hat{y}_i$ is the model prediction value, y_i is the actual value, and N is the number of samples in the validation set.

Through Bayesian optimization, find the combination of hyperparameters that minimizes $f(\theta)$:

$$\theta^* : \theta^* = \underset{\theta}{\mathrm{argmin}}\, f(\theta) \tag{16}$$

(3) Initialize the Gaussian process, which is used to model the distribution of the objective function. It constructs an approximate model based on historical observations $(\theta, f(\theta))$ and predicts the performance of new parameter combinations based on this model. The Gaussian process assumes that the values of the objective function follow a multivariate Gaussian distribution:

$$f(\theta) \sim \mathcal{GP}(\alpha(\theta), k(\theta, \theta')) \tag{17}$$

where $\alpha(\theta)$ is the predicted mean, and $k(\theta, \theta')$ is the kernel function (RBF kernel selected), which describes the similarity between parameter combinations.

(4) In Gaussian process optimization, a sampling function is used to determine the parameter combination for the next sampling step. The sampling function used is the upper confidence bound (UCB) sampling function, which takes the following form:

$$\mathrm{UCB}(\theta) = \mu(\theta) + \kappa \cdot \sigma(\theta) \tag{18}$$

where $\mu(\theta)$ represents the current estimated error value, $\sigma(\theta)$ represents the predicted uncertainty (variance), and κ represents the mean value predicted by the model for the Gaussian process. The key to optimizing the Gaussian process is to strike a balance between exploring new parameter combinations and utilizing existing optimal parameters through the UCB function.

(5) Iterative update. First, select a parameter combination θ_n and calculate its corresponding prediction error $f(\theta_n)$; then, update the Gaussian process model with the new observation value $(\theta_n, f(\theta_n))$; next, maximize the acquisition function UCB(θ) based on the updated model to find the next set of evaluation parameters θ_{n+1}: $\theta_{n+1} = \underset{\theta}{\text{argmax}}\text{UCB}(\theta)$; finally, stop when the maximum number of iterations or convergence criteria is reached.

Introducing the hierarchical Bayesian optimization (HBO) algorithm to improve the efficiency and accuracy of parameter search. Specific steps:

(1) Define the hierarchical parameter space: All model hyperparameters are divided into two levels:

Control layer parameters θ_{control}: Used to control the search space or candidate set of the base layer parameters, specifically including the learning rate range:

$$\theta_{\text{control}} = \left\{\text{lr range} = [1e^{-5}, 1e^{-1}], \text{batchsize candidate set} = \{32, 64, 128\}, \text{dmodel candidate set} = \{16, 32, 64\}\right\}$$

Base layer parameters θ_{base}: Under the constraints given by the control layer parameters, the actual parameter combinations used for model training, specifically including

$$\theta_{\text{base}} = \{lr = 0.0032, d\_ff = 0.1, train\_epochs = 50, batch\_size = 64, d\_model = 32\}$$

The total parameter space is represented as $\Theta = \{\theta_{\text{control}}, \theta_{\text{base}}\}$

(2) A two-level Bayesian model is adopted. The first level uses GP to model the control parameters and predict the optimal objective function value under this combination. The second level models the underlying parameter space under the control parameters and searches for the optimal underlying layer parameters within the control layer range.

(3) Each round of the optimization process includes the following steps: first, sample $\theta^{(i)}_{\text{control}}$ from the control layer parameter space; then, under the constraints specified by $\theta^{(i)}_{\text{control}}$, sample $\theta^{(i)}_{\text{base}}$ from the underlying layer parameter space using the GP-UCB strategy to minimize loss; finally, obtain the objective function value $f(\theta^{(i)}_{\text{control}}, \theta^{(i)}_{\text{base}})$ and update the two-layer GP models.

(4) The optimization objective function is expressed as

$$\theta^* = \arg\min_{\theta_{\text{control}}} \mathbb{E}_{\theta_{\text{basc}}}[f(\theta_{\text{control}}, \theta_{\text{base}})] \tag{19}$$

That is, search for the combination of control parameters that minimizes the expected loss of the base layer in the control layer space.

Figure 3 illustrates the Gaussian process parameter tuning process, and Table 1 shows the impact of different hyperparameter combinations on the test loss (test_loss) during model training. The impact of different hyperparameter combinations on model performance is not the result of a single parameter but rather the combined effect of multiple parameters. θ* denotes the optimized hyperparameter configuration. For example, a higher learning rate (lr) combined with a smaller model dimension (d_model) may lead to a higher test loss, but if other parameters (such as batch size and training epochs) are appropriately tuned, the loss can still remain low. As shown in Table 1, the first hyperparameter combination (lr = 0.0162674, d_model = 16, d_ff = 8, etc.) achieved the lowest test loss (0.008477), indicating that this combination performs best in the model.

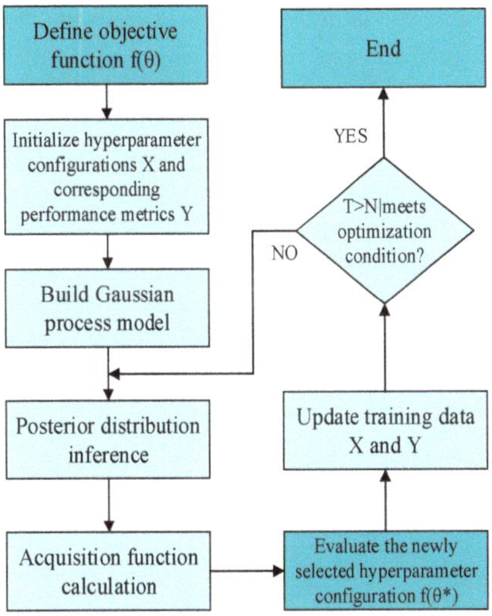

Figure 3. Parameter optimization process using the Bayesian algorithm.

Table 1. Selected parameters and errors from Bayesian optimization (partial).

Parameters of the SOFTS Deep Learning Model						
lr	**d_model**	**d_ff**	**dropout**	**Train_Epochs**	**Batch_Size**	**Test_Loss**
0.01627	16	8	0.06194	16	89	0.008477
0.08974	8	8	0.12194	17	90	0.024948
0.00013	32	8	0.12194	15	90	0.036025
0.00014	8	16	0.00195	8	88	0.096082
0.08635	32	32	0.07821	16	112	0.164758

5. Case Study

5.1. Data Source and Preprocessing

This study employs Python 3.11 for simulation analysis, utilizing real operational data from the main transformer of a hydropower station in Yunnan, China, collected hourly between August 2023 and July 2024. A total of 8484 data samples were obtained and subsequently partitioned into training, validation, and test sets with a ratio of 7:1:2. The training data were used to train the SOFTS model, while the test set served for model validation.

5.2. Performance Metrics

Predictions of the transformer top-oil temperature were generated for the test set, and the mean square error (MSE), mean absolute error (MAE), and root mean square error (RMSE) were computed by comparing these predictions with the corresponding actual values. The calculation formulas are presented as follows:

$$\begin{aligned} e_{\mathrm{MSE}} &= \frac{1}{N}\sum_{i=1}^{N}\left(Y_{\mathrm{oil}} - \hat{Y}_{\mathrm{oil}}\right)^2 \\ e_{\mathrm{MAE}} &= \frac{1}{N}\sum_{i=1}^{N}\left|Y_{\mathrm{oil}} - \hat{Y}_{\mathrm{oil}}\right| \end{aligned} \tag{20}$$

$$e_{\mathrm{RMSE}} = \sqrt{\frac{1}{N}\sum_{i=1}^{N}\left(Y_{\mathrm{oil}} - \hat{Y}_{\mathrm{oili}}\right)^2} \tag{21}$$

where e_{MSE} is the mean square error, e_{MAE} is the mean absolute error, and e_{RMSE} is the root mean square error.

5.3. Data Processing

Figure 4 shows the five IMFs of the transformer top-oil temperature signal obtained via VMD. The selection of the number of IMFs was further justified by analyzing the relationship between the IMF number and the associated error/residual energy. The optimal number of IMFs, K, was determined by considering both the stability of the center frequencies and the residual information content. When K was less than five, some modes exhibited frequency mixing, whereas for K greater than five, redundant modes appeared, and the reduction in residual energy became marginal.

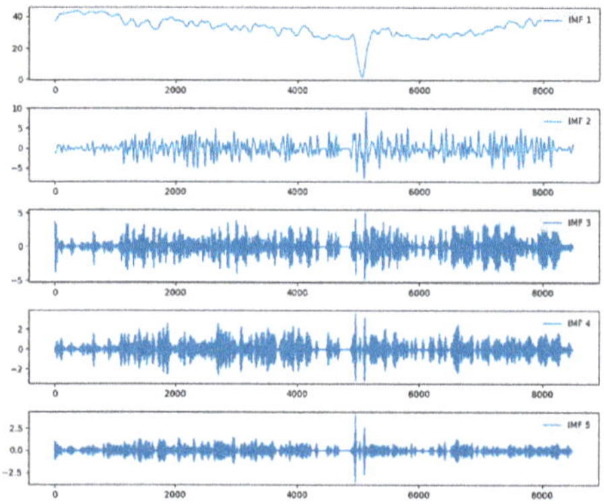

Figure 4. VMD_IMF decomposition results.

As shown in Table 2, the effect of different IMF numbers on the results, the IMF number was chosen to balance center frequency stability and residual information: fewer than five caused frequency mixing, while more than five introduced redundancy with limited residual energy reduction. VMD decomposes the signal into components across different frequency bands, with IMF 1 capturing the lowest-frequency trend and IMF 5 the highest-frequency fluctuations. Figure 5 presents the Kernel PCA projection of these IMFs onto the first two principal components.

Principal Component 1 and Principal Component 2 are the two dominant components derived from the kernel principal component analysis (KPCA), representing the primary projection directions of the original high-dimensional data in the transformed space. The projection reveals pronounced non-linear characteristics of the data, forming an L-shaped distribution. This pattern suggests that some data points cluster near the origin, while others extend linearly along a specific direction. Such a distribution implies underlying structures within the IMF components, where certain components exhibit high concentration with minimal variation, whereas others display substantial variability. The IMF components obtained via VMD demonstrate complex non-linear relationships, which KPCA effectively captures and visualizes in a reduced-dimensional space. This visualization is crucial for further analysis of the features and interrelations among the IMF components.

Table 2. The effect of different IMF numbers on the results.

Number of IMFs K	Residual Energy (%)	RMSE	MAE
3	12.4	0.72	0.57
4	7.1	0.65	0.52
5	4.8	0.63	0.50
6	4.6	0.65	0.51
7	4.5	0.66	0.52
8	4.5	0.67	0.53

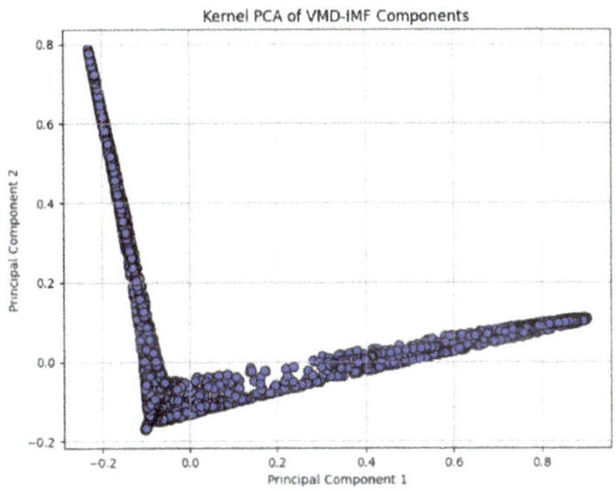

Figure 5. Kernel PCA components of VMD-IMFs.

VMD was adopted as it adaptively determines mode frequency and bandwidth to mitigate mixing and boundary effects, allows predefined IMF numbers for stable decomposition, and produces more stationary IMFs that support Kernel PCA and TSHAP-MLP-based feature analysis.

5.4. Multivariate Input Feature Selection

Table 3 lists the SHAP values for each feature, and Figure 6 shows the relationship between different features and the model output, where the X-axis represents the SHAP value of the feature. The larger the SHAP value, the greater the influence of the feature on the output result, and vice versa. Positive values indicate that the feature increases the predicted value, while negative values indicate that it decreases the predicted value. The Y-axis includes the abbreviations for each feature, and the color of the points represents the feature's value, ranging from blue (low value) to red (high value), i.e., the feature's value range. The redder the color, the higher the feature's value, while blue indicates a lower value. The horizontal distribution shows how the SHAP values of each feature are distributed across different samples. The more scattered the points, the greater the variation in the feature's impact on different samples. For example, the 1F feature (main variable A winding temperature) has a significant impact, with SHAP values ranging from −30 to 10 and red and blue points interspersed, indicating that the feature has a substantial positive or negative impact on the model output across different samples. It is worth noting that some features exhibit both positive and negative SHAP contributions with a wide spread. This reflects their bidirectional physical role in transformer thermal behavior: for instance, ambient temperature and load variations may exert opposite influences under different cooling or seasonal conditions. Such wide distributions highlight context-dependent dynamics rather than model instability. The wide SHAP value distributions observed for certain features indicate that their influence on transformer top-oil temperature is not unidirectional but varies with operating context. This aligns with physical understanding, where the same factor may act differently under varying load and cooling conditions.

Table 3. SHAP values of individual features.

Abbreviation	Feature Description	SHAP Value
1F	1F: Main transformer phase A winding temperature	6.158109
LCU13A-1-A-Ia	LCU13A: No. 1 main transformer HV side power factor cos	4.533494
LCU13A-1-Q	1F: Outlet A phase voltage Ua	4.246660
LCU13A-1-A-Ua	LCU13A: Active power on the high-voltage side of No. 1 main transformer	4.204933
1F-A-1	1F: Active power (transmitter 1)	2.454016
1F-A	LCU13A: No. 1 main transformer HV side phase A current Ia	2.163187
LCU13A-1-cos	1F: Outlet A phase current Ia	1.013335
1F-Ua	LCU13A: No. 1 main transformer high-voltage side reactive power Q	0.874439
LCU13A-1-P	1F: Reactive power	0.454664
1F-A-Ia	LCU13A: No. 1 main transformer HV side power factor cos	0.206889

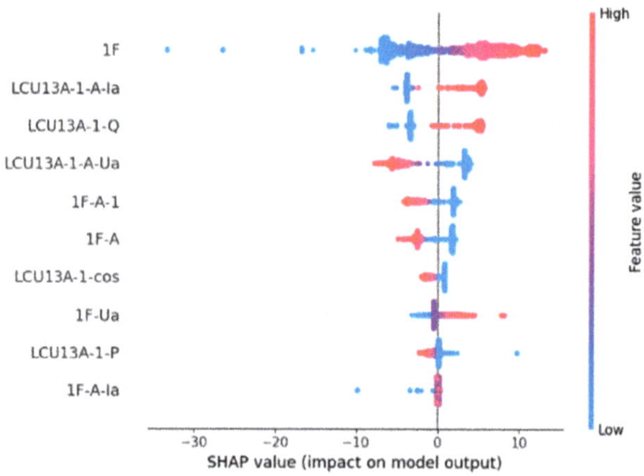

Figure 6. SHAP value impact analysis diagram.

Based on the SHAP values in Figure 6 and Table 3, the larger the SHAP value, the greater the influence of that feature on the model's prediction. Remove features with SHAP values less than 1 and select other features with high importance as model inputs. Table 4 (effect of SHAP threshold on results) summarizes the influence of different thresholds. A threshold of $|SHAP| > 1$ was adopted for feature retention in this study. This value was determined by comparing predictive performance under different thresholds. When the threshold is too low (e.g., 0.2), many redundant features are retained, potentially leading to overfitting; when it is too high (e.g., 4), some critical features are discarded, resulting in insufficient information for the model.

Table 4. Effect of SHAP threshold on results.

SHAP Threshold	Number of Retained Features	RMSE	MAE
0.2	10	0.69	0.56
1.0	7	0.63	0.50
4	4	0.66	0.53

6. Results Comparison

6.1. Analysis of Transformer Top Oil Temperature Prediction Results Based on SOFTS

A SOFTS training program was developed using Python to train the 5918 samples in the training set, validate the 850 samples in the validation set, and test the 1696 samples in the test set. While dedicated noise experiments were not performed, the integrated framework of VMD decomposition, SHAP-guided feature selection, and SOFTS adaptation, complemented by standard preprocessing of outliers and missing data, inherently enhances the model's robustness to noise and incomplete sensor measurements.

Figure 7 illustrates the comparison between the actual top-oil temperature of the transformer and the corresponding predictions generated by the SOFTS model. The blue line reflects the true temperature trend, characterized by periodic fluctuations, while the orange dotted line represents the SOFTS model's predicted values. Overall, the model's predictions exhibit strong agreement with the actual data, particularly in the later phase (right side of the figure), where it accurately captures the steady upward trend and subtle temperature variations.

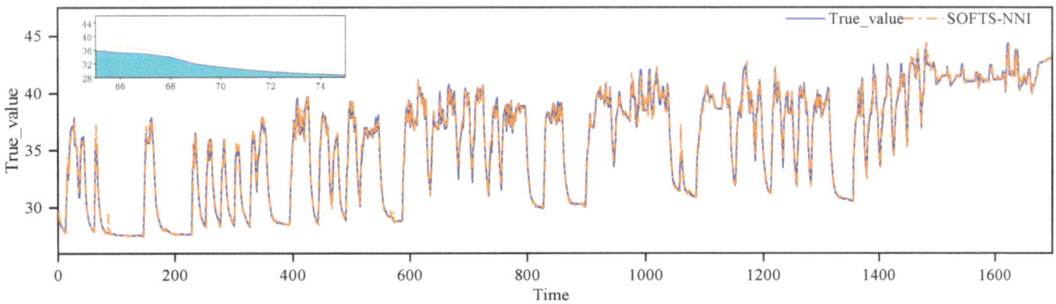

Figure 7. Comparison of SOFTS predicted values and ground truth.

As time progresses, the SOFTS model gradually demonstrates its advantage in capturing long-term and short-term dependencies, particularly excelling in the more stable fluctuations in the later stages, where the predicted results closely align with the actual values. Unlike earlier hybrid frameworks that mainly combine signal decomposition with traditional neural models, our method uniquely integrates VMD, Kernel PCA, and time-aware SHAP with a self-organized forecasting backbone optimized by hierarchical Bayesian search. This synergy enables both higher accuracy and greater adaptability under varying operating conditions.

Table 5 presents the RMSE results obtained from the SOFTS model's prediction of the transformer's top-layer oil temperature, representing the standard deviation between predicted and actual values. The RMSE is 0.632, indicating a low level of prediction error. The MAE, which measures the average absolute error between predicted and actual values, is 0.396, suggesting that the average deviation per prediction point is relatively small. This low MAE further confirms the model's strong predictive capability, with consistently minimal errors in forecasting top-oil temperature.

Table 5. Data of evaluation metrics.

Error_Name	SOFTS_Error_Value
RMSE	0.63221
MAE	0.39568
R^2	0.98195
MSE	0.39969
sMAPE	0.01069
NSE	0.98204

To demonstrate the advantages of the model proposed in this experiment from multiple perspectives, Table 5 also presents the coefficient of determination (R^2), symmetric mean absolute percentage error (sMAPE), and Nash–Sutcliffe efficiency coefficient (NSE) of the experimental results obtained using the model. R^2 reflects the model's explanatory power, with a value of 0.982, indicating that 98.2% of the fluctuations in top-layer oil temperature can be well explained by the model. sMAPE is a symmetric percentage error metric that considers the relative difference between predicted and actual values. The closer the value is to 0, the smaller the relative error of the model. Here, sMAPE is 0.0107, meaning that the model's prediction error accounts for only 1.07% of the actual value. NSE is used to evaluate the accuracy of model predictions, with a higher value indicating better predictive performance. Here, NSE is 0.982, indicating that the model's predictive accuracy is very high, approaching perfect prediction. The comprehensive performance of these indicators shows that the SOFTS model can effectively perform accurate predictions of transformer top-oil temperature and that the model has high robustness and stability.

6.2. Comparison of SOFTS Parameter Tuning Results

Figure 8 presents a comparison of the SOFTS model's prediction performance before and after parameter tuning for transformer top-oil temperature. The blue line denotes the actual top-oil temperature, while the orange line illustrates the predicted values generated by the SOFTS model prior to parameter optimization. Although the orange curve generally follows the trend of the actual values, it exhibits considerable discrepancies in fluctuation amplitude at several points, particularly in regions characterized by rapid temperature changes, where it fails to accurately capture peak values. In contrast, the red line reflects the prediction outcomes after parameter tuning. Compared to the orange line, it can be observed that the red line shows a significant improvement in agreement with the actual values after parameter tuning. After parameter tuning, the model's prediction accuracy has significantly improved, especially during the initial phase of intense fluctuations, where the model can better fit the actual data. This indicates that by adjusting parameters such as the learning rate, feature dimension, and regularization parameters, the model's ability to capture temperature time-series fluctuations can be effectively enhanced.

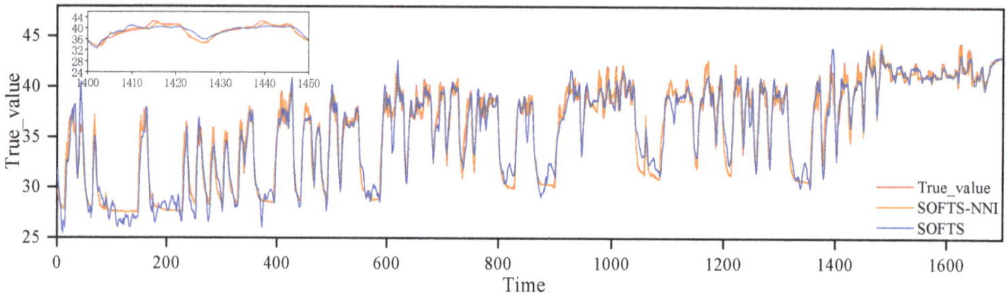

Figure 8. SOFTS model prediction comparison with and without parameter optimization.

6.3. Comparative Analysis of Different Deep Learning Models

To evaluate the performance of the SOFTS model in forecasting transformer top-oil temperature, this study compares it with several deep learning approaches, including transformer [33], TimesNet [34], LSTM [35], and the traditional BP algorithm [36]. The same test samples as described in Section 3.1 are utilized. For the transformer and TimesNet models, the dataset is split into training, validation, and test sets in a ratio of 7:1:2. In contrast, for LSTM and BP, the data are divided into training and test sets using an 8:2 split.

LSTM and BP follow their original 8:2 training–test split, whereas transformer, TimesNet, and SOFTS-NNI use a 7:1:2 chronological split. All models share an identical, temporally non-overlapping test set. The validation portion available to transformer, TimesNet, and SOFTS-NNI (10% of the data) is employed for early stopping and hyperparameter tuning, ensuring that every architecture leverages the same total amount of data for learning while being evaluated on the same strictly held-out samples. And to ensure fairness, all baseline models were tuned using standard Bayesian optimization, with the search space covering key hyperparameters such as network depth, number of hidden units, learning rate, and dropout ratio, with a maximum of 100 iterations. The SOFTS model was further refined using the proposed hierarchical Bayesian optimization for more precise tuning. The computational evaluation shows that the training phase required less than 60 s and the inference phase less than 5 s, indicating that the overall computational burden remains low and fully meets the requirements of real-time monitoring applications.

Figure 9 compares the top-oil temperature prediction results of the transformer using five different machine learning models. As shown in Figure 10, an analysis of the prediction errors reveals that the SOFTS model demonstrates strong consistency with actual temperature fluctuations, especially during the mid-term period characterized by frequent peaks and troughs, where it accurately captures the dynamic variations in the data. It is observed that transformer shows clustered error spikes, which stem from its reliance on historical attention when facing regime shifts. In contrast, SOFTS-NNI produces smoother error distributions due to multi-scale decomposition and adaptive pooling. In contrast, both the transformer and TimesNet models show delayed responses to changes, particularly at extreme points, leading to noticeable deviations from the actual values and an inability to effectively capture fluctuation patterns. The LSTM model exhibits substantial prediction errors during the early and mid-term stages, reflecting poor responsiveness to sudden changes; although its performance improves in the later stage, it still lags behind the other models. Notably, during periods of rapid top-oil temperature variation, the SOFTS model responds more promptly, whereas the BP model displays a delayed reaction in certain instances.

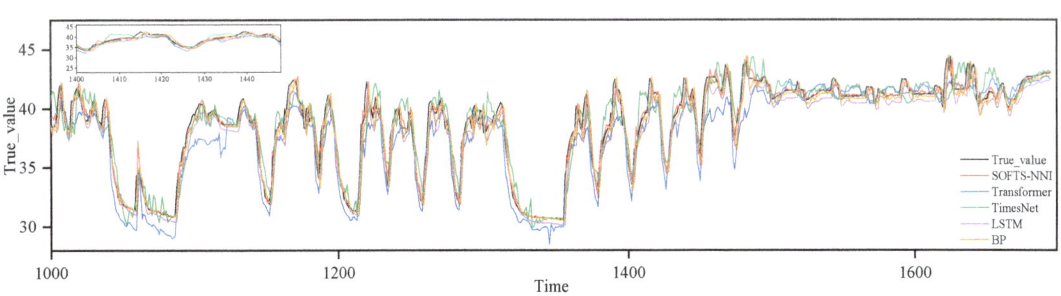

Figure 9. Comparison of temperature prediction results from different models.

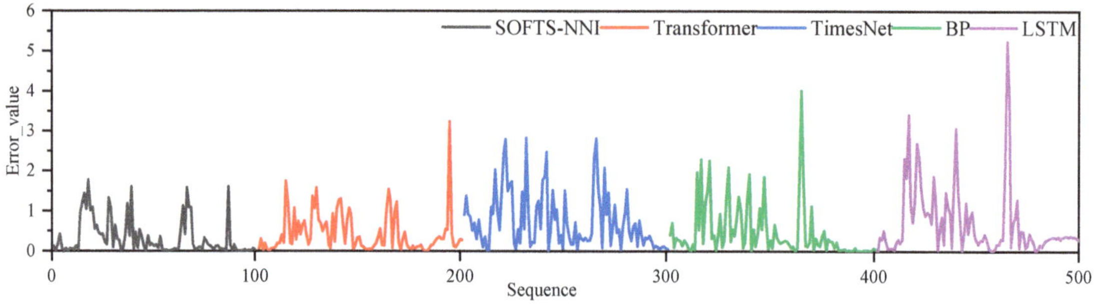

Figure 10. Comparison of error values across different models.

The error sensitivity of different models to sudden input variations reflects their architectural characteristics. While BP and LSTM tend to show abrupt error spikes, transformer-based models may cluster errors when historical contexts differ from the current regime. In contrast, SOFTS combines decomposition, feature selection, and adaptive pooling to alleviate this sensitivity, leading to smoother error distributions under sudden changes. The clustered high errors in the transformer reflect the intrinsic sensitivity of attention-based models to abrupt regime changes. By contrast, SOFTS-NNI integrates decomposition, feature selection, and adaptive pooling, which smooths error distribution and enhances robustness under dynamic conditions. The BP model performs relatively well under stable load conditions due to the monotonicity and regularity of top-oil temperature data, allowing shallow networks to capture the load–temperature mapping. However, BP errors may increase significantly under abrupt fluctuations and seasonal changes, indicating a possible lack of robustness. In contrast, SOFTS demonstrates superior stability and generalization under complex conditions, highlighting the benefits of deep architectures and dynamic feature modeling.

Overall, the computational analysis demonstrates that despite the integration of additional modules, efficiency is well preserved, with preprocessing contributing minimally, SOFTS forward propagation remaining the main cost, and only marginal overhead observed in both training and real-time inference.

Table 6 shows the evaluation indicators obtained by the five different models when performing top-layer oil temperature calculations, and Figure 11 shows the evaluation indicator analysis chart corresponding to the five different models.

Table 6. Performance evaluation metrics for various models.

Models	e_{RMSE}	e_{MSE}	e_{MAE}
SOFTS-NNI	0.6358	0.3997	0.3990
Transformer	1.1733	1.3768	0.9009
TimesNet	1.1986	1.4367	0.8975
LSTM	1.0806	1.1679	0.8089
BP	0.8229	0.6772	0.5422

Based on the results presented in Table 6 and Figure 11, the SOFTS-NNI model—optimized using the Bayesian optimization algorithm—achieves strong performance across all three evaluation metrics, particularly exhibiting notably low RMSE and MSE values. Compared with the best-performing alternative (the BP model), SOFTS-NNI achieves reductions of 18.71% in RMSE, 27.75% in MSE, and 14.31% in MAE, indicating a significant improvement in predictive accuracy. The relatively low MAE further reflects the model's ability to minimize average prediction errors. Overall, the SOFTS model demonstrates outstanding performance across all evaluation indicators, achieving the lowest values of RMSE, MSE, and MAE, thereby confirming its superiority in forecasting transformer top-oil temperature. Although this study is based on a single operational dataset, the observed error distribution patterns primarily reflect the intrinsic modeling capabilities of the different architectures. Therefore, while absolute errors may vary across other datasets, the relative trends in error distribution are expected to remain consistent. Although no dedicated experiments on seasonal or abrupt changes were conducted, the framework is inherently designed to handle such scenarios. VMD decomposes the signal into IMFs capturing long-term seasonal trends and short-term fluctuations, TSHAP dynamically adjusts the relative importance of seasonal and load-dependent features, and the SOFTS STAR module enhances adaptability via stochastic pooling. Collectively, these mechanisms improve the framework's robustness against both seasonal variations and abrupt load changes.

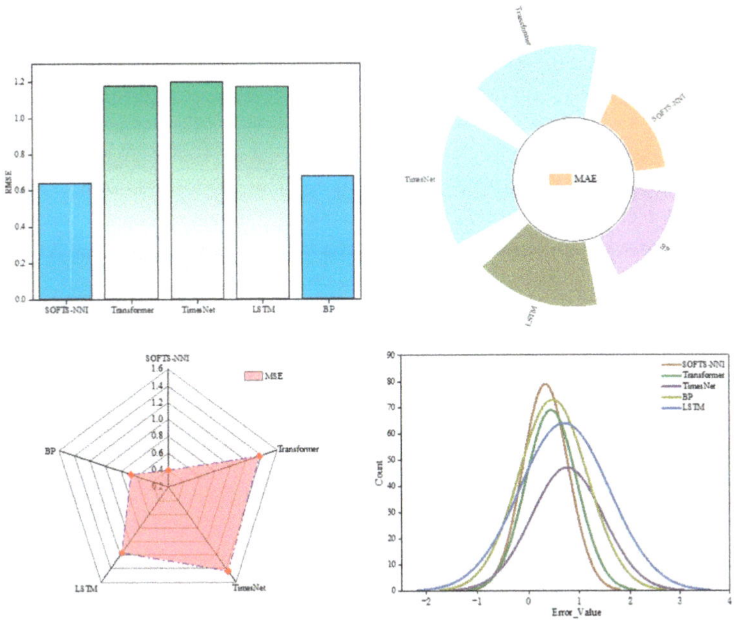

Figure 11. Visualization of evaluation metrics for various models.

7. Conclusions

To address the issues of low accuracy and weak interference resistance in transformer oil temperature prediction, this paper proposes a transformer oil temperature prediction method that combines VMD, Kernel PCA, SHAP-MLP feature selection methods, and the SOFTS deep learning model. Using historical oil temperature data from the main transformer of a hydropower station in Yunnan, China, as a case study, this paper conducts experimental research and draws the following conclusions:

(1) When compared with traditional neural network models such as BP and other deep learning models like LSTM, transformer, and TimesNet, the SOFTS model demonstrated significant advantages in terms of prediction accuracy and model stability. Experimental results show that the SOFTS model, optimized using Bayesian optimization algorithms, achieved the smallest evaluation metrics (RMSE, MSE, and MAE) in top-level oil temperature prediction, indicating higher prediction accuracy and lower error rates.

(2) This study combines VMD and Kernel PCA methods for feature extraction of top-level oil temperature signals and uses the SHAP-MLP feature selection method to optimize input features, significantly improving the model's prediction accuracy. The combination of these techniques effectively reduces the model's input dimension, removes redundant features, and enhances the model's generalization capability and efficiency.
(3) This study provides an effective deep learning method for transformer top-layer oil temperature prediction, demonstrating its practical application value in power equipment monitoring. Future research will continue to explore other optimization algorithms and model structures to further enhance the prediction performance of transformer operating conditions, particularly in terms of stability and robustness in complex environments.
(4) This study demonstrates the effectiveness of the SOFTS framework on hydropower transformer data. Leveraging the universal thermal–electrical load coupling, VMD and KPCA extract transferable features, TSHAP-MLP identifies key variables across operating conditions, and the STAR module adapts to diverse load patterns. Future work will validate the method on transformers with varying capacities, cooling methods, and regional grid conditions and explore transfer learning to enhance cross-device generalization.
(5) The distinctiveness of this study lies in the multi-level integration of decomposition, non-linear reduction, temporal feature selection, and adaptive forecasting, which has not been explored in previous transformer temperature prediction works. Future work will explicitly evaluate the proposed framework under noisy and incomplete data, which are common in real-world monitoring systems, and will extend validation to datasets exhibiting stronger seasonal cycles and more frequent abrupt load fluctuations, thereby providing a more quantitative assessment of its adaptability.
(6) Beyond methodological improvements, advances in transformer insulating liquids are also reshaping the thermal environment of power equipment. Khelifa et al. [37] showed that adding ZrO_2 nanoparticles to mineral, synthetic, and natural esters can significantly enhance AC breakdown voltage at optimal concentrations, improving insulation reliability. Koutras et al. [38] further reported that semiconducting nanoparticles (SiC, TiO_2) improve the initial thermal and dielectric performance of natural esters but may accelerate agglomeration and property degradation with aging. These findings highlight that material-driven changes affect heat dissipation and insulation stability, underscoring the need for accurate top-oil temperature forecasting in next-generation transformers.

Author Contributions: Z.T.: conceptualization, data curation, and methodology; Y.X.: writing—review and editing; X.M.: software and visualization; Y.Z.: methodology, investigation, software, and writing—original draft; T.P.: conceptualization and methodology; C.Z.: conceptualization and writing—review and editing. All authors have read and agreed to the published version of the manuscript.

Funding: This work is supported by China Yangtze Power Co., Ltd. (CYPC) Sponsored Project (Z522302017).

Data Availability Statement: The data presented in this study are available on request from the corresponding author.

Conflicts of Interest: Authors Zhixiang Tong and Yan Xu were employed by the company China Yangtze Power Co., Ltd. Authors Xianyu Meng, Yongshun Zheng, Tian Peng and Chu Zhang were employed by the company NARI Engineering Technology Co., Ltd.

Abbreviations

VMD	Variational Mode Decomposition
BiGRU	Bidirectional Gated Recurrent Unit
Kernel PCA	Kernel Principal Component Analysis
BWO	Black Widow Optimization
ELM	Extreme Learning Machines
IMF	Intrinsic Mode Function
SVM	Support Vector Machines
MLP	Multilayer Perceptron
RMSE	Root Mean Square Error
CNN	Convolutional Neural Network
BiLSTM	Bidirectional Long Short-Term Memory
LSTM	Long Short-Term Memory
RNN	Recurrent Neural Network
CEEMDAN	Complete Ensemble Empirical Mode Decomposition with Adaptive Noise
SA	Self-Attention
EEMD	Ensemble Empirical Mode Decomposition
PSO	Particle Swarm Optimization
TOT	Top Oil Temperature
HKELM	Hybrid Kernel Extreme Learning Machine
IWOA	Improved Whale Optimization Algorithm
TCN	Temporal Convolutional Network
SOFTS	Self-organized Time Series Forecasting System
NSE	Nash–Sutcliffe Efficiency
MAE	Root Mean Absolute Error
sMAPE	Symmetric Mean Absolute Percentage Error
EMD	Empirical Mode Decomposition
R^2	Coefficient of Determination
MSE	Mean Squared Error
KELM	Kernel Extreme Learning Machine

References

1. Li, S.; Xue, J.; Wu, M.; Xie, R.; Jin, B.; Wang, K. Prediction of transformer top oil temperature based on improved weighted support vector regression based on particle swarm optimization. In Proceedings of the 2021 International Conference on Advanced Electrical Equipment and Reliable Operation (AEERO), Beijing, China, 15–17 October 2021; IEEE: Piscataway, NJ, USA, 2021; pp. 1–5.
2. Huang, X.; Zhuang, X.; Tian, F.; Niu, Z.; Chen, Y.; Zhou, Q.; Yuan, C. A Hybrid ARIMA-LSTM-XGBoost Model with Linear Regression Stacking for Transformer Oil Temperature Prediction. *Energies* **2025**, *18*, 1432. [CrossRef]
3. Zhang, C.; Ma, H.; Hua, L.; Sun, W.; Nazir, M.S.; Peng, T. An evolutionary deep learning model based on TVFEMD, improved sine cosine algorithm, CNN and BiLSTM for wind speed prediction. *Energy* **2022**, *254*, 124250. [CrossRef]
4. Peng, T.; Zhang, C.; Zhou, J.; Nazir, M.S. An integrated framework of Bi-directional long-short term memory (BiLSTM) based on sine cosine algorithm for hourly solar radiation forecasting. *Energy* **2021**, *221*, 119887. [CrossRef]
5. Guo, Y.; Chang, Y.; Lu, B. A review of temperature prediction methods for oil-immersed transformers. *Measurement* **2025**, *239*, 115383. [CrossRef]
6. Dong, N.; Zhang, R.; Li, Z.; Cao, B. Prediction model of transformer top oil temperature based on data quality enhancement. *Rev. Sci. Instrum.* **2023**, *94*, 074707. [CrossRef] [PubMed]
7. Deng, W.; Yang, J.; Liu, Y.; Wu, C.; Zhao, Y.; Liu, X.; You, J. A Novel EEMD-LSTM Combined Model for Transformer Top-Oil Temperature Prediction. In Proceedings of the 2021 8th International Forum on Electrical Engineering and Automation (IFEEA), Xi'an, China, 3–5 September 2021; IEEE: Piscataway, NJ, USA, 2021; pp. 52–56.
8. Cao, Y. An Improved Prediction Method of Transformer Oil Temperature. In Proceedings of the 2023 IEEE 6th Information Technology, Networking, Electronic and Automation Control Conference (ITNEC), Chongqing, China, 24–26 February 2023; IEEE: Piscataway, NJ, USA, 2023; pp. 987–991.
9. Hu, J.; Liu, Y.; Peng, Y.; Chen, C.; Wan, J.; Cao, B. Data-Driven short-term prediction for top oil temperature of Residential Transformer. In Proceedings of the 2022 China International Conference on Electricity Distribution (CICED), Changsha, China, 7–8 September 2022; IEEE: Piscataway, NJ, USA, 2022; pp. 303–309.

10. Chang, J.; Duan, X.; Tao, J.; Ma, C.; Liao, M. Power Transformer Oil Temperature Prediction Based on Spark Deep Learning. In Proceedings of the 2022 IEEE International Conference on High Voltage Engineering and Applications (ICHVE), Chongqing, China, 25–29 September 2022; IEEE: Piscataway, NJ, USA, 2022; pp. 1–4.
11. Wang, K.; Zhang, H.; Wang, X.; Li, Q. Prediction method of transformer top oil temperature based on VMD and GRU neural network. In Proceedings of the 2020 IEEE International Conference on High Voltage Engineering and Application (ICHVE), Beijing, China, 6–10 September 2020; IEEE: Piscataway, NJ, USA, 2020; pp. 1–4.
12. Wang, X.; Fu, C.; Zhao, Z.; Fu, B.; Jiang, T.; Xu, Y.; Zhang, L.; Hou, Y. Prediction of transformer top oil temperature based on AC-BiLSTM model. In Proceedings of the 2023 4th International Conference on Smart Grid and Energy Engineering (SGEE), Zhengzhou, China, 24–26 November 2023; IEEE: Piscataway, NJ, USA, 2023; pp. 390–394.
13. Lei, L.; Liao, F.; Liu, B.; Tang, P.; Zheng, Z.; Qi, Z.; Zhang, Z. Research on Transformer Temperature Rise Prediction and Fault Warning Based on Attention-GRU. In Proceedings of the 2023 5th International Conference on Smart Power & Internet Energy Systems (SPIES), Shenyang, China, 1–4 December 2023; IEEE: Piscataway, NJ, USA, 2023; pp. 94–99.
14. Wang, Y.Q.; Yue, G.L.; He, J.; Liu, H.L.; Bi, J.G.; Chen, S.F. Research on Top-Oil Temperature Prediction of Power Transformer Based on Kalman Filtering Algorithm. *High Volt. Appar.* **2014**, *50*, 74–79+86.
15. Dong, X.; Jing, L.; Tian, R.; Dong, X. Transformer Top-Oil Temperature Prediction Method Based on LSTM Model. *Proc. China Electr. Power Soc.* **2023**, *38*, 38–45.
16. Yang, J.; Lu, W.; Liu, X. Prediction of Top Oil Temperature for Oil-immersed Transformers Based on PSO-LSTM. In Proceedings of the 2021 4th International Conference on Energy, Electrical and Power Engineering (CEEPE), Chongqing, China, 23–25 April 2021; IEEE: Piscataway, NJ, USA, 2021; pp. 278–283.
17. Zou, D.; Xu, H.; Quan, H.; Yin, J.; Peng, Q.; Wang, S.; Dai, W.; Hong, Z. Top-Oil Temperature Prediction of Power Transformer Based on Long Short-Term Memory Neural Network with Self-Attention Mechanism Optimized by Improved Whale Optimization Algorithm. *Symmetry* **2024**, *16*, 1382. [CrossRef]
18. Liu, J.; Hou, Z.; Liu, B.; Zhou, X. Mathematical and Machine Learning Innovations for Power Systems: Predicting Transformer Oil Temperature with Beluga Whale Optimization-Based Hybrid Neural Networks. *Mathematics* **2025**, *13*, 1785. [CrossRef]
19. Hao, Y.; Zhang, Z.; Liu, X.; Yang, Y.; Liu, J. Inversion Method for Transformer Winding Hot Spot Temperature Based on Gated Recurrent Unit and Self-Attention and Temperature Lag. *Sensors* **2024**, *24*, 4734. [CrossRef]
20. Zhang, Z.; Kong, W.; Li, L.; Zhao, H.; Xin, C. Prediction of transformer oil temperature based on an improved pso neural network algorithm. In *Recent Advances in Electrical & Electronic Engineering (Formerly Recent Patents on Electrical & Electronic Engineering)*; Bentham Science Publishers: Sharjah, United Arab Emirates, 2024; Volume 17, pp. 29–37.
21. Li, K.; Xu, Y.; Wei, B.; Hua, H.; Qi, X. Transformer Top-Oil Temperature Prediction Model Based on PSO-HKELM. *High Volt. Eng.* **2018**, *44*, 2501–2508.
22. Li, R. Research on Transformer Top-Oil Temperature Prediction Method Based on Bayesian Network. Master's Thesis, Southwest Jiaotong University, Chengdu, China, 2018.
23. Qi, X.; Li, K.; Yu, X.; Lou, J. Transformer Top-Oil Temperature Interval Prediction Based on Kernel Extreme Learning Machine and Bootstrap Method. *Proc. Chin. Soc. Electr. Eng.* **2017**, *37*, 5821–5828+5860.
24. Li, S.; Xue, J.; Wu, M.; Xie, R.; Jin, B.; Zhang, H.; Li, Q. Transformer Top-Oil Temperature Prediction Based on Particle Swarm Optimization Improved Weighted Support Vector Regression. *High Volt. Appar.* **2021**, *5*, 103–109.
25. Patro, S.; Sahu, K.K. Normalization: A preprocessing stage. *arXiv* **2015**, arXiv:1503.06462. [CrossRef]
26. Dragomiretskiy, K.; Zosso, D. Variational Mode Decomposition. *IEEE Trans. Signal Process.* **2014**, *62*, 531–544. [CrossRef]
27. Zhang, C.; Li, Z.; Ge, Y.; Liu, Q.; Suo, L.; Song, S.; Peng, T. Enhancing short-term wind speed prediction based on an outlier-robust ensemble deep random vector functional link network with AOA-optimized VMD. *Energy* **2024**, *296*, 131173. [CrossRef]
28. Schölkopf, B.; Smola, A.; Müller, K.-R. Kernel principal component analysis. In *Artificial Neural Networks—ICANN'97*; Gerstner, W., Germond, A., Hasler, M., Nicoud, J.-D., Eds.; Springer: Berlin/Heidelberg, Germany, 1997; Volume 1327, pp. 583–588.
29. Lundberg, S.M.; Lee, S.-I. A unified approach to interpreting model predictions. In Proceedings of the 31st International Conference on Neural Information Processing Systems (NeurIPS 2017), Long Beach, CA, USA, 4–9 December 2017; Curran Associates, Inc.: Red Hook, NY, USA, 2017; Volume 30.
30. Taud, H.; Mas, J.F. Multilayer Perceptron (MLP). In *Geomatic Approaches for Modeling Land Change Scenarios*; Camacho Olmedo, M.T., Paegelow, M., Mas, J.-F., Escobar, F., Eds.; Springer International Publishing: Cham, Switzerland, 2018; pp. 451–455.
31. Han, L.; Chen, X.-Y.; Ye, H.-J.; Zhan, D.-C. Softs: Efficient multivariate time series forecasting with series-core fusion. *arXiv* **2024**, arXiv:2404.14197. [CrossRef]
32. Snoek, J.; Larochelle, H.; Adams, R.P. Practical Bayesian optimization of machine learning algorithms. In Proceedings of the 25th International Conference on Neural Information Processing Systems (NeurIPS 2012), Lake Tahoe, NV, USA, 3–6 December 2012; Curran Associates, Inc.: Red Hook, NY, USA, 2012; Volume 25.

33. Vaswani, A.; Shazeer, N.; Parmar, N.; Uszkoreit, J.; Jones, L.; Gomez, A.N.; Kaiser, Ł.; Polosukhin, I. Attention is all you need. In Proceedings of the 31st International Conference on Neural Information Processing Systems (NeurIPS 2017), Long Beach, CA, USA, 4–9 December 2017; Curran Associates, Inc.: Red Hook, NY, USA, 2017; Volume 30.
34. Wu, H.; Hu, T.; Liu, Y.; Zhou, H.; Wang, J.; Long, M. Timesnet: Temporal 2d-variation modeling for general time series analysis. *arXiv* **2022**, arXiv:2210.02186.
35. Hochreiter, S.; Schmidhuber, J. Long Short-Term Memory. *Neural Comput.* **1997**, *9*, 1735–1780. [CrossRef]
36. Rumelhart, D.E.; Hinton, G.E.; Williams, R.J. Learning representations by back-propagating errors. *Nature* **1986**, *323*, 533–536. [CrossRef]
37. Khelifa, H.; Vagnon, E.; Beroual, A. Effect of zirconia nanoparticles on the AC breakdown voltage of mineral oil, synthetic, and natural esters. *IEEE Trans. Dielectr. Electr. Insul.* **2024**, *31*, 731–737. [CrossRef]
38. Koutras, K.N.; Peppas, G.D.; Tegopoulos, S.N.; Kyritsis, A.; Yiotis, A.G.; Tsovilis, T.E.; Gonos, I.F.; Pyrgioti, E.C. Aging impact on relative permittivity, thermal properties, and lightning impulse voltage performance of natural ester oil filled with semiconducting nanoparticles. *IEEE Trans. Dielectr. Electr. Insul.* **2023**, *30*, 1598–1607. [CrossRef]

processes

MDPI

Article

Intelligent Deep Learning Modeling and Multi-Objective Optimization of Boiler Combustion System in Power Plants

Chen Huang [1], Yongshun Zheng [2], Hui Zhao [1], Jianchao Zhu [1], Yongyan Fu [2], Zhongyi Tang [2], Chu Zhang [2] and Tian Peng [2,*]

[1] Huaneng Huaiyin No. 2 Power Generation Co., Ltd., Huai'an 223300, China; hc3794@163.com (C.H.); m13151337218_1@163.com (H.Z.); zhujianchao1234@163.com (J.Z.)
[2] Faculty of Automation, Huaiyin Institute of Technology, Huai'an 223003, China; zys19352399371@163.com (Y.Z.); 17696933553@163.com (Y.F.); 001tzy@163.com (Z.T.); zhangchuhust@foxmail.com (C.Z.)
* Correspondence: husthydropt@126.com; Tel.: +86-150-0523-8786

Abstract

The internal combustion process in a boiler in power plants has a direct impact on boiler efficiency and NOx generation. The objective of this study is to propose an intelligent deep learning modeling and multi-objective optimization approach that considers NOx emission concentration and boiler thermal efficiency simultaneously for boiler combustion in power plants. Firstly, a hybrid deep learning model, namely, convolutional neural network–bidirectional gated recurrent unit (CNN-BiGRU), is employed to predict the concentration of NOx emissions and the boiler thermal efficiency. Then, based on the hybrid deep prediction model, variables such as primary and secondary airflow rates are considered as controllable variables. A single-objective optimization model based on an improved flow direction algorithm (IFDA) and a multi-objective optimization model based on NSGA-II are developed. For multi-objective optimization using NSGA-II, the average NOx emission concentration is reduced by 5.01%, and the average thermal efficiency is increased by 0.32%. The objective functions are to minimize the boiler thermal efficiency and the concentration of NOx emissions. Comparative analysis of the experiments shows that the NSGA-II algorithm can provide a Pareto optimal front based on the requirements, resulting in better results than single-objective optimization. The effectiveness of the NSGA-II algorithm is demonstrated, and the obtained results provide reference values for the low-carbon and environmentally friendly operation of coal-fired boilers in power plants.

Keywords: boiler combustion system; hybrid deep learning model; multi-objective optimization; CNN-BiGRU; improved flow direction algorithm

1. Introduction

The issue of controlling pollutant emissions from thermal power plants has received increasing attention and focus [1]. Coal-fired boilers in thermal power plants produce large amounts of nitrogen oxides (NOx) during combustion. NOx, comprising nitric oxide (NO) and nitrogen dioxide (NO_2), significantly contributes to severe air pollution [2]. NOx is an acidic gaseous pollutant that can cause significant damage to the human respiratory system when emitted into the air in large quantities. It forms fine particles of nitric acid and nitrate, causing atmospheric pollution and, in severe cases, acid rain, which can affect the surrounding animals and plants [3]. The use of coal-fired boilers in power plants to generate electricity has caused environmental pollution problems, and many power plants have sought to reduce NOx emission concentrations. However, incomplete combustion in boilers results in wasted fuel calorific values, which, in turn, leads to lower combustion efficiency and increased fuel consumption, thus increasing costs [4]. Measures to reduce NOx emissions often reduce boiler efficiency and affect the economic efficiency [5]. Therefore, a balance between NOx emission and boiler efficiency must be considered to find a solution that will reduce NOx emissions without unduly sacrificing boiler efficiency and plant economics.

As society continues to develop, many power plants have been equipped with plant-level real-time monitoring information systems (SISs), which store much of the operational data during boiler operation, such as the operating conditions and parameters, and provide data support for developing boiler combustion optimization technology [6]. The ever-evolving field of artificial intelligence and machine learning has recently introduced several techniques to enhance boiler combustion optimization technology. NOx emission levels and boiler thermal efficiency are crucial performance indicators for evaluating boiler operation. A machine learning-based predictive model can be created for the NOx emission concentration and boiler thermal efficiency using historical data on boiler operations [7]. In addition, an intelligent evolutionary algorithm can be employed to optimize controllable operational parameters of the boiler combustion system. This will help lower the NOx emission concentration and enhance boiler thermal efficiency, ultimately improving the economic efficiency of power plants [8].

NOx production and thermal efficiency during boiler combustion depend on various factors, including the coal feed, boiler load, primary and secondary airflow, etc. Therefore, the boiler combustion model is a complex multivariable and nonlinear system, making it challenging to develop an optimal model using traditional methods. Therefore, many scholars have conducted intelligent modeling through historical data to predict NOx concentration emissions and boiler combustion efficiency [9]. Currently, the main approaches for data modeling are machine learning and deep learning models.

Li et al. [10] proposed a model of Extreme Learning Machine (ELM) based on an artificial bee colony algorithm for predicting the thermal efficiency of boilers. After experimental verification, the model is capable of markedly increasing the reliability of forecasting results and provides strong support for boiler operation. Tan et al. [11] found through experiments that the long short-term memory (LSTM) model outperformed support vector machines (SVMs) in predicting NOx emissions from coal-fired boilers. Liu et al. [12] employed a generalized correlation (GC) entropy-based hybrid LSTM model to predict NOx emission concentrations, which yielded satisfactory outcomes. Wang et al. [13] developed a hybrid model by utilizing an attention mechanism with an LSTM model and also introduced the complete ensemble empirical mode decomposition with adaptive noise (CEEMDAN) method for the decomposition of raw historical data. Wang et al. [14] introduced a new regression method, the convolutional neural network (CNN), which utilizes convolutional neural networks to enhance NOx emission prediction accuracy and efficiency in both steady-state and transient scenarios for coal-fired boilers. The CNN-BiGRU model was proposed by Ma et al. [15] for the NOx concentration prediction of a 660 MW coal-fired boiler. The model displays excellent prediction performance and accurately predicts NOx concentrations. Based on this, this paper will use the CNN-BiGRU model to construct a NOx concentration and boiler combustion efficiency prediction model and, subsequently, to conduct research. Li et al. [16] developed a CNN-LSTM model augmented with a channel-wise attention module to predict NOx emissions from coal-fired boilers, achieving improved accuracy and robustness over standard deep learning models. Wan et al. [17] applied a

CNN-LSTM-attention framework to forecast the boiler load in CHP plants, reporting a MAPE of below 1%.

Based on the application of machine learning algorithms to model and simulate boiler combustion processes, some researchers have used intelligent evolutionary algorithms to optimize controllable boiler operating parameters under different operating conditions, aimed at boosting boiler efficiency and lowering nitrogen oxide emissions. According to the boiler operation requirements, there are two main methods for optimizing coal-fired boilers [18]: One is to use single-objective optimization to improve boiler efficiency or to reduce the NOx emission concentration. The other is to use multi-objective optimization to balance the NOx concentration and improve boiler efficiency. The single-objective optimization methods for boiler combustion systems include the simulated annealing genetic algorithm (SAGA) [19], genetic algorithm (GA) [20], ant colony optimization algorithm (ACO) [21], and other algorithms [22]. Dirik [20] developed a NOx emission model using a neuro–fuzzy inference system. The model used the GA algorithm to optimize boiler combustion by determining optimal values for the operating parameters. Wei et al. [19] proposed a support vector regression (SVR) model to predict the concentration of NOx emissions and employed SAGA for optimizing the operating parameters of coal-fired boilers. Yao et al. [23] implemented a genetic algorithm to improve system operations, aiming to suppress NOx emission levels. Zhou et al. [21] obtained optimal operating parameters based on the ACO algorithm for reducing NOx emissions.

During the last several years, many multi-objective optimization algorithms have been introduced to solve the problem of optimizing the control of systems such as boiler combustion and denitrification systems [9]. A predictive model is built based on the raw data, and multiple objective functions are established, such as the system efficiency, pollutant emission concentration, operating cost, etc. By solving the objective functions using multi-objective optimization algorithms, multiple Pareto optimal fronts are found, and feasible solutions for the corresponding controllable parameters are obtained, providing reliable recommendations for the operation and production of the equipment.

The most commonly used multi-objective optimization algorithms include non-dominated sorting genetic algorithm II (NSGA-II) [24], multi-objective particle swarm optimization (MOPSO) [25], multi-objective differential evolution (MODE) [26], etc. Xu et al. [25] proposed an improved multi-objective particle swarm optimization (IMOPSO) to optimize boiler combustion to simultaneously enhance thermal efficiency and reduce NOx emissions. Zheng et al. [26] suggested employing a multi-objective evolutionary algorithm (MOEA/D) in combination with a kernel extreme-learning machine (KELM) prediction model for optimal control. The aim was to reduce pollutant emissions and enhance energy efficiency, and the simulation results confirmed the efficacy of this approach. Wu et al. [27] proposed a design for a steam power system (SPS) that incorporated pollutant reduction technology. The NSGA-II algorithm was used to obtain the Pareto optimization curve, and simulation results verified the applicability of NSGA-II. In this study, the NSGA-II multi-objective optimization algorithm is applied to enhance the boiler combustion performance. Tang et al. [28] developed DYNA-A3C, a dynamic-weight reinforcement learning model, to optimize combustion performance under deep peak-shaving scenarios, effectively improving NOx control. Meanwhile, advanced multi-objective optimization algorithms such as MOEA/D and SPEA2 have been employed to handle the trade-off between emission reduction and efficiency. Zheng et al. [26] applied MOEA/D integrated with a kernel extreme learning machine (KELM) for combustion optimization, yielding a diverse and efficient set of solutions.

Motivated by the preceding techniques, this paper proposes a research method for intelligent modeling and multi-objective optimization of boiler combustion based on deep hybrid learning. The scientific contributions made by this study are set out below:

(1) A hybrid deep learning model, known as CNN-BiGRU, is constructed using historical operational data to accurately predict the NOx emission concentration and to optimize the boiler combustion efficiency of a 300 MW coal-fired boiler. This advanced model integrates convolutional neural networks (CNNs) and bidirectional gated recurrent units (BiGRUs) to capture complex temporal and spatial patterns in boiler combustion. Compared with prior works using LSTM, GRUs, or basic CNN models, the CNN-BiGRU model demonstrates superior prediction accuracy.

(2) The NOx emission concentration and boiler combustion efficiency are identified as the optimization objectives. Boiler combustion is optimized with the single goal of using the improved flow direction algorithm (IFDA). The IFDA incorporates iterative local search (ILS) and chaotic initialization to enhance the optimization capabilities, outperforming the original FDA algorithm. The controllable variables and the optimization interval are determined, and the optimal solution is found by continuously adjusting the values of the controllable variables to reduce the NOx emission concentration or to improve the thermal efficiency of the boiler.

(3) To optimize the combustion performance of coal-fired boilers, the non-dominated sorting genetic algorithm II (NSGA-II) is employed as a multi-objective optimization tool. The NSGA-II algorithm is utilized to balance NOx emission reduction and boiler efficiency improvement, generating a Pareto optimal front that provides practical solutions for real-world applications. This approach overcomes the limitations of the single-objective optimization methods previously used. The resulting Pareto front is compared against single-objective optimization outcomes, thereby validating the superiority and practicality of the multi-objective approach. The NSGA-II algorithm achieves a significant reduction in NOx emissions and an increase in boiler thermal efficiency, demonstrating the model's potential for promoting the low-carbon and environmentally sustainable operation of coal-fired boilers.

In this study, Section 2 provides an introduction to the case study and data sources. Section 3 describes the improved FDA algorithm, the NSGA-II algorithm, and the general process of this paper. The results of the prediction model for the boiler combustion system is presented in Section 4. Next, Section 5 compares and analyzes the simulation results obtained from single-objective and multi-objective optimization approaches applied to coal-fired boilers. A summary of the experimental results is presented in Section 6.

2. Methods

2.1. Improved Flow Direction Algorithm (IFDA)

Karami et al. [29] developed the flow direction algorithm (FDA), which is based on tracking the flow movement to the basin outlet. The flow direction algorithm (FDA) employs a specific policy for allocating search duration to balance global and local searches. This is accomplished through two key components: gradually reducing the neighborhood radius to negligible values and implementing a sink-filling process that aids in escaping local solutions. The steps involved in the FDA can be summarized as follows:

$$F1\_\boldsymbol{I} = \begin{bmatrix} i_1^1 & i_2^1 & \cdots & i_s^1 \\ i_1^2 & i_2^2 & \cdots & i_s^2 \\ \vdots & \vdots & \ddots & \vdots \\ i_1^w & i_2^w & \cdots & i_s^w \end{bmatrix} \tag{1}$$

where $F1\_I$ denotes the location of the flow, s denotes the dimension of the problem, and the population size is w.

Each flow undergoes objective function evaluation, with the exit point determined by identifying the most favorable value. The corresponding matrix representation of the objective functions is shown as follows:

$$F1\_fit = \begin{bmatrix} fit1 \\ fit2 \\ \vdots \\ fitw \end{bmatrix} \tag{2}$$

A set of neighboring flows is constructed for each stream element based on a predefined neighborhood radius Δ. The position of the neighborhood is calculated as follows:

$$N1\_\boldsymbol{I}(n) = F1\_\boldsymbol{I}(m) + \boldsymbol{randn}^{*}\Delta \tag{3}$$

where $N1\_\boldsymbol{I}$ denotes the nth position of the neighbor and $\boldsymbol{randn}$ is a random value. $F1\_\boldsymbol{I}(m)$ denotes the m th position in the flow direction.

The value of the objective function is computed for each neighbor, and the best neighbor is determined using the following approach:

$$\Delta = \boldsymbol{rand}^{*}\boldsymbol{Irand} - \boldsymbol{rand}^{*}F1_{I}(m)^{*}BestX - F1_{I}(m)^{*}\boldsymbol{W} \tag{4}$$

$$\mathbf{r_d} = \sum_{a=1}^{A} R_a - \varphi\Delta t \tag{5}$$

where $\boldsymbol{rand}$ stands for a random location and $\boldsymbol{Irand}$ is a random location determined by the direct runoff. $\boldsymbol{W}$ is a non-linear weight, $\mathbf{r_d}$ denotes the direct runoff, R_a denotes the rainfall, Δt denotes the time interval, and m denotes the time step.

An update of the flow velocity vector is performed using Equation (6) when the optimal neighbor exhibits a better objective function value than the current flow. A new flow position is generated according to the relations outlined in Equation (7). The expression for the flow velocity vector can be found in Equation (8):

$$F1\_newX(m) = f\_X(m) + V^{*}\frac{F1\_X(m) - N1\_X(n)}{\|F1\_X(m) - N1\_X(n)\|} \tag{6}$$

$$V = \begin{bmatrix} v_1\, v_2 \cdots v_s \end{bmatrix} \tag{7}$$

$$V = \boldsymbol{randm}^{*}S_0 \tag{8}$$

where $F1\_newX(m)$ represents the location of the new flow, $\boldsymbol{randm}$ represents a randomly generated number, and the variable S_0 indicates the slope linking the neighbor to the current flow.

The position of the flow is updated according to the following formula:

$$\begin{cases} if \quad F1\_fit(\boldsymbol{r}) < F1\_fit(m) \\ F1\_newX(m) = F1\_X(i) + rand\boldsymbol{m}^{*}(F1\_X(r) - F1\_X(m)) \\ else \\ F1\_newX(m) = F1\_X(m) + 2randm^{*}(Best\_X - F1\_X(m)) \end{cases} \tag{9}$$

where $\boldsymbol{r}$ is a random integer.

The final part calculates the objective function for the new flows. The resulting results are compared with the existing flows, and if the effect improves, the objective function and the positions of the flows are updated, and the optimal answer is returned. Conversely, the number of neighborhoods with neighborhood radius Δ is recreated for each flow, and the subsequent steps are repeated.

The original FDA algorithm has yielded relatively excellent results in the study of many optimization problems, but it also suffers from some drawbacks. For this reason, Ma et al. [15] added iterative local search (ILS) strategies and chaotic initialization to improve the FDA algorithm.

The FDA algorithm is frequently employed for swarm intelligence optimization and relies on random distribution for the initial population generation. However, this approach can compromise population diversity, leading to premature convergence, especially for more intricate problems where the algorithm may rapidly converge to a suboptimal local solution. Escaping local optima can present a significant challenge when seeking better solutions. Various strategies have been proposed to overcome the tendency of the FDA algorithm to fall into local optimization during location updates. One such method involves implementing a chaos strategy to determine the initial population distribution [29], which leverages the random behavior of chaotic systems to increase the initial population diversity. An alternative strategy to combat early local optimum convergence and improve the global search ability of the FDA algorithm involves utilizing an ILS technique [30]. Essentially, the ILS approach introduces perturbations to the current local optimum solution and conducts a new search to identify a superior global optimum solution [31]. A key advantage of the ILS approach is its ability to rapidly jump out of local optima, facilitating the solution exploration capabilities of the FDA. An ILS strategy is adopted and integrated at the end of each iteration to counter the issue of local-optimum-solution trapping in the FDA algorithm. This strategy enables the algorithm to more effectively bypass local optima, resulting in a quicker convergence rate.

The improved flow direction algorithm has an optimization search flow, as shown in Figure 1.

2.2. Non-Dominated Sorting Genetic Algorithm-II (NSGA-II)

Deb et al. [32] introduced the non-dominated sorting genetic algorithm-II (NSGA-II), which incorporates crowding distance calculations to identify the complete Pareto optimal front. This approach also ensures that solutions along the front are evenly distributed, thereby maintaining population diversity throughout the optimization process. The NSGA-II algorithm avoids the loss of the best individuals in the population by introducing an elite strategy and expanding the sampling space while also increasing the computational speed of the algorithm. The NSGA-II algorithm steps are described below:

Step 1: Produce an initial population with N number of individuals;

Step 2: Determine whether to generate new individuals by calculating fitness values and rank them;

Step 3: Generate a second-generation population through strategies such as cross and variation, and use an elite strategy to merge the first- and second-generation populations to form a new population;

Step 4: Perform non-dominated sorting on the new population and calculate the crowd level. Select the best individuals to form a new population again, and keep repeating the above operations until the end conditions are met.

Figure 2 shows the flow of the NSGA-II algorithm.

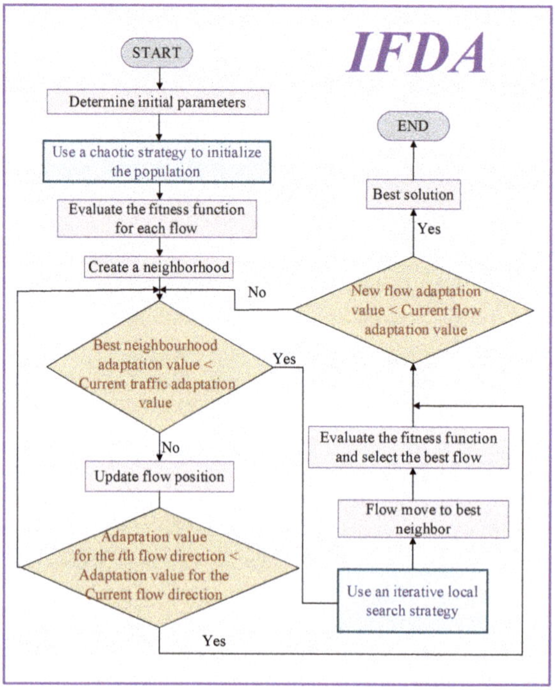

Figure 1. Flow chart of the improved flow direction algorithm.

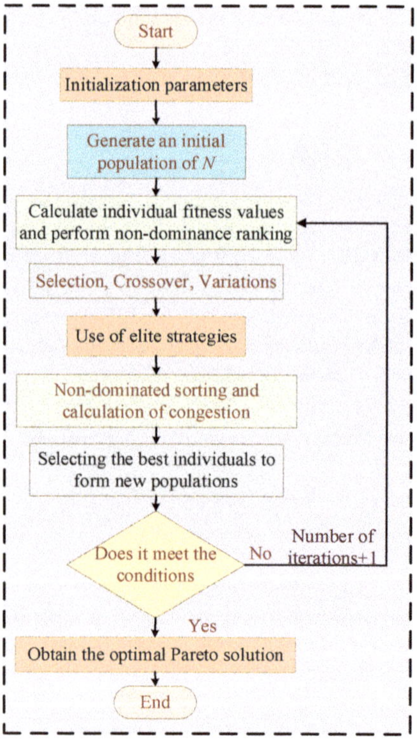

Figure 2. NSGA-II algorithm calculation flow chart.

2.3. The CNN-BiGRU Prediction Model

Ma et al. [15] employed a convolutional neural network (CNN) to extract deep features from a dataset before utilizing the BiGRU model to predict NOx. Then, the hyperparameters of the CNN-BiGRU are optimized using the IFDA algorithm. In this paper, based on Ma et al. [15], a NOx concentration and coal-fired boiler efficiency prediction model will be constructed based on 300 MW boiler operation data as the research object. Figure 3 shows the final construction of the CNN-BiGRU model.

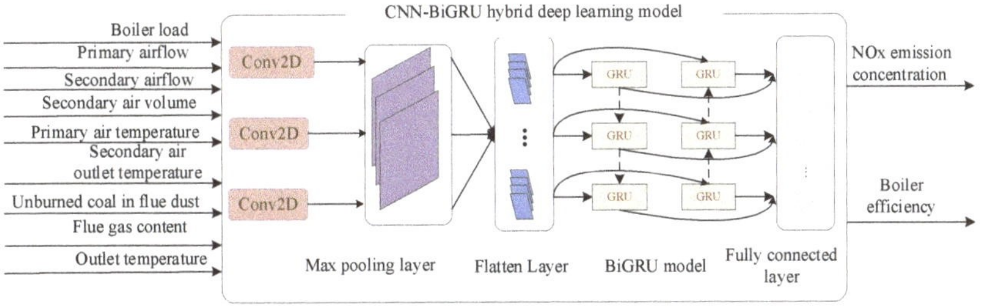

Figure 3. A boiler combustion prediction model based on the CNN-BiGRU.

The CNN-BiGRU model comprises two convolutional layers with kernel sizes of 3 and 5, respectively, followed by ReLU activation. The BiGRU layer contains 128 hidden units, followed by a dense output layer. The model is trained for 100 epochs with a batch size of 64, using the Adam optimizer and a learning rate of 0.001. The IFDA algorithm is applied to optimize key hyperparameters, including the learning rate, number of filters, and hidden units by minimizing the validation RMSE.

2.4. Boiler Efficiency Calculations

Boiler efficiency is defined as the ratio of useful heat output to the total heat input during the coal combustion process in a boiler [33]. Boiler efficiency is used to assess the economics of coal-fired boilers. Improving boiler efficiency increases energy efficiency and is one of the main objectives of this study for optimal control. Two main methods for calculating boiler thermal efficiency are currently used in the actual generation process, namely the positive- and negative-balance method [34].

When the boiler efficiency is calculated using the positive balance method in the actual production process, it is necessary to obtain the total heat input value of the boiler QH_2 and the effective heat value QH_1 and to obtain the boiler efficiency from the ratio of the two. The calculation formula is shown in Equation (10):

$$\eta = \frac{QH_1}{QH_2} \tag{10}$$

where QH_1 and QH_2 represent the effective heat value per kilogram of fuel and the total heat input, respectively.

However, in practice, the errors in the effective heat obtained and the total heat input are too large to accurately calculate boiler efficiency, so plants generally use the negative balance method to calculate boiler efficiency. The negative balance method differs from the forward balance method in that the negative balance method indirectly calculates boiler efficiency by obtaining various heat loss values for coal-fired boilers. This calculation method not only gives a more accurate picture of the boiler efficiency but also allows the current operating conditions of the boiler to be known from the heat loss values. The calculation formula for the negative balance method is shown in Equation (11).

$$\eta = 100 - \sum_{j=3}^{7} QH_j \tag{11}$$

where QH_j represents various heat loss values, mainly for the following types of heat loss: exhaust heat loss QH_3, gas incomplete combustion heat loss QH_4, solid incomplete combustion heat loss QH_5, heat loss QH_6, and ash physical heat loss QH_7.

Exhaust heat loss QH_3 refers to the heat in the flue gas emitted from the boiler; more heat is lost the higher exhaust temperature is. Gas incomplete combustion heat loss QH_4 refers to the combustible gases remaining in the flue gas emitted from the boiler; such gases do not burn and release heat during the boiler operation, resulting in heat loss. The solid incomplete combustion heat loss QH_5 and ash physical heat loss QH_7 refer to the coal in the combustion process; some slag and leakage of coal contain unburned coal material, thus causing heat loss; the more unburned coal material generated by the heat loss, the more significant this will be. Heat loss QH_6 is the heat loss caused by the heat dissipation of the equipment to the outside due to the combustion process of the boiler; this heat loss generally accounts for a relatively small amount.

3. General Flowchart of the CNN-BiGRU and NSCA-II for Boiler Combustion Modeling

The general process of the CNN-BiGRU and NSCA-II (CNN-BiGRU-NSCA-II)-based study is shown below:

(1) Firstly, the study object was identified as a 300 MW circulating fluidized bed boiler, and 3000 samples were selected as the raw data from its historical operation data, of which 70% were classified as the training set and 30% as the test set. There were 25 input variables, while the model output variables mainly involved boiler thermal efficiency and NOx emission concentration.

(2) A CNN-BiGRU-based boiler combustion prediction model was constructed and employed to forecast the concentrations of NOx emissions and the thermal efficiency of the boiler.

(3) A total of five variables, namely the primary airflow rate (y_6~y_7), secondary airflow rate (y_{12}~y_{13}), and flue gas oxygen content (y_{20}), were selected as controllable variables according to the actual situation, and the optimization interval was determined.

(4) The boiler thermal efficiency and NOx emission concentration were denoted as the optimization objectives.

(5) The NSGA-II algorithm was employed to identify the optimal solution for controllable variables within the defined optimization range, with the best parameter set corresponding to the minimum fitness value achieved by the algorithm.

(6) Simulation analyses were performed to evaluate and compare the outcomes of the IFDA under single-objective and multi-objective optimization scenarios.

The specific process based on the CNN-BiGRU-NSGA-II is shown in Figure 4.

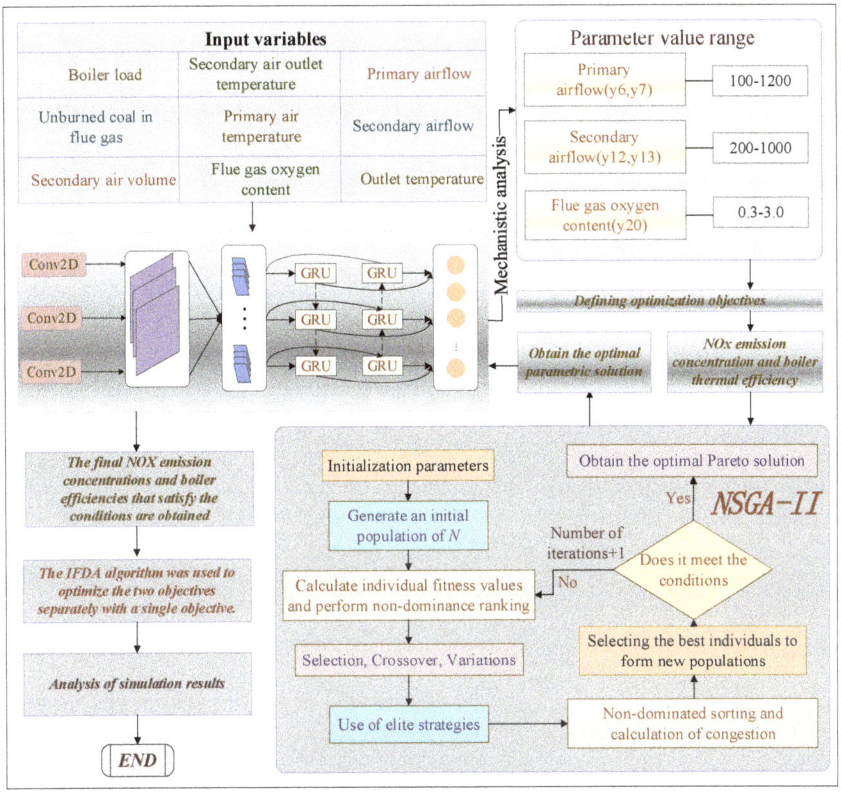

Figure 4. General flowchart of the CNN-BiGRU and NSCA-II for boiler combustion modeling.

4. Case Study

4.1. Data Source and Preprocessing

In this study, operational data were collected from a 300 MW circulating fluidized bed (CFB) boiler, based on historical combustion operation records. A total of 3000 samples were selected as the original dataset after preprocessing. The dataset was divided into 70% for training and 30% for testing. Twenty-five key variables were selected as inputs, including the boiler load (y_1), coal feed rates (y_2–y_5), primary airflow at the burner inlets (y_6–y_7), primary air temperature (y_8–y_9), primary fan inlet temperature (y_{10}–y_{11}), secondary airflow (y_{12}–y_{13}), secondary air outlet temperature (y_{14}–y_{15}), motor currents (y_{16}–y_{17}), flue gas oxygen content (y_{18}), flue gas temperature (y_{19}), and outlet temperature of slag cooler (y_{20}–y_{23}). The two output variables were NOx emission concentration and efficiency. The dataset was provided by Hu et al. [26] and was a real dataset from the monitoring information system (SIS) of circulating fluidized bed boilers (CFBBs).

To guarantee the reliability of the combustion model used in this study, a meticulous selection process was employed to screen and process accurate operating data from the power plant for testing and training purposes. The primary factors influencing NOx generation in coal-fired boilers included the direct airflow, secondary airflow, coal feed, load, and oxygen content. A comprehensive analysis was conducted on these factors to determine the input variables for subsequent experiments. Based on the generation mechanism of NOx, 23 variables were selected as input variables, and NOx emission concentration and boiler thermal efficiency were selected as the output variables of the model. The initial selection of 23 variables is presented in Table 1 below:

Table 1. Coal-fired boiler inputs and outputs.

	Variable Name	Unit	Variable
Input	Load	%	y_1
Ambient pressure	Coal quantity of the coal feeder	t/h	y_2–y_5
Ambient humidity	Primary airflow of the burner inlet	kNm^3/h	y_6–y_7
Air-filter pressure difference	Primary air temperature of the burner inlet	°C	y_8, y_9
Gas-turbine exhaust pressure	Primary air fan inlet temperature	°C	y_{10}, y_{11}
Turbine inlet temperature	Distribution flow of the secondary air	kNm^3/h	y_{12}, y_{13}
Turbine after temperature	Distribution flow of secondary air	kNm^3/h	
Compressor discharge pressure	Air preheater secondary air outlet air temperature	°C	y_{14}, y_{15}
Turbine energy yield	Limestone powder conveying motor current	AMP	y_{16}, y_{17}
Carbon monoxide	Flue gas oxygen content	%	y_{18}
Nitrogen oxides	Flue gas temperature	°C	y_{19}
	Outlet temperature of slag cooler	°C	Y_{20}–y_{23}
Output	NOx emission concentration	mg/m^3	
	efficiency	%	

4.2. Performance Metrics

Four evaluation metrics were used in this study to assess the effectiveness of the prediction model: Root Mean Square Error (*RMSE*), Correlation Coefficient (*R*), Mean Absolute Error (*MAE*), and Mean Absolute Percentage Error (*MAPE*).

$$RMSE = \sqrt{\frac{1}{p}\sum_{a=1}^{p}(Yc_a - Ge_a)^2} \tag{12}$$

$$R = \frac{\sum_{a=1}^{p}(Yc_a - \overline{Yc_a})(Ge_a - \overline{Ge_a})}{\sqrt{\sum_{a=1}^{p}(Yc_a - \overline{Yc_a})^2 \times \sum_{j=1}^{p}(Ge_a - \overline{Ge_a})^2}} \tag{13}$$

$$MAE = \frac{1}{p}\sum_{n=1}^{p}|Yc_n - Ge_n| \tag{14}$$

$$MAPE = \frac{1}{p}\sum_{n=1}^{p}\left|\frac{Yc_n - Ge_n}{Ge_n}\right| \times \frac{100}{p} \tag{15}$$

where Ge_a is the observed value and p is the total number of predicted samples. Yc_a is the predicted value, and $\overline{Y_{p_i}}$ and $\overline{O_i}$ represent the average observed and average predicted values, respectively.

5. Comparison Results

5.1. Analysis of the NOx Prediction Results

In this research study, a novel hybrid deep learning model dubbed CNN-BiGRU was utilized in constructing boiler combustion prediction models. Table 2 displays the NOx emission concentration prediction-error evaluation index values for each model. To present the NOx prediction outcomes more transparently, this subsection employs a sample size of 60 consecutive forecast results to generate comparative graphs depicting the model prediction performance, as shown in Figures 5 and 6.

Table 2. Evaluation index of the prediction error for the NOx emission concentration in each model.

NOx Emission Prediction Model	RMSE	MAE	R	MAPE
LSTM	4.7415	3.8592	0.9953	0.0266
GRU	4.2426	3.4098	0.9964	0.0274
BiGRU	3.8799	3.3472	0.9977	0.0228
CNN-BiGRU	2.8784	2.4829	0.9991	0.0199
CNN-FDA-BiGRU	2.1745	1.7395	0.9985	0.0137
CNN-IFDA-BiGRU	1.3396	1.064	0.9992	0.0078

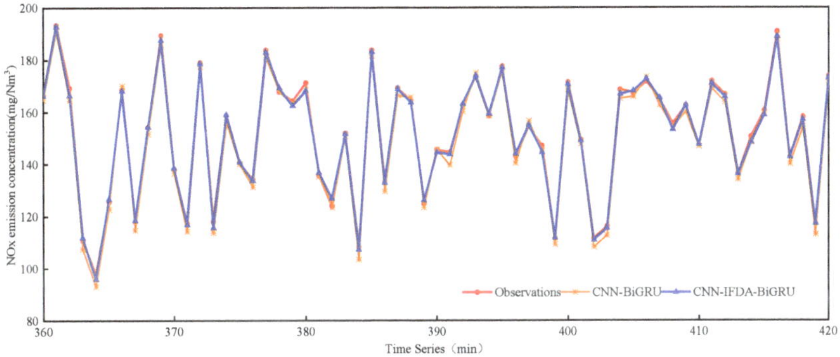

Figure 5. Comparison of NOx emission concentration prediction results from the hybrid deep learning model.

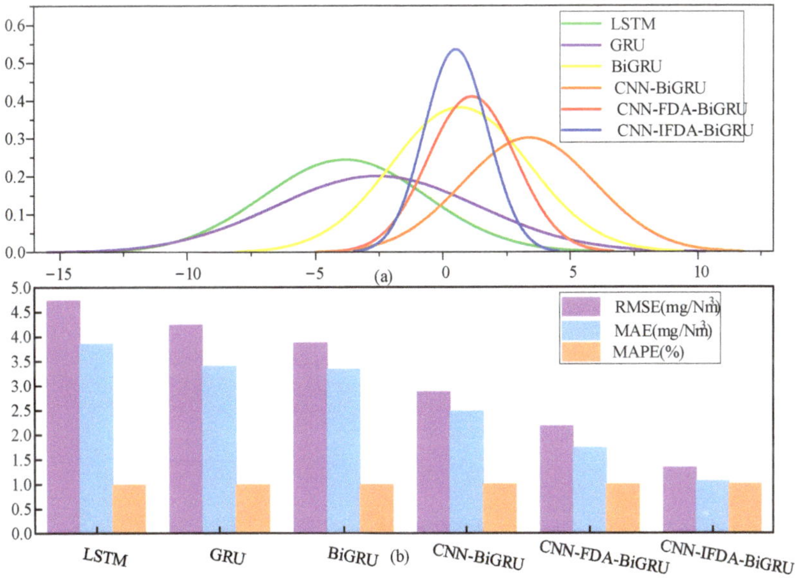

Figure 6. Probability distribution diagram and error index comparison. (**a**) Probability density of the prediction errors. (**b**) Bar-chart comparison of RMSE, MAE and MAPE for each model.

For the NOx emission concentration prediction experiment, the CNN-IFDA-BiGRU model achieved RMSE, MAE, R, and MAPE values of 1.3396, 1.064, 0.9992, and 0.0078, respectively, as detailed in Table 2. This model demonstrates the lowest error metrics—including for the RMSE, MAE, and MAPE—while also attaining the highest correlation coefficient R. When comparing the CNN-BiGRU model to the standalone BiGRU model, the former clearly outperforms the latter, exhibiting reduced RMSE, MAE, and MAPE values, which indicates its superior predictive capability. Furthermore, as illustrated in Figure 5, the IFDA-optimized model surpasses the FDA-optimized model in prediction accuracy. These results confirm the effectiveness of the IFDA algorithm in minimizing prediction errors and enhancing the overall performance of the CNN-BiGRU model. To provide a more comprehensive evaluation, Figure 6 presents the normal distribution of the NOx prediction results alongside a comparative analysis of the error metrics across all the models.

Observing the probability distribution in Figure 6, it becomes evident that the CNN-IFDA-BiGRU model exhibits a higher frequency of error values close to zero compared with other models. On the other hand, the LSTM model displays the smallest error values. This further indicates that the proposed model has better results in predicting NOx emission concentrations. The error indicator bar chart clearly illustrates a consistent decreasing trend in the simulation results of each model. This trend is strong evidence of the effectiveness of the methods proposed in this study.

5.2. Analysis of the Thermal Efficiency of Coal-Fired Boilers Prediction Results

Like in Section 4.1, the boiler combustion prediction models were constructed using the hybrid deep learning model CNN-BiGRU. Table 3 shows the values of the boiler thermal efficiency prediction-error evaluation metrics for each model. The thermal efficiency prediction results for coal-fired boilers are shown in Figures 7 and 8.

Table 3. Comparison of evaluation indicators for the prediction error for the thermal efficiency of coal-fired boilers in various models.

Thermal Efficiency Prediction Model	RMSE	MAE	R	MAPE
LSTM	0.0573	0.0453	0.9935	0.00050
GRU	0.0527	0.0404	0.9927	0.00045
BiGRU	0.0470	0.0348	0.9938	0.00038
CNN-BiGRU	0.0421	0.0365	0.9984	0.00040
CNN-FDA-BiGRU	0.0273	0.0189	0.9981	0.00021

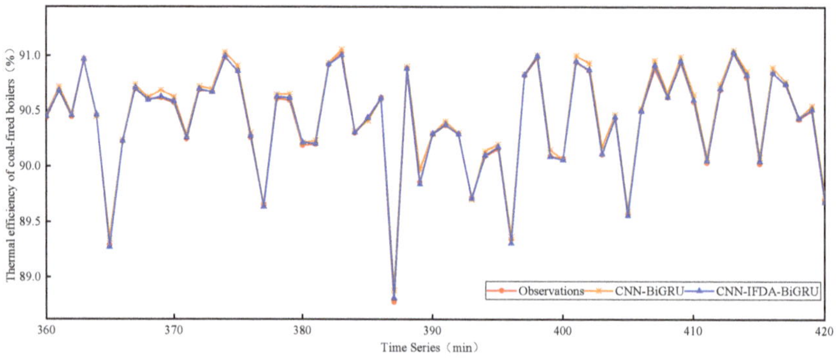

Figure 7. Comparison of boiler thermal efficiency prediction results of the hybrid deep learning model.

In the boiler thermal efficiency prediction experiments, the CNN-IFDA-BiGRU model achieved RMSE, MAE, R, and MAPE values of 0.0144, 0.0113, 0.9996, and 0.00012, respectively. Among all the control groups, this model exhibits the lowest error metrics—including for the RMSE, MAE, and MAPE—along with the highest correlation coefficient R. As illustrated in Figure 7, the IFDA-optimized model outperforms the FDA-optimized model in terms of prediction accuracy. To further compare the performance of each model in boiler thermal efficiency prediction, a plot of the normal distribution of the NOx prediction results and a comparison of error indicators for each model were plotted, as shown in

Figure 8. The probability distribution plot in Figure 8 shows that the CNN-IFDA-BiGRU model has the most error values close to zero and the LSTM model has the least, further indicating that the proposed model has the better performance in predicting boiler thermal efficiency. The histogram of error indicators demonstrates a consistent decreasing trend in the simulation results of each model, implying the effectiveness of the proposed methods in predicting the thermal efficiency of boilers.

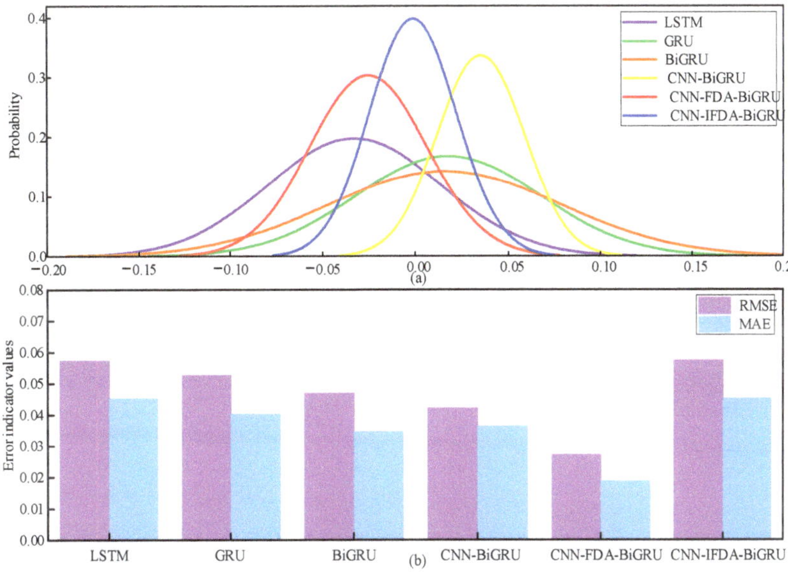

Figure 8. Performance comparison in boiler thermal-efficiency prediction. (**a**) Probability-density distribution of the prediction errors. (**b**) Bar-chart comparison of RMSE and MAE for each model.

5.3. Coal-Fired Boilers with Multiple Objectives

In practice, reducing NOx emissions from boilers can reduce the boiler thermal efficiency, thus affecting the efficiency of the power plant. Therefore, during boiler combustion operations, it is imperative to consider not only the limits of NOx emission concentrations but also the efficiency of the boiler. This study starts with these two objectives. In combination with the previous prediction model of the boiler combustion system based on the CNN-BiGRU, a single-objective optimization of these two objectives is carried out first. Then a multi-objective optimization model considering NOx emission concentration and boiler combustion efficiency as optimization objectives are constructed using NSGA-II, and the simulation results are analyzed.

This study utilizes real-world boiler operating data as the foundation for the model, employing an algorithm to identify the optimal boiler operating parameters that fulfil the requirements. This approach facilitates optimal control over boiler efficiency and NOx emission concentration. In this study, only the primary airflow (y_6~y_7), secondary airflow (y_{12}~y_{13}), and flue gas oxygen content (y_{20}) parameters are optimized in the given optimization range. In contrast, the other operating parameters are kept constant during optimization.

5.3.1. Single-Objective Optimization of NOx Emission Concentrations from Coal-Fired Boilers

Single-objective optimization is the optimization of only one objective function, within a given range, by finding the optimal solution for the parameter variables to satisfy the objective function. Single-objective optimization can be expressed by the following Equation (16):

$$\begin{gathered} Min(orMax) \quad fun(x) \\ \mathbf{s1.} \\ x = [x_1, x_1, \cdots, x_m, \cdots, x_M] \\ x_{m\min} \le x_m \le x_{m\max} \end{gathered} \tag{16}$$

where $Min(orMax)\ fun(x)$ is the approximate objective function, $x = [x_1, x_1, \cdots, x_m, \cdots, x_M]$ is the constraint condition, and $x_{m\min} \le x_m \le x_{m\max}$ is the range of the parameter variables.

Using Equation (17), a single-objective optimization model of the NOx emission concentration using the FDA algorithm and the IFDA algorithm is constructed as

$$\begin{gathered} Minfun(y) \\ s.t. \\ y = [y_1, y_1, \cdots, y_m, \cdots, y_{13}] \\ y_{m\min} \le y_m \le y_{m\max} \end{gathered} \tag{17}$$

where $fun(y)$ is the NOx emission concentration and $y = [y_1, y_1, \cdots, y_m, \cdots, y_{13}]$ represents the input variables.

Certain parameter variables are not adjustable during the practical operation of boilers [32]. The dataset in this paper shows that boiler load and temperature are uncontrollable parameter variables. Accordingly, based on practical considerations, five variables were chosen as controllable parameters in this study: primary airflow (y_6–y_7), secondary airflow (y_{12}–y_{13}), and flue gas oxygen content (y_{20}). The optimal ranges for these controllable variables are presented in Table 4.

Table 4. The optimization interval of the selected controllable variables.

Controllable Variables	Flue Gas Oxygen Content (%a)	Primary Airflow (KNm^3/h)	Secondary Airflow (KNm^3/h)
Controllable interval	0.3–3.0	1200–1000	100–1200

This study first used the CNN-BiGRU model to predict NOx emission concentrations. Then the FDA algorithm and the modified FDA algorithm were used to continuously search for the best values within the selected optimization intervals of the five variables until their optimal values were found. The populations of the IFDA algorithm were set to 20, the neighbors to 5, and the maximum number of iterations to 100. The minimum value of the NOx emission concentration obtained during the optimization process was taken as the optimal solution. Finally, optimal values of the controllable variables were obtained based on this optimal solution. In this section, the original FDA algorithm with the same parameter settings is used as a comparison to demonstrate the optimality-finding capability of the IFDA algorithm. The optimization process is depicted in Figure 9, with the corresponding results shown in Table 5.

Table 5. Optimization results of controllable variables for NOx emission concentration.

Variables	Primary Airflow		Secondary Airflow		Flue Gas Oxygen Content
	y_6	y_7	y_{12}	y_{13}	y_{18}
Observations	866.15	864.27	540.70	947.79	1.06
FDA	853.03	793.89	521.58	896.28	1.01
IFDA	847.51	680.05	526.79	877.91	0.86

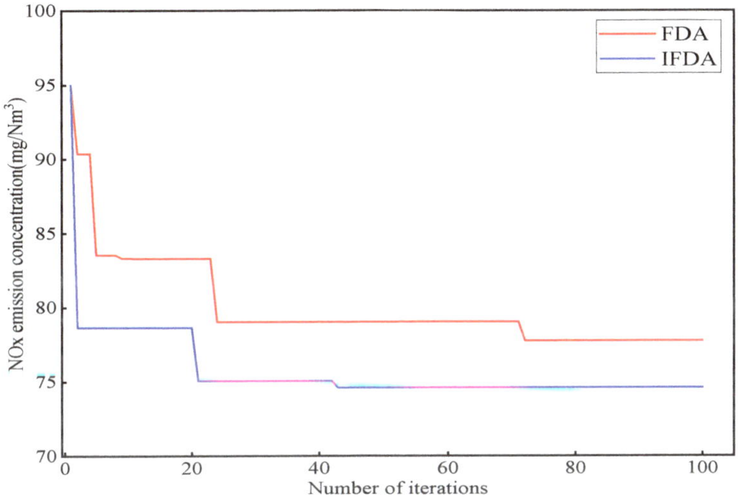

Figure 9. NOx emission concentration optimization process.

As can be seen from the information in Table 5 and Figure 9, the distribution of the blue and red curves represents the optimization process of the IFDA and FDA for the pair of NOx emission concentrations. This indicates that the IFDA algorithm has a better optimization capability and can find more suitable parameter variables for the boiler; thus, it can effectively achieve the optimization goal of reducing NOx emission concentrations.

5.3.2. Single-Objective Optimization of Thermal Efficiency in Coal-Fired Boilers

As in Section 5.3.1, this subsection uses the IFDA algorithm to perform a single-objective optimization of the boiler thermal efficiency, with the single-objective optimization function constructed as

$$\begin{gathered} Minfun(y) \\ s.t. \\ y = [y_1, y_1, \cdots, y_m, \cdots, y_{13}] \\ y_{m\min} \leq y_m \leq y_{m\max} \end{gathered} \tag{18}$$

where $fun(y)$ is the thermal efficiency of the boiler and $y = [y_1, y_1, \cdots, y_m, \cdots, y_{13}]$ represents the input variables.

To ensure the accuracy and validity of the experiment, this summary also uses the sample data from Section 5.3.1. Also, it selects the five variables of primary airflow (y_6~y_7), secondary airflow (y_{12}~y_{13}), and flue gas oxygen content (y_{20}) as control variables, with the variable optimization range shown in Table 4. After determining the variables to be optimized and the optimization target, the IFDA algorithm is used to continuously search for the optimum value within the optimization interval of these five variables based on the boiler thermal efficiency prediction model until the best value is found. Throughout the optimization procedure, the optimal solution is identified by maximizing the thermal efficiency of the coal-fired boiler. Subsequently, the optimal values of the controllable variables are derived based on this solution. The algorithm's optimization process is illustrated in Figure 10, while the corresponding results are summarized in Table 6.

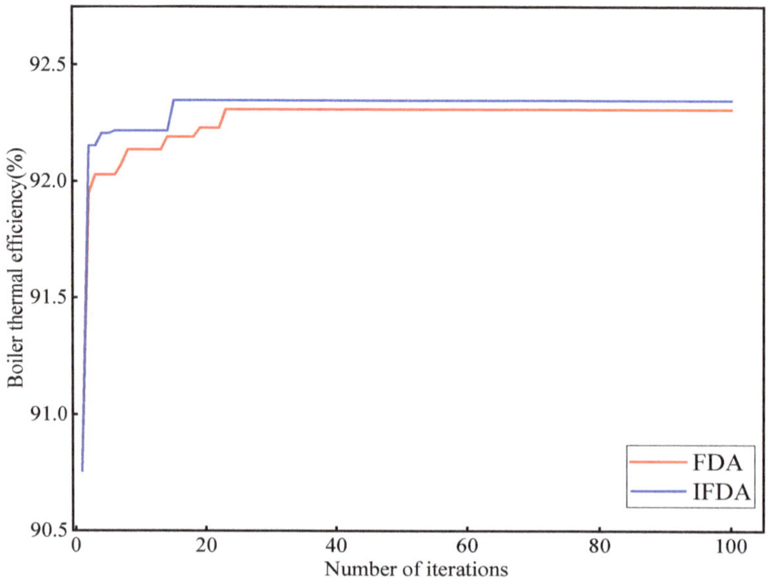

Figure 10. Thermal efficiency optimization process for coal-fired boilers.

Table 6. Optimization results of controllable variables for boiler thermal efficiency.

Variables	Primary Airflow		Secondary Airflow		Flue Gas Oxygen Content
	y_6	y_7	y_{12}	y_{13}	y_{20}
Observations	866.15	864.27	540.70	947.79	1.06
FDA	896.72	868.02	560.38	952.09	1.31
IFDA	921.51	980.05	596.79	977.91	2.09

As illustrated in Table 6 and Figure 10, the blue and red curves correspond to the IFDA and FDA optimization processes for the boiler thermal efficiency, respectively. Following optimization of the boiler's controllable parameters by the FDA algorithm, the thermal efficiency increased from 90.75% to 92.31%. With the improved IFDA algorithm, the thermal efficiency similarly rose from 90.75% to 92.31%. These results further confirm that the IFDA optimization algorithm is more effective in identifying optimal parameter values to enhance boiler thermal efficiency.

5.3.3. Multi-Objective Optimization for Boiler Combustion

The main objective of coal-fired boiler combustion optimization is to minimize NOx emissions while increasing the boiler efficiency as much as possible. Therefore, this study uses the multi-objective optimization algorithm NSGA-II to find the optimal solution for the controllable parameter variables within the optimization range. The optimal parameter solution is obtained when the algorithm is minimally adaptive. Through this optimal parameter, the operation process of the coal-fired boiler is controlled, and finally, the NOx emission concentration and boiler efficiency satisfying the conditions are obtained.

The multi-objective model constructed in this study is shown in Equation (19).

$$\begin{cases} Min \;\; 100 - fun_1(y) \\ Min \;\; fun_2(y) \\ s.t. \\ y = [y_1, y_1, \cdots, y_m, \cdots, y_{13}] \\ a \leq y_m \leq b \end{cases} \tag{19}$$

where $100 - fun_1(y)$ is the boiler thermal efficiency loss value and $fun_2(y)$ is the NOx emission concentration.

Among the Pareto optimal solutions, the middle point is selected as the recommended solution, balancing NOx emission and thermal efficiency. This approach assumes equal preference for emission reduction and energy performance, aligning with practical decision-making scenarios in power plants. Choosing endpoints on the Pareto front, in contrast, would prioritize either strict emission control or maximum efficiency, potentially sacrificing the other objective.

The range for the controllable parameter variables used in this study is the same as when single-objective optimization was carried out. To verify the effectiveness of the NSGA-II algorithm, three samples—with low NOx emission concentration, medium NOx emission concentration, and high NOx emission concentration—were selected as experimental control groups in the original data, named Sample A, Sample B, and Sample C, respectively. The optimized results are shown in Figure 11. The two endpoints and the middle point of the Pareto optimal front for samples A, B, and C were selected for analysis, and the specific results are shown in Table 7.

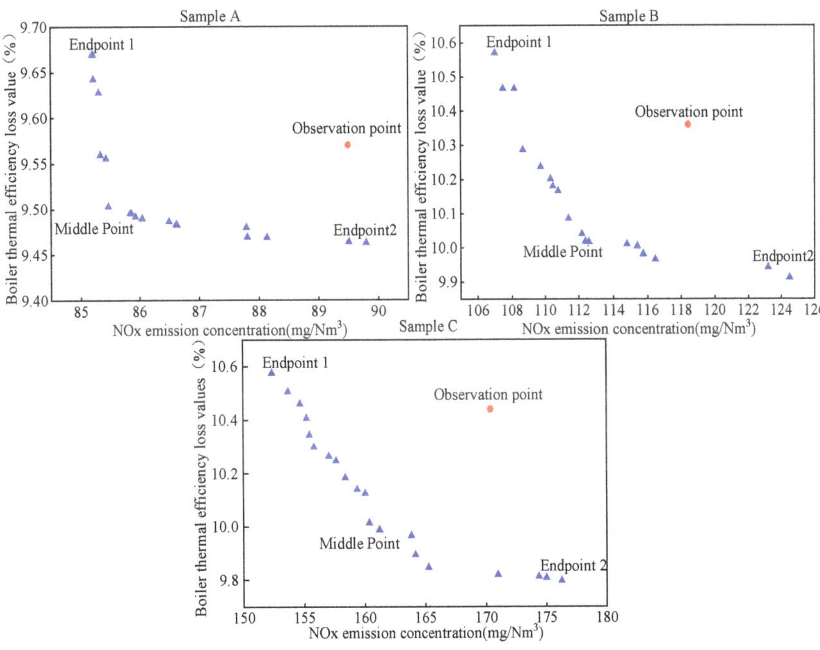

Figure 11. Pareto optimal front of the different samples.

Following the application of the NSGA-II algorithm for optimization, significant reductions in NOx emission concentrations for the coal-fired boiler are evident when comparing the data in Figure 11 and Table 7 with pre-optimization results. Concurrently, an improvement in boiler thermal efficiency is also observed. Analysis of endpoints 1 and 2 from the three sample datasets reveals a trade-off: when NOx emissions reach their minimum, boiler combustion efficiency declines relative to observed values, whereas the highest combustion efficiency corresponds to an increase in NOx emissions compared with observations. This indicates that single-objective optimization can only satisfy one goal at a time. To overcome this limitation, a multi-objective optimization algorithm is employed to generate the Pareto optimal front, from which a recommended intermediate solution is selected.

Table 7. Optimization results of the samples.

Sample	Optimization Objectives	NOx Emission Concentration (mg/Nm^3)	Thermal Efficiency (%)
Sample A	Observations	89.49	90.43
	Endpoints 1	85.19	90.34
	Endpoints 2	89.79	90.54
	Middle Point	85.46	90.50
Sample B	Observations	118.43	89.64
	Endpoints 1	107.01	89.43
	Endpoints 2	124.50	90.09
	Middle Point	112.37	89.98
Sample C	Observations	170.36	89.56
	Endpoints 1	152.33	89.42
	Endpoints 2	176.27	90.20
	Middle Point	161.17	90.01

The middle point of the Pareto optimal front, selected from samples A, B, and C, is chosen as the recommended point for analysis. The original data for sample A shows a NOx emission concentration of 89.49 mg/Nm^3 and a thermal efficiency of 90.43%. After optimization, the values obtained are 85.46 mg/Nm^3 for the NOx emission concentration and 90.50% for thermal efficiency. This represents a 4.5% reduction in the NOx emission concentration and a 0.07% increase in boiler thermal efficiency. For sample B, the original data shows a NOx emission concentration of 118.43 mg/Nm^3 and a thermal efficiency of 89.64%. After optimization, the values obtained are 112.37 mg/Nm^3 for the NOx emission concentration and 89.98% for thermal efficiency. This represents a 5.12% reduction in the NOx emission concentration and a 0.38% increase in boiler thermal efficiency. For sample C, the original data shows a NOx emission concentration of 170.36 mg/Nm^3 and a thermal efficiency of 89.64%. After optimization, the values obtained are 161.17 mg/Nm^3 for the NOx emission concentration and 90.01% for thermal efficiency. This represents a 5.40% reduction in the NOx emission concentration and a 0.50% increase in boiler thermal efficiency. By calculating the average of the middle points of the three samples, it is determined that the average reduction in NOx emission concentration is 5.01%. In contrast, the average increase in boiler thermal efficiency is 0.32%. These results demonstrate that the NSGA-II algorithm can effectively optimize and control combustion in boilers.

6. Conclusions

This study develops a predictive model for boiler combustion systems by employing a hybrid deep learning approach to accurately forecast NOx emission concentrations and boiler thermal efficiency. Following this, the NSGA-II multi-objective optimization algorithm is applied to optimize the controllable parameters within specified ranges, identifying the optimal values and deriving the Pareto optimal front. The proposed method demonstrates superior predictive accuracy and robust generalization capabilities, successfully achieving NOx emission levels and boiler efficiencies that satisfy the targeted criteria. The key contributions of this research are summarized as follows:

(1) A prediction model for the boiler combustion system was developed based on the operating data generated by a 300 MW boiler, which was used to predict the NOx emission concentration and thermal efficiency of the boiler. For the CNN-IFDA-BiGRU model, the MAPE for predicting the NOx emission concentration is 0.78%, while the MAPE for predicting the boiler thermal efficiency is 0.012%.
(2) The boiler combustion system utilizes the FDA and IFDA for single-objective optimization of the boiler thermal efficiency and NOx emission concentration, respectively. Simulation results indicate that the IFDA possesses superior optimization capabilities, effectively achieving the optimization goals of reducing the NOx emission concentration and improving the thermal efficiency of coal-fired boilers separately and efficiently.
(3) The NSGA-II algorithm was employed to optimize the boiler combustion process through an analysis of simulation results. The primary goals were to reduce NOx emission concentrations, improve boiler thermal efficiency, and achieve effective control over the boiler's combustion system. Utilizing operational data from coal-fired boilers, the proposed approach generated a Pareto optimal front tailored to specific objectives. The findings provide valuable insights for promoting the low-carbon and environmentally sustainable operation of coal-fired boilers. The effectiveness of the NSGA-II algorithm was validated by comparing NOx emission concentrations and boiler thermal efficiencies across various samples before and after optimization, supported by detailed simulation analyses.

Author Contributions: C.H.: Conceptualization, Data curation, Methodology. Y.Z.: Methodology, Investigation, Software, Writing—original draft. H.Z.: Writing—review and editing. J.Z.: Software, Visualization. Y.F.: Visualization, Writing—review and editing. Z.T.: Writing—review and editing. C.Z.: Conceptualization, Writing—review and editing. T.P. Conceptualization, Methodology. All authors have read and agreed to the published version of the manuscript.

Funding: This work is supported by the National Natural Science Foundation of China (NSFC) (No. 62303191 and No. 62306123), the Post-graduate Research & Practice Innovation Program of Jiangsu Province (SJCX25_2187), the Postgraduate Science & Technology Innovation Program of Huaiyin Institute of Technology (HGYK202509), the Huai'an Science and Technology Project-Frontier Technology R&D Grant (HAG202410), the Natural Science Foundation of the Jiangsu Higher Education Institutions of China (No. 23KJD480001), and the Double-innovation Doctor Program of Jiangsu province (No. JSSCBS20201033 and No. JSSCBS20201037).

Data Availability Statement: The data presented in this study are available on request from the corresponding author.

Acknowledgments: Special thanks are given to the "Qinglan Project" and "333 project" of Jiangsu Province.

Conflicts of Interest: Authors Chen Huang, Hui Zhao and Jianchao Zhu were employed by the company Huaneng Huaiyin No. 2 Power Generation Co., Ltd. The remaining authors declare that the research was conducted in the absence of any commercial or financial relationships that could be construed as a potential conflict of interest.

Abbreviations

ICGO	Improved Chaos Game Optimization	CGO	Chaos Game Optimization
BiGRU	Bidirectional Gated Recurrent Unit	PSO	Particle Swarm Optimization
NO	Nitric Oxide	NOx	Nitrogen Oxides
ML	Machine Learning	ACO	Ant Colony Optimization Algorithm
ELM	Extreme Learning Machines	IGOA	Improved Grasshopper Optimization Algorithm
SAGA	Simulated Annealing Genetic Algorithm	NSGA-II	Non-dominated Sorting Genetic Algorithm II
SVM	Support Vector Machine	MOPSO	Multi-objective Particle Swarm Optimization
NO_2	Nitrogen Dioxide	MODE	Multi-objective Differential Evolution
RMSE	Root Mean Square Error	ILS	Iterative Local Search
CNN	Convolutional Neural Network	MAE	Root Mean Absolute Error
ANNSVR	Support Vector Regression Artificial Neural Network	MAPE	Mean Absolute Percentage Error
LSTM	Long Short-Term Memory	EMD	Empirical Mode Decomposition
RNN	Recurrent Neural Network	R	Correlation Coefficient
CEEMDAN	Complete Ensemble Empirical Mode Decomposition with Adaptive Noise	LSH	Locality Sensitive Hashing
SPS	Steam Power System	FDA	Flow Direction Algorithm

References

1. Yang, G.; Wang, Y.; Li, X. Prediction of the NOx emissions from thermal power plant using long-short term memory neural network. *Energy* **2020**, *192*, 116597. [CrossRef]
2. Resitoglu, I.A.; Keskin, A.; Karaman, B.; Ozarslan, H. Suppressing Calcium Deactivation in Selective Catalytic Reduction of NO_x from Diesel Engines Using Antimony. *Processes* **2025**, *13*, 1914. [CrossRef]
3. Ren, M.; Liang, Y.; Chen, J.; Xu, X.; Cheng, L. Fault detection for NOx emission process in thermal power plants using SIP-PCA. *ISA Trans.* **2023**, *140*, 46–54. [CrossRef] [PubMed]
4. Ke, X.; Zhu, S.; Huang, Z.; Zhang, M.; Lyu, J.; Yang, H.; Zhou, T. Issues in deep peak regulation for circulating fluidized bed combustion: Variation of NOx emissions with boiler load. *Environ. Pollut.* **2023**, *318*, 120912. [CrossRef] [PubMed]
5. Bian, C.; Huang, J.; Sun, R. Numerical optimization of combustion and NOX emission in a retrofitted 600MWe tangentially-fired boiler using lignite. *Appl. Therm. Eng.* **2023**, *226*, 120228. [CrossRef]
6. Li, S.; Xu, T.; Hui, S.; Wei, X. NOx emission and thermal efficiency of a 300MWe utility boiler retrofitted by air staging. *Appl. Energy* **2009**, *86*, 1797–1803. [CrossRef]
7. Wen, X.; Li, K.; Wang, J. NOx emission predicting for coal-fired boilers based on ensemble learning methods and optimized base learners. *Energy* **2023**, *264*, 126171. [CrossRef]
8. Qiao, J. A novel online modeling for NOx generation prediction in coal-fired boiler. *Sci. Total Environ.* **2022**, *847*, 157542. [CrossRef] [PubMed]
9. Liu, X.; Bansal, R.C. Integrating multi-objective optimization with computational fluid dynamics to optimize boiler combustion process of a coal fired power plant. *Appl. Energy* **2014**, *130*, 658–669. [CrossRef]
10. Li, G.; Niu, P.; Ma, Y.; Wang, H.; Zhang, W. Tuning extreme learning machine by an improved artificial bee colony to model and optimize the boiler efficiency. *Knowl.-Based Syst.* **2014**, *67*, 278–289. [CrossRef]
11. Tan, P.; He, B.; Zhang, C.; Rao, D.; Li, S.; Fang, Q.; Chen, G. Dynamic modeling of NOx emission in a 660 MW coal-fired boiler with long short-term memory. *Energy* **2019**, *176*, 429–436. [CrossRef]
12. Liu, X.; Qi, H.; Ren, M.; Cheng, L. A prediction method of NOx in thermal power plants using GC-LSTM neural network. In Proceedings of the 2020 Chinese Automation Congress (CAC), Shanghai, China, 6–8 November 2020. [CrossRef]
13. Wang, X.; Liu, W.; Wang, Y.; Yang, G. A hybrid NOx emission prediction model based on CEEMDAN and AM-LSTM. *Fuel* **2022**, *310*, 122486. [CrossRef]
14. Wang, Z.; Peng, X.; Cao, S.; Zhou, H.; Fan, S.; Li, K.; Huang, W. NOx emission prediction using a lightweight convolutional neural network for cleaner production in a down-fired boiler. *J. Clean. Prod.* **2023**, *389*, 136060. [CrossRef]

15. Ma, H.; Peng, T.; Zhang, C.; Ji, C.; Li, Y.; Nazir, M.S. Developing an evolutionary deep learning framework with random forest feature selection and improved flow direction algorithm for NOx concentration prediction. *Eng. Appl. Artif. Intell.* **2023**, *123*, 106367. [CrossRef]
16. Li, N.; Lv, Y.; Hu, Y. Prediction of NOx Emissions from a Coal-Fired Boiler Based on Convolutional Neural Networks with a Channel Attention Mechanism. *Energies* **2023**, *16*, 76. [CrossRef]
17. Wan, A.P.; Yang, J.; Miao, X. Boiler Load Forecasting of CHP Plant Based on Attention Mechanism and Deep Neural Network. *J. Shanghai Jiao Tong Univ.* **2023**, *57*, 316–325. [CrossRef]
18. Deb, K.; Agrawal, S.; Pratap, A.; Meyarivan, T. A Fast Elitist Non-dominated Sorting Genetic Algorithm for Multi-objective Optimization: NSGA-II. In Proceedings of the International Conference on Parallel Problem Solving from Nature, Paris, France, 18–20 September 2000. [CrossRef]
19. Wei, Z.; Li, X.; Xu, L.; Cheng, Y. Comparative study of computational intelligence approaches for NOx reduction of coal-fired boiler. *Energy* **2013**, *55*, 683–692. [CrossRef]
20. Dirik, M. Prediction of NOx emissions from gas turbines of a combined cycle power plant using an ANFIS model optimized by GA. *Fuel* **2022**, *321*, 124037. [CrossRef]
21. Zhou, H.; Pei Zhao, J.; Gang Zheng, L.; Lin Wang, C.; Fa Cen, K. Modeling NOx emissions from coal-fired utility boilers using support vector regression with ant colony optimization. *Eng. Appl. Artif. Intell.* **2012**, *25*, 147–158. [CrossRef]
22. Song, J.; Romero, C.E.; Yao, Z.; He, B. Improved artificial bee colony-based optimization of boiler combustion considering NOX emissions, heat rate and fly ash recycling for on-line applications. *Fuel* **2016**, *172*, 20–28. [CrossRef]
23. Yao, Z.; Romero, C.; Baltrusaitis, J. Combustion optimization of a coal-fired power plant boiler using artificial intelligence neural networks. *Fuel* **2023**, *344*, 128145. [CrossRef]
24. Wei, L.; Yan, F.; Hu, J.; Xi, G.; Liu, B.; Zeng, J. Nox conversion efficiency optimization based on NSGA-II and state-feedback nonlinear model predictive control of selective catalytic reduction system in diesel engine. *Appl. Energy* **2017**, *206*, 959–971. [CrossRef]
25. Xu, W.; Huang, Y.; Song, S.; Chen, B.; Qi, X. A novel online combustion optimization method for boiler combining dynamic modeling, multi-objective optimization and improved case-based reasoning. *Fuel* **2023**, *337*, 126854. [CrossRef]
26. Zheng, W.; Wang, C.; Liu, D. Data-driven based multi-objective combustion optimization covering static and dynamic states. *Expert Syst. Appl.* **2022**, *210*, 118531. [CrossRef]
27. Xiao, W.; Cheng, A.; Li, S.; Jiang, X.; Ruan, X.; He, G. A multi-objective optimization strategy of steam power system to achieve standard emission and optimal economic by NSGA-II. *Energy* **2021**, *232*, 120953. [CrossRef]
28. Tang, W.J.; Wang, H.T.; Liu, P.J.; Qian, F.L. Combustion Optimization Under Deep Peak Shaving Based on DYNA-A3C With Dynamic Weight. *Front. Energy Res.* **2023**, *10*, 953387. [CrossRef]
29. Karami, H.; Anaraki, M.V.; Farzin, S.; Mirjalili, S. Flow Direction Algorithm (FDA): A Novel Optimization Approach for Solving Optimization Problems. *Comput. Ind. Eng.* **2021**, *156*, 107224. [CrossRef]
30. Chen, H.; Li, W.; Yang, X. A whale optimization algorithm with chaos mechanism based on quasi-opposition for global optimization problems. *Expert Syst. Appl.* **2020**, *158*, 113612. [CrossRef]
31. Gan, C.; Cao, W.; Wu, M.; Chen, X. A new bat algorithm based on iterative local search and stochastic inertia weight. *Expert Syst. Appl.* **2018**, *104*, 202–212. [CrossRef]
32. Kalyanmoy, D. A fast and elitist multi-objective genetic algorithm: NSGA-II. *IEEE Trans. Evol. Comput.* **2002**, *6*, 182–197.
33. Toksari, M.D. A hybrid algorithm of Ant Colony Optimization (ACO) and Iterated Local Search (ILS) for estimating electricity domestic consumption: Case of Turkey. *Int. J. Electr. Power Energy Syst.* **2016**, *78*, 776–782. [CrossRef]
34. Zhang, Q.; Yi, H.; Yu, Z.; Gao, J.; Wang, X.; Lin, H.; Shen, B. Energy-exergy analysis and energy efficiency improvement of coal-fired industrial boilers based on thermal test data. *Appl. Therm. Eng.* **2018**, *144*, 614–627. [CrossRef]

processes

Article

Design and Simulation of a Feedback Controller for an Active Suspension System: A Simplified Approach

Vasileios Provatas and Dimitris Ipsakis *

School of Production Engineering and Management, Technical University of Crete, 73100 Chania, Greece; vasilisprovatas1998@gmail.com
* Correspondence: dipsakis@tuc.gr; Tel.: +30-28210-37362

Abstract: The concept of controlling vehicle comfort is a common problem that is faced in most under- and postgraduate courses in Engineering Schools. The aim of this study is to provide a simplified approach for the feedback control design and simulation of active suspension systems, which are applied in vehicles. Firstly, the mathematical model of an active suspension system (a quarter model of a car) which consists of a passive spring, a passive damper and an actuator is provided. In this study, we chose to design and compare the following controllers: (a) conventional P, PI and PID controllers that were tuned through two conventional methodologies (Ziegler–Nichols and Tyreus–Luyben); (b) an optimal PID controller that was tuned with a genetic algorithm (GA) optimization framework in terms of the minimization of certain performance criteria and (c) an internal model controller (IMC) based on the process transfer function. The controllers' performance was assessed in a series of realistic scenarios that included set-point tracking with and without disturbances. In all cases, the IMC controller and the optimal PID showed superior performance. On the other hand, the P and PI controllers showed a rather insufficient behavior that involved persistent errors, overshoots and eventually, uncomfortable ride oscillations. Clearly, a step-by-step approach such as this, that includes modeling, control design and simulation scenarios can be applied to numerous other engineering examples, which we envisage to lead more students into the area of automatic control.

Keywords: active car suspension; PID controller; model-based controller; modeling and simulation; tuning methods; controller performance criteria; optimal PID

1. Introduction

One of the modern applications of automatic control technology in mobile applications is vehicle control (either with or without a driver) that ensures road-handling ability and ride comfort. One of these fascinating applications is the cruise control system that adjusts the car's speed based on data received by a set of sensors and is implemented with advanced control techniques [1–3]. Another interesting area is lane-free control systems that use sensors and cameras to steer vehicles and retain them inside their lane without any action from the driver. Furthermore, the blind-spot-monitoring control system can detect when a car or a random object is out of sight. The latter can be combined with anti-collision warning control systems that warn the driver if such an obstacle is detected within the vehicle direction. Clearly, the above applications have received significant attention from researchers [4,5]. Still, however, the suspension control system, which is responsible for controlling and adjusting the ride comfort with an aim of reducing oscillations, is always at the forefront of research and lectures in engineering schools (mainly in undergraduate studies).

The core aim of the suspension system (see Figure 1) is to isolate the vehicle's cabin from external road disturbances (e.g., it separates the car body from the wheels) in order to maximize the driving comfort for the passengers and ensure car stability. Another important role of the suspension system is to keep the wheel in contact with the road surface, as vibrations may be harmful to the passengers and the vehicle itself. These requirements for improved driving comfort and passenger safety have pushed car manufacturers to develop new active suspension systems. This type of suspension has the ability to adapt to harsh road surface conditions and is realized through the ability to store, dissipate and utilize energy within the system. Compared with the passive suspension (open loop without a control action), the active suspension performs better due to the force/power actuator. The actuator is a mechanical control system, which is mounted on the suspension, and can be pneumatic, hydraulic, electromagnetic, etc. The controller collects data from various sensors mounted on the vehicle that can analyze the road profile before any control action.

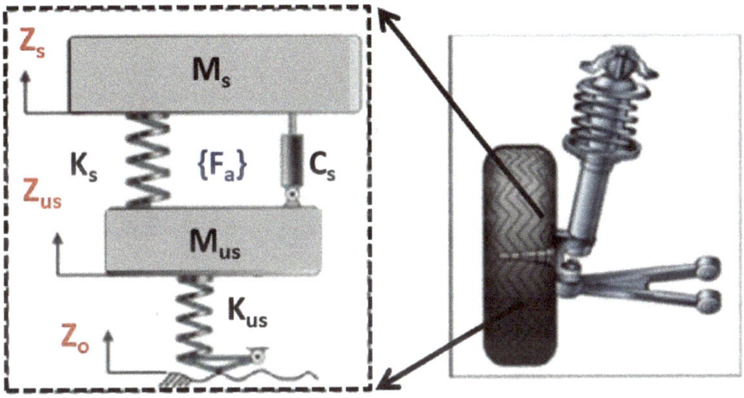

Figure 1. A quarter-car model of the suspension system.

The design and simulation of a car suspension can be found in various forms in the literature. Most studies take a mathematical model for the active suspension into account and simulate it along with a (properly tuned) PID controller. In most case studies, simulations using a hydraulic actuator and tested in various road surface scenarios (uneven road, potholes and random road entrances) prove to be an established line of work. Such results have proven that the active suspension system with PID control improves driving comfort [6]. In [7], a significant effort was made to reduce the motion of a rail car body and the damper. The PID controller was tuned through the Ziegler–Nichols methodology, and the simulated results between the passive and the active suspension system proved that the active suspension system with PID control improves ride comfort, gives a low peak overshoot (amplitude) and a faster settling time. Another group of studies discussed the performance of a PID controller (tuned with the Ziegler–Nichols method or with fuzzy methods) in a semi-active suspension of a vehicle. Again, the results proved that the active suspension system significantly improves driving comfort [8–10]. An interesting approach was reported in a study by A. Al-Zughaibi and H. Davies [11]. The controller was designed using the Routh–Hurwitz stability criterion and the respective simulations were carried out in a system with unknown time delays, different masses and road disturbance. The simulation results showed a significant improvement in performance and disturbance rejection. A quite interesting study using a combination of commercial software (used for extracting vehicle responses) and control design investigated the performance of a K-seat-based PID that is able to enhance motion comfort [12]. Along the same trend, J. Zhu et al. [13] worked on the nonlinear modeling and control of emergency rescue vehicles (e.g., active suspension systems in fire trucks) that lack body stability. Their disturbance rejection control design provided a complementary perspective to the results derived from feedback control implementation.

Moving on to modern control techniques, a neural-network-based model predictive controller (NN-MPC), designed for a non-linear servo-hydraulic vehicle suspension system, was compared to a PID controller. The analysis revealed the superior performance of the NN-MPC over the PID, as it showed a better tracking of the set-point, despite the disturbances that emerged [14]. J. Kim et al. [15] developed an MPC algorithm for semi-active suspension systems with road preview that managed to improve both the ride comfort and the road-handling performance. In particular, the damping force (input variable) was minimized towards achieving predefined trajectories. A detailed mathematical methodology based on multi-agent communication topology and MPC was presented by N. Zhang et al. [16] with an aim to minimize the vertical acceleration, pitch acceleration and roll acceleration of the vehicle body. Another meaningful case study dealt with the development of controllers in motors [17]. The control system was realized and tuned by an artificial intelligence strategy that is based on heuristic optimization techniques (particle swarm optimization, PSO). At first, the development of the mathematical model of the DC motor was carried out. Then, a feedback control system was applied under the adaptive PSO and the conventional Ziegler–Nichols method. The simulation and comparison of the above controller with the PSO and Ziegler–Nichols tuning methods showed that the PSO algorithm can be effectively used to tune the PID controller. In [18], three tuning methods were used for a PID controller, (heuristic tuning, Ziegler–Nichols tuning and the iterative learning algorithm). Afterwards, the simulation and comparison of the above-mentioned tuning methods, as well as of the comparison between the passive and active suspension system, were carried out for scenarios of three different road disturbances (bump and hole, sinusoidal and random input). The outcome of the study is that the active suspension system performs better than the passive suspension system, whereas the PID controller that was tuned by the iterative learning algorithm performed better than the other two tuning methods. In a more detailed study regarding the development of a full vehicle model, an adaptive PID control was designed [19]. Specifically, this study is worth mentioning because it developed an adaptive controller for active suspension systems with parameter uncertainties. Their aim was to improve the transient response of the body and to save the communication resources of the in-vehicle network. An interesting study that tried to address problems relating to gain estimation was presented by Y. Jeong et al. [20]. This group designed a linear-quadratic (LQ) static output feedback with a 2-DOF quarter-car model for an active suspension system. This type of control uses available sensor signals measured in real vehicles and formulates an optimization problem that can be solved via a heuristic optimization method. It was concluded that the proposed controller showed the best performance in terms of ride comfort.

Several other studies can also be found in the literature and include control approaches such as the linear quadratic regulator [21–23], adaptive sliding control [24], H∞ control [25,26], advanced fuzzy control [27,28], optimal control [15,29] and neural-network-based methods for tuning [30,31].

Aside from vehicles, specifically those in train suspension systems, the study by X. He et al. in [32] studied the riding comfort of high-speed trains (which affect the travel experience of passengers) and aimed at reducing the vibrations through the application of a vertical dynamics model of railway vehicles. Furthermore, a unique study with actual implementation scenarios was reported by K. Ikeda et al. [33]. The purpose of this study was to estimate the psychological state of an occupant based on biometric information. Based on the research findings, the ride quality of occupants while driving can be recorded and control algorithms can be retrofitted in real time.

Since several studies dealing with vehicle suspensions have been carried out in the literature, the focus of this study is not claimed to be unique among such interesting and useful studies. The vision of the present study is to provide a simplified step-by-step approach that can be used as an applied example of modeling and control for engineering students' lectures. The concept of active suspension systems has significant benefits, as it can be comprehensively taught via the following consecutive educational requirements: (a) the set-up of the mathematical model of a quarter of a car (1/4) active suspension in the dual form of differential equations, state-space model and transfer functions, (b) the design and application of a feedback control system based on P, PI and PID controllers, (c) the design and application of a feedback control system based on internal model control (IMC) and (d) the evaluation of the controllers' performance under different road conditions (including disturbances). The above are accompanied by applied tuning techniques such as the methodologies of Ziegler–Nichols and Tyreus–Luyben and optimal tuning through genetic algorithms (GA).

The structure of the paper is as follows: Section 2 presents the mathematical model of the active suspension system. In this section, open loop simulations are used to evaluate the system dynamics under scenarios of varying input force and by applying random disturbances. Next, Section 3 designs the feedback control system that is realized in the form of P, PI and PID controllers (tuned via the Ziegler–Nichols and Tyreus–Luyben methodologies), in the form of an optimal PID controller tuned via a GA optimization framework and in the form of a feedback internal model controller. The comparison of the performance of all controllers is evaluated (Section 4) in different scenarios that include set-point variations with or without the presence of road disturbances.

2. Mathematical Model of the Active Suspension System

2.1. 2nd and 1st (Reduced) Order Differential Equations

Figure 2 shows a typical active suspension system that consists of a passive spring, a passive damper and an actuator. The actuator can be hydraulic, pneumatic or electromagnetic. Through this element, the controller can decide whether to add or dissipate energy with the help of measurements (sensors that could provide road profile data). When the mechanical spring, damper and actuator are connected, a hydraulic actuator can be used and is able to control both the mass of the tire (M_{us}) and the vehicle mass or vehicle body (M_s).

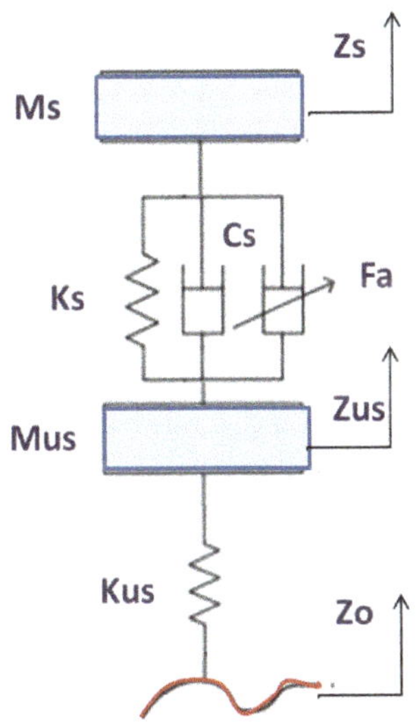

Figure 2. Active suspension system of a (1/4) quarter car.

The equations of motion can be written as follows (In the case of F_a = 0, the system reduces to the open loop passive suspension with no control actions):

$$M_s\frac{d^2Z_s}{dt^2} = F_a(t) - K_s(Z_s(t) - Z_{us}(t)) - C_s(\frac{dZ_s}{dt} - \frac{dZ_{us}}{dt}) \quad (1)$$

$$M_{us}\frac{d^2Z_{us}}{dt^2} = -F_a(t) + K_{us}(Z_o(t) - Z_{us}(t)) + K_s(Z_s(t) - Z_{us}(t)) + C_s(\frac{dZ_s}{dt} - \frac{dZ_{us}}{dt}) \quad (2)$$

where F_a [N] is the control force from the hydraulic actuator, M_s [kg] is the sprung mass, M_{us} [kg] is the unsprung mass, Z_s [m] is the sprung mass position, Z_{us} [m] is the unsprung mass position, Z_o [m] is the road surface anomaly, K_s [N/m] is the suspension spring (stiffness), K_{us} [N/m] is the tire spring element and C_s [N·s/m] is the damping co-efficient of the variable shock absorber.

Due to the fact that Equations (1) and (2) are second-order differential equations, auxiliary variables (x_1, x_2, x_3, x_4) can be used, so that the reduction process to differential equations of a first-order magnitude can be carried out. Based on the above, there are four (4) state variables (x_1, x_2, x_3, x_4) and one (1) output is defined as $y = (x_1 - x_3) = (Z_s - Z_{us})$:

$$\begin{array}{ll} x_1 = Z_s & \Rightarrow \frac{dx_1}{dt} = x_2 \\ x_2 = \frac{dZ_s}{dt} & \frac{dx_2}{dt} = \left(\frac{F_a}{M_s}\right) - \left(\frac{K_s}{M_s}\right)\cdot x_1 + \left(\frac{K_s}{M_s}\right)\cdot x_3 - \left(\frac{C_s}{M_s}\right)\cdot x_2 + \left(\frac{C_s}{M_s}\right)\cdot x_4 \\ x_3 = Z_{us} & \frac{dx_3}{dt} = x_4 \\ x_4 = \frac{dZ_{us}}{dt} & \frac{dx_4}{dt} = -\left(\frac{F_a}{M_{us}}\right) + \left(\frac{K_{us}}{M_{us}}\right)\cdot Z_0 - \left(\frac{K_{us}}{M_{us}}\right)\cdot x_3 + \left(\frac{K_s}{M_{us}}\right)\cdot x_1 - \left(\frac{K_s}{M_{us}}\right)\cdot x_3 + \left(\frac{C_s}{M_{us}}\right)\cdot x_2 - \left(\frac{C_s}{M_{us}}\right)\cdot x_4 \\ \Rightarrow & \\ & y = (x_1 - x_3) \end{array} \quad (3)$$

2.2. State-Space form Equations

From the above set of four (4) differential equations, the desired state-space form can be formulated and used thereafter in the following form $\{x(t_0) = x_0 = 0\}$:

$$\begin{array}{l} \dot{x}(t) = A\cdot x(t) + B\cdot u(t) + E\cdot d(t) \\ y(t) = C\cdot x(t) + D\cdot u(t) \end{array} \quad (4)$$

The state-space model of matrices A, B, C, D, E can be realized:

$$\begin{bmatrix} \frac{dx_1}{dt} \\ \frac{dx_2}{dt} \\ \frac{dx_3}{dt} \\ \frac{dx_4}{dt} \end{bmatrix} = \begin{bmatrix} 0 & 1 & 0 & 0 \\ -\frac{K_s}{M_s} & -\frac{C_s}{M_s} & \frac{K_s}{M_s} & \frac{C_s}{M_s} \\ 0 & 0 & 0 & 1 \\ \frac{K_s}{M_{us}} & \frac{C_s}{M_{us}} & -\frac{(K_s+K_{us})}{M_{us}} & -\frac{C_s}{M_{us}} \end{bmatrix} \cdot \begin{bmatrix} x_1 \\ x_2 \\ x_3 \\ x_4 \end{bmatrix} + \begin{bmatrix} 0 \\ \frac{1}{M_s} \\ 0 \\ -\frac{1}{M_{us}} \end{bmatrix} \cdot F_a + \begin{bmatrix} 0 \\ 0 \\ 0 \\ \frac{K_{us}}{M_{us}} \end{bmatrix} \cdot Z_o \quad (5a)$$

$$y = [1\,0\, -1\,0]\cdot \begin{bmatrix} x_1 \\ x_2 \\ x_3 \\ x_4 \end{bmatrix} + [0]\cdot F_a \quad (5b)$$

2.3. Mathematical Model in the Form of Transfer Functions

Based on the above, the transfer functions of the process (G_p) and of the disturbance (G_d) can be derived from the following relationships:

$$\begin{array}{l} G_p(s) = \frac{Y(s)}{U(s)} = \frac{Z_s(s)-Z_{us}(s)}{F_a(s)} = C\cdot(sI - A)^{-1}\cdot B + D \\ G_p(s) = \frac{(M_{us}+M_s)\cdot s^2+K_{us}}{(M_s\cdot s^2+C_s\cdot s+K_s)\cdot(M_{us}\cdot s^2+C_s\cdot s+K_s+K_{us})-(C_s\cdot s+Ks)^2} \end{array} \quad (6a)$$

$$\begin{array}{l} G_d(s) = \frac{Y(s)}{D(s)} = \frac{Z_s(s)-Z_{us}(s)}{Z_o(s)} = C\cdot(sI - A)^{-1}\cdot E \\ G_d(s) = \frac{-K_{us}\cdot M_s\cdot s^2}{(M_s\cdot s^2+C_s\cdot s+K_s)\cdot(M_{us}\cdot s^2+C_s\cdot s+K_s+K_{us})-(C_s\cdot s+Ks)^2} \end{array} \quad (6b)$$

$$Y(s) = G_p(s) \cdot U(s) + G_d(s) \cdot D(s) \tag{6c}$$

Before proceeding to the series of open loop simulations, the main parameters of the active suspension need to be defined and are shown in Table 1.

Table 1. Active suspension system parameters.

System Parameters	Values	Units
M_s	243	kg
M_{us}	40	kg
C_s	370	N·s/m
K_s	14,671	N/m
K_{us}	124,660	N·s/m
Initial Conditions	0 at t = 0 s	

2.4. Open Loop Simulations

Open loop simulations show the dynamic performance of the suspension system when no control action is applied. Such simulations highlight the need for a proper control design since oscillations, high overshoots and prolonged settling times are always reported. Certainly, closed loop simulations are more important, but for a simplified approach that is trying to force engineering students to delve into control systems, it is worth devoting a few lines to the open loop performance of the suspension system.

Based on the mathematical model of Section 2 (in either form of the differential equations, state-space and Laplace domain), we performed the following simulation scenarios regarding road disturbances (see also Figure 3a,b regarding scenarios C and D):

- (A) Constant force of F_a = 3500 N with no road disturbance Z_o = 0 m.
- (B) Variable force with no road disturbance Z_o = 0 m:

$$\begin{aligned} F_a &= \{0 \text{ N}, \quad \text{for} \quad 0 < t < 5 \text{ s}\} \\ F_a &= \{3500 \text{ N}, \text{ for} \quad 5 \le t \le 12 \text{ s}\} \\ F_a &= \{2000 \text{ N}, \text{ for} \quad t > 12 \text{ s}\} \end{aligned}$$

- (C) Constant force of F_a = 3500 N and variable (short time bump and hole) road disturbance:

$$\begin{aligned} Z_o &= \{0 \text{ m}, \quad \text{for} \quad 0 < t < 20 \text{ s}\} \\ Z_o &= \{0.04 \text{ m}, \quad \text{for} \quad 20 \le t \le 40 \text{ s}\} \\ Z_o &= \{0.02 \text{ m}, \quad \text{for} \quad 40 < t < 60 \text{ s}\} \\ Z_o &= \{-0.02 \text{ m}, \text{ for} \quad 60 \le t \le 75 \text{ s}\} \\ Z_o &= \{-0.05 \text{ m}, \text{ for} \quad t > 75 \text{ s}\} \end{aligned}$$

- (D) Constant force of F_a = 3500 N and random (continuous bump and hole) road disturbance: Z_o = {random variation for 0 < t < 100 s}

Figure 4a–d show the displacement deviation (with the output variable of our model as $y = x_1 - x_3 = Z_s - Z_{us}$) under the four (4) different and arbitrarily selected scenarios A to D. As can be seen in Figure 4a, by applying a constant force equal to 3500 N, the suspension system oscillates for a short period of nearly 8–10 s and stabilizes at a new operation point (0.24 m displacement with a peak overshoot of 0.4 m). Similarly, Figure 4b shows the effect of applying a variable force of 0, 2000 and 3500 N. Initially, the system is at the desired peaceful state (no oscillations). At t = 5 s, a 3500 N force is applied for a short period of time (until the 12th second) and then a sudden force reduction at 2000 N is applied. In scenario B, the steady-state displacement is reduced from 0.24 m to 0.135 m, with descending and detrimental (for the vehicle and the passengers) oscillations. Proceeding to scenario C and D in Figure 4c,d, the effects of short- and long-term bump and hole disturbances (under a constant force of 3500 N) are observed, respectively. During the effect of the disturbance, the suspension system oscillates (the higher the disturbance → the higher the oscillation magnitude) and stabilizes at the displacement of 0.24 m in both scenarios. Clearly, scenario D is a worst-case scenario, where it is seen that inside the vehicle, the passengers suffer significantly from road anomalies.

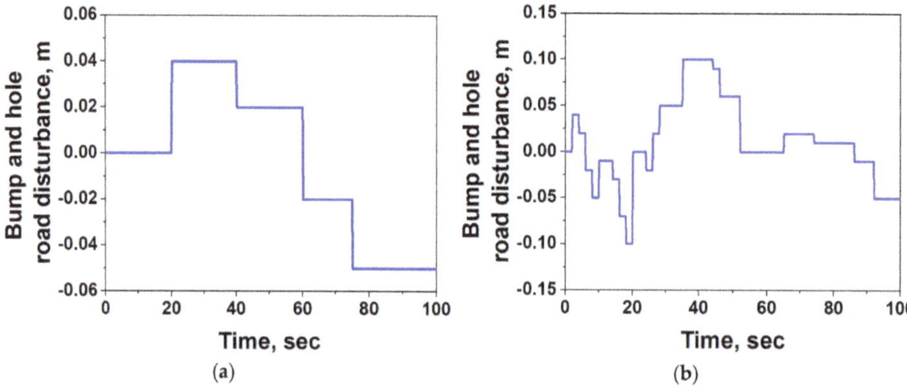

Figure 3. Bump and hole road disturbance (road anomaly) (**a**) short-term and (**b**) long-term.

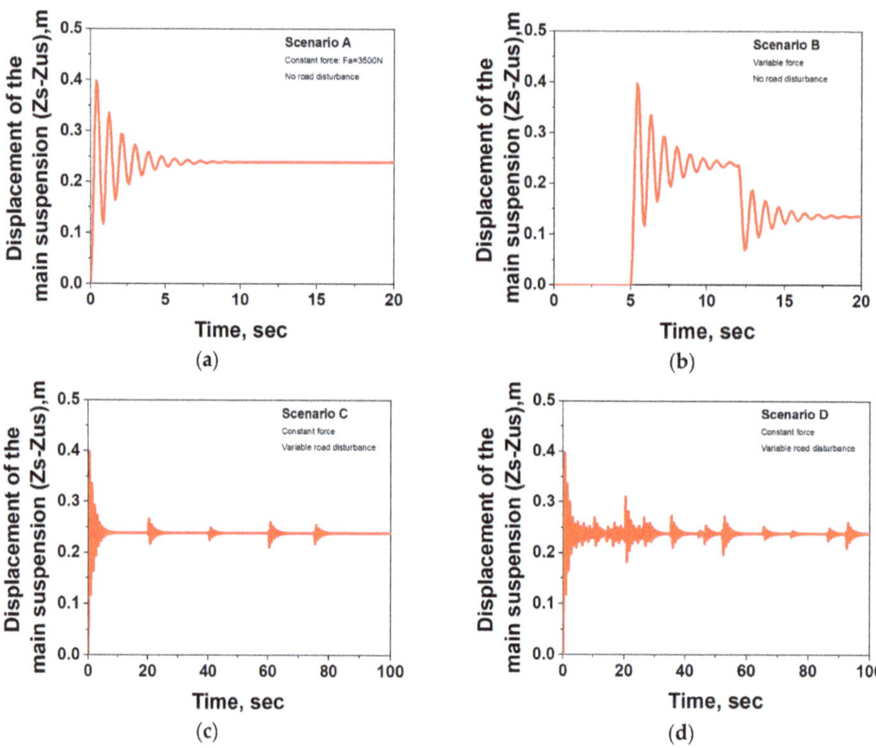

Figure 4. Open loop simulations for scenarios A–D: displacement of the suspension system as $y = Z_s - Z_{us}$ (**a**) constant force without disturbance, (**b**) variable force without disturbance, (**c**) constant force with short-term disturbance and (**d**) constant force with long-term disturbance.

3. Feedback Control System Design

The open loop simulation scenarios presented above show that a control system is needed if we want to provide ride comfort for the vehicle passengers. Its aim should be quick disturbance rejection (e.g., from road anomalies) in a way that will not stress the hydraulic actuator of the feedback controller. With this in mind, this section is devoted to the investigation of the suitability of two types of controllers, (a) conventional P, PI and PID controllers and (b) internal model control, IMC. Figure 5 shows the proposed feedback control system. The transfer functions of each element involved, are shown afterwards in Table 2.

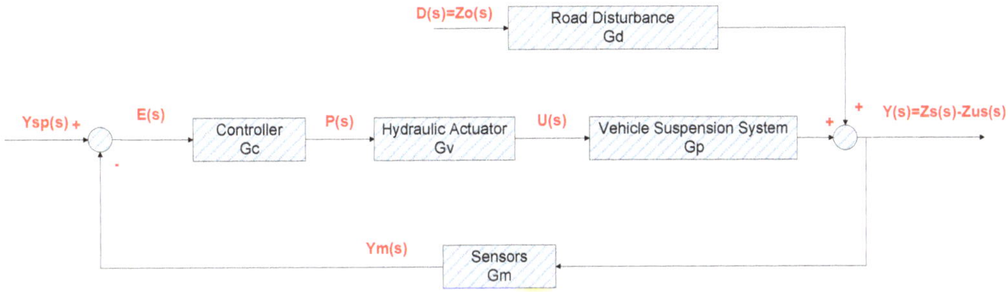

Figure 5. Feedback control for the active suspension system.

Table 2. Closed loop transfer functions (based on Table 1).

Transfer Function	
$G_p(s)$	$\frac{283\cdot s^2+124660}{(243\cdot s^2+370\cdot s+14671)\cdot(40\cdot s^2+370\cdot s+139331)-(370\cdot s+14671)^2}$
$G_d(s)$	$\frac{-30292380\cdot s^2}{(243\cdot s^2+370\cdot s+14671)\cdot(40\cdot s^2+370\cdot s+139331)-(370\cdot s+14671)^2}$
$G_v(s)$	$\frac{1}{10s+1}$
$G_m(s)$	$\frac{1}{30s+1}$
$G_c(s)$	Equations (7) and (10)

3.1. PID Controller

The PID controller takes the following form:

$$\textbf{\textit{PID Controller}}: G_c(s) = K_c\cdot(1+\frac{1}{\tau I\cdot s}+\tau D\cdot s) \tag{7}$$

In order to apply and evaluate a feedback control system, the following steps are implemented and discussed in detail afterwards (see also Figures 6 and 7 for the GA framework):

- **Step 1:** Tuning of the different (P, PI and PID) controllers via the Ziegler–Nichols and Tyreus–Luyben methods and a by using genetic algorithms (PID (OPT)).
- **Step 2:** Implementation of the selected type of controllers (either, P, PI or PID) and fine tuning (only if necessary for the cases using Ziegler–Nichols and Tyreus–Luyben tuning) and verification via realistic simulation scenarios.
- **Step 3:** Selection of the best type of controller to be compared with the IMC afterwards.

The methodologies of Ziegler–Nichols and Tyreus–Luyben tuning have some limitations in terms of applications. Some of these refer to (a) the tuned parameters of a PID are not optimal and require fine tuning prior to implementation, (b) the output of the controller has its own limitations (e.g., a maximum force to be applied) and these limits cannot be surpassed due to mechanical reasons and (c) PID control might end up with saturation issues that can only be solved through anti-wind-up techniques (it is beyond the scope of this study to include their implementation).

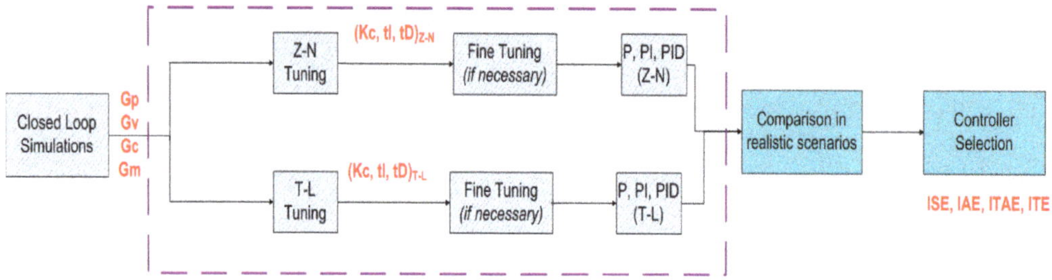

Figure 6. Conventional tuning methodology for PID controllers.

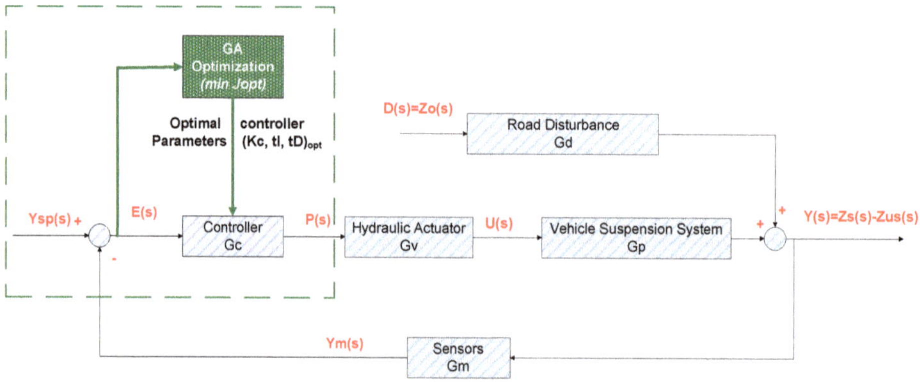

Figure 7. Optimal tuning methodology for PID controllers.

3.2. Optimal Tuning Via Genetic Algorithms

The evaluation of optimal PID controllers is also an interesting topic of research in the literature. M. Kaldas et al. [34] estimated controller parameters by using a gradient-based optimization routine with an objective function that includes both ride comfort and road holding. The ride performance of the road preview control strategy was evaluated through a stochastic road profile and the results highlighted the importance of the road preview control strategy, which significantly improves ride performance. A similar approach was recently published by G. Abbas et al. [35], who used a similar optimization pattern with the integral squared error (ISE) as their objective function. In that particular study, a low overshoot and fast settling times have been reported.

In this study, we selected the MATLAB optimization toolbox for the optimal PID tuning [36]. Specifically, we selected the genetic algorithm (GA) to be our optimization solver since it is an established method for both constrained and unconstrained problems. The algorithm modifies a population of individual solutions. For each step, the GA randomly selects individuals from the current population and uses them as parents to produce the children for the next generation. Over successive generations, the population evolves and the best point in the population approaches an optimal solution [36]. For our problem, the optimal tuning was performed by seeking the optimal solution (equivalently the pairing of the PID controller parameters Kc, τI and τD) that minimizes the performance criterion of the squared error:

$$J_{opt} = \min[\int_{0}^{t} (Y(t) - Y_{sp}(t))^2 dt] \quad (8)$$

where $Y_{sp}(t)$ is the set-point in m and $Y(t)$ is the $Z_s(t) - Z_{us}(t)$ system output variable in m.

Table 3 shows the controller parameters (Kc gain, τI integral time and τD derivative time) for steps 1 and 2. In the cases of Ziegler–Nichols and Tyreus–Luyben tuning, it was observed that all controllers led to a very aggressive control action and we had to reduce Kc by a factor of 100 and increase τI by a factor of 10 (as respective to the critical gain and period that were found through root locus analysis at the values of Kcr = 1.9×10^7 and Pcr = 2.85 s). The implementation and comparison of the above controllers will be compared and discussed in Section 4.

Table 3. Selection of controllers based on the Ziegler-Nichols (Z-N), Tyreus-Luyben (T-L) and optimal (OPT) tuning method.

Controller	Kc	τI	τD
P (Z-N)	98,500	-	-
PI (Z-N)	88,650	23.79	-
PID (Z-N)	118,200	14.27	0.357
PI (T-L)	197,000	62.8	-
PID (T-L)	197,000	62.8	0.453
PID (OPT)	19,461	40	7.39

3.3. IMC Controller

The IMC controller can be easily designed based on the system's Gp function (see Table 2) via the following equation [37]:

$$G_P(s) = G^{-}{}_p(s) \cdot G^{+}{}_p(s) \quad (9)$$

where $G^{-}{}_p(s)$ is the minimum phase and $G^{+}{}_p(s)$ is the non-minimum phase (which will be equal to 1 in our case). In our case that we have the transfer functions of G_c and G_m, the theory allows as to add them (unless it is mentioned otherwise).

The $Gc(s)$ transfer function of the controller can be derived through the following equation:

$$G_c(s) = \frac{1}{G^{-}{}_p(s)} \cdot \frac{1}{(\lambda \cdot s + 1)^r - G^{+}{}_p(s)} \quad (10)$$

where r is the difference between the order of the denominator and the order of the nominator and λ is a time constant that can be defined from the user (we will use $\lambda = 5$ in our case).

3.4. Controller Performance Criteria

In order to evaluate and screen the performance of the controllers, a series of controller performance criteria have been used [37,38]:

Integral Squared Error (ISE)	**Integral time Squared Error (ITSE)**
$ISE = \int_0^\infty e^2(t)dt$	$ITSE = \int_0^\infty t \cdot e^2(t)\, dt$
Integral Absolute Error (IAE)	**Integral Time Absolute Error (ITAE)**
$IAE = \int_0^\infty \lvert e(t) \rvert\, dt$	$IATE = \int_0^\infty t \cdot \lvert e(t) \rvert\, dt$

4. Evaluation and Comparison of Feedback Controllers

Based on the analysis performed in the previous section, the evaluation of all feedback controllers is to be compared under two road scenarios:

(a) Scenario 1: the displacement $Z_s - Z_{us}$ is 0.05 m (set-point) and no disturbances are applied.

(b) Scenario 2: the displacement $Z_s - Z_{us}$ is 0.05 m (set-point) and disturbances are applied as shown in Figure 3b (with the difference that the time period will be prolonged to 300 s).

4.1. Simulation Scenario 1: Set-Point Tracking under No Disturbances

Based on the above two scenarios, the P, PI and PID controllers that were tuned via Z-N and T-L methods, and presented in Table 3, are compared in the following figures. The force actuator represents the exit variable from the hydraulic actuator, U(s), shown in Figure 5. Table 4 summarizes control parameters such as overshoot (%), peak time, rise time, settling time and steady-state error for Scenario 1 (the IMC will be discussed in Section 4.3).

Table 4. Performance criteria for the comparison of controllers during simulation scenario 1.

	Overshoot, %	Peak Time, s	Rise Time, s	Settling Time (±5%), s	Steady-State Error, m
P	198.6	9	1.8	--	0.0062
PI (Z-N)	245.6	10	1.9	90.5	0
PI (T-L)	400	4	0.8	82.5	0
PID (Z-N)	340.4	9	1.3	118.5	0
PID (T-L)	370	6	0.3	59.5	0
PID (OPT)	59	0.5	0.3	4.5	0
IMC	74	16.5	8.5	46	0

As shown in Figure 8a, the P controller (red color) fails to reach the set-point of 0.05 m and after 90–100 s reaches its steady state value at 0.0435 (featuring a persistent steady-state error of 13%). On the other hand, both PI controllers reach the desired set-point at nearly the same settling point of ~80–90 s after a few oscillations. The PI controller tuned via the T-L method shows a more aggressive action since an overshoot of 0.2 m (~400%) occurs, whereas the PI controller tuned via the Z-N method shows a lower, but still high, overshoot of 0.1 m (~245%). Figure 8b shows the respective dynamic performance of the control actuator for the three (3) controllers. As expected, PI tuned via the T-L method shows the most aggressive action, and the P controller the least aggressive.

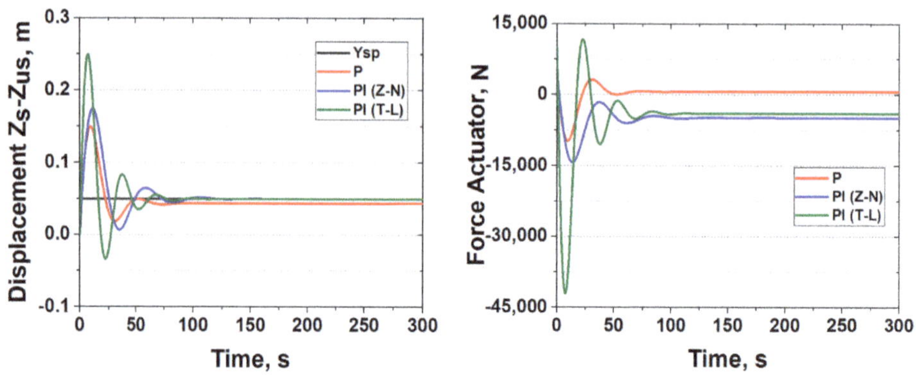

(a) (b)

Figure 8. Comparison of P and PI controllers during scenario 1: (**a**) displacement of the active suspension system and (**b**) force actuator control action.

Following the above analysis, Figure 9a shows the performance of the three (3) PID controllers. As can be seen, the optimal PID controller shows an extremely superior performance since the values of overshoot (59%) and settling time (<5 s) are extremely low. On the other hand, the two PID controllers (Z-N and T-L) present a descending oscillation that is more evident in the PID (Z-N) with a settling time of ~120 s as compared to the lower settling time of the PID (T-L) at ~60 s. Meanwhile, both these controllers indicate a significant overshoot of 0.17–0.18 m (~340–370%). Figure 9b supplements the analysis by depicting the control actuator dynamics for the three PID controllers. As expected, the least action is provided by the optimal PID and the highest by the PID (Z-N) and the PID (T-L).

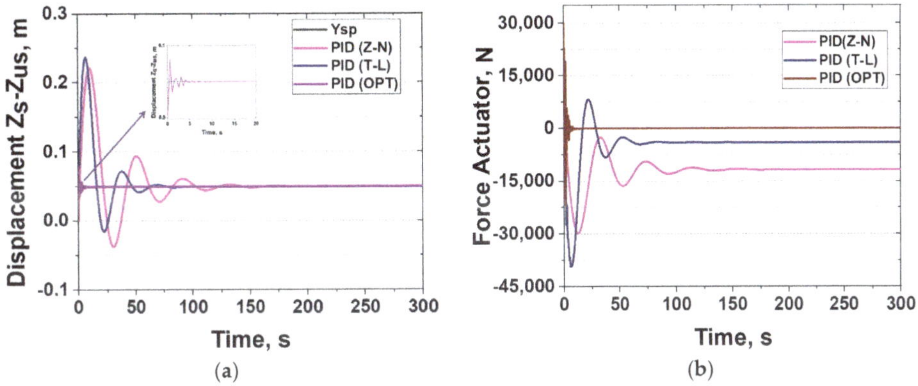

Figure 9. Comparison of PID controllers during scenario 1: (**a**) displacement of the active suspension system and (**b**) force actuator control action.

4.2. Simulation Scenario 2: Set-Point Tracking and Disturbance Rejection

Following the previous analysis, this subsection presents the control simulation of scenario 2. Since we have both a constant set-point tracking and a simultaneous disturbance rejection, we skipped the information regarding overshoot, peak time, rise time, etc., that are more or less similar to scenario 1 and shown in Table 4. As can be seen in Figure 10a, the P controller fails to reach set-point at 0.05 m, and after the effect of disturbances (see Figure 3b for the road anomaly profiles) it reaches its steady state value at 0.0435 (featuring a constant error of 13%). On the other hand, the other two PI controllers reach the desired set-point at nearly the same settling point after oscillating significantly during the disturbance emergence. The PI (T-L) controller shows more aggressive action in all time periods where a bump or hole is encountered, followed by a slightly more comfortable performance by the PI (Z-N). Figure 10b shows the respective dynamic performances of the control actuator for the three (3) controllers and, as expected, the PI (T-L) shows the most aggressive action and the P controller the least aggressive. Overall, the controller performances are similar to the ones in scenario 1, and the spikes refer to the emerged road anomalies.

Figure 11a shows the performance of the three (3) PID controllers. Similar to the previous case in scenario 1, the optimal PID controller shows an extremely superior performance as it exhibits the lowest oscillations (in magnitude and duration) among the three controllers. The two PID controllers (Z-N and T-L) present a more aggressive oscillation which is persistent over time. Among the two, the PID (T-L) shows a better performance in terms of low peaks and settling times. Figure 11b depicts the control actuator dynamics for the three PID controllers. As expected, the least overall action is provided by the optimal PID and the highest by the PID (Z-N). It is worth mentioning that the optimal PID presents a very aggressive action during the imminent disturbance effect, and this needs to be taken into account during implementation.

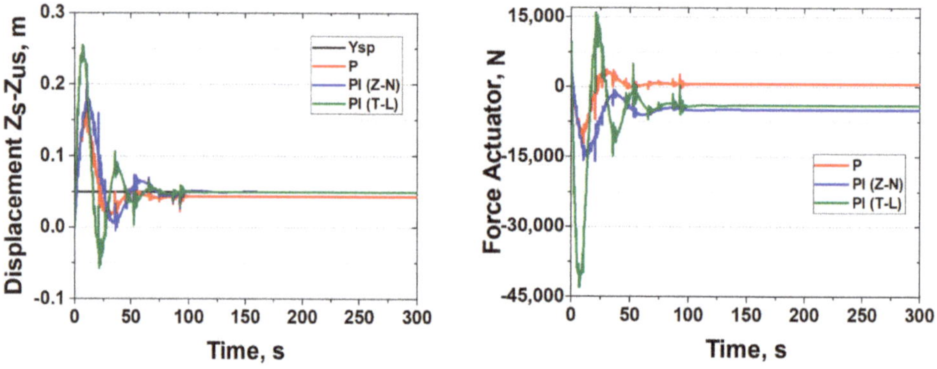

(a) (b)

Figure 10. Comparison of P and PI controllers during scenario 2: (**a**) displacement of the active suspension system and (**b**) force actuator control action.

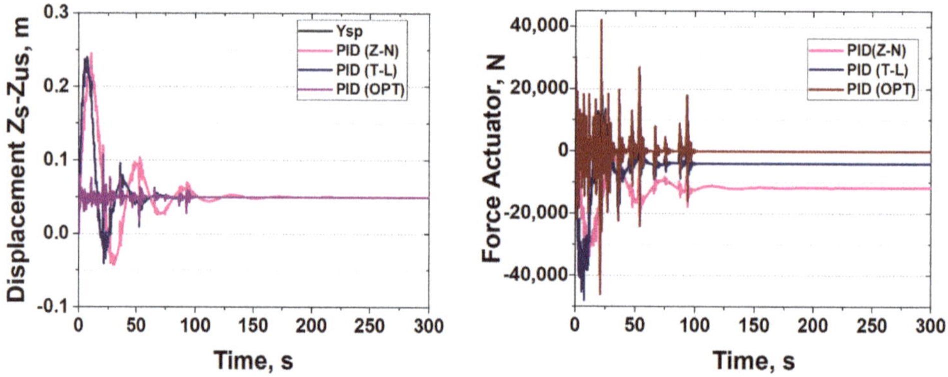

(a) (b)

Figure 11. Comparison of PID controllers during scenario 2: (**a**) displacement of the active suspension system and (**b**) force actuator control action.

4.3. Simulation Scenarios 1 and 2: Comparison of IMC and Optimal PID Controllers

Based on the previous analyses in Sections 4.1 and 4.2, it was revealed that the optimal PID is the best choice among all controllers. Hence, in this section, the IMC controller and the optimal PID will be compared in the same scenario 1 and in a modified scenario 2 (here, the disturbance shown in Figure 3b will be applied, not from t = 0, but from t = 100 s and thereafter). Regarding scenario 1, as shown in Figure 12a and Table 4, the optimal PID controller presents the lowest overshoot (59%) and settling time (4.5 s). The IMC presents an overshoot of 74% (see also Table 4) that last for about 45–50 s (settling time), but in a very smooth way. If we compare the control actions of both controllers in Figure 12b, we observe the minimum effort provided by the IMC (nearly zero action). This result shows that the actuator in the IMC controller design will face the least pressure during road riding and this is considered an asset of this feedback controller.

Regarding the modified scenario 2 shown in Figure 13a, the optimal PID controller presents the lowest overshoot and settling time during the disturbance emergence at t = 100 s and afterwards. The IMC indicates a very smooth performance in the same time period. If we compare the control actions of both controllers in Figure 13b, we can again observe the minimum effort provided by the IMC (nearly zero action). This result shows that the actuator in the IMC controller design will face the least pressure during road riding compared to the optimal PID. Clearly, this should be a concern during implementation, since min/max hard constraints on the actuator could lead to a problematic control action.

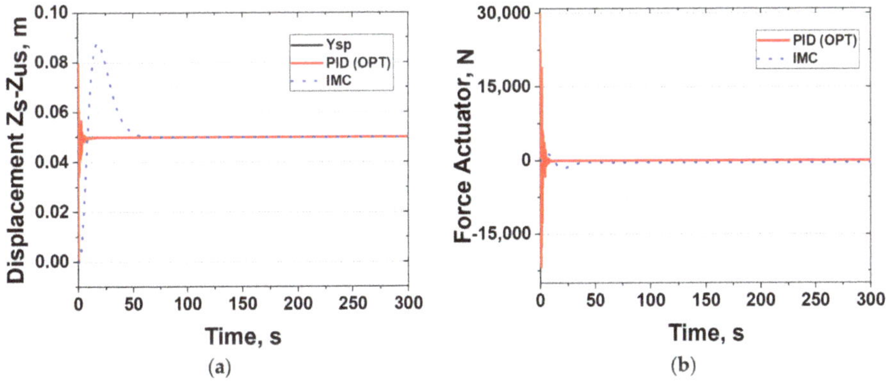

Figure 12. Comparison of optimal PID and IMC controllers during scenario 1: (**a**) displacement of the active suspension system and (**b**) force actuator control action.

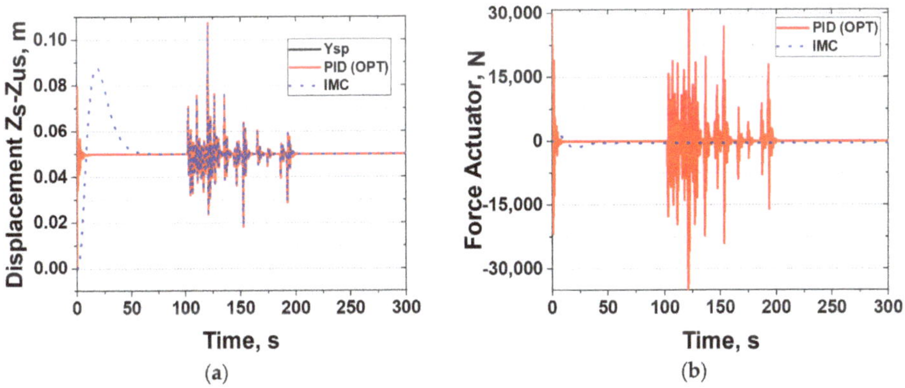

Figure 13. Comparison of optimal PID and IMC controllers during modified scenario 2: (**a**) displacement of the active suspension system and (**b**) force actuator control action.

4.4. Performance Criteria and Critical Analysis

Tables 5 and 6 show the controller performance criteria (see Section 3.4) for both scenarios and for all controllers. The green color represents the two lowest (best) values (which are attributed to the optimal PID and IMC controllers. The optimally tuned PID shows superior performance for scenario 1 that depicts set-point tracking. Similarly, Table 6, which depicts the results of scenario 2 with the inclusion of disturbance rejection, shows that the optimally tuned PID again shows an excellent performance. As can be inferred, the PID controller tuned optimally via genetic algorithms (see Section 3.2 and Figure 7) can be applied in a vehicle active suspension system, provided that the operating and mechanical limits of the actuator can handle the performance as shown earlier (high force is required during control). The latter is crucial, as any violation of the mechanical constraints will lead to disruptive control actions. On the other hand, the IMC controller is also an excellent choice as it exhibits (a) a smooth operation with low overshoot, (b) a peaceful settling to the desired set-point and (c) the lowest effort from the force actuator.

Table 5. Performance control criteria for scenario 1.

Performance Criterion	P	PI (Z-N)	PI (T-L)	PID (Z-N)	PID (Z-N)	PID (OPT)	IMC
ISE	0.239	0.406	0.716	0.780	0.592	0.004	0.065
IAE	6.721	5.692	6.532	8.911	5.550	0.138	2.126
ITSE	6.355	6.116	7.591	13.551	5.386	0.0012	0.916
ITAE	622.37	139.4	121.13	276.1	87.97	0.618	38.39

Table 6. Performance control criteria for scenario 2.

Performance Criterion	P	PI (Z-N)	PI (T-L)	PID (Z-N)	PID (Z-N)	PID (OPT)	IMC
ISE	0.247	0.413	0.720	0.787	0.597	0.0167	0.079
IAE	6.899	5.802	6.633	8.879	5.693	0.967	2.431
ITSE	6.672	6.424	7.895	13.539	5.697	0.397	1.348
ITAE	631.44	147.79	130.12	274.39	99.77	31.01	56.5

To conclude the analysis, it worth highlighting that the presented material is suited to educational courses such as "Fundamentals/Introduction in Control Theory" and includes the following parts that could be taught in a step-by-step process: (a) mathematical modeling of mechanical systems, (b) linear models and transfer functions, (c) PID tuning via conventional and optimization methods, (d) model-based control, (e) performance criteria and (f) evaluation in realistic scenarios.

5. Conclusions

The core aim of this study was the establishment of a feedback control system for an active suspension system. The open loop simulations revealed that the presence of disturbances induces high overshoots and continuous oscillations that are persistent in time. With this indication in mind, the need for a control system design was verified. Next, the evaluation and comparison of three sets of feedback controllers took place: (a) P, PI and PID tuned controllers via the Ziegler–Nichols and Tyreus–Luyben methods, (b) an optimally tuned PID controller via genetic algorithms and (c) an internal model controller. Through a set of simulation scenarios that involved set-point tracking with and without the effect of bumps and holes (road disturbances), it was shown that the optimally tuned PID and IMC controllers exhibit a superior performance in terms of ride comfort and minimum control actions. Among the two, the IMC controller exhibited the lowest (overall) action from the actuator, whereas the optimal PID controller presented the best results for the performance criteria of errors, peak overshoots and settling and rise times, but with extreme mechanical effort from the actuator elements.

Future studies regarding the effective control of active suspension systems should involve the application of advanced methodologies such as model predictive control. This way, the optimal action from the actuator could be applied in order to sustain a predefined trajectory despite emerged disturbances. Furthermore, the application of the most crucial research findings could be implemented using lab hardware equipment, which could involve a similar mechanical structure accompanied by automation utilities.

Author Contributions: All authors contributed equally to the main contributions to this paper. All authors have read and agreed to the published version of the manuscript.

Funding: This research received no external funding.

Data Availability Statement: The data presented in this study are available on request from the corresponding author. The data are not publicly available due to confidentiality.

Conflicts of Interest: The authors declare no conflict of interest.

Nomenclature

A, B, C, D, E	State-space matrices
C_s	the damping co-efficient of the variable shock absorber, N·s/m
E	the error to the controller
F_a	the control force from the hydraulic actuator, N
G_c, G_p, G_v, G_m, G_d	transfer functions denoting controller (c), process (p), actuator (v), sensor (m), disturbance (d)
K_c	controller gain
K_s	the suspension spring (stiffness), N/m
K_{us}	the tire spring element, N/m
M_s	the sprung mass, kg
M_{us}	the unsprung mass, kg
t	time, s
tI	controller integral time, s
tD	controller derivative time, s
P	output variable from the controller to the actuator
U	output variable from the actuator to the active suspension system
x_1, x_2, x_3, x_4	auxiliary variables used in model reduction
$Y_{sp}(t)$, $Y(t)$	set-point and system output variable, respectively, m
Y_m	ouput variable from the measurement element
Z_s	the sprung mass position, m
Z_{us}	the unsprung mass position, m
Z_o	the road surface (anomaly), m

References

1. Petri, A.-M.; Petreuș, D.M. Adaptive Cruise Control in Electric Vehicles with Field-Oriented Control. *Appl. Sci.* **2022**, *12*, 7094. [CrossRef]
2. Brugnolli, M.M.; Angelico, B.A.; Lagana, A.A.M. Predictive Adaptive Cruise Control Using a Customized ECU. *IEEE Access* **2019**, *7*, 55305–55317. [CrossRef]
3. Chen, C.; Guo, J.; Guo, C.; Chen, C.; Zhang, Y.; Wang, J. Adaptive Cruise Control for Cut-In Scenarios Based on Model Predictive Control Algorithm. *Appl. Sci.* **2021**, *11*, 5293. [CrossRef]
4. Yang, W.; Chen, Y.; Su, Y. A Double-Layer Model Predictive Control Approach for Collision-Free Lane Tracking of On-Road Autonomous Vehicles. *Actuators* **2023**, *12*, 169. [CrossRef]
5. Tang, B.; Hu, Z.; Jiang, H.; Yin, Y.; Yang, Z. Dynamic Lane Tracking Control of the Commercial Vehicle Based on RMPC Algorithm Considering the State of Preceding Vehicle. *Machines* **2022**, *10*, 534. [CrossRef]
6. Shafiei, B. A Review on PID Control System Simulation of the Active Suspension System of a Quarter Car Model While Hitting Road Bumps. *J. Inst. Eng. Ser. C* **2022**, *103*, 1001–1011. [CrossRef]
7. Daniyan, I.A.; Mpofu, K.; Osadare, D.F. Design and simulation of a controller for an active suspension system of a rail car. *Cogent. Eng.* **2018**, *5*, 1–15. [CrossRef]
8. Gowda, D.V.; Chakrasali, S. Comparative Analysis of Passive and Semi-Active Suspension System for Quarter Car Model Using PID Controller. 2014. Available online: https://www.researchgate.net/publication/276276510 (accessed on 1 September 2023).
9. Ding, X.; Li, R.; Cheng, Y.; Liu, Q.; Liu, J. Design of and Research into a Multiple-Fuzzy PID Suspension Control System Based on Road Recognition. *Processes* **2021**, *9*, 2190. [CrossRef]
10. Han, S.-Y.; Dong, J.-F.; Zhou, J.; Chen, Y.-H. Adaptive Fuzzy PID Control Strategy for Vehicle Active Suspension Based on Road Evaluation. *Electronics* **2022**, *11*, 921. [CrossRef]
11. Al-Zughaibi, A.I. Controller Design for Active Suspension System of 1/4 Car with Unknown Mass and Time-Delay. Available online: https://www.researchgate.net/publication/280934598 (accessed on 1 September 2023).
12. Papaioannou, G.; Ning, D.; Jerrelind, J.; Drugge, L. A K-Seat-Based PID Controller for Active Seat Suspension to Enhance Motion Comfort. *SAE Int. J. Connect. Autom. Veh.* **2022**, *5*, 189–199. [CrossRef]
13. Zhu, J.; Zhao, D.; Liu, S.; Zhang, Z.; Liu, G.; Chang, J. Integrated Control of Spray System and Active Suspension Systems Based on Model-Assisted Active Disturbance Rejection Control Algorithm. *Mathematics* **2022**, *10*, 3391. [CrossRef]
14. Dahunsi, O.A.; Dahunsi, O.A.; Pedro, J.O. Neural Network-Based Identification and Approximate Predictive Control of a Servo-Hydraulic Vehicle Suspension System Ergonomic Evaluation of Dynamic and Post Effects of Vibration on Earthmoving EQUIPMENT Operators View Project Neural Network-Based Identification and Approximate Predictive Control of a Servo-Hydraulic Vehicle Suspension System. 2010. Available online: https://www.researchgate.net/publication/49586591 (accessed on 1 September 2023).

15. Kim, J.; Lee, T.; Kim, C.J.; Yi, K. Model predictive control of a semi-active suspension with a shift delay compensation using preview road information. *Control Eng. Pract.* **2023**, *137*, 105584. [CrossRef]
16. Zhang, N.; Yang, S.; Wu, G.; Ding, H.; Zhang, Z.; Guo, K. Fast Distributed Model Predictive Control Method for Active Suspension Systems. *Sensors* **2023**, *23*, 3357. [CrossRef]
17. Payakkawan, P.; Klomkarn, K.; Sooraksa, P. Dual-line PID controller based on PSO for speed control of DC motors. In Proceedings of the 2009 9th International Symposium on Communications and Information Technology, ISCIT 2009, Incheon, Republic of Korea, 28–30 September 2009; pp. 134–139. [CrossRef]
18. Mat, M.H.; Darns, I.Z.M. Self-tuning PID controller for active suspension system with hydraulic actuator. In Proceedings of the IEEE Symposium on Computers and Informatics, ISCI 2013, Langkawi, Malaysia, 7–9 April 2013; IEEE Computer Society: Piscataway, NJ, USA, 2013; pp. 86–91. [CrossRef]
19. Zeng, Q.; Zhao, J. Event-triggered controller design for active suspension systems: An adaptive backstepping method with error-dependent gain. *Control Eng. Pract.* **2023**, *136*, 105547. [CrossRef]
20. Jeong, Y.; Sohn, Y.; Chang, S.; Yim, S. Design of Static Output Feedback Controllers for an Active Suspension System. *IEEE Access* **2022**, *10*, 26948–26964. [CrossRef]
21. Liu, J.; Liu, J.; Li, Y.; Wang, G.; Yang, F. Study on Multi-Mode Switching Control Strategy of Active Suspension Based on Road Estimation. *Sensors* **2023**, *23*, 3310. [CrossRef]
22. Kozek, M.; Smoter, A.; Lalik, K. Neural-Assisted Synthesis of a Linear Quadratic Controller for Applications in Active Suspension Systems of Wheeled Vehicles. *Energies* **2023**, *16*, 1677. [CrossRef]
23. Alvarez-Sánchez, E. A Quarter-Car Suspension System: Car Body Mass Estimator and Sliding Mode Control. *Procedia Technol.* **2013**, *7*, 208–214. [CrossRef]
24. Nguyen, T.A. Research on the Sliding Mode—PID control algorithm tuned by fuzzy method for vehicle active suspension. *Forces Mech.* **2023**, *11*, 100206. [CrossRef]
25. Fu, Z.-J.; Dong, X.-Y. $H\infty$ optimal control of vehicle active suspension systems in two time scales. *Automatika* **2021**, *62*, 284–292. [CrossRef]
26. Chen, L.; Xu, X.; Liang, C.; Jiang, X.-W.; Wang, F. Semi-active control of a new quasi-zero stiffness air suspension for commercial vehicles based on H_2H_∞ state feedback. *J. Vib. Control.* **2023**, *29*, 1910–1926. [CrossRef]
27. Ahmad, I.; Ge, X.; Han, Q.-L. Communication-Constrained Active Suspension Control for Networked In-Wheel Motor-Driven Electric Vehicles with Dynamic Dampers. *IEEE Trans. Intell. Veh.* **2022**, *7*, 590–602. [CrossRef]
28. Merah, A.; Hartani, K.; Yazid, N.E.H.; Chikouche, T.M. New Integrated Full Vehicle Suspension System for Improvements in Vehicle Ride Comfort and Road Holding. *SAE Int. J. Veh. Dyn. Stab. NVH* **2022**, *6*, 267–281. [CrossRef]
29. Azmi, R.; Mirzaei, M.; Habibzadeh-Sharif, A. A novel optimal control strategy for regenerative active suspension system to enhance energy harvesting. *Energy Convers. Manag.* **2023**, *291*, 117277. [CrossRef]
30. Liu, Y.J.; Zeng, Q.; Liu, L.; Tong, S. An Adaptive Neural Network Controller for Active Suspension Systems with Hydraulic Actuator. *IEEE Trans. Syst. Man Cybern. Syst.* **2020**, *50*, 5351–5360. [CrossRef]
31. Qu, Z.; Liu, J.; Li, Y.; Yang, F.; Liu, J. Study on Multi-Mode Switching Control of Intelligent Suspension under Full Road Section. *Processes* **2023**, *11*, 1776. [CrossRef]
32. He, X.L.; Chen, J.; Tang, D.Y.; Peng, S.; Tang, B.B. Using an Inerter-Based Suspension to Reduce Carbody Flexible Vibration and Improve Riding-Comfort. *SAE Int. J. Veh. Dyn. Stab. NVH* **2023**, *7*, 137–151. [CrossRef]
33. Ikeda, K.; Kuroda, J.; Uchino, D.; Ogawa, K.; Endo, A.; Kato, T.; Kato, H.; Narita, T. A Study of a Ride Comfort Control System for Ultra-Compact Vehicles Using Biometric Information. *Appl. Sci.* **2022**, *12*, 7425. [CrossRef]
34. Kaldas, M.M.; Soliman, A.M.A.; Abdallah, S.A.; Mohammad, S.S.; Amien, F.F. Road Preview Control for Active Suspension System. *SAE Int. J. Veh. Dyn. Stab. NVH* **2022**, *6*, 371–383. [CrossRef]
35. Abbas, G.; Asad, M.U.; Gu, J.; Alelyani, S.; Balas, V.E.; Hussain, M.R.; Farooq, U.; Awan, A.B.; Raza, A.; Chang, C. Multivariable unconstrained pattern search method for optimizing digital pid controllers applied to isolated forward converter. *Energies* **2021**, *14*, 77. [CrossRef]
36. The MathWorks Inc. *Optimization Toolbox*, version 9.4 (R2022b); The MathWorks Inc.: Natick, MA, USA, 2022.
37. Bequette, B. *Process Control: Modeling, Design, and Simulation*; Prentice Hall Press: Upper Saddle River, NJ, USA, 2002.
38. Nise, N.S. *Control Systems Engineering*, 8th ed.; Wiley: Hoboken, NJ, USA, 2019.

processes

MDPI

Article

The Influence of Blade Tip Clearance on the Flow Field Characteristics of the Gas–Liquid Multiphase Pump

Yuxuan Deng *, Yanna Li, Jing Xu, Chunyan Kuang and Yanli Zhang

Bailie School of Petroleum Engineering, Lanzhou City University, Lanzhou 730071, China
* Correspondence: dengyuxuan@lzcu.edu.cn; Tel.: +86-159-0815-5884

Abstract: Gas–liquid multiphase pumps are critical transportation devices in the petroleum and chemical engineering industries, and improving their conveyance efficiency is crucial. This study investigates the influence of blade tip clearance variations on the flow characteristics within a multiphase pump. Numerical simulations were conducted using Eulerian two-phase and SST k-ω turbulence models with four distinct tip clearance sizes (0 mm, 0.3 mm, 0.6 mm, and 0.9 mm). The performance curve, tip leakage flow (TLF), and internal gas distribution were subjected to analysis. The results indicate that the TLF is linearly related to the clearance size and traverses multiple flow passages, resulting in energy losses and a reduced pump head coefficient. Larger tip clearances (0.6 mm and 0.9 mm) exhibited a more uniform flow pattern, contrasting the irregularities seen with a 0.3 mm clearance. Compared to no tip clearance (0 mm), gas holdup within the impeller passages decreased by 18.39%, 39.62%, and 58.53% for clearances of 0.3 mm, 0.6 mm, and 0.9 mm, respectively, leading to decreased overall system efficiency. This study highlights the connection between tip clearance size and flow dynamics in multiphase pumps, offering insights for optimal tip clearance selection during multiphase pump design.

Keywords: multiphase pump; two-phase flow; tip clearance; numerical simulation

1. Introduction

Gas–liquid multiphase pumps, characterized by their simple design, extensive flow capability, and minimal sensitivity to solid particle entrainment, are integral to the transportation of petroleum. Since the 21st century's onset, offshore oil and gas exploration and development have surpassed onshore activities, leading to a steady increase in production and storage volumes. As a result, offshore regions are now viewed as strategic reserves for global oil and gas supplies [1]. The evolution of gas–liquid multiphase pump technology is anticipated to further enhance production from both marginal and deep-water fields.

Despite the considerable advancements, the internal dynamics of gas–liquid multiphase pumps remain intricately complex, leaving various flow mechanisms unexplored. Previous studies have examined the influence of different flow parameters on pump performance, including inlet gas void fraction (IGVF), rotational speed, etc. Cheng et al. [2] introduced a novel cylindrical vane pump, whose efficiency was corroborated through experimental and numerical methods. The Computational Fluid Dynamics (CFD) model of the pump exhibited a maximum error rate of 5.7%, validating its successful design as a positive-displacement pump. Furthermore, the pump maintained consistent efficiency across diverse conditions. Xu et al. [3] tested the multiphase pump under various flow rates and IGVF conditions, finding that increasing IGVF corresponded to a decreasing stable operational range, peaking stability at 50% IGVF. Saadawi et al. [4] studied a blade-type gas–liquid multiphase pump in an oil field, noting enhanced pressure performance with higher rotational speeds. This conclusion was echoed by Ali and Zhang [5,6] through both numerical and experimental research. The medium's viscosity significantly influences the performance of multiphase pumps. Specifically, increased viscosity results in reduced pressure during the transport of high-viscosity media [7–9]. In the realm of optimization, Liu et al. [10,11] thoroughly investigated pump–turbine behavior during power-off scenarios, blending numerical simulations with empirical methods. Their work clarifies critical aspects such as transient flow characteristics, pressure fluctuations, and operational shifts, providing valuable theoretical and practical insights into the field. Zhang et al. [12] proposed a three-dimensional design methodology suitable for high-flow pumps based on the impeller passage mesh of a gas–liquid multiphase pump, a method verified experimentally. Cao et al. [13] combined the three-dimensional design method with computational fluid dynamics (CFD) to design a multiphase pump and conducted experiments using this approach. Zhang et al. [14] merged a backpropagation (BP) neural network with a multi-objective genetic algorithm, achieving optimal outcomes in numerical simulations under distinct operating scenarios with varying IGVF. Kim et al. [15] optimized specific structural pump parameters, determining that their enhanced design minimized flow separation, thus improving operational efficiency.

Currently, a cohesive and mature design theory for multiphase pumps is absent. With the progression of computer technology, the field is evolving towards a more multi-objective and intelligent paradigm. Zhang et al. [16] performed a numerical simulation to study the flow field within a multistage gas–liquid multiphase pump under various operational conditions. Li et al. [17,18] employed the Eulerian model to predict the bubble distribution and diameter in the pump, providing a better understanding of bubble motion characteristics. Zhang et al. [19,20] investigated the effects of flow parameters on gas distribution and interphase forces in single-stage and multistage pumps under different flow rates, IGVF, and viscosity using numerical simulations. Sun et al. [21] analyzed the impact of blade tip clearance on vortex structure, gas distribution, and energy characteristics, concluding that a wider clearance leads to substantial energy losses and alters gas accumulation patterns. Zhang et al. [22] studied the resistance variation under different IGVF conditions using a non-uniform bubble model with different drag coefficients. Another investigation by Zhang et al. [23] empirically assessed experimental analyses of pressure pulsation under different IGVF and rotational speed conditions. Their findings indicate that both dynamic and static interference lead to heightened pressure pulsation. With decreasing rotational speed, the pressure pulsation initially declines, but later increases with elevated IGVF. This suggests that controlled air ingress into the pump can optimize pressure pulsation. However, elevated IGVF levels tend to amplify pressure pulsation. Shi et al. [24] used numerical simulations to study the influence of the clearance at the blade tip on the pressure and velocity fields on the S2 flow surface of the impeller in a multiphase pump.

In summation, prevailing research on the impact of tip clearance on flow characteristics largely emphasizes leakage vortex cavitation and velocity distribution within the clearance gap. However, a more comprehensive examination is required, especially concerning leakage flow patterns and gas holdup distribution in the gap region of gas–liquid multiphase pumps. This investigation employs the Eulerian two-fluid model for numerical simulations of internal flow in a gas–liquid multiphase pump with different clearances. This methodology provides insights into the effects of tip clearance on the pump's external characteristics, gap leakage flow, and gas distribution in the impeller. The findings from this study can be crucial references in optimizing the design of gas–liquid multiphase pumps.

2. Numerical Model

2.1. Computational Method

In multiphase pumps, the flow of two-phase fluids is inherently non-uniform. Although factors such as slip velocity and momentum transfer can often be neglected, the non-uniform assumption offers a more accurate depiction of the pump's internal flow dynamics compared to a uniform model. This study employs the non-uniform Eulerian two-fluid model for its calculations. While both fluid phases operate within a shared pressure field, they maintain individual velocity fields. The Reynolds-averaged Navier–

Stokes equation for a steady, incompressible fluid is solved. The governing equations are as follows:

Continuity Equation:

$$\nabla \cdot (\alpha_k \rho_k V_k) = 0 \tag{1}$$

Momentum Equation:

$$\nabla \cdot (\alpha_k \rho_k V_k V_k) = -\alpha_k \nabla p + \nabla \cdot (\alpha_k \tau_{ij}) + M_k + f_k \tag{2}$$

where the subscript k represents any phase (l is liquid, g is gas). α_k is the k-phase volume fraction, satisfying $\alpha_g + \alpha_l = 1$. ρ_k is the k-phase density, V_k is the k-phase absolute velocity, p is the pressure, τ is the viscous stress tensor, M_k is the interphase force on k-phase, and f_k is the mass force.

The steady flow within the multiphase pump was calculated using ANSYS CFX 17.0, a commercial computational fluid dynamics software. We employed the SST k-ω turbulence model to address the Reynolds stress. This model combines the characteristics of the k-ω and k-ε models, and can accurately predict flow separation under adverse pressure gradients. Additionally, it exhibits specific adaptability in predicting the movement of clearance leakage vortices [25,26].

2.2. Computational Model

The computational model presented in this paper consists of four primary components: the inlet pipe, impeller, diffuser, and outlet pipe, as depicted in Figure 1. Key design parameters for the multiphase pump are a design flow rate Q_d = 50 m^3/h, a design head H = 15 m, and a rotational speed n = 2950 r/min. Other main technical parameters can be found in Table 1.

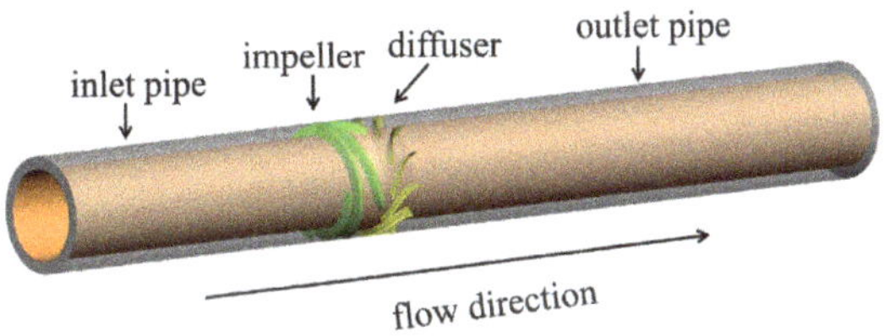

Figure 1. Computational model.

Table 1. Main technical parameters.

Parameter	Impeller	Diffuser
Blade number	4	11
Shroud radius (mm)	75	75
Inlet hub radius (mm)	58.96	67
Outlet hub radius (mm)	67	60
Axial length (mm)	55	65

The tip clearance is defined as the gap between the blade wheel edge and the inner wall of the pump casing, as depicted in Figure 2. The presence of this clearance allows a portion of the fluid to flow from the high-pressure side of the blade to the low-pressure suction side, resulting in TLF. This phenomenon significantly alters the flow dynamics within the impeller passage, subsequently impacting the pump's performance. In this study, four tip clearance sizes, δ_1 = 0 mm, δ_2 = 0.3 mm, δ_3 = 0.6 mm, and δ_4 = 0.9 mm, were selected to investigate the influence of the tip clearance on the flow characteristics of the multiphase pump.

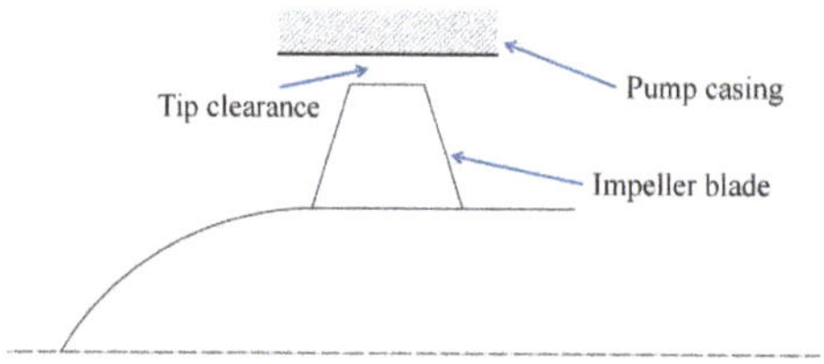

Figure 2. Schematic diagram of tip clearance.

The computational domains were discretized using a high-precision structured grid, making use of both ICEM and TurboGrid tools. For components with slender geometries, including diffusers and impellers, an "O-shaped" grid was adopted for localized refinement, ensuring a smooth transition between component grids and capturing flow intricacies effectively. Additionally, grid refinement was focused on the near-wall region to capture the flow dynamics within the impeller passages with greater accuracy. Specifically, a gap grid of 12 layers was designated for $\delta_2 = 0.3$ mm, while 26 and 40 layers were assigned for $\delta_3 = 0.6$ mm and $\delta_4 = 0.9$ mm, respectively. Figure 3 illustrates both the computational domains and the local grid. For details of the tip clearance grid, refer to Supplementary Figure S1 in the Supplementary Materials.

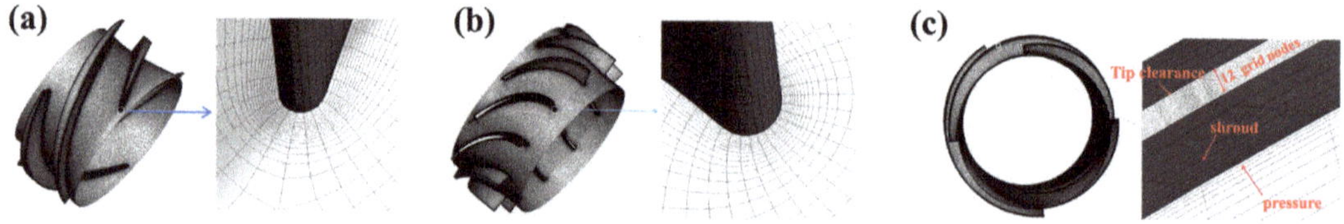

Figure 3. Computational grid. (**a**) Blade; (**b**) diffuser; (**c**) tip clearance.

A grid independence verification analysis was executed to confirm the quality of the grid. This verification process was based on calculations performed on five distinct grid sets, each varying by specific parameters, with the IGVF set to 0%. The head and efficiency were used as evaluation metrics. The data suggests consistency in both head and efficiency across the last three grid sets, as illustrated in Figure 4. To achieve a balance between computational efficiency and result accuracy, the third grid set was selected for subsequent numerical calculations.

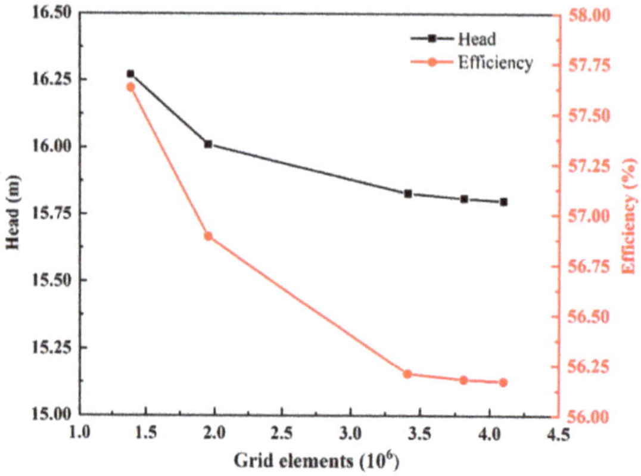

Figure 4. Grid independence verification.

2.3. Boundary Conditions and Setting

The selected flow medium consists of a two-phase mixture, with water serving as the continuous phase and air as the dispersed phase. The water has a density of 997 kg/m^3, and the bubble diameter is consistently 0.1 mm. Volume fractions for both water and air are determined by operating conditions, ensuring their combined sum always equals 1. The inlet boundary condition is defined by the total mass flow rate, whereas the boundary condition for the outlet corresponds to the average static pressure. The pump operates at a rotational speed of 2950 rpm. Since the impeller is a rotating domain, the interface between the inlet extension pipe and the impeller and the impeller and the diffuser were modeled using the frozen rotor approach. The simulations were deemed converged when the residuals for all equations fell below a threshold of 1×10^{-5}.

2.4. Gas–Liquid Multiphase Pump Test Rig

The experimental setup is illustrated in Figure 5. Air was supplied by an air compressor, while water was provided by a water tank. The water and air were individually channeled into the buffer, where they were thoroughly mixed before entering the gas–liquid multiphase pump. Subsequently, the water was recirculated back into the tank, while the air was discharged into the atmosphere. Flowmeters and pressure sensors were strategically positioned on the relevant pipes to monitor the gas and liquid flow rates, as well as the pump's inlet and outlet pressures, respectively. The main sensor parameters are shown in Table S1 of the Supplementary Materials.

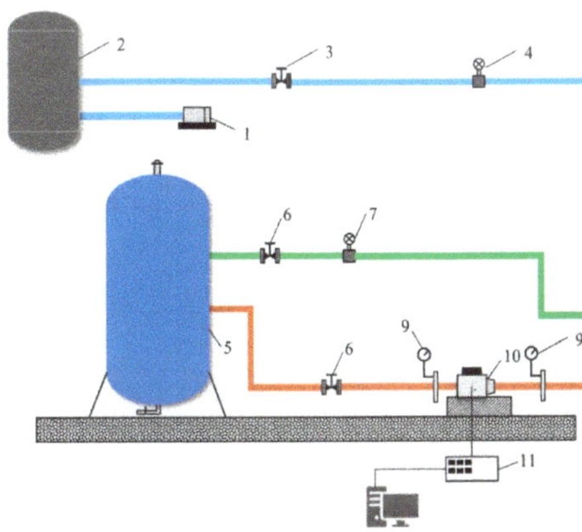

Figure 5. Experimental schematic diagram. 1, air compressor; 2, gas tank; 3, valve; 4, gas flowmeter; 5, water tank; 6, valve; 7, liquid flowmeter; 8, buffer; 9, pressure sensor; 10, multiphase pump; 11, data collector.

3. Results and Discussion

3.1. Validation of Numerical Calculation Methods

To assess the accuracy of the numerical calculations, a comparison was conducted between the experimental and numerically calculated results for both head and efficiency, as depicted in Figure 6. The efficiency exhibited a maximum error of 7.95% at IGVF = 20%, while the head presented a minimum error of 3.48% at IGVF = 10%. Overall, both experimental and calculated data showed congruent trends, affirming the accuracy of the numerical approach. However, it is noteworthy that as IGVF escalates, discrepancies in results become more apparent. This phenomenon primarily arises from the need to equalize pressures in the gas and liquid pipelines, ensuring the gas–liquid mixture does not enter the contrasting phase's pipeline before reaching the buffer. During the tests, maintaining identical pressures in the gas and liquid pipes conduits grows more arduous as the IGVF increases, leading to fluctuations in the measured data and larger errors.

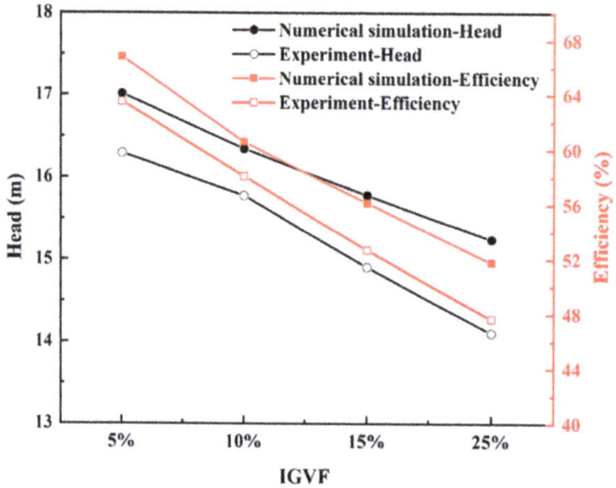

Figure 6. Validation of the numerical calculation.

3.2. External Characteristics Prediction

In order to analyze the impact of tip clearance on the head of the gas–liquid multiphase pump, the head is dimensionless. The head coefficient φ_h is defined as:

$$\varphi_h = \frac{p_2 - p_1}{\rho u_2^2} \tag{3}$$

where p_1 and p_2 are the inlet and outlet pressures of the pump, respectively, and u_2 is the circumferential velocity at the impeller outlet rim.

The values of the pump head coefficient, calculated under various tip clearances, are presented in Table 2. These values were derived for a pump with an IGVF of 10%. Analysis of Table 2 reveals a consistent decline in the pump head coefficient with the progressive increase in tip clearance from 0 mm to 0.9 mm. When comparing the results with the case of a 0 mm tip clearance, the declines recorded are 1.82% at 0.3 mm, 3.21% at 0.6 mm, and 4.83% at 0.9 mm. The decrease in the head coefficient is attributed to increased liquid recirculation when the IGVF is at a specific condition. As the impeller processes the fluid, an increasing tip clearance prompts a fraction of this fluid to circulate back to the impeller's suction side. This recirculation negatively impacts the fluid's effective inflow, subsequently reducing the pump head, leading to a reduced coefficient. Additionally, the quantitative relationship between the gap and head coefficient is elaborated upon in Supplementary Figure S2.

Table 2. Head coefficient under different tip clearance.

Tip Clearance (mm)	Head Coefficient (φ_h)
0	0.3023
0.3	0.2968
0.6	0.2926
0.9	0.2877

3.3. Influence of Tip Clearance on the TLF

Figure 7 illustrates the relationship between TLF and varying tip clearances. A nearly linear correlation is evident between gap leakage and clearance size, with leakage surging markedly as the tip clearance expands. As the tip clearance increases from 0 mm to 0.9 mm by increments of 0.3 mm, the corresponding TLF volumes are 2.23 kg/s, 4.65 kg/s, and 7.24 kg/s, accounting for 17.89%, 37.31%, and 58.09% of the inlet mass flow rate, respectively. Data analysis reveals an ongoing surge in leakage with increasing tip clearance, peaking at 58.09% for 0.9 mm. This trend is explained by Figure 8a, which demonstrates that a significant portion of the leakage flow passing through the first tip clearance (marked as I) proceeds to subsequent tip clearance (marked as II), with portions even advancing to further clearances (marked as III, IV). This leads to the fluid navigating multiple flow paths, incurring cumulative losses. In addition, leakage can substantially diminish energy efficiency due to the associated decrease in fluid pressure within the pump, directly impacts its performance. Enhanced tip clearance offers a wider flow route, intensifying fluid losses. The compact streamlines depicted in Figure 8b corroborate this, illustrating an increased level of fluid leakage with broader tip clearance.

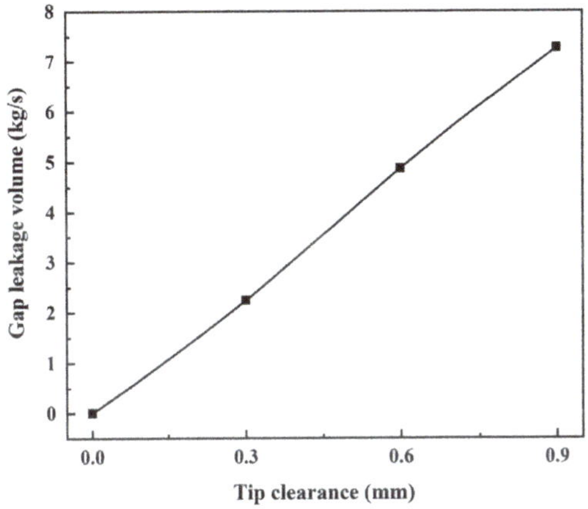

Figure 7. TLF volume at different tip clearance.

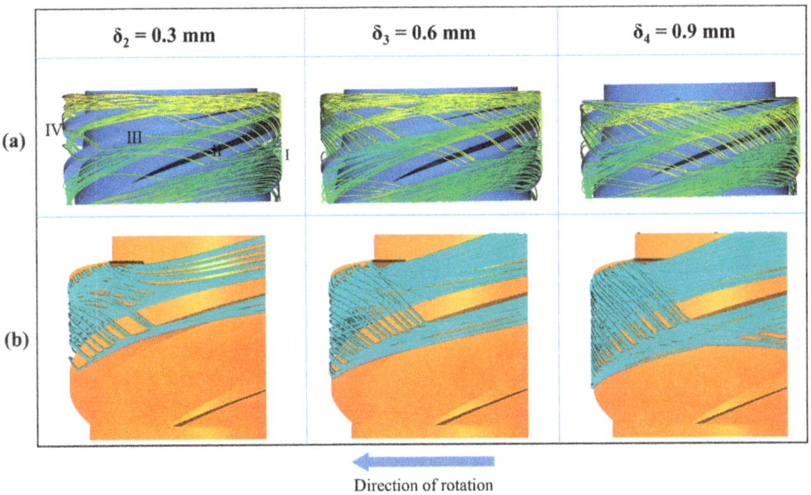

Figure 8. TLF streamlines at different tip clearances. (**a**) Leakage flow in runners; (**b**) leakage flow in the clearance.

The investigation of external characteristics reveals a steady decrease in the head coefficient of the gas–liquid multiphase pump as the tip clearance increases. This phenomenon is intricately related to TLF, which interrupts mainstream motion and induces turbulence within the flow field. Figure 9 represents the vorticity distribution at 90% of the impeller blade height. At tip clearance of 0 mm, the absence of tip clearance eliminates any potential TLF, yielding a distinctly different vorticity distribution than in situations with tip clearance. Predominantly, vorticity in such situations stems from a fluid collision with the blade surface, followed by backflow. With greater tip clearance, both the range of vorticity distribution and the vortex intensity rise. This trend is due to the leakage flow through the clearance, potentially triggering a vortex effect and consequent energy loss. Incremental expansion of the tip clearance amplifies the leakage flow, resulting in increased energy consumption. It is evident that tip clearance significantly affects the overall performance of the pump. This influence is dynamic, changing with variations in the tip clearance and follows a consistent trend: as the tip clearance widens, the pump's head coefficient consistently reduces. This insight enhances our understanding of fluid dynamics and provides critical guidance for designing and optimizing gas–liquid multiphase pumps.

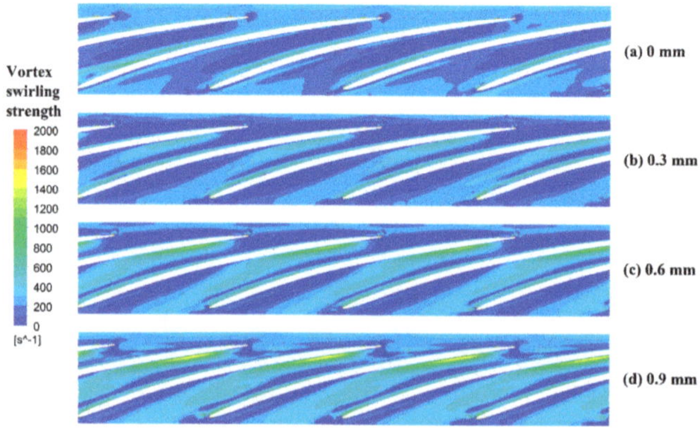

Figure 9. Vortex swirling strength distribution of the impeller.

Flow leakage through the clearance gaps significantly influences the flow distribution within the impeller passages and their corresponding mass flow rates. Figure 10 illustrates the mass flow rate variation across each impeller passage with different tip clearances. Notably, Passage 2, with no tip clearance (0 mm), exhibits much lower flow rates compared to other passages. At a tip clearance of 0.3 mm, the flow rates in both Passages 2 and 3 diminish. In contrast, clearances of 0.6 mm and 0.9 mm showcase more consistent flow distribution. This inconsistency is largely attributed to the complexities of gap fluid dynamics. Tip leakage flow (TLF) is particularly pronounced in passages with smaller clearances, exemplified by Passage 2. Such flow perturbations disturb the adjacent flow, causing non-uniform flow velocities within the passage, compromising impeller stability and efficiency, potentially leading to vibrations and jeopardizing the pump's operational robustness. Passages with larger clearances typically demonstrate more balanced flow rates, usually augmenting impeller performance. In summary, the magnitude of the tip clearance critically determines the flow and mass flow rate distribution within the passages. Strategic design and optimization of this clearance are vital to boost pump efficiency and reliability, mitigating undesirable leakages, refining flow distribution, and enhancing overall performance.

Figure 11 illustrates the gas void fraction (GVF) distribution within the flow passage at a blade height of 0.5. Figure 11 reveals pronounced gas accumulation in Passage 2 with a tip clearance of 0 mm and in Passages 2 and 3 with a clearance of 0.3 mm. This accumulation diminishes the overflow of liquid, leading to a notably uneven mass flow distribution. Nonetheless, when the tip clearance increases to 0.6 mm and 0.9 mm, there is a marked change in gas distribution. In these scenarios, gas disperses more uniformly throughout the passages, resulting in a more equitable mass flow distribution. This enhanced uniformity, which presumably augments pump performance, corresponds with the data in Figure 10. Uniform gas distribution at broader tip clearances implies a pivotal threshold beyond which gas obstruction in the impeller passage can be effectively reduced. This revelation holds considerable relevance for the design and strategic optimization of gas–liquid multiphase pumps.

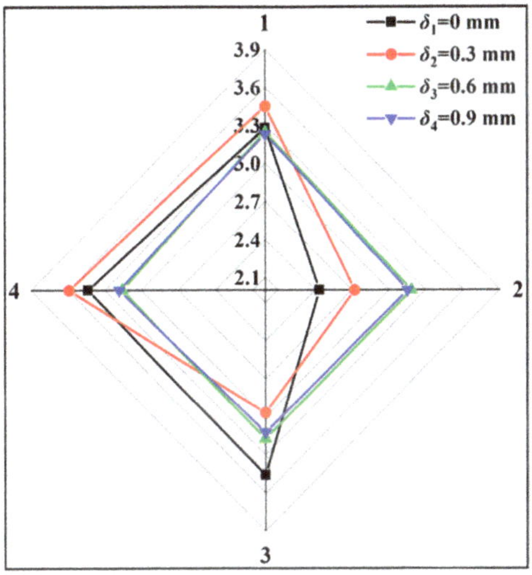

Figure 10. Flow distribution in each flow passage of the impeller.

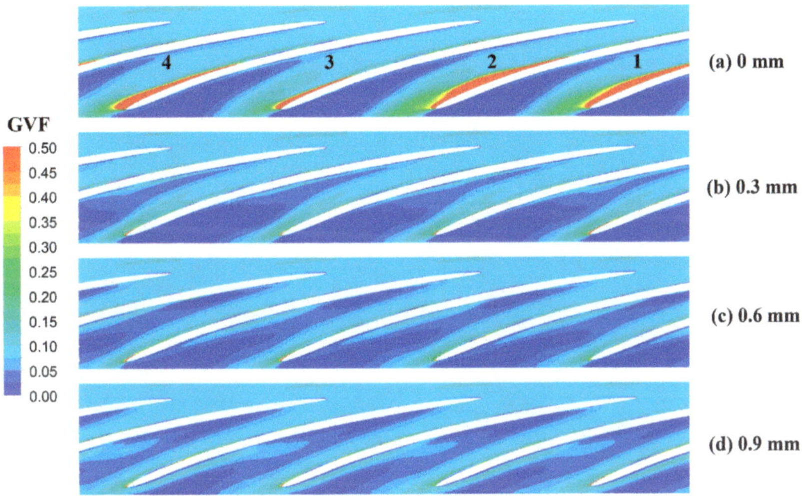

Figure 11. Distribution of gas void fraction in impeller.

3.4. Influence of Blade Tip Clearance on Gas Void Fraction Distribution

In the impeller, a randomly selected flow passage is analyzed to examine the distribution of the gas void fraction near the tip clearance. Cross-sections are designated every 15° around the rotational axis, extending from the inlet to the outlet. These sections are labeled sequentially from A1 to A6, as shown in Figure 12.

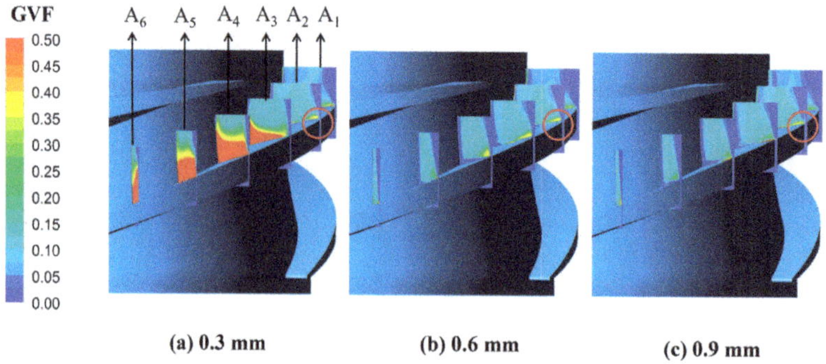

Figure 12. Distribution of gas void fraction around tip clearances.

In Figure 12, it is evident that under various tip clearance conditions, there is no significant accumulation of gas either in the gap region or on the blade's pressure surface adjacent to the gap. The absence of this accumulation is likely due to the increased flow velocity at the gap, which causes the gas to become entrained in the liquid flow. The reduced relative pressure on the nearby suction surface results in gas accumulation. Moreover, at a tip clearance of 0.3 mm, the GVF in each section notably increases, compared to clearances of 0.6 mm or 0.9 mm. This trend can be linked to the expansion of the TLF area with an increasing tip clearance which, in turn, reduces the primary flow and boosts the liquid flow velocity. As shown in Figure 13, with the rise in tip clearance, there is a proportional increase in the liquid phase's velocity. This elevated velocity encourages the entrainment of a larger gas volume by the liquid flow. The gas entrainment effect is especially prominent at a tip clearance of 0.3 mm, confirming our previous observations. However, as the clearance expands further to 0.6 mm and 0.9 mm, the amount of entrained gas starts to decrease, possibly due to a higher liquid flow velocity resulting in gas–liquid separation. Further examination of the figure shows a denser region of gas concentration at a tip clearance of 0.3 mm. Such concentrated gas aggregations might compromise the liquid flow's stability, subsequently impacting blade performance and efficiency.

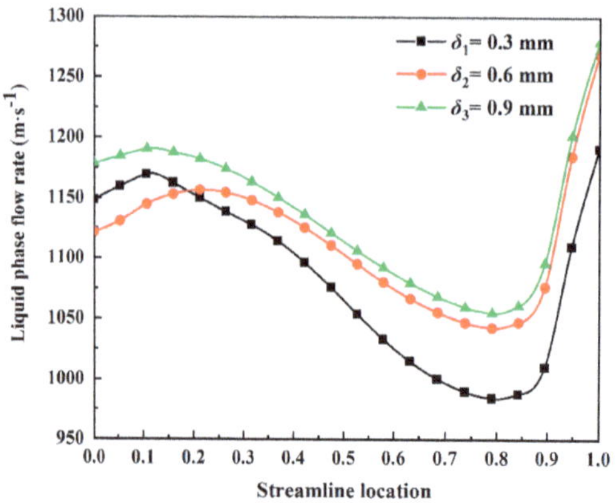

Figure 13. Distribution of liquid phase velocity in impeller passage.

Figure 14 illustrates the variations in the average GVF within the impeller passages across different tip clearances. The GVF is at its peak with no tip clearance and decreases as the clearance expands. Notably, the GVF experiences a significant decline when the clearance reaches 0.9 mm, consistent with the observations in Figure 11. The average GVF in the impeller passages transitions sharply from 15.12% at 0 mm clearance to 12.34%,

9.13%, and then 6.27%. These percentages correspond to reductions of 18.39%, 39.62%, and 58.53%, respectively, compared to the condition with no clearance. The pronounced reduction in GVF with increasing clearance can be ascribed to several factors. Firstly, a larger clearance may introduce heightened turbulence within the impeller passages, facilitating the more efficient dispersion of gas bubbles and thereby decreasing the local gas concentration. Secondly, a larger clearance might enhance gas–liquid separation, allowing gas to evacuate more easily from the impeller area, resulting in a decreased GVF. However, while increased tip clearance reduces gas concentration in the impeller passages, amplifying its gas transport capability, it also augments the potential for leakage through the gap. Such leakage could negate the advantages of diminished gas concentration, leading to a decline in the system's overall efficiency.

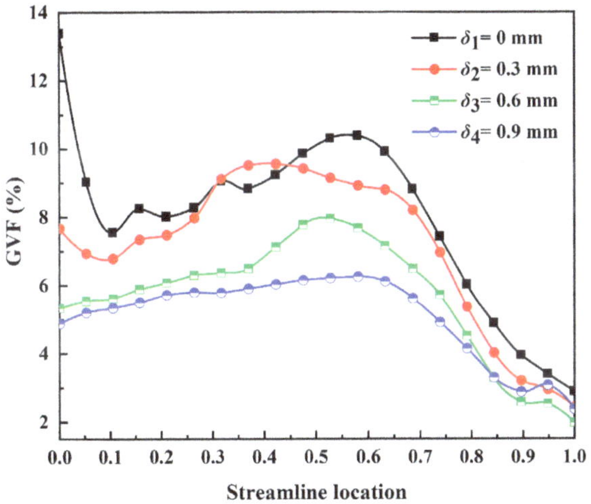

Figure 14. Distribution of average gas void fraction in impeller passage.

4. Conclusions

Using the Eulerian non-uniform two-fluid model, this study investigates the effects of varying blade tip clearances on the external characteristics of the multiphase pump, the leakage flows across the clearance gap, and the internal gas distribution inside the impeller. The results provide guidance for selecting the optimal tip clearance during multiphase pump design.

(1) The presence of a tip clearance in the multiphase pump leads to leakage flow within the gap. This leakage flow has a near-linear relationship with the size of the clearance. As the clearance increases, the leakage volume grows, dispersing across multiple flow passages, resulting in energy loss and a notable reduction in the pump head coefficient.

(2) The tip clearance plays a pivotal role in influencing the flow distribution within the passages. Clearances of 0 mm and 0.3 mm show marked irregularities in flow distribution. However, larger clearances, such as 0.6 mm and 0.9 mm, promote a more uniform flow distribution. This suggests that a greater tip clearance can reduce the often-seen flow irregularities or "clogging" inside the impeller passage.

(3) While the tip clearance does not lead to substantial gas accumulation, a discernible concentration is present near the clearance on the blade's suction side. A notable observation is the diminishing gas content in the impeller passage with increasing tip clearance. Compared to a scenario with no tip clearance (0 mm), there is a significant 58.53% decrease in gas content at a clearance of 0.9 mm.

Supplementary Materials: The following supporting information can be downloaded at: https://www.mdpi.com/article/10.3390/pr11113170/s1, Figure S1: Tip clearance grid. Figure S2: Relationship between the gap and head coefficient. Table S1: Main sensor parameters.

Author Contributions: Conceptualization, Y.D.; Methodology, Y.D.; Writing—Original Draft Preparation, J.X.; Writing—Review and Editing, Y.L.; Visualization, Y.Z.; Supervision, Y.D.; Software, C.K. All authors have read and agreed to the published version of the manuscript.

Funding: This research was funded by the Open Research Subject of Key Laboratory of Fluid and Power Machinery, Ministry of Education (Grant No. LTDL2020-005), Gansu Provincial Key R&D Program (Grant No. 22YF7GA129).

Data Availability Statement: All the data are already in the article.

Acknowledgments: The authors would like to thank all staff of the treatment plants for making this study possible.

Conflicts of Interest: The authors declare no conflict of interest.

References

1. Hu, W.; Bao, J.; Hu, B. Trend and progress in global oil and gas exploration. *Pet. Explor. Dev.* **2013**, *40*, 439–443. [CrossRef]
2. Cheng, Y.; Wang, X.; Ur Rehman, W.; Sun, T.; Shahzad, H.; Chai, H. Numerical simulation and experimental performance research of cylindrical vane pump. *Proc. Inst. Mech. Eng. Part C J. Mech. Eng. Sci.* **2021**, *236*, 1698–1720. [CrossRef]
3. Xu, Y.; Cao, S.L.; Reclari, M.; Wakai, T.; Sano, T. Multiphase performance and internal flow pattern of helico-axial pumps. *IOP Conf. Ser. Earth Environ. Sci.* **2019**, *240*, 032029. [CrossRef]
4. Saadawi, H. Operating Multiphase Helicoaxial Pumps in Series To Develop a Satellite Oil Field in a Remote Desert Location. *SPE Proj. Facil. Constr.* **2008**, *3*, 1–6. [CrossRef]
5. Ali, A.; Yuan, J.; Deng, F.; Wang, B.; Liu, L.; Si, Q.; Buttar, N.A. Research Progress and Prospects of Multistage Centrifugal Pump Capability for Handling Gas–Liquid Multiphase Flow: Comparison and Empirical Model Validation. *Energies* **2021**, *14*, 896. [CrossRef]
6. Zhang, J.; Cai, S.; Li, Y.; Zhu, H.; Zhang, Y. Visualization study of gas–liquid two-phase flow patterns inside a three-stage rotodynamic multiphase pump. *Exp. Therm. Fluid Sci.* **2016**, *70*, 125–138. [CrossRef]
7. Maleki, A.; Ghorani, M.M.; Haghighi, M.H.S.; Riasi, A. Numerical study on the effect of viscosity on a multistage pump running in reverse mode. *Renew. Energy* **2020**, *150*, 234–254. [CrossRef]
8. Liu, M.; Tan, L.; Xu, Y.; Cao, S. Optimization design method of multistage multiphase pump based on Oseen vortex. *J. Pet. Sci. Eng.* **2020**, *184*, 106532. [CrossRef]
9. Shamsuddeen, M.M.; Ma, S.-B.; Kim, S.; Yoon, J.-H.; Lee, K.-H.; Jung, C.; Kim, J.-H. Flow Field Analysis and Feasibility Study of a Multistage Centrifugal Pump Designed for Low-Viscous Fluids. *Appl. Sci.* **2021**, *11*, 1314. [CrossRef]
10. Liu, J.; Liu, S.; Sun, Y.; Wu, Y.; Wang, L. Three Dimensional Flow Simulation of Load Rejection of a Prototype Pump-Turbine. *Eng. Comput.* **2013**, *29*, 417–426. [CrossRef]
11. Liu, J.; Liu, S.; Sun, Y.; Jiao, L.; Wu, Y.; Wang, L. Three-Dimensional Flow Simulation of Transient Power Interruption Process of a Prototype Pump-Turbine at Pump Mode. *J. Mech. Sci. Technol.* **2013**, *27*, 1305–1312. [CrossRef]
12. Cao, S.; Peng, G.; Yu, Z. Hydrodynamic Design of Rotodynamic Pump Impeller for Multiphase Pumping by Combined Approach of Inverse Design and CFD Analysis. *J. Fluids Eng.* **2004**, *127*, 330–338. [CrossRef]
13. Zhang, J.; Zhu, H.; Yang, C.; Li, Y.; Wei, H. Multi-objective shape optimization of helico-axial multiphase pump impeller based on NSGA-II and ANN. *Energy Convers. Manag.* **2011**, *52*, 538–546. [CrossRef]
14. Kim, J.H.; Lee, H.C.; Kim, J.H.; Choi, Y.S.; Yoon, J.Y.; Yoo, I.S.; Choi, W.C. Improvement of Hydrodynamic Performance of a Multiphase Pump Using Design of Experiment Techniques. *J. Fluids Eng.* **2015**, *137*, 081301. [CrossRef]
15. Zhang, Y.; Zhang, J.; Zhu, H.; Cai, S. 3D Blade Hydraulic Design Method of the Rotodynamic Multiphase Pump Impeller and Performance Research. *Adv. Mech. Eng.* **2014**, *6*, 803972. [CrossRef]
16. Zhang, J.; Li, Y.; Vafai, K.; Zhang, Y. An investigation of the flow characteristics of multistage multiphase pumps. *Int. J. Numer. Methods Heat Fluid Flow.* **2018**, *28*, 763–784. [CrossRef]
17. Li, Y.; Yu, Z.; Zhang, W.; Yang, J.; Ye, Q. Analysis of bubble distribution in a multiphase rotodynamic pump. *Eng. Appl. Comput. Fluid Mech.* **2019**, *13*, 551–559. [CrossRef]
18. Li, Y.; Yu, Z. Distribution and motion characteristics of bubbles in a multiphase rotodynamic pump based on modified non-uniform bubble model. *J. Pet. Sci. Eng.* **2020**, *195*, 107569. [CrossRef]
19. Zhang, W.; Yu, Z.; Li, Y. Analysis of flow and phase interaction characteristics in a gas-liquid two-phase pump. *Oil Gas Sci. Technol.-Rev. IFP Energ. Nouv.* **2018**, *73*, 69. [CrossRef]
20. Zhang, W.; Xie, X.; Zhu, B.; Ma, Z. Analysis of phase interaction and gas holdup in a multistage multiphase rotodynamic pump based on a modified Euler two-fluid model. *Renew. Energy* **2021**, *164*, 1496–1507. [CrossRef]

21. Sun, W.; Yu, Z.; Zhang, K.; Liu, Z. Analysis of Tip Clearance Effect on the Transportation Characteristics of a Multiphase Rotodynamic Pump Based on the Non-Uniform Bubble Model. *Fluids* **2022**, *7*, 58. [CrossRef]
22. Zhang, W.; Yu, Z.; Li, Y. Application of a non-uniform bubble model in a multiphase rotodynamic pump. *J. Pet. Sci. Eng.* **2019**, *173*, 1316–1322. [CrossRef]
23. Zhang, J.; Cai, S.; Zhu, H.; Qiang, R. Characteristic Analysis of Pressure Fluctuation in a Three-Stage Rotodynamic Multiphase Pump. *ASME Int. Mech. Eng. Congr. Expo.* **2018**, *140*, 101107. [CrossRef]
24. Shi, G.; Liu, Z.; Xiao, Y.; Li, H.; Liu, X. Velocity characteristics in a multiphase pump under different tip clearances. *Proc. Inst. Mech. Eng. Part A J. Power Energy* **2020**, *235*, 454–475. [CrossRef]
25. Li, X.; Li, Z.; Zhu, B.; Wang, W. Effect of Tip Clearance Size on Tubular Turbine Leakage Characteristics. *Processes* **2021**, *9*, 1418. [CrossRef]
26. Kock, F.; Herwig, H. Local Entropy Production in Turbulent Shear Flows: A High-Reynolds Number Model with Wall Functions. *Int. J. Heat Mass Transf.* **2004**, *47*, 2205–2215. [CrossRef]

processes MDPI

Article

A Graphical User Interface for Calculating Exergy Destruction for Combustion Reactions

M. Özgün Korukçu

Department of Mechanical Engineering, University of Bursa Uludag, Gorukle, 16059 Bursa, Turkey; ozkorukcu@uludag.edu.tr; Tel.: +90-2242941927

Abstract: The combustion of fuels has been studied by many researchers as it is used in a wide range of engineering applications. The chemical equilibrium approach served as the foundation for the investigation of combustion reactions. This article presents a software application designed to facilitate the calculation of combustion processes by calculating the combustion of 16 fuels among the common alkanes (C_nH_{2n+2}) and alcohols ($C_nH_{2n+1}OH$). The Ozan Combustion Calculator (OCC) offers a user-friendly and efficient graphical user interface (GUI) that allows users to easily input data and obtain results. The program was developed using MATLAB 2021a and LaTeX software, ensuring its reliability and accuracy. To perform these calculations, the program utilizes calculations of the thermophysical properties of fuels and water obtained from tables. The program consists of five modules, each serving a specific purpose. These modules calculate various parameters, such as the Adiabatic Flame Temperature, Exergy of Combustion with Dry Air, Exergy of Combustion with Moist Air, Energy of Combustion with Dry Air, and Energy of Combustion with Moist Air. Additionally, the program can be used to investigate the impact of relative humidity on the adiabatic flame temperature and exergy destruction. The results obtained from the calculations reveal that the adiabatic flame temperature exhibits a linear decrease as the relative humidity increases. On the other hand, exergy destruction demonstrates a quadratic increase with higher relative humidity values. The program derives mathematical relationships for the adiabatic flame temperature and exergy destruction with respect to relative humidity values, with a high regression coefficient ($r^2 = 0.999$). The versatility of OCC makes it suitable for various applications. It can be utilized in university settings for both undergraduate- and graduate-level courses, providing students with a practical tool for studying combustion processes. Additionally, it finds applications in industrial settings for the design and optimization of combustors, gas turbines, and burners. The user-friendly interface and accurate calculations make OCC a valuable resource in the field of combustion engineering.

Keywords: combustion; chemical equilibrium; fuel; adiabatic flame temperature; exergy

1. Introduction

The utilization of energy heavily relies on the combustion of hydrocarbons, which can occur in liquid, solid, or gaseous states. This combustion process serves various purposes, including power generation, production within the process industry, heating systems for both residential and industrial use, implementation of safety measures, control of pollutant emissions, and active management of combustion processes [1]. In general, the primary parameter for premixed gases is the equivalence ratio (ϕ). When the equivalence ratio is greater than 1 ($\phi > 1$), it signifies rich combustion where the fuel is in excess. Conversely, a lean regime ($\phi < 1$) indicates that the oxidizer is in excess. Combustion in practical burners predominantly occurs at or below the stoichiometric mixture ratio ($\phi = 1$) [2]. In the field of combustion, the adiabatic flame temperature (AFT) holds great significance. It refers to the temperature at which the combustion process occurs adiabatically, without any heat transfer or alterations in kinetic or potential energy. In this context, the AFT represents the maximum temperature that the reactants can attain. However, it is important to emphasize that the highest AFT can only be achieved when the mixture is stoichiometric (ϕ = 1). To regulate the AFT, one can adjust the equivalence ratio (ϕ) to maintain the temperature values of the components within specific allowable limits [3]. AFT is directly influenced by pressure, the initial temperature, and the initial composition of the system [4].

In their study, Ufot et al. [5] investigated the AFT computations for gas turbine combustors employing conventional combustors. The analysis focused on a predetermined energy input and a stoichiometric ratio (ϕ) of 1.4. Glaude et al. [6] conducted a study to examine the utilization of biodiesels derived from rapeseed, soybean, and sunflower, as well as conventional fuels such as gas oil and natural gas. The focus of their investigation was to analyze the impact of these fuels on NOx emissions and the AFT. An analysis of the combustion products revealed that biodiesel products exhibit AFT and NOx indices that are comparable to those of gas oil but higher than those of natural gas. Kayadelen [7] conducted a study to examine the influence of natural gas components on the AFT, equilibrium products, and thermodynamics. The findings indicated that the AFT values were predominantly affected by CO_2, with isobutane (C_4H_{10}), propane (C_3H_8), and ethane (C_2H_6) also playing a role. Kayadelen [8] enhanced a comprehensive model to accurately predict the thermodynamic properties, AFT, and equilibrium products of various fuels, fuel blends, and additives. Law et al. [9] conducted a study to explore the potential enhancement of AFT values through off-stoichiometric combustion. The findings of their research indicated that a reduction in heat release, coupled with the occurrence of product dissociation peaks, leads to an increase in AFT values. In their study, Ozsari et al. [10] conducted a comparison of AFT, energy, and emission values for natural gas samples obtained from various countries, including Russia, the USA, Iran, and Australia. The researchers examined the performance of oxy-fuel combustion and conventional combustion with air for these natural gas samples. The findings of their investigation revealed that Russian natural gas exhibited the lowest emission values, while Australian natural gas demonstrated the lowest entropy production. The investigation conducted by Movileanu et al. [11] focused on analyzing the adiabatic flame temperature (AFT) of different fuel–air mixtures in combustion processes that were either isobaric or isochoric. The examination carried out by Sakhrieh [12] delved into the impact of the equivalence ratio, pressure, and initial temperature parameters on the AFT and laminar flame speed. The results of the study indicated that the optimal values for the AFT and laminar flame speed were attained when the equivalence ratio was maintained at ϕ = 1. In their study, Wu et al. [13] made significant advancements in a newly developed model aimed at determining the upper flammability limit of alkane combustion in the presence of air-CO_2 mixtures. To validate their model, the researchers conducted experiments utilizing mixtures of n-butane-CO_2 and n-isopentane-CO_2. An empirical model to calculate the heat of combustion of saturated carbons contained in power-generating fuels was conducted by Sagadeev and Sagadeev [14]. An additive scheme was used to compute the heat values of the fuels.

Numerous programs and methodologies are available for the determination of the AFT, energy, and chemical equilibrium in combustion processes. Reynolds made significant improvements to the STANJAN program by integrating the element potential method for the analysis of chemical equilibrium [15]. Furthermore, Gordon and McBride developed a computer program named CAE, which facilitates the calculation of complex chemical equilibrium compositions and their respective applications [16]. A computer program was introduced by Olikara [17] to facilitate the computation of the gas constant, enthalpy, and internal energy, as well as the partial derivatives of combustion products with temperature. On a different note, Eriksson developed a MATLAB graphical user interface (GUI) for a chemical equilibrium program. Program users can choose different fuels and products for combustion reactions. This program is designed to calculate the thermal properties of reactants and products involved in a combustion reaction [18]. By employing six major gas species and simplifying the governing equations into fourth-degree polynomials, Jarungthammachote successfully developed a quartic combustion model (QRC). This model allows for the determination of the AFT, molar fractions of combustion products, and properties [19]. GASEQ ver 0.79 is a software application that provides a solution for a range of calculations such as the AFT, compositions, adiabatic compression and expansions, equilibrium constants, and shock calculations [20]. The work of Frenklach et al. [21] highlights an optimized chemical reaction mechanism that pertains to the combustion of methane. By utilizing this model, one can effectively determine the ignition and flame properties associated with various methane combustion combinations. Cantera is a web-based application that computes the adiabatic flame temperature (AFT) based on the pressure, initial temperature, fuel type, and equivalence ratio [22]. EQLBRM, developed by Pratt [23], is a Fortran IV-based text-oriented program that draws inspiration from DOS/UNIX. Yazar and Demir [24] proposed a graphical user interface (GUI) for calculating the combustion, emission, and engine performance of internal combustion engines using the Java programming language. The effects of the in-cylinder pressure, temperature, mass ratios of gas components, and engine performance characteristics were obtained for different excess air ($\phi > 1$) ratios. Jakoubek et al. [25] developed online control algorithms to control flue gas emissions in power plants. The developed model is in communication with a real system to prevent unwanted emission values and to sustain energy production in power plants. Yildiz [26] used the chemical equilibrium-based combustion model and evaluated the effects of H_2 addition to biogases with different CO_2 contents. According to their study, the addition of H_2 increases the adiabatic flame temperature.

Efficiency evaluation of a thermal system heavily depends on the energy release derived from combustion. Although energy is predominantly employed in thermodynamic calculations and analyses, its quality cannot be adequately assessed solely based on these factors. The occurrence of irreversibilities during a process results in energy losses, consequently leading to the generation of entropy. Exergy, also known as available energy, holds significant importance in the design of thermal systems due to its association with exergy destruction or irreversibility [27]. The investigation carried out by Bouras and Khaldi focused on the exergy analysis of diverse fuel compositions. Their objective was to identify the ideal fuel composition that would result in both efficient combustion and economic benefits [28]. The exergy analysis performed by Koyun delved into the combustion of diverse octane and hydrogen blends within internal combustion engines. The study encompassed both complete and partial incomplete combustion scenarios, providing valuable insights into the energy efficiency of these processes [29]. Ozsari and Ust provided a detailed exergy analysis of the oxy-combustion process for diesel, gasoline, methane, and natural gas [30]. An exergy analysis of methane combustion was performed by Silva and Rouboa [31]. Som and Datta provided a comprehensive explanation regarding the occurrence of irreversibilities and exergy destruction during the combustion of gaseous, liquid, and solid fuels. Their study concluded that the primary cause of irreversibilities stemmed from the internal thermal exchange, which was directly associated with the presence of high-temperature gradients resulting from heat transfer in combustion reactions [32]. In their research, Taniguchi et al. [33] conducted a comprehensive investigation into the exergy and energy of combustion, power generation, and heat pump processes.

The present study aims to provide a comprehensive understanding of the combustion of alkanes and alcohols through the development of a graphical user interface (GUI) program. The developed program can calculate fuels of 12 alkanes (C_nH_{2n+2}) and 4 alcohols ($C_nH_{2n+1}OH$), which are not calculated in other programs. This program allows for the calculation of the AFT, energy, and exergy using both dry and moist air for a range of different fuels. Displaying the combustion equation on the screen, which was not available in previous studies, is an important feature.

Another aspect of this study involves investigating the correlation between the relative humidity and adiabatic flame temperature, as well as exergy destruction during combustion reactions. It is important to highlight that the impact of relative humidity on these parameters has not been thoroughly examined in previous research, as evidenced by the existing literature. Therefore, the present study aims to address this gap and contribute valuable insights to this research area.

2. Method

Within a combustion reaction, the fuel undergoes oxidation, resulting in the release of heat energy from the system. The primary oxidizer employed in this reaction is predominantly air. Air is modeled as a mixture of O_2 and 3.76 kmol of N_2. Combustion reactions of alkanes that are in the C_nH_{2n+2} form can be written for stoichiometric combustion (ϕ = 1), rich combustion (ϕ > 1), and excess air combustion (ϕ < 1) as Equations (1)–(3).

$$C_nH_{2n+2} + \varnothing(O_2 + 3.76\ N_2)\ \rightarrow\ a\ CO_2 + b\ H_2O + c\ N_2\ \varnothing = 1 \tag{1}$$

$$C_nH_{2n+2} + \varnothing(O_2 + 3.76\ N_2)\ \rightarrow\ a\ CO_2 + b\ H_2O + c\ N_2 + d\ O_2\ \varnothing > 1 \tag{2}$$

$$C_nH_{2n+2} + \varnothing(O_2 + 3.76\ N_2)\ \rightarrow\ a\ CO_2 + b\ H_2O + c\ N_2 + d\ CO\ \varnothing < 1 \tag{3}$$

Combustion reactions of alcohols that are in the $C_nH_{2n+1}OH$ form are modeled in Equations (4) and (5).

$$C_nH_{2n+1}OH + \varnothing(O_2 + 3.76\ N_2)\ \rightarrow\ a\ CO_2 + b\ H_2O + c\ N_2\ \varnothing = 1 \tag{4}$$

$$C_nH_{2n+1}OH + \varnothing(O_2 + 3.76\ N_2)\ \rightarrow\ a\ CO_2 + b\ H_2O + c\ N_2 + d\ O_2\ \varnothing > 1 \tag{5}$$

The air–fuel ratio (AF) can be defined as the proportion of the mass of air (m_{air}) to the mass of fuel (m_{fuel}), as depicted in Equation (6).

$$AF = \frac{m_{air}}{m_{fuel}} \tag{6}$$

In cases where the air is not dry, the partial pressure of water vapor can be calculated by taking into account the relative humidity (φ_{air}) of the air and the saturation pressure at the given air temperature ($P_{sat@Tair}$).

$$P_{v,air} = \varphi_{air} P_{sat@Tair} \tag{7}$$

Under ideal gas assumptions, the mole number of moist air ($N_{v,air}$) can be found by using Equation (8).

$$N_{v,air} = \left(\frac{P_{v,air}}{P_{total}}\right) N_{total} \tag{8}$$

Combustion reactions of alkanes that are in the C_nH_{2n+2} form with moist air can be written for stoichiometric combustion (ϕ = 1), rich combustion (ϕ > 1), and excess air combustion (ϕ < 1) as Equations (9)–(11).

$$C_nH_{2n+2} + \varnothing(O_2 + 3.76\ N_2) + N_{v,air}H_2O \rightarrow\ a\ CO_2 + b\ H_2O + c\ N_2\ \varnothing = 1 \tag{9}$$

$$C_nH_{2n+2} + \varnothing(O_2 + 3.76\ N_2) + N_{v,air}H_2O\ \rightarrow\ a\ CO_2 + b\ H_2O + c\ N_2 + d\ O_2\ \varnothing > 1 \tag{10}$$

$$C_nH_{2n+2} + \varnothing(O_2 + 3.76\ N_2) + N_{v,air}H_2O\ \rightarrow\ a\ CO_2 + b\ H_2O + c\ N_2 + d\ CO\ \varnothing < 1 \tag{11}$$

Combustion reactions of alcohols that are in the $C_nH_{2n+1}OH$ form with moist air are modeled in Equations (12) and (13).

$$C_nH_{2n+1}OH + \varnothing(O_2 + 3.76\ N_2) + N_{v,air}H_2O\ \rightarrow\ a\ CO_2 + b\ H_2O + c\ N_2\ \varnothing = 1 \tag{12}$$

$$C_nH_{2n+1}OH + \varnothing(O_2 + 3.76\ N_2) + N_{v,air}H_2O\ \rightarrow\ a\ CO_2 + b\ H_2O + c\ N_2 + d\ O_2\ \varnothing > 1 \tag{13}$$

According to the first law of thermodynamics, the steady-state energy balance can be written as Equation (14).

$$Q - W = \Delta H \tag{14}$$

The total enthalpy change between the reactants and products can be used to determine the energy released (Q_{out}) from the combustion system when there is no work present in the system. Equation (15) incorporates various factors such as the number of moles (N), the formation enthalpy ($\overline{h}^{\circ}_{for}$), the enthalpy of the component at a defined temperature ($\overline{h}_T$), and the reference enthalpy of the component at 1 atm and 25 °C ($\overline{h}^{\circ}_{ref}$).

$$Q_{out} = \underbrace{\sum N_R\left(\overline{h}^{\circ}_{for} + \overline{h}_T - \overline{h}^{\circ}_{ref}\right)_R}_{\substack{Energy\ in\ by\ mass \\ per\ mole\ of\ fuel}} - \underbrace{\sum N_P\left(\overline{h}^{\circ}_{for} + \overline{h}_T - \overline{h}^{\circ}_{ref}\right)_P}_{\substack{Energy\ out\ by\ mass \\ per\ mole\ of\ fuel}} \tag{15}$$

$$Q = W = 0 \tag{16}$$

In an adiabatic combustion process, the enthalpies of reactants (H_R) and products (H_P) will be equal, as expressed in Equations (17) and (18).

$$H_R = H_P \tag{17}$$

Equation (18) shows the energy balance between reactants and products. By using Equation (18), the enthalpy of the products ($\overline{h}_{T,P}$) and the adiabatic flame temperature (T_{af}) can be found. Since there is only one equation and multiple unknowns, T_{af} can be solved by trial and error or numerically.

$$\sum N_R\left(\overline{h}^{\circ}_{for} + \overline{h}_T - \overline{h}^{\circ}_{ref}\right)_R = \sum N_P\left(\overline{h}^{\circ}_{for} + \overline{h}_T - \overline{h}^{\circ}_{ref}\right)_P \tag{18}$$

The net change in the entropy of a system can be calculated with Equation (19).

$$\underbrace{S_{in} - S_{out}}_{\substack{Net\ entropy \\ transfer\ by \\ heat\ and\ mass}} + \underbrace{S_{gen}}_{\substack{Entropy \\ generation}} = \underbrace{\Delta S_{system}}_{\substack{Change\ in \\ entropy}} \tag{19}$$

The entropy generation for closed or steady-flow systems can be found in Equation (20).

$$S_{gen} = S_P - S_R + \sum \frac{Q_{out}}{T_{ambient}} \tag{20}$$

If the process is adiabatic ($Q_{out} = 0$), Equation (21) can be used.

$$S_{gen,\ adiabatic} = S_P - S_R \geq 0 \tag{21}$$

In the context of the combustion reaction, both the reactants and products are considered to be ideal gas mixtures. The entropy per mole ($\overline{s}_n$) for any component within the mixture can be determined using Equation (22). This equation takes into account various factors, including the absolute entropy ($\overline{s}^{\circ}_n$); the universal gas constant (R_u), which is typically valued at 8.3144 7 kJ/kmol K; the mole fraction of the component (y_n); the total pressure of the mixture (P_m); and the pressure of the ambient environment ($P_{ambient}$).

$$\overline{s}_n(T, P_n) = \overline{s}^{\circ}_n(T, P_0) - R_u ln \frac{y_n P_m}{P_{ambient}} \tag{22}$$

The entropy of a component can be found in Equation (23). Here, N_n is the mole number of the component.

$$S_n = N_n \overline{s}_n(T, P_n) = N_n\left[\overline{s}^{\circ}_n(T, P_0) - R_u ln \frac{y_n P_m}{P_{ambient}}\right] \tag{23}$$

Using Equations (21)–(23), entropy generation is obtained as in Equation (24).

$$S_{gen} = \sum N_P \bar{s}_P - \sum N_R \bar{s}_R + \sum \frac{Q_{out}}{T_{ambient}} \quad (24)$$

The exergy destruction in the combustion system can be found in Equation (25).

$$X_{destroyed} = T_{ambient}\, S_{gen} \quad (25)$$

Equations (1)–(25) were taken from Cengel and Boles [27].

The following assumptions were made in the calculations in this study:

- The air and products involved in the combustion process are considered to be ideal gases.
- The pressure during combustion and the pressure of the surrounding environment remain constant.

Figure 1 illustrates the flow diagram of the AFT calculation. Once the fuel type and equivalence ratio are selected, the program proceeds to calculate the combustion equilibrium constants and calculate enthalpy values [27,34]. Subsequently, the initial guess for the AFT is obtained, followed by another enthalpy calculation retrieval and the actual AFT calculation. If the calculated error falls below the tolerance limit ($\Delta\varepsilon = 1 \times 10^{-15}$), the program terminates the loop and displays the results on the screen.

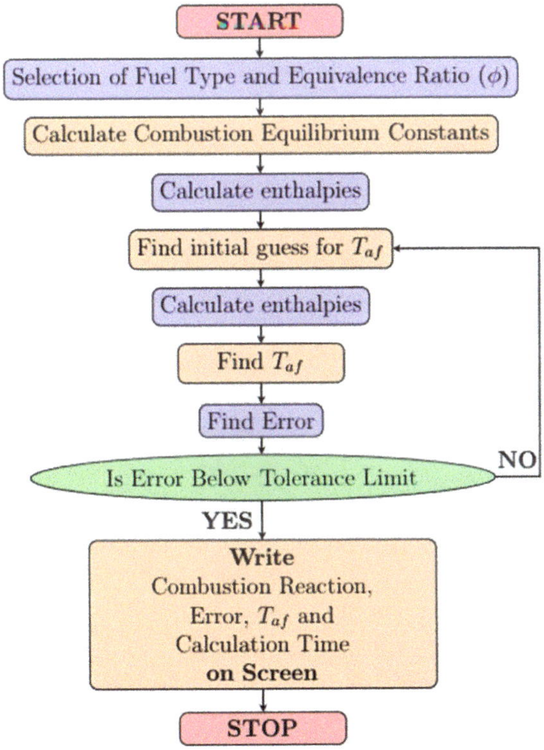

Figure 1. Flow diagram of adiabatic flame temperature (AFT) calculation.

The flow diagram depicted in Figure 2 provides a visual representation of the steps involved in calculating the AFT and conducting exergy analysis for dry and moist air. The AFT calculation procedure remains unchanged from the first flow diagram. However, when dealing with moist air, the temperature of the air (T_{air}) and the relative humidity ratio of the air (φ_{air}) are selected as initial parameters. Tables are then consulted to determine the moles of water vapor. Following the AFT calculation, the program proceeds to compute the entropy generation and exergy destruction and ultimately displays the results on the screen.

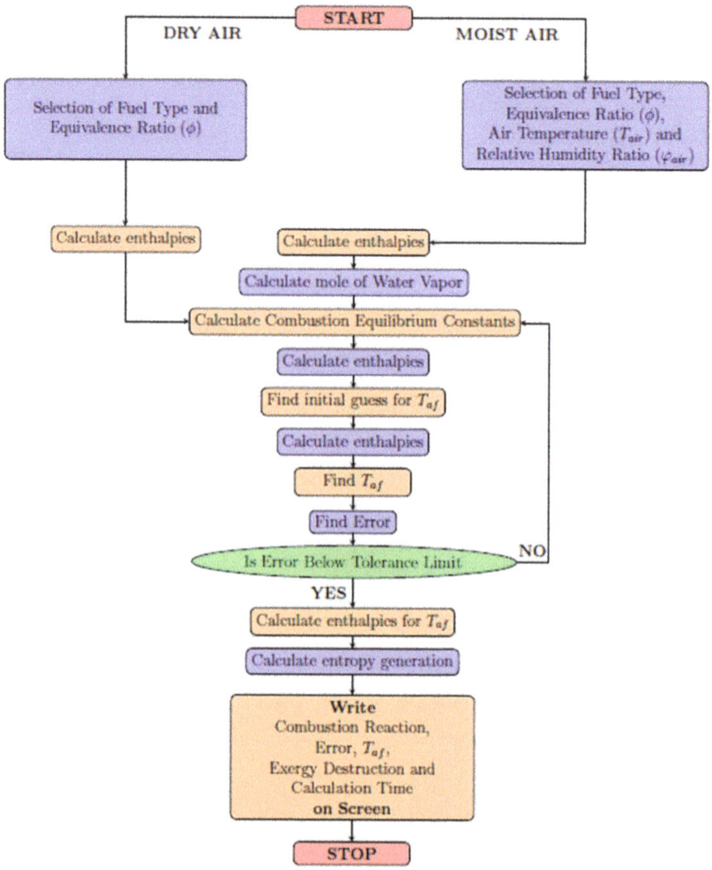

Figure 2. Flow diagram of adiabatic flame temperature (AFT) calculation and exergy analysis for dry and moist air.

Figure 3 depicts a flow diagram illustrating the process of conducting energy and exergy analyses for both dry and moist air. The user is required to make several selections and input various parameters. These include choosing the fuel type, specifying the equivalence ratio (ϕ), providing the air temperature (T_{air}), indicating the relative humidity ratio of the air (φ_{air}), and, if moist air is being used, entering the mass flow rate of fuel and the combustion temperature. In the case of moist air, additional steps are taken. Tables are consulted to determine the mole number of water vapor, which is then calculated. Subsequently, the combustion equilibrium constants are computed, followed by the determination of the energy, exergy destruction, air–fuel ratio, and mass flow rate of air. Finally, the obtained results are displayed on the screen.

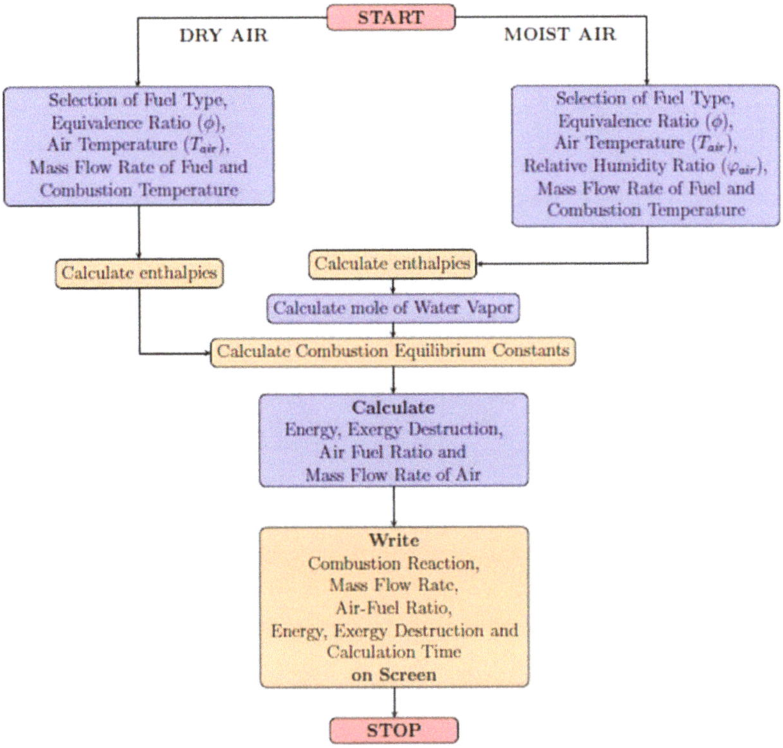

Figure 3. Flow diagram of energy and exergy analyses for dry and moist air.

3. Verification of the Calculations

The software was validated by calculating the adiabatic flame temperature of liquid octane ($C_8H_{18\ (liquid)}$) combustion for stoichiometric combustion ($\phi = 1$), rich combustion ($\phi > 1$), and excess air combustion ($\phi < 1$). The stoichiometric combustion ($\phi = 1$) of liquid ethanol ($C_2H_5OH_{\ (liquid)}$) was compared with experimental data [35]. The validation results are presented in Table 1.

Table 1. Comparison of adiabatic flame temperature values with Refs. [3,35]

.

Fuel	Equivalence Ratio	Ref. [3]	Ref. [27]	Ref. [35]	Present Study	Error (%)
$C_8H_{18\ (liquid)}$	$\phi = 0.9$	-	2395 K	-	2394.5 K	0.02
$C_8H_{18\ (liquid)}$	$\phi = 1$	-	2286 K	-	2285.9 K	0.004
$C_8H_{18\ (liquid)}$	$\phi = 4$	961.8 K	962 K	-	962.03 K	0.003
$C_2H_5OH_{\ (liquid)}$	$\phi = 1$	-	-	2242 K	2292 K	2.2

The adiabatic flame temperature and exergy destruction for the combustion of methane (CH_4) at 25 °C and 1 atm with an equivalence ratio of $\phi = 1.5$ were compared with those given by the developed software, and the results are presented in Table 2.

Table 2. Comparison of adiabatic flame temperature and exergy destruction values with Ref. [3].

Fuel	Equivalence Ratio	Ref. [3]	Present Study	Error (%)
		Adiabatic Flame Temperature		
CH_4	$\phi = 1.5$	1789 K	1789.04 K	0.002
		Exergy Destruction		
CH_4	$\phi = 1.5$	288,000 kJ/kmol	288,247 kJ/kmol	0.08

The heat transfer rate and exergy destruction for the combustion of liquid propane ($C_3H_{8(liquid)}$) at 25 °C and 0.4 kg/min mass flow rate with air at 12 °C and an equivalence ratio of $\phi = 2.5$ were compared with those given by the developed software, and the results are presented in Table 3.

Table 3. Comparison of heat transfer rate and exergy destruction values with Ref. [3].

Fuel	Equivalence Ratio	Ref. [3]	Present Study	Error (%)
		Heat Transfer Rate		
$C_3H_{8(liquid)}$	$\phi = 2.5$	28.853 kJ/s	28.858 kJ/s	0.017
		Exergy Destruction		
$C_3H_{8(liquid)}$	$\phi = 2.5$	1,120,510 kJ/kmol	1,118,672 kJ/kmol	0.16

As can be seen from Tables 1–3, the results from the Ozan Combustion Calculator (OCC) are in good agreement with the results obtained from references. In light of these results, it can be concluded that the developed graphical user interface (GUI) exhibits a high level of accuracy and reliability.

4. Implementation

The primary interface of the OCC program empowers users to select one of five problems through its main screen. Figure 4 visually presents this main screen, where users can conveniently choose their desired problem to calculate. The selection of fuels is made by using the predefined list of fuels. In addition to fuel selection, users are prompted to input various parameters, including the equivalence ratio (ϕ), air temperature (T_{air}), relative humidity ratio of the air (φ_{air}), mass flow rate of fuel, and combustion temperature. Figure 5 showcases the comprehensive list of fuels available in the AFT calculation program.

Figure 6 shows the results from the AFT calculation of liquid octane ($C_8H_{18\,(liquid)}$) combustion with an equivalence ratio of $\phi = 1.5$ (Adiabatic Flame Temperature Calculation Module).

Figure 7 shows the AFT result and exergy analysis of benzene (C_6H_6) combustion with an equivalence ratio of $\phi = 1$ (Adiabatic Flame Temperature and Exergy Calculation Module).

The AFT result and exergy analysis of butane (C_4H_{10}) combustion with an equivalence ratio of $\phi = 0.8$ at 300 K and $\varphi_{air} = 0.9$ is presented in Figure 8 (Adiabatic Flame Temperature of Combustion With Moist Air and Exergy Calculation Module).

In Figure 9, energy and exergy analyses of methyl alcohol (CH_3OH) combustion with an equivalence ratio of $\phi = 1.2$ at 288 K, fuel mass flow rate of 12 kg/s, and combustion temperature of 1250 K are presented (Energy and Exergy Module).

Figure 10 shows the results of energy and exergy analyses of ethyl alcohol (C_2H_5OH) combustion with an equivalence ratio of $\phi = 1.2$ at 300 K and $\varphi_{air} = 0.95$, fuel mass flow rate of 5 kg/s, and combustion temperature of 1250 K (Fuel Combustion With Moist Air Energy and Exergy Calculation Module).

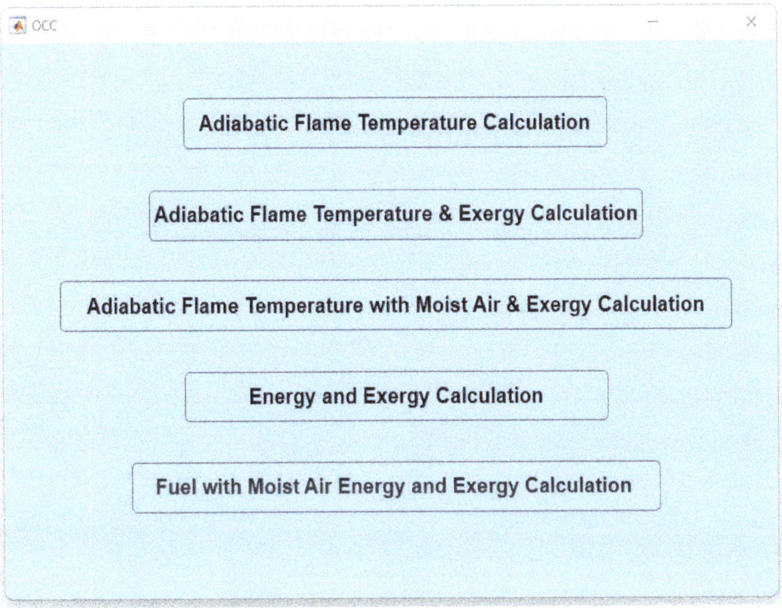

Figure 4. The main screen of the OCC.

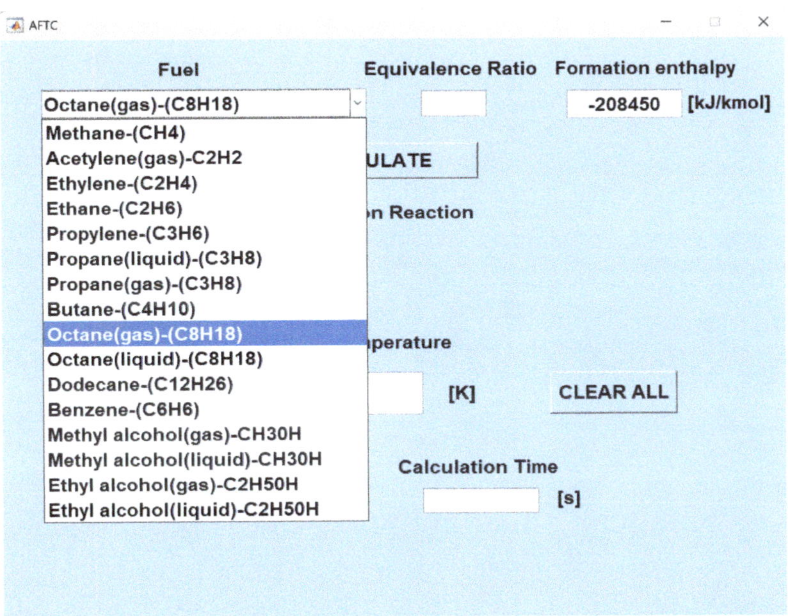

Figure 5. Selection of fuels from the AFT calculation program.

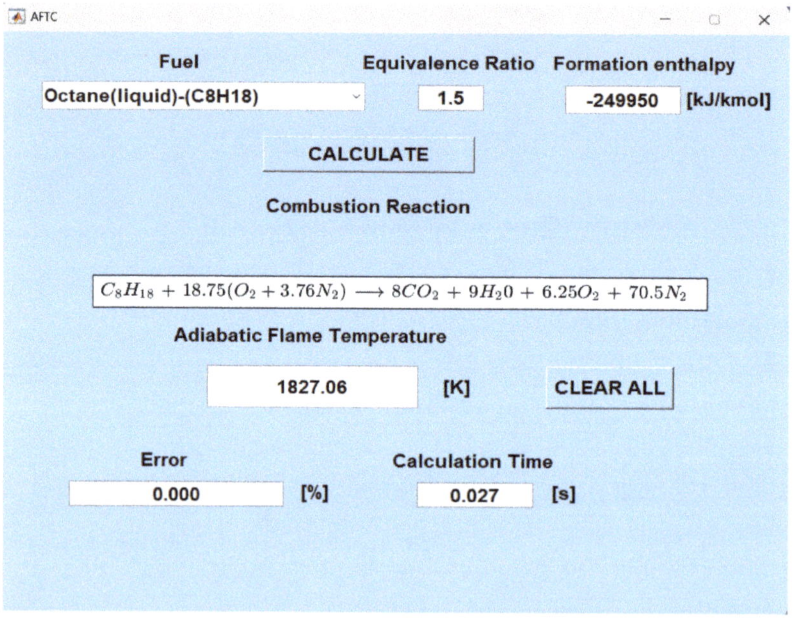

Figure 6. AFT calculation of liquid octane ($C_8H_{18\,(liquid)}$) combustion with an equivalence ratio of ϕ = 1.5.

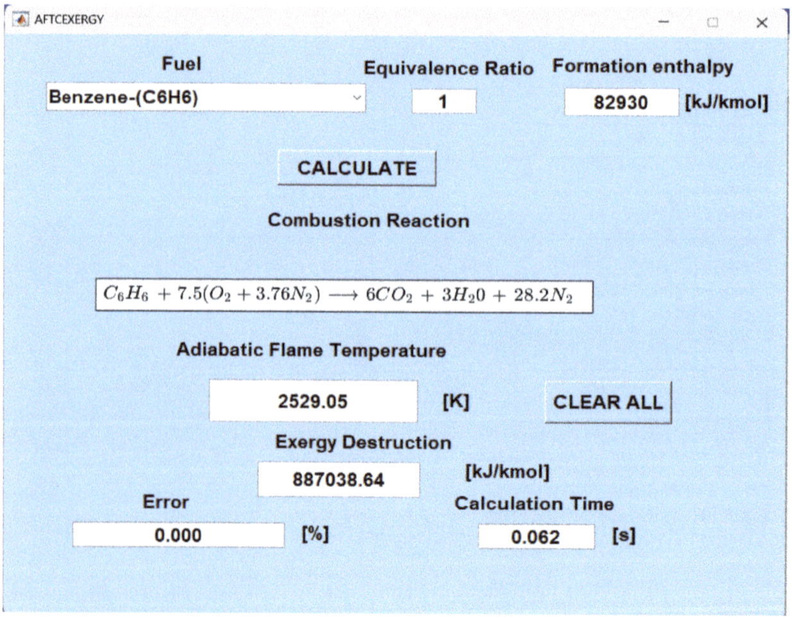

Figure 7. AFT and exergy analysis of benzene (C_6H_6) combustion with an equivalence ratio of ϕ = 1.

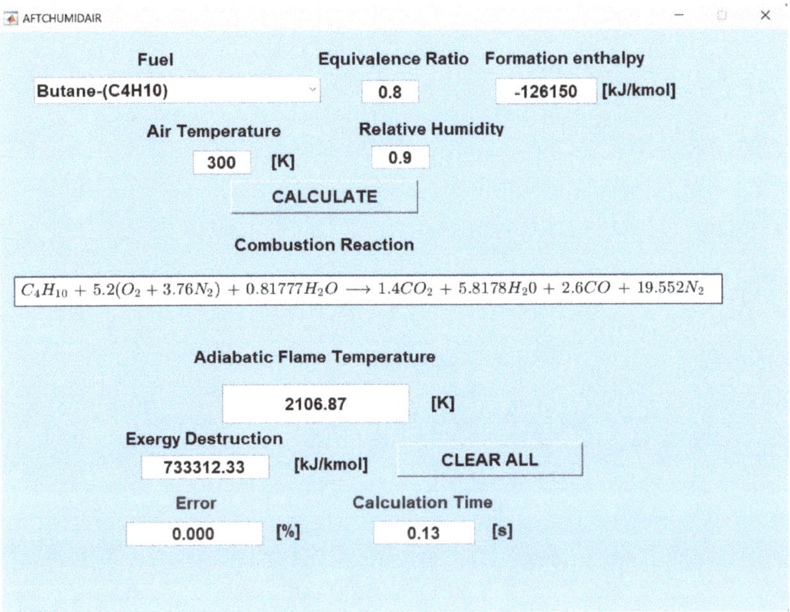

Figure 8. AFT and exergy analysis of butane (C_4H_{10}) combustion with an equivalence ratio of ϕ = 0.8 at 300 K and $\varphi_{air} = 0.9$.

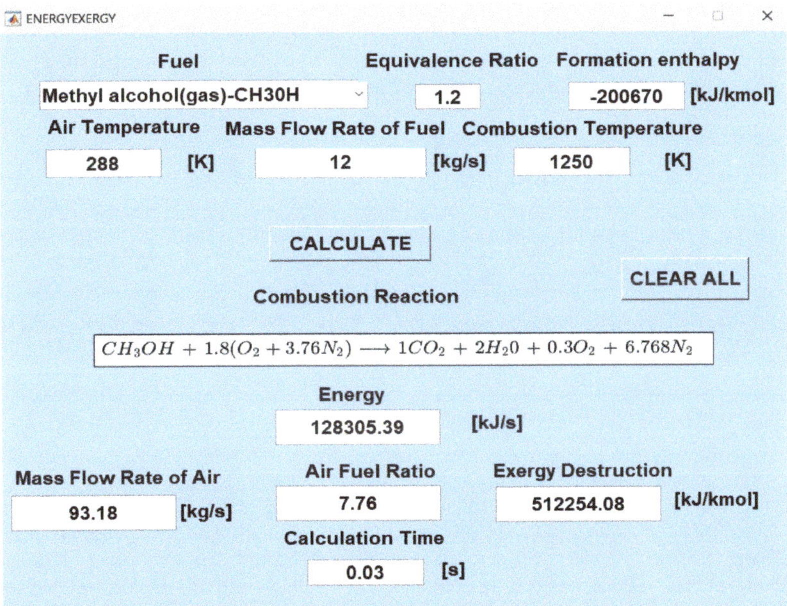

Figure 9. Energy and exergy analyses of methyl alcohol (CH_3OH) combustion with an equivalence ratio of ϕ = 1.2 at 288 K and a fuel mass flow rate of 12 kg/s.

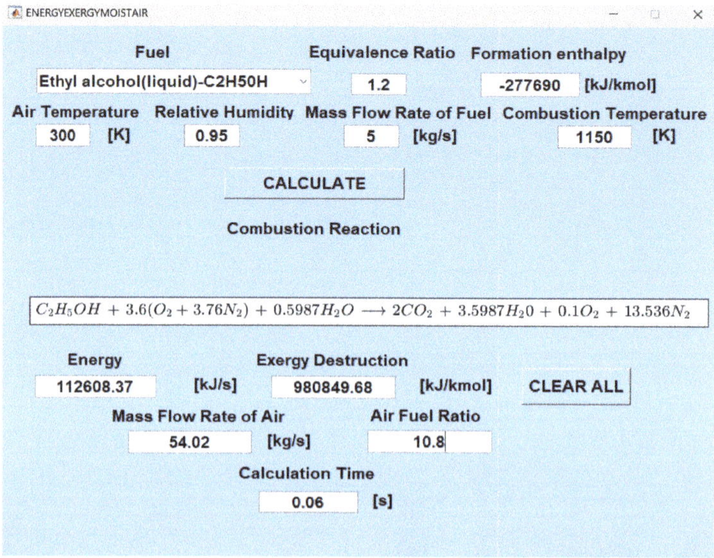

Figure 10. Energy and exergy analyses of ethyl alcohol (C_2H_5OH) combustion with an equivalence ratio of ϕ = 1.2 at 300 K and $\varphi_{air} = 0.95$, with a fuel mass flow rate of 5 kg/s.

5. Effect of Relative Humidity on Adiabatic Flame Temperature and Exergy Destruction

This study aimed to examine the impact of relative humidity on the adiabatic flame temperature and exergy destruction in combustion systems. The calculations were conducted while keeping the equivalence ratio at 1 and the air temperature at 300 K. Figure 11 provides a graphical representation of the variations in adiabatic flame temperature with the relative humidity for all the fuels analyzed in this research.

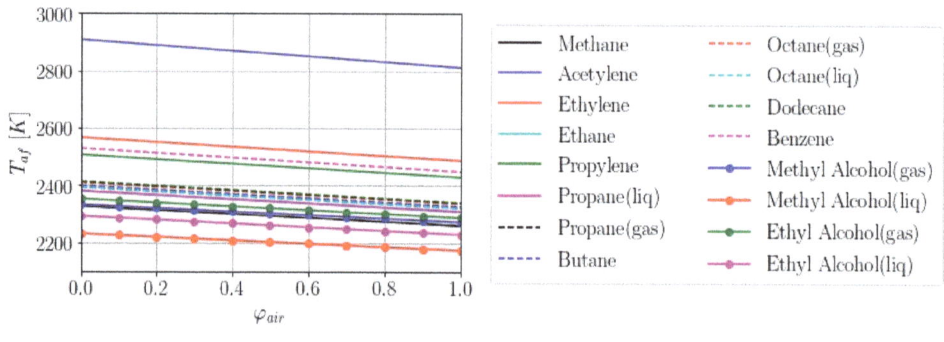

Figure 11. Variation in the adiabatic flame temperature (AFT) with relative humidity (φ_{air}).

The derivation of an equation to express the change in adiabatic flame temperature (AFT) with relative humidity (φ_{air}) was accomplished through the application of linear regression analysis. The findings from both Figure 11 and Equation (21) demonstrate a clear negative linear relationship between the adiabatic flame temperature (AFT) and relative humidity (φ_{air}). The coefficients of Equation (26), which are specific to the fuels examined in this study, are provided in Table 4.

$$T_{af} = a\varphi_{air} + b \tag{26}$$

Table 4. Coefficients for Equation (26).

Fuel	a	b	r^2
Methane (CH_4)	−68.284	2328.5	0.999
Acetylene (C_2H_2)	−96.605	2909.7	0.999
Ethylene (C_2H_4)	−80.18	2567.7	0.999
Ethane (C_2H_6)	−71.182	2382	0.999
Propylene (C_3H_6)	−77.774	2507.7	0.999
Propane (Liquid) (C_3H_8)	−72.011	2381.7	0.999
Propane (Gas) (C_3H_8)	−72.534	2395.6	0.999
Butane (C_4H_{10})	−72.946	2400.5	0.999
Octane (Gas) (C_8H_{18})	−73.701	2411.1	0.999
Octane (Liquid) (C_8H_{18})	−73.123	2395.6	0.999
Dodecane ($C_{12}H_{26}$)	−73.967	2414.8	0.999
Benzene (C_6H_6)	−80.915	2530.2	0.999
Methyl Alcohol (Gas) (CH_3OH)	−61.366	2334.1	0.999
Methyl Alcohol (Liquid) (CH_3OH)	−57.982	2231.9	0.999
Ethyl Alcohol (Gas) (C_2H_5OH)	−66.645	2354.8	0.999
Ethyl Alcohol (Liquid) (C_2H_5OH)	−64.404	2293.5	0.999

Another expression for exergy destruction ($X_{destroyed}$) variation during the combustion of the studied fuels with relative humidity (φ_{air}) is given in Equation (27). According to Equation (27), there is a quadratic positive correlation between exergy destruction ($X_{destroyed}$) and relative humidity (φ_{air}). The coefficients of Equation (27) are represented in Table 5.

$$X_{destroyed} = c\varphi_{air}{}^2 + d\varphi_{air} + e \tag{27}$$

Table 5. Coefficients for Equation (27).

Fuel	c	d	e	r^2
Methane (CH_4)	1187.1	2067.5	235,602	0.999
Acetylene (C_2H_2)	1442.5	4584.1	284,419	0.999
Ethylene (C_2H_4)	1780.5	4137.2	354,237	0.999
Ethane (C_2H_6)	2075.4	3967	428,564	0.999
Propylene (C_3H_6)	2683.4	5959.5	543,265	0.999
Propane (Liquid) (C_3H_8)	2958.2	5794.6	639,774	0.999
Propane (Gas) (C_3H_8)	2972.4	5831.9	621,081	0.999
Butane (C_4H_{10})	3881.4	7678.6	813,235	0.999
Octane (Gas) (C_8H_{18})	7431	15220	2,000,000	0.999
Octane (Liquid) (C_8H_{18})	7392.8	15087	2,000,000	0.999
Dodecane ($C_{12}H_{26}$)	10964	22786	2,000,000	0.999
Benzene (C_6H_6)	4271.5	11620	887,039	0.999
Methyl Alcohol (Gas) (CH_3OH)	896.75	1439.9	214,778	0.999
Methyl Alcohol (Liquid) (CH_3OH)	887.72	1318.1	243,462	0.999
Ethyl Alcohol (Gas) (C_2H_5OH)	1826.9	3163.6	404,904	0.999
Ethyl Alcohol (Liquid) (C_2H_5OH)	1793.9	3044.4	435,784	0.999

6. Conclusions

A novel software application, utilizing a graphical user interface (GUI), was developed to facilitate the analysis of combustion processes based on chemical equilibrium. This software offers users the ability to compute the adiabatic flame temperature and determine the equilibrium products resulting from the combustion of commonly used alkanes and alcohols. Notably, this software surpasses its predecessors by providing a visual representation of the combustion equation directly on the screen, a feature that was previously unavailable in combustion calculation software. Furthermore, the software's functionality extends to calculating the energy released to the ambient surroundings and evaluating the exergy destruction associated with the combustion reaction.

This study also delved into the influence of relative humidity on the adiabatic flame temperature and exergy destruction. The results indicate a negative linear correlation between relative humidity and the adiabatic flame temperature. Furthermore, a positive quadratic relationship was identified between exergy destruction and relative humidity. The expressions describing the adiabatic flame temperature and exergy destruction with respect to relative humidity demonstrate a strong correlation coefficient value ($r^2 = 0.999$).

The easy-to-use graphical interface allows the user to control the input parameters to see the combustion equation and the analysis results. It is also helpful for lecturers when teaching the subject of combustion, giving homework, and preparing or correcting for exams.

This software application is suitable for implementation in both undergraduate- and graduate-level courses that encompass combustion reactions. The utilization of this program effectively reduces the occurrence of calculation errors. Furthermore, the fuel library within the program can be expanded to include additional fuel blends and incorporate other functionalities, thereby paving the way for potential commercialization in the future. Moreover, this program finds applications in various industrial sectors, such as the design of combustion chambers, internal combustion engines, gas turbines, and burners.

Funding: This research received no external funding.

Data Availability Statement: Data are contained within the article.

Conflicts of Interest: The author declares no conflict of interest.

Nomenclature

AF	Air–fuel ratio
$\bar{h}^{\circ}_{for}$	Formation enthalpy (kJ/kmol)
$\bar{h}^{\circ}_{ref}$	Reference enthalpy (kJ/kmol)
$\bar{h}_T$	Enthalpy at given temperature (kJ/kmol)
H	Enthalpy (kJ)
m_{air}	Mass of air (kg)
m_{fuel}	Mass of fuel (kg)
N	Mole number of component
N_{total}	Total mole number
$N_{v,air}$	Mole number of moist air
P_{total}	Total pressure (kPa)
$P_{v,air}$	Partial pressure of water vapor (kPa)
Q	Heat energy (kJ)
R_u	Universal gas constant (kJ/kmol K)
$\bar{s}^{\circ}_n$	Absolute entropy (kJ/kmol K)
S	Entropy (kJ/K)
T	Temperature (K)
W	Mechanical energy (kJ)
y_n	Mole fraction of component
X	Exergy destruction (kJ/kmol)
Greek Letters	
φ	Relative humidity
ϕ	Equivalence ratio
Subscripts	
gen	Generated
n	Component
R	Reactants
P	Products

References

1. Kuo, K.K. *Principles of Combustion*; John Wiley&Sons, Inc.: Hoboken, NJ, USA, 2005.
2. Poinsot, T.; Veynante, D. *Theoretical and Numerical Combustion*, 2nd ed.; RT Edwards, Inc.: Morningside, Australia, 2005; Volume 53.
3. Van Wylen, G.J.; Borgnakke, C.; Sonntag, R.E. *Fundamentals of Thermodynamics*; Wiley, John Wiley&Sons, Inc.: Hoboken, NJ, USA, 2009.
4. Williams, F.A. *Combustion Theory*; CRC Press: Boca Raton, FL, USA, 1985.
5. Ufot, E.; Lebele-Alawa, B.T.; Douglas, I.E. On the adiabatic flame temperature in gas-turbine combustors. *Eur. J. Sci. Res.* **2011**, *64*, 387–393.
6. Glaude, P.A.; Fournet, R.; Bounaceur, R.; Molière, M. Adiabatic flame temperature from biofuels and fossil fuels and derived effect on NOx emissions. *Fuel Process. Technol.* **2010**, *91*, 229–235. [CrossRef]
7. Kayadelen, H.K. Effect of natural gas components on its flame temperature, equilibrium combustion products and thermodynamic properties. *J. Nat. Gas Sci. Eng.* **2017**, *45*, 456–473. [CrossRef]
8. Kayadelen, H.K. A multi-featured model for estimation of thermodynamic properties, adiabatic flame temperature and equilibrium combustion products of fuels, fuel blends, surrogates and fuel additives. *Energy* **2018**, *143*, 241–256. [CrossRef]
9. Law, C.K.; Makino, A.; Lu, T.F. On the off-stoichiometric peaking of adiabatic flame temperature. *Combust. Flame* **2006**, *145*, 808–819. [CrossRef]
10. Ozsari, I.; Ust, Y.; Kayadelen, H.K. Comparative Energy and Emission Analysis of Oxy-Combustion and Conventional Air Combustion. *Arab. J. Sci. Eng.* **2021**, *46*, 2477–2492. [CrossRef]
11. Movileanu, C.; Mitu, M.; Brinzea, V.; Musuc, A.; Mocanu, M.; Razus, D.; Oancea, D. Adiabatic flame temperature of fuel-air mixtures in isobaric and isochoric combustion processes. *Rev. Chim.* **2011**, *62*, 376–379.
12. Sakhrieh, A. The adiabatic flame temperature and laminar flame speed of methane premixed flames at varying pressures. *Acta Period. Technol.* **2019**, *50*, 220–227. [CrossRef]
13. Wu, M.; Shu, G.; Chen, R.; Tian, H.; Wang, X.; Wang, Y. A new model based on adiabatic flame temperature for evaluation of the upper flammable limit of alkane-air-CO_2 mixtures. *J. Hazard. Mater.* **2018**, *344*, 450–457. [CrossRef]
14. Sagadeev, E.V.; Sagadeev, V.V. Calculation of the heats of combustion of aromatic hydrocarbons contained in power-generating fuel. *High Temp.* **2006**, *44*, 530–534. [CrossRef]
15. Reynolds, W.C. *The Element Potential Method for Chemical Equilibrium Analysis: Implementation in the Interactive Program STANJAN*; Stanford University: Stanford, CA, USA, 1986.
16. Gordon, S.; McBride, B.J. *Computer Program for Calculation of Complex Chemical Equilibrium Compositions and Applications I. Analysis*; NASA RP-1311; NASA: Washington, DC, USA, 1994; p. 58.
17. Olikara, C.; Borman, G.L. A computer program for calculating properties of equilibrium combustion products with some applications to I.C. engines. *SAE Tech. Pap.* **1975**, 7191. [CrossRef]
18. Eriksson, L. CHEPP—A chemical equilibrium program package for Matlab. *SAE Tech. Pap.* **2004**, *113*, 730–741. [CrossRef]
19. Jarungthammachote, S. Simplified model for estimations of combustion products, adiabatic flame temperature and properties of burned gas. *Therm. Sci. Eng. Prog.* **2020**, *17*, 100393. [CrossRef]
20. Morley, C. Gaseq Chemical Equilibrium Program. Available online: http://www.gaseq.co.uk/ (accessed on 15 December 2023).
21. Frenklach, M.; Wang, H.; Goldenberg, M.; Smith, G.P.; Golden, D.M.; Bowman, C.T.; Hanson, R.K.; Gardiner, W.C.; Lissianski, V. *GRI-Mech—An Optimized Detailed Chemical Reaction Mechanism for Methane Combustion*; SRI International: Menlo Park, CA, USA, 1995.
22. Cantera. Available online: http://elearning.cerfacs.fr/combustion/tools/adiabaticflametemperature/index.php (accessed on 15 December 2023).
23. Pratt, D.T. Available online: https://aste-classes.usc.edu/utility/flame_temperature.php (accessed on 15 December 2023).
24. Yazar, O.; Demir, B. Development of the Software Program That Can Account the Combustion, Emission and Engine Performance Values of Internal Combustion Engines. *Eur. J. Technol.* **2022**, *12*, 129–138. [CrossRef]
25. Jakoubek, P.; Hanzal, L. Matlab As a Tool for Fossil Fuel Power Plant Combustion Control. Available online: https://dsp.vscht.cz/konference_matlab/MATLAB10/full_text/044_Jakoubek.pdf (accessed on 15 December 2023).
26. Yıldız, M. Chemical equilibrium based combustion model to evaluate the effects of H_2 addition to biogases with different CO_2 contents. *Int. J. Hydrogen Energy* **2023**, *52*, 1334–1344. [CrossRef]
27. Cengel, Y.; Boles, M.A. *Thermodynamics an Engineering Approach*, 5th ed.; McGraw-Hill: New York, NY, USA, 2011.
28. Bouras, F.; Khaldi, F. Optimization of Combustion Efficiency Using a Fuel Composition Based on CH_4 and/or H_2. *Russ. J. Appl. Chem.* **2020**, *93*, 1954–1959. [CrossRef]
29. Koyun, T. Exergy Analysis of Different Blends of Hydrogen and Octane for Combustion Conditions of Internal Combustion Engine. *J. Nanoelectron. Optoelectron.* **2018**, *13*, 1656–1668. [CrossRef]
30. Ozsari, I.; Ust, Y. Effect of varying fuel types on oxy-combustion performance. *Int. J. Energy Res.* **2019**, *43*, 8684–8696. [CrossRef]
31. Silva, V.; Rouboa, A. Methane combustion: An exergy analysis. *AIP Conf. Proc.* **2011**, *1389*, 179–182. [CrossRef]
32. Som, S.K.; Datta, A. Thermodynamic irreversibilities and exergy balance in combustion processes. *Prog. Energy Combust. Sci.* **2008**, *34*, 351–376. [CrossRef]
33. Taniguchi, H.; Mouri, K.; Nakahara, T.; Arai, N. Exergy analysis on combustion and energy conversion processes. *Energy* **2005**, *30*, 111–117. [CrossRef]

34. Stull, D.R.; Prophet, H. *JANAF Thermochemical Tables*, 2nd ed.; National Institute of Standards and Technology: Gaithersburg, MD, USA, 1971.
35. Zhou, J.; Lu, C.; Xu, C.; Yu, Z. Experimental and Numerical Study on the Effect of Hydrogen Addition on Laminar Burning Velocity of Ethanol–Air Mixtures. *Energies* **2022**, *15*, 3114. [CrossRef]

processes

Article

Simulation and Experimental Design of Magnetic Fluid Seal Safety Valve for Pressure Vessel

Zhenggui Li [1,2,*], Ziyue Wang [2,3,4], Changrong Shen [3], Wangxu Li [2], Yanxiong Jiao [1], Chuanshi Cheng [5], Jie Min [5] and Yuanyuan Li [5]

1 School of Engineering, Qinghai Institute of Technology, Xining 810016, China
2 Key Laboratory of Fluid and Power Machinery, Ministry of Education, Xihua University, Chengdu 610039, China
3 Jiangsu Water Resources Co., Ltd., East Route of South-to-North Water Diversion, Nanjing 215341, China
4 Sichuan Huadian Luding Hydropower Co., Ltd., Luding 610041, China
5 China Yangtze Power Co., Ltd., Maintenance Plant, Yichang 443000, China
* Correspondence: lzhgui@mail.xhu.edu.cn

Abstract: This article focuses on the safety valve of pressure vessels, and a new ferrofluid sealing device for pressure vessel safety valves is developed based on a special magnetic circuit. A combined method of numerical calculation and experimental analysis is used to study the relationship between seal clearance, number of seals, pole slot width, pole tooth height, pole tooth width, and the sealing pressure of the ferrofluid sealing device. The research results show that seal clearance and pole tooth width have a significant impact on the sealing performance, and as the dimensions increase, the sealing pressure decreases. As the number of seals, pole tooth height, and slot width increase, the sealing performance initially improves and then decreases. This phenomenon is attributed to the increase in magnetic reluctance in the magnetic circuit. In experimental studies, when the excitation current of the electromagnet is 240 mA and the coil turns number 30, the sealing capacity is 61.22 kPa. When the excitation current is 200 mA and the coil turns number 80, the sealing capacity is 168.24 kPa. The experiments demonstrate the compensating ability of magnetic fluid seals in combination with safety valve seals, confirming that combined seals have higher reliability compared to conventional mechanical seals.

Keywords: pressure vessel safety valve; ferrofluid sealing device; polar tooth parameters; sealing capacity

1. Introduction

Pressure vessels, as process equipment for storing and carrying high-pressure gases or liquids, play a significant role in the industrial field. The sealing performance of pressure vessels is crucial to the reliability, stability, and safety of energy systems. Therefore, there is widespread attention being paid to exploring long-term effective sealing methods [1–3]. Traditional sealing methods used in process equipment like pressure vessels, such as flange connections, threaded seals, and rubber seals, often struggle to achieve high reliability in sealing.

Issues such as the loosening of flange seals during prolonged operation, minute leaks due to imprecise machining of threaded seals, and deformations of rubber seals in corrosive environments make achieving reliable sealing challenging. Once these sealing problems occur, they can result in unnecessary economic losses, and even threaten human safety.

Ferrofluid is a new type of intelligent liquid that can control the shape of the magnetic fluid without contact using a magnetic field to achieve the desired effect. By guiding it into the sealing clearance, a liquid sealing ring can be formed. Compared to rubber seals and mechanical end-face seals, ferrofluid seals have better corrosion resistance and reliability. It has been reported that magnetic fluid seals in tank periscopes have a service life of up to 10 years [4]. The excellent performance of ferrofluid seals has also been demonstrated in industrial products such as chemical equipment reactors [5], transportation equipment centrifugal pumps, and other applications [6]. With advancements in ferrofluid preparation processes and an understanding of the dynamic mechanism of ferrofluids, high-performance hydrophobic ferrofluid and high-temperature-resistant ferrofluids can now be mass-produced [7,8]. The theory of magnetic particle chaining under field conditions has been gradually accepted [9], the mechanisms of magnetic adhesion effects under magnetic and thermal fields have been revealed [10], and improvements have been made to the ferrofluid transport equations at small-micro-gap scales [11]. These new technologies and insights provide theoretical references for optimizing ferrofluid seals. However, under liquid environments or high-pressure conditions, ferrofluid seals still fall short of ideal performance. A new approach to resolving this issue involves combining ferrofluid seals with traditional seals in a combined sealing structure. For instance, Li [12] integrated ferrofluid with threaded seals in hydraulic equipment to achieve long-term effective dynamic sealing in liquid environments. Yang [13] proposed structural improvements by combining ferrofluid with labyrinth seals, using the groove part of the labyrinth seal as the magnetic assembly of the ferrofluid seal, significantly enhancing the pressure resistance of the ferrofluid seal. Wang [14] introduced an embedded ferrofluid seal combination structure based on mechanical end face seals, achieving tight leak-free sealing at 500 rpm and 0.5 MPa conditions. Other improvements combining general mechanical structures, such as adding grooved shafts [15] and buffer channels [16], have positively contributed to improving the stability of ferrofluid seal interfaces. For safety valve sealing, the drawbacks of traditional sealing forms have led to many potential hazards. Exploring new sealing methods has become an urgent technological challenge. With the development of the basic theory and application technology of ferrofluid seals, utilizing ferrofluid seals to supplement mechanical end face seals has emerged as a new approach to achieving the tight sealing of safety valves.

This article focuses on the research of ferrofluid seals for safety valves. It starts by selecting the type of ferrofluid based on high-temperature and high-pressure sealing environments. The stable mechanism of ferrofluids is analyzed from the perspectives of ferrofluid preparation and performance, considering thermodynamic parameters of ferrofluids to theoretically calculate pressure resistance. Subsequently, a coupling structure between mechanical end faces and ferrofluid seals is designed based on magnetic principles. Numerical methods are employed to study the impact of structural parameters on the performance of ferrofluid seals. Finally, experimental verification is conducted using a test rig for safety valve ferrofluid seals to confirm the feasibility and reliability of the seal structure. This study aims to provide technical guidance for addressing stability sealing issues in safety valves.

2. Preparation and Performance Analysis of Ferrofluids

The general temperature range that ferrofluids can withstand is between −20 °C and 105 °C; temperatures that are too high or too low can lead to the failure of the ferrofluid. According to investigations, the highest heat resistance values for ferrofluids based on diester, perfluoropolyether oil, and silicone oil carriers are 200 °C, 300 °C, and 350 °C, respectively. The safety valves developed in this study are mostly intended for high-temperature and high-pressure conditions. Although the temperature resistance values of perfluoropolyether oil and silicone oil are equally superior, perfluoropolyether oil is more expensive in terms of material cost. After comprehensive consideration, choosing silicone oil as the base fluid for preparing ferrofluids better meets production needs [17–21]. Chemical coprecipitation is a common method for synthesizing Fe_3O_4 nanoparticles. By applying chemical coprecipitation, the required magnetic particles can be rapidly obtained. The chemical reaction process is represented as $Fe^{2+} + 2Fe^{3+} + 8OH^- \rightarrow Fe_3O_4 + 4H_2O$. Taking 40.5 g of $FeCl_3 \cdot 6H_2O$ and 18.6 g of $FeCl_2 \cdot 7H_2O$, we dissolve them together in 250 mL of deionized water at 65 °C, then raise the temperature to 70 °C under constant stirring.

Next, we add 25.6 g of sodium hydroxide solution to the mixed solution while maintaining a stirring speed of 300 rpm, and continue stirring for 1 h to produce black Fe_3O_4 magnetic particles.

Subsequently, we maintain a constant reaction temperature of 50 °C using a water bath and slowly add hydrochloric acid in increments to the magnetic particles, adjusting the pH to the range of 4–5. We then add 90 mL of anhydrous ethanol and 6 mL of tetraethyl orthosilicate (TEOS) to the solution to initiate the hydrolysis and condensation reactions. Under continuous and uniform stirring, we let the mixture react for 1.5 h (hydrolysis process), and then gradually add sodium hydroxide solution during the reaction process to adjust the pH to 9–10 (condensation process) to ensure a thorough reaction progress. After completion, we place the solution in a magnetic field for separation. Firstly, we wash the separated material with anhydrous ethanol, followed by rinsing five to seven times with deionized water, to eventually obtain core–shell nano-magnetic particles of $Fe_3O_4@SiO_2$ (as shown in Figure 1a).

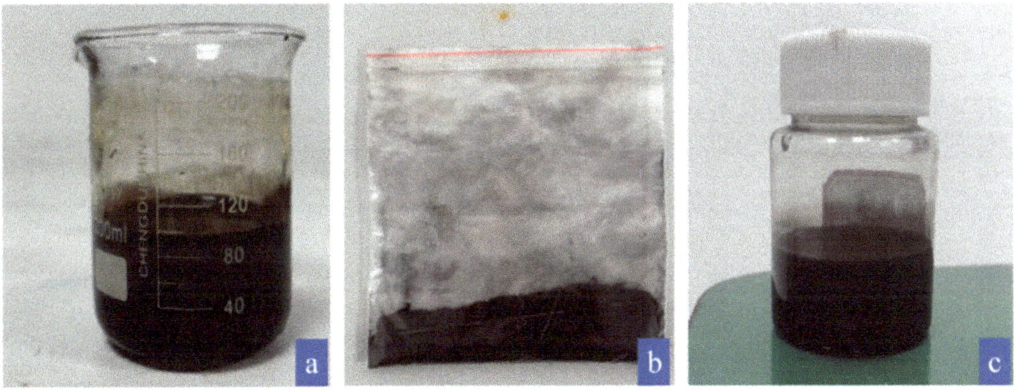

Figure 1. Samples of ferrofluid preparation process ((**a**) core–shell nano-magnetic particles; (**b**) dried nano-magnetic particles; (**c**) silicone-based ferrofluid).

We successively add 10 mL of deionized water, 10 mL of silane coupling agent A1120, and 180 mL of anhydrous ethanol, and gradually add hydrochloric acid to adjust the pH to the range of 4–5, stirring continuously for 2 h to ensure secondary coating on the $Fe_3O_4@SiO_2$. Upon completion of the secondary coating, we wash the sample three to five times with anhydrous ethanol and deionized water, then dry the sample at 80 °C for 12 h. Finally, we grind the sample, pack it, and store it to obtain dried $Fe_3O_4@SiO_2@A1120$ particles (Figure 1b).

According to reference materials [22], the magnetic particles are mixed with the base carrier liquid in a ratio of 1:3 (10 g of coated particles to 30 g of dimethyl silicone oil) at a temperature of 70 °C. We then stir uniformly at a mechanical stirring speed of 500 rpm for 2 h to ensure the even dispersion of the magnetic particles in the base liquid. After removing the undissolved large particles, a clear upper liquid is obtained, which is the stably dispersed silicone-based ferrofluid (Figure 1c). The hydrolysis and condensation processes of tetraethyl orthosilicate are depicted in Figure 2.

The crystalline structures of Fe_3O_4, $Fe_3O_4@SiO_2$, and $Fe_3O_4@SiO_2@A120$ particles were analyzed using the X-ray diffractometer Ultima IV (Rigaku Corporation, Tokyo, Japan), as shown in Figure 3. The consistency in diffraction peak positions indicates that the particle coating behavior does not alter the crystal structure of Fe_3O_4. Fourier transform infrared spectroscopy analysis of surface-treated nano-magnetic particles was conducted using the Nicolet iS20 Fourier Transform Infrared Spectrometer (Thermo Scientific, Waltham, MA, USA).

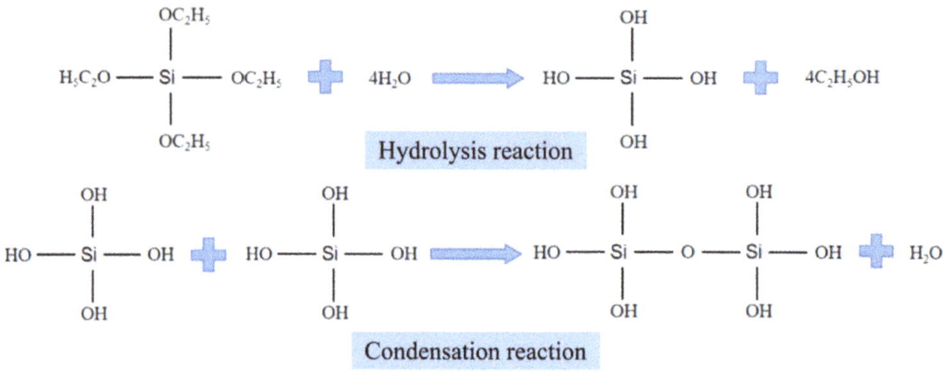

Figure 2. Hydrolysis and polycondensation of ethyl orthosilicate.

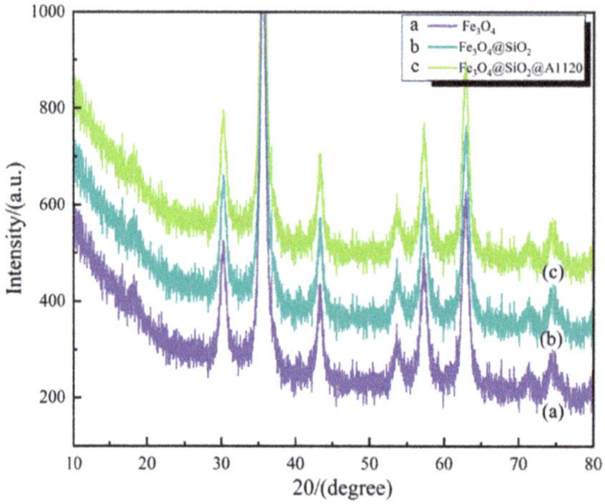

Figure 3. XRD of Fe_3O_4@SiO_2@A1120 nanoparticles.

Figure 4 presents the infrared spectra of Fe_3O_4@SiO_2 particles (a) and Fe_3O_4@SiO_2@A1120 particles (b). The data reveal two peaks corresponding to Si–O bonds at 1117.86 cm^{-1} and 1032.18 cm^{-1} on curve b, with additional peaks such as 1204.68 cm^{-1} corresponding to C-N stretching vibrations and 1449.57 cm^{-1} corresponding to C-H bending vibrations, matching the characteristic peaks of A1120. On curve a, only one peak corresponding to a Si–O bond is observed at 1018.67 cm^{-1}, along with characteristic hydroxyl group peaks at 1625.94 cm^{-1} and 1621.05 cm^{-1}.

The thermogravimetric analysis (TGA) of surfactant-modified particles was performed using the Mettler TGA2 Thermal Analyzer (Zurich, Switzerland), generating the weight loss curve depicted in Figure 5. It is observed that before 160 °C, the mass reduction of the particles primarily results from the evaporation of residual water attached to the surface of nanoparticles. As the temperature increases further, the decrease in thermal weight curve occurs due to the decomposition and evaporation of the surfactant enveloping the particle's surface. Additionally, within a temperature range of room temperature to 600 °C, Fe_3O_4@SiO_2@A1120 particles exhibit a weight loss 6.4% higher than that of Fe_3O_4@SiO_2 particles, confirming the successful encapsulation of A1120 on the surface of Fe_3O_4@SiO_2 particles.

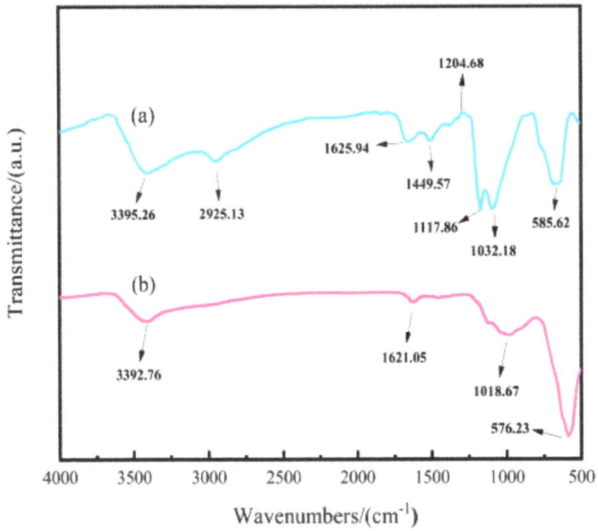

Figure 4. Infrared spectrum of Fe_3O_4@SiO_2@A1120 nanoparticles.

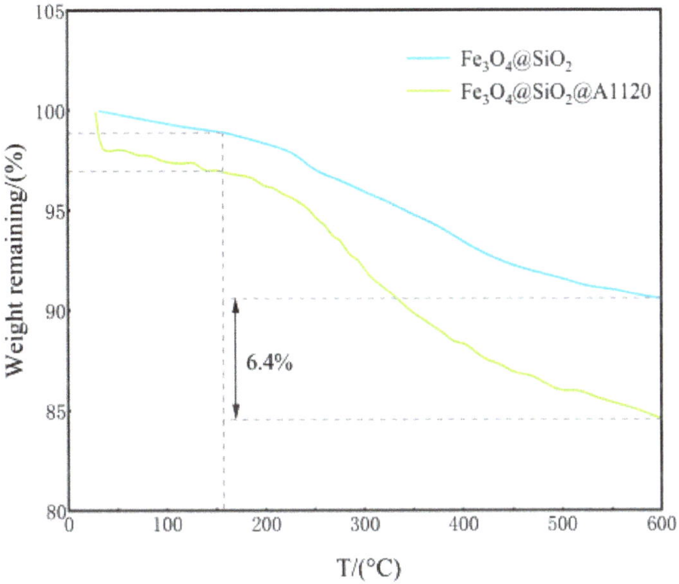

Figure 5. TG diagram of Fe_3O_4@SiO_2@A1120 nanoparticles.

Finally, after allowing the silicone-based ferrofluid to stand undisturbed in a gravitational field for 1 to 4 weeks, it was observed that the magnetic particles maintained a stable dispersed state with the base carrier liquid, without any signs of separation (Figure 6). In conclusion, the silicone-based ferrofluid prepared by chemical coprecipitation exhibits good stability and heat resistance.

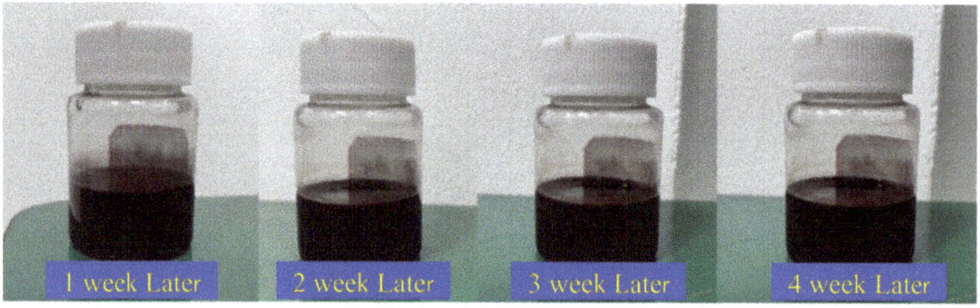

Figure 6. A static diagram of a silicone oil-based ferrofluid over a period of time.

3. Theoretical Analysis

Under the magnetic field environment, the ferrofluid present in the clearance forms a liquid sealing ring, as shown in Figure 7. The Bernoulli equation for ferrofluids derived from the general fluid Navier–Stokes equations is presented as Equation (1) [23–25]. Solving the physical problem depicted in Figure 7 yields the following: in a static seal configuration where the motion velocity v of the ferrofluid is 0 and the sealing medium and ferrofluid are at the same horizontal level, the relative elevation h is 0. Consequently, the expression for the maximum pressure Δp_{max} in the ferrofluid seal is obtained as given in Equation (2), where the minimum value of magnetic field intensity H_{min} in the seal clearance approaches 0, simplifying Equation (2) to Equation (3). In a multi-stage sealing device, the total sealing pressure is the sum of the individual single-stage sealing pressures, expressed as in Equation (4) [26–28].

$$p + \frac{1}{2}\rho v^2 + \rho g h - \mu_0 \int_0^H M dH = C \tag{1}$$

$$\Delta p_{max} = \mu_0 \int_{H_1}^{H_2} M dH \tag{2}$$

$$\Delta p_{max} = \mu_0 M_s H_{max} \tag{3}$$

$$\Delta p_{max} = N \mu_0 M_s H_{max} \tag{4}$$

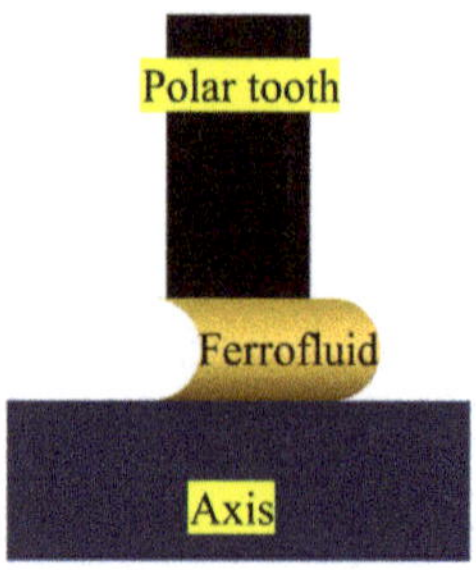

Figure 7. Distribution of magnetic liquid in the seal gap under static pressure seal.

In the equations, p represents fluid pressure, ρ indicates fluid density, g is the gravitational acceleration, μ_0 stands for the relative magnetic permeability of the ferrofluid, M denotes the magnetization intensity of the ferrofluid, M_s is the saturation magnetization intensity of the ferrofluid, H represents magnetic field intensity, H_{max} and H_{min} denote the maximum and minimum values of magnetic field intensity, and N signifies the number of sealing stages.

The transport equation for magnetic fluid in the gap is derived from Newton's second law. Ferrofluids working inside a safety valve must satisfy the Navier–Stokes (N-S) equations for ferrofluids. The significant difference between the N-S equations for ferrofluids and those for conventional fluids lies in the inclusion of the magnetic body force μ_0MH as a source term (Equations (5)–(7)).

$$\frac{\partial \mu_0 MH}{\partial x} - \frac{1}{\rho}(\frac{\partial p}{\partial x}) + \eta \nabla^2 v_x = \frac{dv_x}{dt} \quad (5)$$

$$\frac{\partial \mu_0 MH}{\partial y} - \frac{1}{\rho}(\frac{\partial p}{\partial y}) + \eta \nabla^2 v_y = \frac{dv_y}{dt} \quad (6)$$

$$\frac{\partial \mu_0 MH}{\partial z} - \frac{1}{\rho}(\frac{\partial p}{\partial z}) + \eta \nabla^2 v_z = \frac{dv_z}{dt} \quad (7)$$

where v_x, v_y and v_z are the velocity components in the x, y, and z directions, and η is the kinematic viscosity of the ferrofluid.

4. Numerical Model

The sealing structure of a safety valve generally employs a mechanical end face seal, as depicted in Figure 8a. The sealing performance deteriorates rapidly when the end face exhibits wear or unevenness. Leveraging the adsorption and flow characteristics of ferrofluids under the influence of a magnetic field, a magnetic circuit structure is designed to effectively distribute the ferrofluid uniformly in the gap between two contacting planes. As illustrated in Figure 8b, the upper part of the mechanical seal end face is designed with a polar tooth structure that perfectly fits the lower end face. With the application of a magnetic field, the polar teeth exhibit magnetization, creating a magnetic field gradient on the sealing end face, thus forming a ferrofluid seal structure. The magnetic field is supplied by an electromagnet; when the circuit is connected, the electromagnet acts as a magnetic source, generating a magnetic field based on the principle of electromagnetic induction. The magnetic field lines will extend along high-permeability materials, following Kirchhoff's magnetic circuit law. The variation in cross-sectional area of the polar tooth structure causes a sharp increase in magnetic induction intensity.

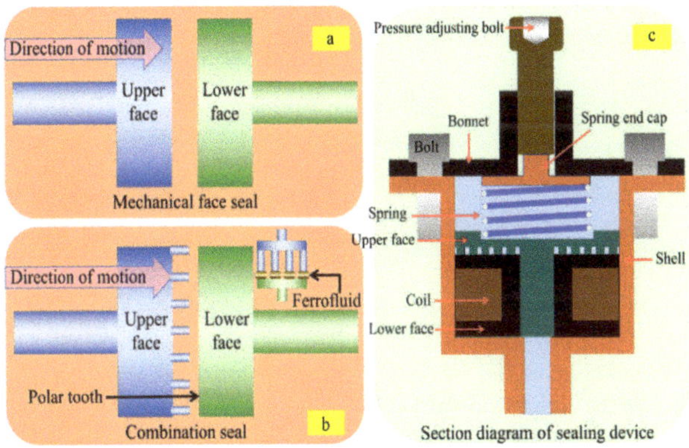

Figure 8. Ferrofluid seal safety valve model diagram ((**a**) mechanical seal concept diagram; (**b**) composite seal concept diagram; (**c**) section diagram of sealing structure).

Iron magnetic particles within the ferrofluid, possessing permanent magnetic dipole moments, react to the external magnetic field, aligning in chains and being fixed in the gap by magnetic forces, forming a liquid "O" ring seal. This effectively isolates the internal medium of the pipeline from the external environment. In the absence of a magnetic field, high-pressure gas enters the pipeline at high temperatures. Once the pressure exceeds the set pressure, the sealing end face is pushed off the electromagnet seat by the medium, resulting in seal failure. With the presence of a magnetic field, the sealing ability of the safety valve comprises both the end face seal and the ferrofluid seal. This dual approach offers an improvement in sealing capability. Additionally, when the mechanical seal end face is not smooth, the ferrofluid can fill all gaps, blocking the passage, thereby achieving a seal. Consequently, there is an enhancement in the reliability and stability of the safety valve seal. Figure 8c illustrates the overall structure of the combined seal.

Using the finite element analysis software ANSYS Maxwell 16.0, numerical calculations are conducted on the ferrofluid sealing device under different seal clearances (l_x), numbers of sealing stages (n), polar tooth groove widths (l_s), polar tooth heights (l_h), and polar tooth widths (l_l). The axially symmetric structure of the ferrofluid sealing device allows for simplification using a two-dimensional model in numerical simulations. According to the requirements of small-pressure equipment safety valves (rated pressure < 0.8 MPa), a ferrofluid seal design for the safety valve is developed.

The ferrofluid seal structure comprises an electromagnet, a sealing top cover, a sealing bottom end, and the ferrofluid, as shown in Figure 9. In the numerical model, the electromagnet is constructed with a copper material, while the sealing top cover and polar teeth are made of 45# steel. The ferrofluid is composed of custom materials with a density of 1.279 g/cm^3, a saturation magnetization intensity of 281.1242 Gs, a thermal conductivity of 0.2 W/(m·K), and a relative magnetic permeability of 1.05. Boundary conditions for the model are set to Balloon (zero magnetic field at infinity, magnetic field tangential to boundaries but allowed to pass through).

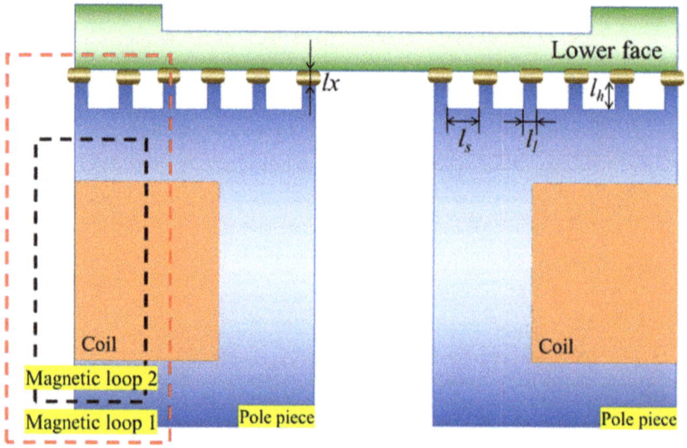

Figure 9. Structural diagram of the magnetic fluid seal part of the safety valve device.

The air domain size is set to 100% of the overall model. The radial bisector line between the sealing top cover and polar teeth is chosen as the study object. By exporting the data obtained from numerical simulations of the magnetic fluid sealing device under various seal clearances, sealing stages, polar tooth heights, polar tooth widths, and polar tooth groove widths, maximum magnetic induction intensity values are derived, then used to calculate the numerical sealing pressure.

The ferrofluid sealing system consists of two magnetic circuits. In the primary magnetic circuit, the magnetic lines generated by the coil start from the north (N) pole, successively passing through the magnetic poles, gaps, sealing top cover, air, and another magnetic pole, before returning to the south (S) pole. In the secondary magnetic circuit, the magnetic lines start from the north pole, pass through a magnetic pole, the outer air, and another magnetic pole, then return to the south pole, forming magnetic leakage on the outside of the device.

5. Result Analysis

5.1. Effect of Sealing Gap on Sealing Pressure

To investigate the relationship between seal clearances and the ferrofluid sealing safety valve device, this section sets the number of sealing stages as five, the polar tooth groove width as 2.0 mm, the polar tooth height as 2.0 mm, and the polar tooth width as 1.0 mm. Within the range of 0.1–0.5 mm, with a variation gradient of 0.1 mm, the study explores the relationship between seal clearances and magnetic sealing pressure. Magnetic field maps of the sealing devices under different clearance conditions are shown in Figure 10.

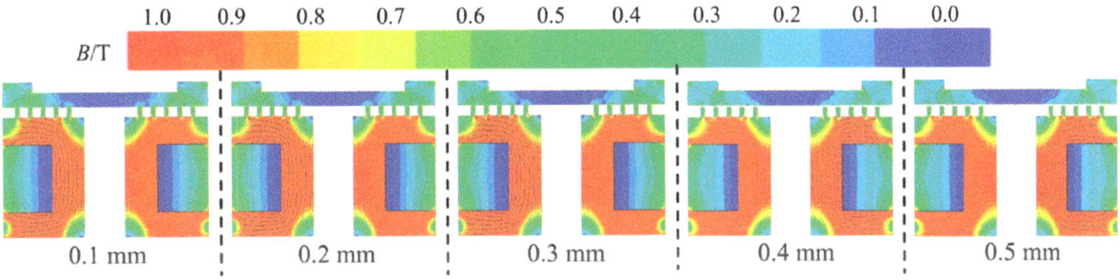

Figure 10. Cloud map of magnetic field distribution of sealing device at different gaps.

The key to the ferrofluid sealing device lies in the effect of the gap magnetic field, where the sealing performance is directly related to the magnetic field gradient. The greater the magnetic field gradient, the better the sealing performance. Therefore, it is crucial to delve into the region of the magnetic field within the gap.

Figure 11 shows Magnetic induction intensity of cut-off line under different seal gap conditions. With an increase in the seal clearance, the air's magnetic reluctance in the magnetic circuit increases. As a result, the positive peak value of the magnetic induction intensity gradually decreases while the negative peak value increases. For example, at gap values of 0.1 mm and 0.5 mm, the positive peak value decreases from 0.3896 T to 0.2291 T, and the negative peak value increases from 0.0387 T to 0.101 T.

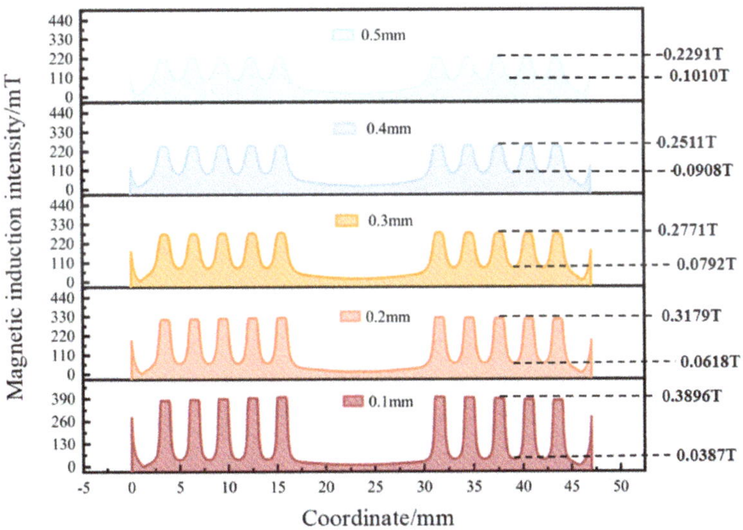

Figure 11. Magnetic induction intensity of cut-off line under different seal gap conditions.

By calculating the difference between the positive and negative peak values shown in Figure 8 to obtain the magnetic induction intensity gradient, and combining it with the sealing pressure formula, it can be inferred that the sealing capability is 35.12 kPa, 30.74 kPa, 23.75 kPa, 19.24 kPa, and 15.38 kPa for seal clearances ranging from 0.1 mm to 0.5 mm. The fundamental reason for the decrease in sealing capability lies in the increased magnetic reluctance at the seal clearance, leading to a reduction in the binding force of the ferrofluid. This manifests in the physical quantities of magnetic induction intensity as a decrease in the positive peak value, an increase in the negative peak value, and a reduction in the magnetic induction intensity gradient.

5.2. Effect of Sealing Stage on Sealing Pressure

Since the ferrofluid seal mainly acts as a compensatory seal in the safety valve device, with the primary sealing of high-pressure gas relying on mechanical seals, this section sets the number of polar teeth in the range of one to seven based on existing research results. The seal clearance is constrained to 0.1 mm, the polar tooth groove width is 2.0 mm, the polar tooth height is 2.0 mm, and the polar tooth width is 1.0 mm. With an increment of one in the number of stages, the relationship between the number of sealing stages and magnetic sealing pressure is studied. Magnetic field maps of the sealing devices under different numbers of sealing stages are shown in Figure 12.

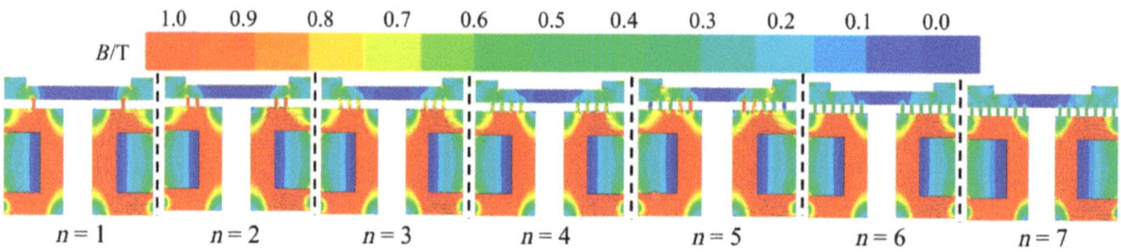

Figure 12. Cloud map of magnetic induction intensity distribution at different sealing stages.

When the number of stages is one, the magnetic induction intensity at the polar teeth portion exceeds 1 T. With more than three stages, the presence of multiple polar teeth generates additional magnetic circuits, each containing air magnetic reluctance, leading to increased magnetic losses and a subsequent decrease in the magnetic induction intensity at the polar teeth portion. However, when considering sealing capability, it is essential to comprehensively evaluate the impact of both the number of stages and the magnetic induction intensity gradient.

Figure 13 shows the distribution curves of magnetic induction intensity at the top and bottom center of the polar teeth. As the number of stages increases, both the positive and negative peak values of the magnetic induction intensity at the seal clearance gradually decrease. The positive peak values for $n = 1$ and $n = 7$ are 1.2897 T and 0.2777 T, while the negative peak values are 0.0923 T and 0.0282 T, respectively. The decrease in positive peak value is detrimental to overall sealing performance improvement, whereas the decrease in negative peak value is beneficial for enhancing overall sealing performance. The effects of these two values are opposite, but the positive peak value changes more rapidly. Overall, the magnetic induction intensity gradient at the seal clearance shows a decreasing trend.

Although the magnetic induction intensity of an individual polar tooth decreases with an increase in the number of stages, the sealing pressure of a multi-stage seal is related to the sum of the magnetic induction intensity gradients of all polar teeth. By summing up the sealing capabilities of each stage, as shown in Figure 14, the comprehensive sealing performance improves with an increase in the number of stages. For example, the sealing pressure is 23.96 kPa for $n = 1$ and 35.12 kPa for $n = 5$. However, blindly increasing the number of stages without considering the total magnetic potential does not consistently enhance sealing performance. Additional stages introduce more magnetic reluctance, including material reluctance and air gap reluctance, leading to a decrease in total sealing capability. For instance, the sealing capability is 35.53 kPa at $n = 6$ and 34.94 kPa at $n = 7$. This result indicates that the optimal number of sealing stages for the designed ferrofluid seal safety valve in this study is 6.

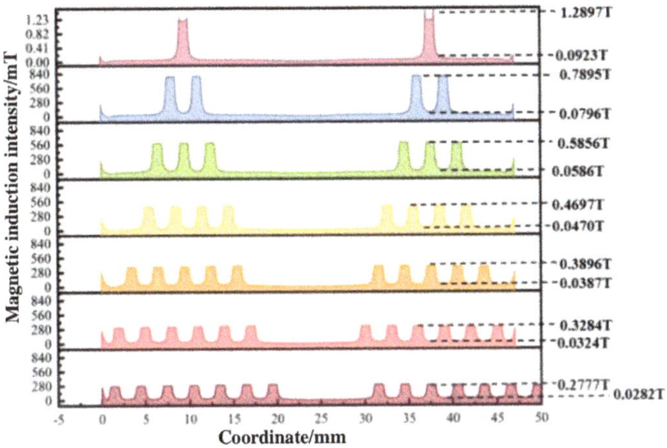

Figure 13. Magnetic induction intensity of truncated line under different sealing stages.

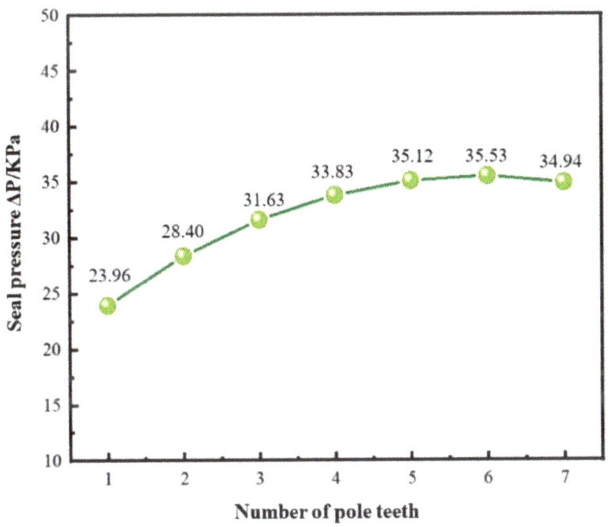

Figure 14. Seal pressure diagram under different seal clearance values.

5.3. Effect of Slot Width on Sealing Pressure

To investigate the relationship between groove width and the ferrofluid seal safety valve device, variables are strictly controlled. The parameters include a seal clearance of 0.1 mm, six polar teeth, a polar tooth height of 2.0 mm, and a polar tooth width of 1.0 mm. The groove is the space between the polar teeth, where a single-stage seal has no grooves, and in a multi-stage seal, the number of grooves is n-1. In this numerical simulation with six polar teeth, there are five grooves. This section sets the groove width in the range of 0.5–2.5 mm with a gradient of 0.5 mm to study the relationship between seal clearance and sealing pressure. Magnetic field maps of the sealing devices under different groove widths are shown in Figure 15.

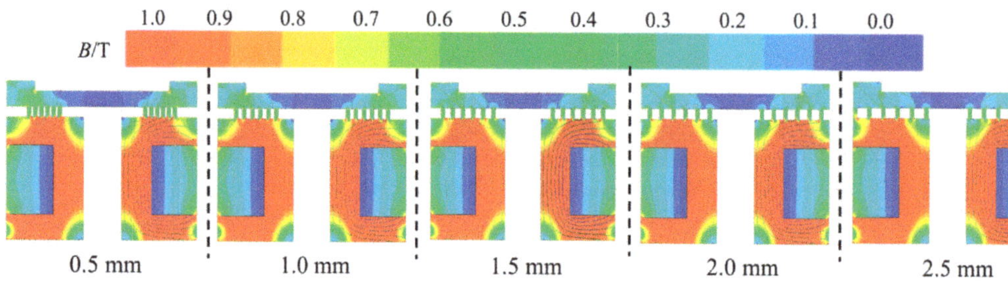

Figure 15. Magnetic field cloud image of a sealing device with different slot widths.

Increasing the groove width does not significantly alter the distribution pattern of the magnetic field. According to the magnetic induction distribution curves at the seal clearance (Figure 16), it is evident that the influence of groove width on the positive peak value of the magnetic induction intensity is relatively small. For groove widths of 0.5 mm and 2.5 mm, the positive peak values are 0.3539 T and 0.3183 T, respectively, showing a slight decrease with increasing groove width. Furthermore, with larger groove widths, there is less mutual influence between the polar teeth, weakening the leakage flux at the roots of the polar teeth and leading to a reduction in the negative peak value of the magnetic induction intensity. At groove widths of 0.5 mm and 2.5 mm, the negative peak values are 0.1262 T and 0.0257 T, respectively.

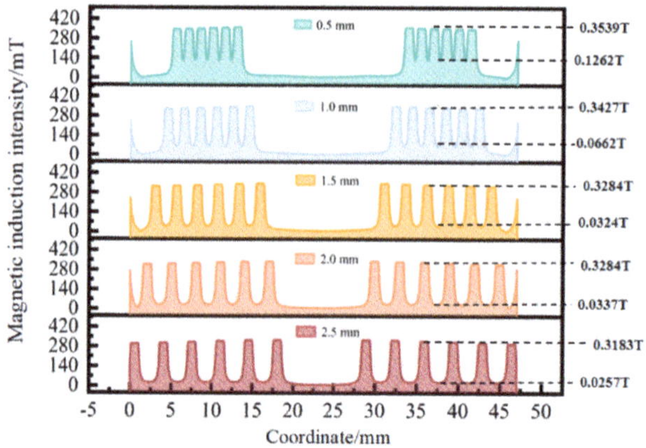

Figure 16. Magnetic induction intensity at different slot widths.

Calculations of the sealing performance at different groove widths yield sealing capabilities of 27.33 kPa, 33.19 kPa, 35.53 kPa, 35.37 kPa, and 35.12 kPa for groove widths of 0.5, 1.0, 1.5, 2.0, and 2.5 mm, respectively. The optimal sealing performance occurs at a groove width of 1.5 mm.

5.4. Influence of Pole Tooth Height on Sealing Pressure

Setting the seal clearance to 0.1 mm, with six polar teeth, a groove width of 1.5 mm, and a polar tooth width of 1.0 mm, this study explores the relationship between polar tooth height and the ferrofluid seal safety valve device. By incrementing in steps of 0.5 mm, the magnetic field and sealing performance are investigated for polar tooth heights ranging from 0.5 to 3.0 mm. Figure 17 displays the magnetic induction intensity distribution maps for different polar tooth heights.

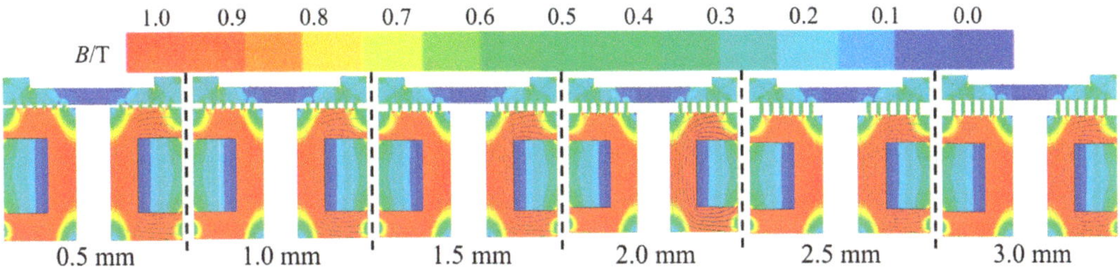

Figure 17. Magnetic field cloud image of the sealing device under different pole tooth heights.

Increasing the polar tooth height accentuates the geometric differences between the polar teeth and the top surface, facilitating the better transmission of magnetic induction along the polar teeth, thereby reducing leakage flux at the ends of the polar teeth and enhancing magnetic aggregation capabilities. When analyzing the magnetic induction intensity distribution curve at the clearance (Figure 18), it is observed that as the polar tooth height increases, the positive peak value initially increases and then decreases, while the negative peak value first decreases and then increases. Overall, there is a trend of the magnetic induction intensity gradient initially increasing with polar tooth height, and then decreasing. For polar tooth heights of 0.5, 1.0, 1.5, 2.0, 2.5, and 3.0 mm, the sealing capabilities are 30.91 kPa, 34.61 kPa, 35.22 kPa, 35.53 kPa, 35.37 kPa, and 35.19 kPa, respectively. The optimal sealing performance is achieved at a polar tooth height of 2.0 mm. The fundamental reason for the improvement in sealing performance with increased polar tooth height is the enhancement in the magnetic aggregation ability. Once the sealing performance reaches its maximum, it indicates efficient utilization of the magnetic potential. Further increasing the polar tooth height introduces additional material (polar teeth) reluctance, leading to a slight decline in sealing performance.

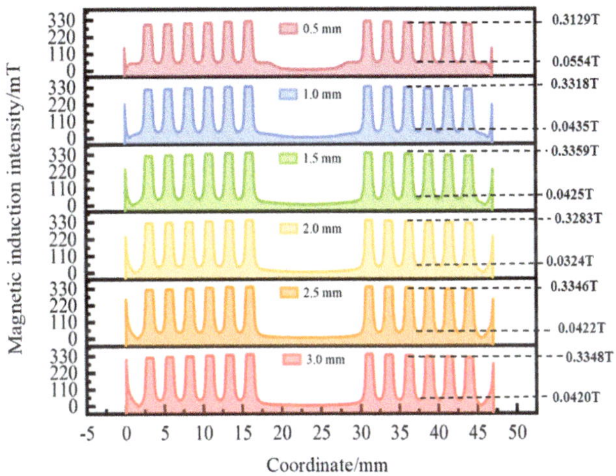

Figure 18. Magnetic induction intensity of the truncated line under different pole tooth heights.

5.5. Influence of Pole Tooth Width on Sealing Pressure

Setting the seal gap at 0.1 mm, with six pole teeth, a groove width of 1.5 mm, and a pole tooth height of 2.0 mm, the effect of pole tooth width on the performance of ferrofluid seals is studied. Widths ranging from 0.5 to 2.5 mm were selected, with a change gradient of 0.25 mm, to investigate the relationship between pole tooth width and ferrofluid seal pressure. The magnetic field maps of the sealing device under different pole tooth widths are shown in Figure 19.

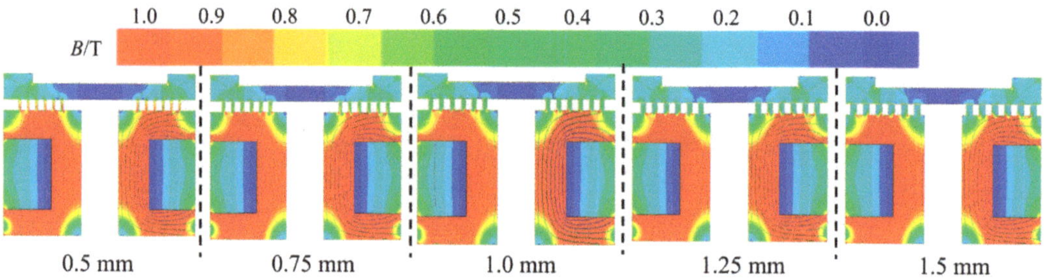

Figure 19. Magnetic field cloud image of a sealing device with different pole tooth widths.

Figure 20 shows Magnetic induction intensity of truncated line under different pole tooth widths. At a pole tooth width of 0.25 mm, the measured magnetic induction intensity on the pole tooth is significantly higher. As the pole tooth width increases, the magnetizing effect on the pole decreases, resulting in a reduction in magnetic induction intensity in certain areas of the pole tooth. Through quantitative analyses of the magnetic induction intensity distribution curve at the seal gap position, an increase in pole tooth width leads to a rapid decrease in the peak positive value of magnetic induction intensity. At pole tooth widths of 0.5 mm and 1.5 mm, the peak positive values of magnetic induction intensity are 0.4971 T and 0.2525 T, respectively. Increasing the pole tooth width by 1 mm reduces the peak positive value by a factor of 2. Although the negative peak value of magnetic induction intensity is closer to zero for large pole tooth widths, its variation can be approximately neglected compared to the variation in positive peak values. The sealing ability for pole tooth widths of 0.5, 0.75, 1.0, 1.25, and 1.5 mm was calculated using the pressure resistance formula, resulting in 52.13, 42.16, 35.53, 30.23, and 26.52 kPa, respectively. The rate of decline in sealing ability is relatively rapid. Among all parameters, the pole tooth width has a significant impact on sealing performance, second only to the seal gap's influence on sealing performance.

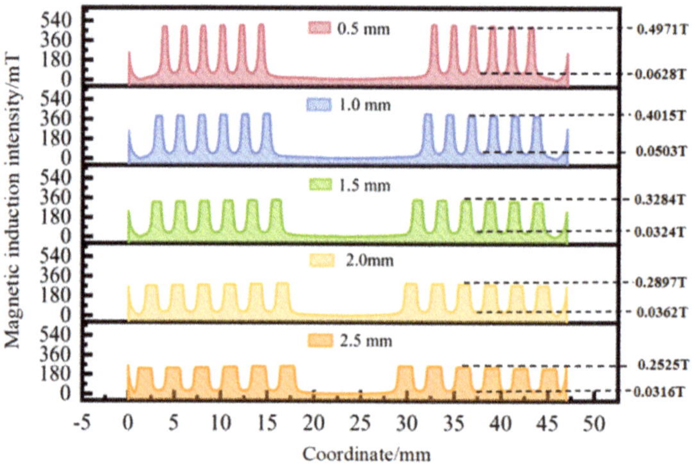

Figure 20. Magnetic induction intensity of truncated line under different pole tooth widths.

6. Test Verification

According to the safety valve technical specification "GB12241-2021" for the allowable deviation of the set pressure of a spring-loaded safety valve [29], when the set pressure is $\geq$0.5 MPa, a 3% deviation in the set pressure is allowed Figure 21 shows the test bench for the ferrofluid seal device of the safety valve. In the experiment, the coil of the electromagnet was set to 10 turns, and the excitation current was 200 mA. The pressurized gas used was nitrogen, and six sets of experiments were conducted. The test results indicate that the sealing device failed at nitrogen pressures of 1.128, 1.125, 1.125, 1.128, 1.125, and 1.125 MPa.

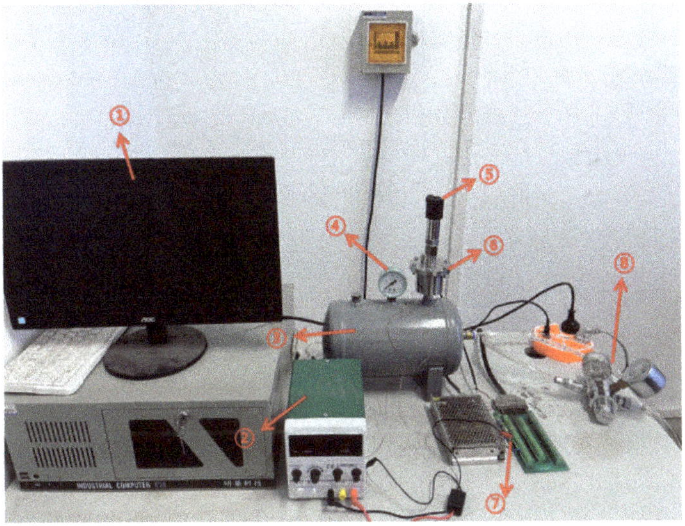

Figure 21. Test site diagram (① data acquisition computer; ②—DC power supply; ③—gas storage tank; ④—pressure gauge; ⑤—pressure sensor; ⑥—magnetic fluid seal safety valve; ⑦—data collection card; ⑧—pressure reducing valve).

Following the technical standard, to ensure seal safety, the reference limit pressure for the seal device was taken as 1.125 MPa. The allowable deviation value for this seal device was calculated to be ±0.034 MPa. The actual deviation value was found to be 0.003 MPa (1.128–1.125 MPa), which is significantly lower than the allowable deviation value. Thus, it can be concluded that the designed ferrofluid seal safety valve exhibits stability at the set pressure.

Upon energizing the DC power supply and inputting current, the ferrofluid is attracted to the gap of the sealing device, forming an "O"-type seal ring [30]. Once the pressure of the medium exceeds the pressure that the seal ring can withstand, the seal ring breaks, causing the valve to open and release the medium, thereby reducing the system pressure. Under the influence of the magnetic force, the valve disc quickly resets to close, re-establishing a stable "O"-type seal ring. As the sealing method of the safety valve combines mechanical sealing with ferrofluid sealing, this study focused on verifying the ferrofluid seal part, hence the removal of the spring part from the device during the experiment.

Figure 22 shows the relationship between the measured seal performance and the excitation current and number of coil turns in the experiment. Initially, with a fixed number of turns of 30, the excitation current increased linearly from 40 mA to 240 mA in increments of 20 mA, resulting in a linear improvement in the seal performance. When the excitation current reached 240 mA, the seal pressure was 61.22 kPa. Subsequently, by fixing the excitation current at 200 mA, the number of coil turns of the device varied linearly from 10 turns to 80 turns, showing a linear relationship between the seal pressure and the number of coil turns. Both results indicate that the seal performance improves linearly with the increase in magnetic field intensity, consistent with theoretical predictions regarding the seal pressure of the ferrofluid seal device, thus confirming the feasibility of this structure.

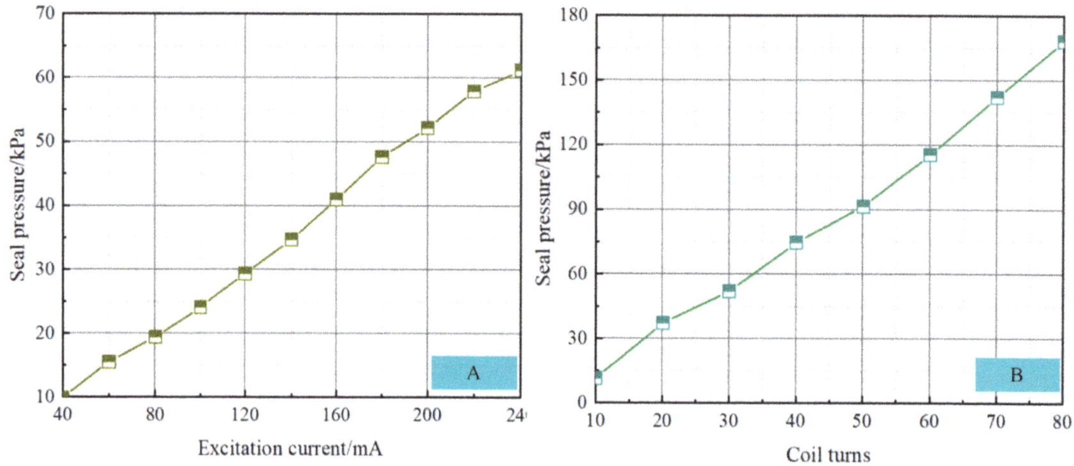

Figure 22. Experimental correlation parameter ((**A**) relation between sealing performance and excitation current; (**B**) relation between sealing performance and coil turns).

7. Conclusions

The study designed and systematically assessed a ferrofluid seal device for use in a safety valve, resulting in some important conclusions that provide preliminary validations of the feasibility of using a ferrofluid seal for safety valves. By conducting simulation calculations on the magnetic circuit, the study analyzed the impacts of key parameters of the device on the sealing capability, and further validated these through experiments, yielding specific conclusions, as follows:

(1) The study combined ferrofluid sealing with mechanical end-face sealing and designed a new type of seal structure for the safety valve. This structure enhances the stability and reliability of the safety valve seal device performance by applying ferrofluid without altering the system of the sealing device;

(2) Using numerical methods, the study investigated the effects of seal clearance, pole tooth height, pole tooth width, groove width, and number of seals on the performance of the magnetic fluid seal. The sealing capacity decreased rapidly with an increase in seal clearance. The trend shows an initial increase followed by a decrease with an increase in pole tooth length, a significant decrease with an increase in pole tooth width, an initial increase followed by a decrease with an increase in groove width, and an initial increase followed by a decrease with an increase in the number of seals. The optimal seal structural dimensions identified were a seal gap of 0.1 mm, a pole tooth height of 2.0 mm, a pole tooth width of 0.5 mm, a groove width of 1.5 mm, and six seals;

Variable	Sealing Performance	
Seal clearance	↑	↓
Pole tooth height	↑	↑↓
Pole tooth width	↑	↑
Groove width	↑	↑↓
Number of seals	↑	↑↓

(3) A novel ferrofluid seal safety valve device was designed as the main mechanical seal, augmented by the magnetic fluid seal as a compensatory seal. This design addresses the shortcomings of traditional spring-loaded safety valves by mitigating leakage issues and ensuring faster opening and closing. Testing conducted based on the manufactured ferrofluid safety valve device demonstrated stable set pressure capability and "zero leakage" features (the leakage rate of the sealed medium was

undetectable using the standard helium mass spectrometry method, with a limit of 1×10^{-11} (Pa·m^3)/s). The pressure resistance of the ferrofluid seal increases with a rise in excitation current and the number of coil turns. At an excitation current of 240 mA and 50 coil turns, the pressure resistance was 61.22 kPa, while at an excitation current of 200 mA and 80 coil turns, the pressure resistance reached 168.24 kPa.

Author Contributions: Conceptualization, Z.L. and Z.W.; methodology, W.L.; software, Y.J.; validation, C.S.; formal analysis, C.C.; investigation, C.S.; resources, J.M. and W.L.; data curation, Y.L.; writing—original draft preparation, Z.L.; writing—review and editing, Z.L.; visualization, W.L.; supervision, W.L.; project administration, W.L.; funding acquisition, Z.L. All authors have read and agreed to the published version of the manuscript.

Funding: This study was supported by the National Natural Science Foundation of China [52379092], "Kunlun Talents" Research Project [2023-QLGKLYCZX-032], Chengdu Science, Technology Bureau project [2024-YF09-0003-GX], Jiangsu South-to-North water transfer technology research and development project [JSNSBD202403], and Project funded by Changjiang Power Co., Ltd. [2324020007].

Data Availability Statement: Data can be obtained by mail from corresponding authors.

Conflicts of Interest: Authors Ziyue Wang and Changrong Shen were employed by the company Jiangsu Water Resources Co., Ltd. Author Ziyue Wang was employed by the company Sichuan Huadian Luding Hydropower Co., Ltd. Authors Chuanshi Cheng, Jie Min and Yuanyuan Li was employed by the company China Yangtze Power Co., Ltd. The remaining authors declare that the research was conducted in the absence of any commercial or financial relationships that could be construed as a potential conflict of interest.

References

1. Pan, S.; Wang, J.; Zhao, X.; Ren, C.; Xin, Y.; Wang, Z. New design and transient flow analysis of a claw vacuum pump with novel gear-claw rotors. *Vacuum* **2023**, *216*, 112470. [CrossRef]
2. Pan, S.; Wang, J.; Tan, Q.; Zhao, F.; Cui, D.; Zhang, C. Analysis and numerical simulation of claw vacuum pumps with novel three-claw rotors. *Vacuum* **2021**, *192*, 110442. [CrossRef]
3. Zhu, S.C.; Guo, J.X.; Zhao, L.F.; Li, H.H.; Zhu, F. Reliability Analysis of Seal Performance of Pressure vessel safety Valve. *China Spec. Equip. Saf.* **2008**, *24*, 68–69.
4. Fan, D. *The Study of Magnetoviscous Effects in High and Low Temperature*; Beijing Jiaotong University: Beijing, China, 2012.
5. Chen, F.; Zhang, J.; Li, Z.; Yan, S.; Li, W.; Yan, Z.; Liu, X. Effect of the surface coating of carbonyl iron particles on the dispersion stability of magnetorheological fluid. *Sci. Rep.* **2024**, *14*, 11358. [CrossRef] [PubMed]
6. Li, W.; Li, Z.; Qin, Z.; Yan, S.; Wang, Z.; Peng, S. Influence of the solution pH on the design of a hydro-mechanical magneto-hydraulic sealing device. *Eng. Fail. Anal.* **2022**, *135*, 106091. [CrossRef]
7. Li, W.; Li, Z.; Han, W.; Li, Y.; Yan, S.; Zhao, Q.; Gu, Z. Pumping-velocity variation mechanisms of a ferrofluid micropump and structural optimization for reflow inhibition. *Phys. Fluids* **2023**, *35*, 052005.
8. Chen, F.; Zhang, J.; Guo, Q.; Liu, Y.; Liu, X.; Ding, W.; Yan, S.; Yan, Z.; Li, Z. Carbonyl Iron Particles' Enhanced Coating Effect Improves Magnetorheological Fluid's Dispersion Stability. *Materials* **2024**, *17*, 4449. [CrossRef]
9. Philipse, A.P.; Maas, D. Magnetic colloids from magnetotactic bacteria: Chain formation and colloidal stability. *Langmuir* **2002**, *18*, 9977–9984. [CrossRef]
10. Li, W.; Li, Z.; Han, W.; Li, Y.; Yan, S.; Zhao, Q.; Chen, F. Measured viscosity characteristics of Fe_3O_4 ferrofluid in magnetic and thermal fields. *Phys. Fluids* **2023**, *35*, 012002. [CrossRef]
11. Li, W.; Li, Z.; Han, W.; Tan, S.; Yan, S.; Wang, D.; Yang, S. Time-mean equation and multi-field coupling numerical method for low-Reynolds-number turbulent flow in ferrofluid. *Phys. Fluids* **2023**, *35*, 125145. [CrossRef]
12. Li, Z.; Qing, J.; Shen, C.; Ma, S.; Wu, D.; Zhu, G.; Wang, Y.; Yang, C.; He, X.; Su, J. Study on performance optimization of double screw-magnetic fluid combination seal. *Phys. Fluids* **2024**, *36*, 022021. [CrossRef]
13. Yang, X.; Li, D.; He, X.; Zhang, H. Numerical and Experimental Studies of Alternative Combined Magnetic Fluid and Labyrinth Seal with Large Gap. *J. Mech. Eng.* **2014**, *50*, 175–179. [CrossRef]
14. Wang, Z.S. *Theoretical and Experimental Research of Mechanical—Magnetic Fluid Combined Seal*; Beijing Jiaotong University: Beijing, China, 2016.
15. Li, W.; Li, Z.; Wang, Z.; Wu, F.; Xu, L.; Peng, S. Turbulence intensity characteristics of a magnetoliquid seal interface in a liquid environment. *Coatings* **2021**, *11*, 1333. [CrossRef]
16. Liu, T.; Cheng, Y.; Yang, Z. Design optimization of seal structure for sealing liquid by magnetic fluids. *J. Magn. Magn. Mater.* **2005**, *289*, 411–414. [CrossRef]

17. Petcharoen, K.; Sirivat, A. Synthesis and characterization of magnetite nanoparticles via the chemical co-precipitation method. *Mater. Sci. Eng. B* **2012**, *177*, 421–427. [CrossRef]
18. Wang, Y.M.; Cao, X.; Liu, G.H.; Hong, R.Y.; Chen, Y.M.; Chen, X.F.; Li, H.Z.; Xu, B.; Wei, D.G. Synthesis of Fe_3O_4 magnetic fluid used for magnetic resonance imaging and hyperthermia. *J. Magn. Magn. Mater.* **2011**, *323*, 2953–2959. [CrossRef]
19. Tsai, C.S.; Chen, H.Y.; Chang, K.P. Using chemical coprecipitation method to prepare magnetite magnetic fluids and explore its applications in hyperthermia. *Key Eng. Mater.* **2011**, *474*, 1776–1780. [CrossRef]
20. Li, Z.; Li, W.; Wang, Q.; Xiang, R.; Cheng, J.; Han, W.; Yan, Z. Effects of medium fluid cavitation on fluctuation characteristics of magnetic fluid seal interface in agricultural centrifugal pump. *Int. J. Agric. Biol. Eng.* **2021**, *14*, 85–92. [CrossRef]
21. Agnihotri, P.; Lad, V.N. Magnetic nanofluid: Synthesis and characterization. *Chem. Pap.* **2020**, *74*, 3089–3100. [CrossRef]
22. Xu, L.F. *Preparation and Characterization of Fe_3O_4@SiO_2 Silicon Oil Based Magnetic Fluid*; Beijing Jiaotong University: Beijing, China, 2017.
23. Liang, A.; Li, H.; Zhang, W.; Yao, Z.; Zhu, B.; Wang, F. Study on pressure fluctuation and rotating stall characteristics in the vaneless space of a pump-turbine in pump mode. *J. Energy Storage* **2024**, *94*, 112385. [CrossRef]
24. Hu, L.; Liang, A.; Li, H.; Zhang, W.; Zhu, B. Impact of rotor-stator axial spacing on the gas-liquid-solid flow characteristics of a multiphase rotodynamic pump based on the Euler multi-fluid model. *Phys. Fluids* **2024**, *36*, 063314. [CrossRef]
25. Li, H.; Yang, J.; Zhang, W.; Hu, L.; Liang, A.; Yao, Z. Energy performance and unsteady gas-liquid flow characteristics of a multiphase rotodynamic pump: An experiment. *Appl. Energy* **2024**, *375*, 124112. [CrossRef]
26. Xu, L.; Kan, K.; Zheng, Y.; Liu, D.; Binama, M.; Xu, Z.; Yan, X.; Guo, M.; Chen, H. Rotating Stall Mechanism of Pump-Turbine in Hump Region: An Insight into Vortex Evolution. *Energy* **2024**, *292*, 130579. [CrossRef]
27. Liu, D.; Li, Z.; Xu, L.; Li, J.; Yang, Y.; Wang, X.; Pang, J.; Liu, X. Vortex Motion in Vaneless Space and Runner Passage of Pump-Turbine in S-Shaped Region. *Phys. Fluids* **2024**, *36*, 025115. [CrossRef]
28. Ren, Z.; Li, D.; Zhou, W.; Li, Z.; Wang, H.; Liu, J.; Li, Y.; Khoo, B.C. Gas–liquid mass-transfer characteristics during dissolution and evolution in quasi-static and dynamic processes. *Int. J. Multiph. Flow* **2024**, *180*, 104970. [CrossRef]
29. *GB/T 12241-2021*; " Safety Valves—General Requirements" Standard. State Administration for Market Regulation, Standardization Administration of China: Beijing, China, 2021.
30. Li, W.; Li, Z.; Han, W.; Li, D.; Yan, S.; Zhou, J. Study of The Flow Characteristics of Pumped Media in The Confined Morphology of A Ferrofluid Pump with Annular Microscale Constraints. *J. Fluids Eng.* **2024**, 1–31. [CrossRef]

processes

MDPI

Article

Evaluation of Advanced Control Strategies for Offshore Produced Water Treatment Systems: Insights from Pilot Plant Data

Mahsa Kashani, Stefan Jespersen and Zhenyu Yang *

AAU Energy, Aalborg University, Esbjerg Campus, Niels Bohrs vej 8, 6700 Esbjerg, Denmark; maka@energy.aau.dk (M.K.); sje@energy.aau.dk (S.J.)
* Correspondence: yang@energy.aau.dk

Abstract

Produced water treatment (PWT) is a critical process in offshore oil and gas production, ensuring compliance with stringent environmental discharge regulations and minimizing environmental impact. This process is characterized by inherent nonlinearities, coupled system dynamics, and the presence of significant disturbances that can impede operational efficiency and separation performance. Effective control strategies are essential to maintain stable operation and high separation efficiency under dynamic and uncertain conditions. This paper presents a comprehensive evaluation of advanced control methods applied to a pilot-scaled PWT facility designed to replicate offshore conditions. Four control solutions are assessed, i.e., (i) baseline approach using PID controllers; (ii) Multi-Input–Multi-Output (MIMO) H_∞ control; (iii) MIMO Model Predictive Control (MPC); and (iv) MIMO Model Reference Adaptive Control (MRAC). The motivation lies in their differing capabilities for disturbance rejection, tracking accuracy, robustness, and computational feasibility. Real-world operational data were used to assess each strategy in regulating critical process variables, the interface water level in the three-phase gravity separator, and the pressure drop ratio (PDR) in the hydrocyclone, both closely linked to de-oiling efficiency. The results highlight the distinct advantages and limitations of each method. In general, the baseline PID solution offers simplicity but limited adaptability, while advanced strategies such as MIMO H_∞, MPC, and MRAC solutions demonstrate enhanced reference-tracking and de-oiling performances subject to diverse operating conditions and disturbances, though different control solutions still exhibit different dynamic characteristics. The findings provide systematic insights into selecting optimal control architectures for offshore PWT systems, supporting improved operational performance and reduced environmental footprint.

Keywords: produced water treatment; de-oiling hydrocyclone; PID control; H_∞ control; MRAC; MPC

1. Introduction

1.1. Motivation

Offshore produced water treatment (PWT) is a crucial process related to sustainable offshore oil and gas production, as it helps mitigate environmental impacts by removing hydrocarbon residuals and other contaminants from the vast produced water before this stream is re-injected into the reservoir (for enhanced oil recovery purposes) or directly discharged to the ocean. This type of process is highly compact and complex due to limited space and weight constraints in offshore applications, as well as inherent nonlinearities, couplings, and diverse and harsh offshore operating conditions. Consequently, different advanced control solutions have been continuously demanded by the relevant industry to comply with increasingly stringent produced water discharge regulations [1–4].

A central technical challenge in PWT arises from the strong interactions and dynamic couplings between multiple subsystems, particularly the upstream three-phase gravity separator (TPGS) and the downstream de-oiling hydrocyclone. The TPGS subsystem plays a critical role by separating the produced fluid into gas, oil, and water phases based on density differences, which inherently involves complex multiphase flow dynamics. Maintaining a stable water interface level in the TPGS is essential, as fluctuations can propagate downstream and adversely affect the hydrocyclone's performance in oil removal. Control challenges include dealing with nonlinearities, time-varying disturbances, and operational constraints such as valve saturation and limited sensor accuracy. These difficulties motivate the exploration of robust and adaptive control strategies that can ensure consistent separation efficiency under dynamic offshore conditions.

One of the key challenges is how to effectively handle the potential conflict caused by the functional coupling between the upstream TPGS system and the downstream de-oiling hydrocyclone system, which has been identified as a key factor in determining the de-oiling efficiency of the entire PWT process [5–7]. This interconnection introduces multivariable interactions and competing control objectives, making it necessary to apply advanced control solutions capable of managing these complexities while optimizing overall system performance.

To explore the feasibility and effectiveness of different advanced control solutions in real-world applications, this study evaluates four different control solutions, i.e., (i) a baseline PID control, (ii) MIMO H_∞ control, (iii) MIMO MPC, and (iv) MIMO MRAC, using a laboratory-scale PWT pilot plant that can mimic various real-world PWT conditions.

1.2. State of the Art

By applying the plant-wide control strategy [7], substantial efforts have been made to address the challenges faced in offshore PWT processes. For example, [2] explored several anti-slugging control strategies aimed at stabilizing the multiphase flow through the riser system, thereby mitigating fluctuations caused by slugging inside the TPGS. The study investigated a range of methods, including feedback linearization, gain scheduling, internal model control, and adaptive PI control, highlighting their potential to improve flow stability and overall system performance.

By focusing on the coupled de-oiling hydrocyclone and TPGS systems, a H_∞ control solution has been proposed in [3], and the system performance is compared against a benchmark PID control. A recent study proposed an H_∞-based fault-tolerant control strategy to mitigate disturbance sensitivity under operational constraints in [8]. The work further introduced a differential game framework to address actuator faults and saturation, and developed a reinforcement learning approach with finite-time convergence to enable rapid fault recovery and maintain safety boundaries.

A nonlinear MPC solution has also been studied and tested in [9]. Moreover, in [10], first-principles and gray-box models were developed to describe dynamic relationships, demonstrating suitability for control design and enabling the integration of nonlinear strategies such as feedback linearization and sliding-mode control, thereby enhancing robustness under uncertainty. Moreover, Vallabhan et al. [11] investigated the application of feedforward, cascade, and MPC algorithms for de-oiling hydrocyclones through a comprehensive simulation study, highlighting their potential to improve separation efficiency and operational reliability.

Though all these advanced control solutions showed better performances compared with a benchmark PI control solution, all these solutions are based on the assumption that the de-oiling efficiency is proportional with one of the controlled variables, named as the pressure drop ratio (PDR), and it has been discovered using (online) oil-in-water (OiW) sensors that this linear relationship is only valid for specific operating ranges subject to a static condition [1,12]. It is clear that a reliable evaluation and design of the dynamic de-oiling efficiency cannot be feasible without any real-time OiW monitoring installed in PWT processes. Addressing this need, in [13], nonlinear control-oriented Hammerstein–Wiener models of an offshore PWT hydrocyclone were identified from OiW concentration data via exhaustive search, and PI control was implemented to maintain discharge concentration tracking under varying operating conditions, with improvements in separation efficiency and reductions in oil discharge demonstrated.

Adaptive control is crucial in real-world applications because it dynamically adjusts to uncertainties and disturbances, ensuring reliable and efficient operation despite changing process conditions [14]. In this context, robust MRAC designs have been developed to address practical challenges such as actuator faults, model uncertainties, and external disturbances [15–17]. Furthermore, advanced frameworks, such as the decentralized state-feedback MRAC proposed in [18–21] with time-varying delays, highlight the scalability and robustness of adaptive control in handling complex, distributed industrial processes. Specifically for PWT applications, a robust multivariable MRAC based on output feedback and LDS decomposition was proposed in [4], demonstrating superior tracking and disturbance rejection compared to optimized PI controllers. Collectively, these studies show that adaptive control techniques, particularly MRAC, substantially improve robustness and performance, underscoring their strong potential for real-world applications.

1.3. Key Contributions

The primary objective of this study is to evaluate and compare advanced control strategies for offshore PWT systems under realistic operating conditions. The assessment focuses on regulating key process variables, namely the water interface level within the TPGS and the PDR in the de-oiling hydrocyclone system, while also measuring dynamic de-oiling efficiency using an advanced online OiW monitoring system. A benchmark PID controller is implemented to represent conventional PWT practices, and its performance is compared with three advanced strategies: H_∞, MPC, and an MRAC solution. Notably, the proposed MRAC-based control is applied for the first time to a PWT system. Furthermore, this study presents a comprehensive experimental comparison across all controllers, incorporating a broader set of performance metrics than our previous work, to provide deeper insights into which control strategy is best suited for real-world offshore applications. In summary, the key contributions of this paper are as follows:

- Real-world Data Application: The study uses real-world pilot-scale data to evaluate control strategies, offering practical insights into their performance under actual operating conditions.
- First-Time MRAC Application in PWT: The study presents the first implementation of an MRAC strategy for the PWT system and examines how well advanced control strategies, particularly MRAC, handle dynamic feed variations and disturbances, showcasing their adaptability and robustness in real time.
- Comprehensive Control Evaluation: A broader set of performance metrics is applied to systematically compare tracking accuracy, de-oiling efficiency, robustness, and control effort between simulation and experimental results.

- Guidelines for Optimal Control Architecture Selection: Based on the evaluation, the study provides actionable guidelines for selecting the most appropriate control strategy depending on system needs, computational resources, and disturbance handling.
- Practical Implications for Plant-wide Control: This work follows the plant-wide control strategy, highlighting the benefits of coordinated control between the upstream and downstream systems to optimize the overall water treatment process.

The rest of the paper is organized as follows. Section 2 introduces the experimental setup and test conditions used in the study. Section 3 presents the relevant mathematical models used for different control developments and the different control solutions selected for this study. Section 4 discusses the results obtained and the comparative analysis of different control solutions. Finally, Section 5 concludes the paper along with recommendations for future work.

2. Experimental Setup and Testing Conditions

The offshore PWT system examined in this study comprises two-stage separation processes as shown in Figure 1, i.e., an upstream TPGS system and a downstream de-oiling hydrocyclone system. This sequential configuration is designed to efficiently remove hydrocarbons and other dispersed impurities from produced water, to ensure compliance with stringent environmental discharge regulations, or to improve the water purity before re-injection into reservoirs. In some cases, additional separation equipment, such as gas flotation units, may be installed downstream of the hydrocyclone system to further enhance oil and solid removal. The dynamics of the considered system are inherently nonlinear and often subject to fluctuating feed conditions, as well as other external disturbances, necessitating robust and adaptable control strategies to maintain optimal operational performance.

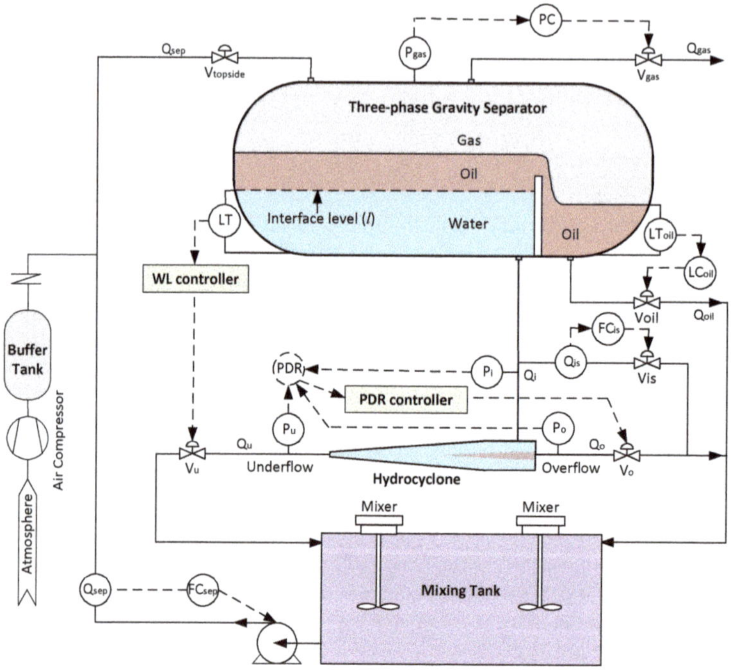

Figure 1. Simplified P&ID of the considered PWT system illustrated by an upstream TPGS, a downstream de-oiling hydrocyclone, and their relevant control loops.

2.1. PWT Pilot Plant

The pilot-scale experimental setup depicted in Figure 2 is designed to emulate an offshore PWT facility and includes the following key components:

- A multiphase flow generation system to mimic the underground reservoir and production well and pipeline systems. Here the oil and water are mixed, and compressed air is injected to mimic gas flow.
- A set of TPGS systems: The multiphase flow is lifted through a riser pipeline and fed into the TPGS system, where the oil, water, and gas are separated according to their density differences. The gas phase ascends to the top and exits through the gas control valve (V_{gas}), which is regulated by a pressure controller. The oil phase skims over a weir plate into the oil compartment, while the sunken water phase is blocked by the weir plate. The level of the interface between oil and water is monitored by a level transmitter (LT) and controlled by a level controller (LC), which adjusts the underflow control valve (V_u) to maintain optimal separation efficiency and stable separator pressure.
- A set of de-oiling hydrocyclone systems: The water separated by the TPGS system is further treated using de-oiling hydrocyclone systems, which employ centrifugal forces to enhance the separation of tiny residual oil from water. Inside the chamber of the hydrocyclone, the dispersed oil is pushed to the axial center, an oil core can be formed subject to proper operating conditions, then the oil core stream exits the hydrocyclone through its overflow port, while the cleaned water is discharged through its underflow port. The de-oiling performance of a hydrocyclone is significantly influenced by its mechanical design and dynamic flow conditions. Achieving optimal separation efficiency requires maintaining a sufficient flow rate to generate a strong centrifugal field within the hydrocyclone. Additionally, an appropriate flow split is essential to ensure effective separation of oil and water phases [22].
- The OiW fluorescence-based Turner designs TD-4100XDC are installed at the inlet C_i and water outlet C_u of the hydrocyclone.

Figure 2. Picture of the PWT pilot plant at AAU offshore laboratory.

2.2. Testing Conditions

In the experimental test, the system operates with a supply tank that continuously mixes oil and water, while the inlet flow rate, Q_{sep}, is regulated by a flow controller FC_{sep}. It is worth mentioning that, to mimic the characteristics of PW, non-detergent SAE30 Midland engine oil dispersed in tap water was used. The separator pressure, P_{gas}, is maintained at 7 bar using a PID controller in conjunction with a control valve V_{gas}. An air compressor and a buffer tank supply air at 8 bar to the system. The side-stream flow rates, Q_{is} and Q_{us}, are maintained at 1.5 L per minute using PID controllers to fulfill the OiW monitor's flow demand. The water level inside the TPGS and the PDR of the hydrocyclone are controlled with reference setpoints of 0.15 m and 2, respectively.

For all experiments, the controller was implemented with a sampling time of $T_s = 0.01$ s, chosen to provide sufficient temporal resolution while maintaining computational efficiency. Each test scenario lasted 3900 s, with data from the startup and shutdown phases excluded to focus on the main experimental phase and capture both transient and steady-state behavior under the given operating conditions. Raw sensor measurements were filtered using a low-pass filter to suppress high-frequency noise prior to analysis. Subsequent data processing, including normalization and statistical evaluation, was carried out in MATLAB to ensure consistent comparison across control strategies. These settings were selected to support reproducibility and to ensure that the results are representative of practical, industrially relevant operations.

Figure 3 illustrates the feeding (liquid) flow rate Q_{sep} to the TPGS system, to mimic a slugging flow pattern between 600 and 3600 s as marked by the vertical lines in Figure 3. It should be noted that the initial 600 s corresponds to the startup phase, while the final 300 s corresponds to the shutdown phase.

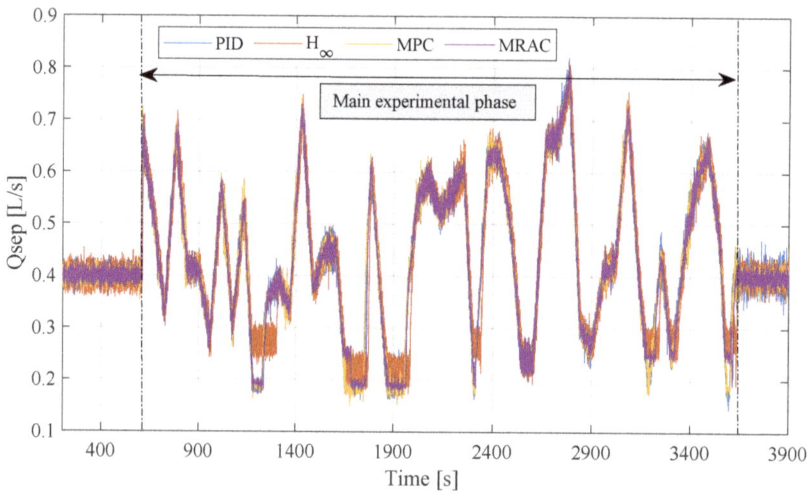

Figure 3. Feeding flow rates (slugging flow pattern) to the TPGS system during different tests subject to different control solutions.

3. Mathematical Models and Control Solutions

3.1. Dynamic Model of TPGS System

The (water interface) dynamic of a TPGS system can be modeled via mass balance analysis. Subject to the assumption of a noncompressible liquid, a simple model describing the water volume inside the TPGS can be formulated as

$$\frac{dV(t)}{dt} = Q_{sep}(t) - Q_{out}(t), \tag{1}$$

where $Q_{sep}(t)$ and $Q_{out}(t)$ denote the volumetric flow rates entering and leaving the TPGS system, respectively. The flow rate $Q_{out}(t)$ can be determined by using the orifice character of the control valve as

$$Q_{out}(t) = C_v f(V_u(t))\sqrt{\frac{\Delta P(t)}{\rho_{water} g}}, \tag{2}$$

where C_v is the orifice coefficient of the control valve, $f(V_u)$ represents the functional relationship between the opening degree of the underflow valve $V_u(t)$ and its effective opening area, $\Delta P(t)$ denotes the pressure drop across the control valve, $\rho_{water})$ is the density of the liquid, and g is the gravitational acceleration.

The water level denoted as $l(t)$ in the TPGS system can be used to substitute the liquid volume $V(t)$ in Equation (1). Then the dynamic model relevant to the water interface inside the TPGS system can be described as

$$\frac{dl(t)}{dt} = \frac{1}{\frac{dV}{dl}}\left(Q_{sep}(t) - C_v f(V_u(t))\sqrt{\frac{\Delta P(t)}{\rho_{water} g}}\right). \tag{3}$$

Under the assumption that the TPGS is a cylinder-shaped vessel, then $V(l(t))$ can be calculated according to the cross-sectional area $A(l(t))$ and the length of the compartment L as [23]

$$V(l(t)) = A(l(t))L = \left(r^2 cos^{-1}(\frac{r - l(t)}{r}) - (r - l(t))\sqrt{2rl(t) - l(t)^2}\right)L, \tag{4}$$

where r represents the radius of the separator.

3.2. De-Oiling Hydrocyclone

One key performance index of the de-oiling hydrocyclone system is the *flow split ratio*, which is defined as

$$F_s \triangleq \frac{Q_o}{Q_i} \tag{5}$$

where Q_o represents the volumetric flow rate in the overflow (e.g., the oil-rich stream), and Q_i is the volumetric flow rate at the inlet of the hydrocyclone. The flow split ratio needs to be properly controlled, such that ideally all oil residuals are expelled through the overflow stream without too much water accompanying, while the pure water goes through the underflow stream. In reality, the oil compositions in the water effluent need to be limited below 30 ppm to fulfill the compliance [24].

As shown in Figure 1, the common cost-effective approach to realize a flow split control is to implement a PDR control over the hydrocyclone system [6,25]. The PDR is defined as

$$PDR \triangleq \frac{\Delta P_{io}}{\Delta P_{iu}} = \frac{P_i - P_o}{P_i - P_u}, \tag{6}$$

where P_i, P_o, and P_u indicate the pressures at the inlet, overflow, and underflow ports of the hydrocyclone system, respectively. The relationship between the PDR and the flow split plays a critical role in regulating the flow dynamics within the hydrocyclone system.

An approximately linear relationship between the flow split and PDR has been indicated in [26]. However, this relationship heavily depends on the specific geometric configuration of the deployed hydrocyclone and specific surface characteristics of its inner chamber materials. Nevertheless, the PDR is an essential parameter for fine-tuning flow split control in de-oiling hydrocyclones. Keeping the PDR constant as throughput varies is a reliable strategy to maintain a steady flow split, thereby stabilizing separation performance. In the following experimental work at the pilot plant, the PDR for the hydrocyclones is recommended within a range of 2–3 to balance separation efficiency and operational stability [3].

As shown in Figure 1, the hydrocyclone system's performance is governed by a PDR control loop, which modulates the overflow valve V_o based on the PDR. This control mechanism ensures efficient separation by maintaining the PDR within desired operational limits. The TPGS and hydrocyclone systems are interconnected through these control loops, enabling coordinated operation to achieve high separation efficiency [3].

3.3. Model of Entire System

A Linear Time-Invariant (LTI) model of the hydrocyclone system was identified in our previous work [3] around a typical operating point with $l = 0.15$ m, $V_u = 0.4168$, and $V_o = 0.1657$. This LTI model is combined with the TPGS model to derive a state-space model (7) of the entire considered system, where the two control inputs are opening degrees of the underflow and overflow control valves, and the two outputs are the PDR and water level inside the TPGS system.

$$\begin{aligned} \dot{X}(t) &= AX(t) + BU(t) + B_d D(t) \\ Y(t) &= CX(t), \end{aligned} \tag{7}$$

here, the state vector of the system is $X(t) = \begin{bmatrix} l(t) & \dot{x}_{Vu}(t) & x_{Vu}(t) & \dot{x}_{Vo}(t) & x_{Vo}(t) \end{bmatrix}^T$, where $l(t)$ is the interface level, while $\dot{x}_{Vu}(t)$, $x_{Vu}(t)$, $\dot{x}_{Vo}(t)$ and $x_{Vo}(t)$ are the states of the black-box PDR. The control input $U(t) = \begin{bmatrix} V_u(t) & V_o(t) \end{bmatrix}^T$. The output $Y(t) = \begin{bmatrix} y_l(t) & y_{\mathrm{PDR}}(t) \end{bmatrix}^T$, $D(t)$ represents system disturbances. The detailed matrices of A, B, and C are

$$\begin{aligned} A &= \begin{bmatrix} -1.23 \times 10^{-5} & 0 & 0 & 0 & 0 \\ 0 & -0.97 & -0.76 & 0 & 0 \\ 0 & 1 & 0 & 0 & 0 \\ 0 & 0 & 0 & -0.93 & -0.65 \\ 0 & 0 & 0 & 1 & 0 \end{bmatrix}, \\ B &= \begin{bmatrix} -1.4 \times 10^{-4} & 0 \\ -1 & 0 \\ 0 & 0 \\ 0 & 1 \\ 0 & 0 \end{bmatrix}, \quad C = \begin{bmatrix} 1 & 0 & 0 & 0 & 0 \\ 0 & 0 & 2.72 & 0 & 1.69 \end{bmatrix}. \end{aligned} \tag{8}$$

3.4. Hydrocyclone's De-Oiling Efficiency

The de-oiling efficiency, denoted as ε_{oil}, of a hydrocyclone system is defined in the following Equation (9). This index is also employed to directly evaluate the de-oiling performance of the system considered.

$$\varepsilon_{oil} \triangleq \frac{Q_o C_o}{Q_i C_i} = 1 - \frac{Q_u C_u}{Q_i C_i}, \tag{9}$$

where C_i, C_o, and C_u represent the volumetric OiW concentrations at the inlet, overflow, and underflow, respectively, and Q_i and Q_u denote the inlet and underflow flow rate, respectively. In many situations, the overflow rate can be very tiny compared with the inlet flow rate Q_i, so the de-oiling efficiency can be approximated using the oil concentration ratio, i.e.,

$$\varepsilon_{red} \approx 1 - \frac{C_u}{C_i}. \tag{10}$$

3.5. Control Solutions

Four different control solutions have been used/developed in the following to perform a systematic comparison.

- *Baseline PID control*: Two PID control loops, which mimic the actual control loops deployed in one of real-life offshore installations, have been developed as detailed in [3]. As shown in Figure 1, this control solution serves as a baseline control to be compared with other advanced control solutions. The level controller manages the water interface by adjusting the underflow valve (V_u) based on feedback from the level transmitter measurements. The PDR is governed by another PID control loop that modulates the overflow valve (V_o) to control the flow split.
- *H_∞ control*: A MIMO robust control solution is developed using H_∞ control theory as shown in Figure 4. The objective of this design is to minimize the worst-case transfer function gain from disturbance $W(t)$ to output $Z(t)$, subject to the condition that the closed-loop system needs to be internally stable. The H_∞ control is naturally formulated to handle MIMO systems, providing coordinated control actions for both the water interface level and the PDR performances.
  The LTI-type controller K is achieved by minimizing the H_∞ norm of the closed-loop system, denoted as $\|\mathcal{F}_l(P,K)\|_\infty < \gamma$ via the D-K iteration design method [3].
- *MPC control*: Based on the same model as used for H_∞ control design, an MPC solution is also developed similar to [9]. To address the rapid dynamics of the PDR loop, the sampling time was set to $T_s = 0.2$ s, while the slow TPGS dynamics necessitated a prediction horizon of $H_p = 600$. The control horizon was tuned to $H_u = 30$ to ensure smooth valve actuation. By solving an optimization problem at each control step, MPC anticipates future disturbances and adjusts control inputs accordingly, effectively handling system constraints and multi-channel interactions.
  The cost function used in the MPC solution is

$$J(k) = \sum_{i=1}^{H_p} \|\hat{Y}(k+i|k) - R(k+i|k)\|^2_{Q(i)} + \sum_{i=0}^{H_u-1} \|\Delta\hat{U}(k+i|k)\|^2_{R(i)}. \tag{11}$$

  where $\hat{Y}(k+i|k)$ represents the predicted outputs (level and PDR) at the $k+i$th step based on the measurement up to the kth step, $R(k+i|k)$ is the reference trajectory, and $Q(i)$ and $R(i)$ are weighting matrices. The optimization determines future control actions to minimize variations caused by inflow Q_{sep} while respecting physical constraints on V_u and V_o. To address the system nonlinearity, the control output is passed through the inverse Hammerstein function given in Equation (12). For more details on the MPC implementation, refer to [9].

$$V_o = \frac{\tan(\frac{V_{o,h}}{k2})}{k_1}. \tag{12}$$

- *MRAC Control*: Different from the robust control strategy, the MRAC solution adapts its control parameters in real time to match a reference behavior provided by a reference model, ensuring robust performance despite impacts due to system uncertainties and external disturbances. An output feedback MRAC is designed in [4,27] as shown in Figure 5. The control signal is described as

$$U(t) = \Theta_1^T(t)\Omega_1(t) + \Theta_2^T(t)\Omega_2(t) + \Theta_3(t)Y(t) + \Theta_4(t)R(t) + \Theta_5(t)D(t), \quad (13)$$

where the adaptive gain matrices are Θ_1, Θ_2, Θ_3, Θ_4, and Θ_5, and $\Omega_1(t)$, $\Omega_2(t)$ are filtered input and output signals, i.e., $\Omega_1(t) = F(s)U(t)$ and $\Omega_2(t) = F(s)Y(t)$, where the filter is defined as $F(s) = \frac{A(s)}{\Lambda(s)}$. Here, $A(s)$ is expressed as $A(s) = I_M, sI_M, ..., s^{\bar{v}-2}I_M]^T$, and a monic stable polynomial $\Lambda(s)$ of degree $\bar{v} - 1$. The parameter $\bar{v}$ presents the observability index of the plant or its upper bound. A diagram of the control system used in this method can be found in Figure 5. Moreover, a more detailed description of this controller can be found in [27].

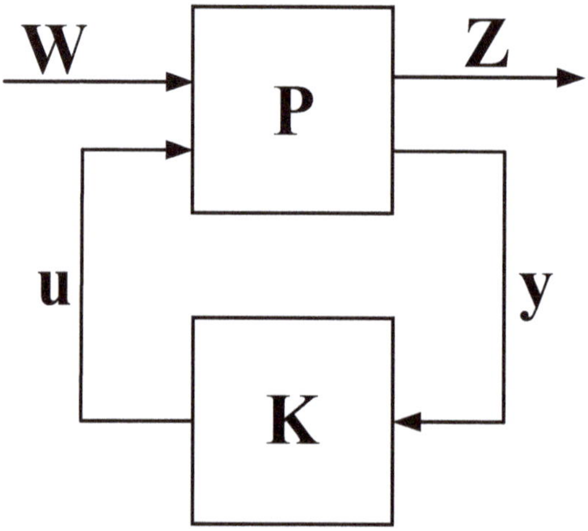

Figure 4. Closed-loop H_∞ control framework.

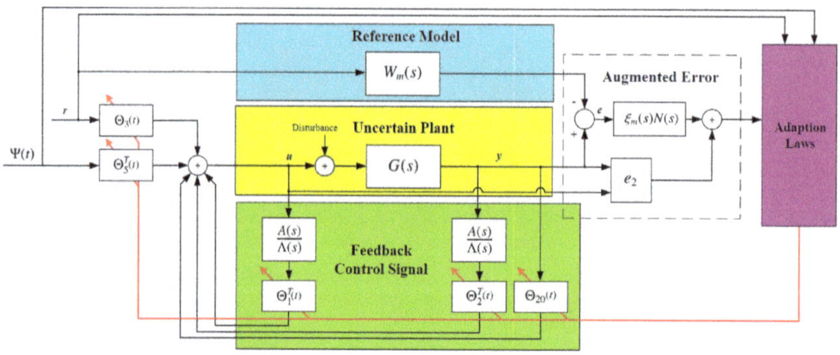

Figure 5. Closed-loop controlled system using output-feedback output-tracking MRAC.

Each control solution represents one type of distinguished features; for instance, the PID solution is the simplest and easiest in terms of control structure and implementation; the H_∞ control is a (5th-order) LTI-structured controller with fixed control parameters; the MRAC is a linear type of controller with adaptive control parameters; while the MPC solution represents the most sophisticated control solution which requires online optimization capability.

Table 1 provides a summary of the control solutions implemented for the PWT system, including their key design parameters and the constraints or assumptions considered in their tuning.

Table 1. Controller design parameters and tuning assumptions.

Controller	Key Parameters	Constraints / Assumptions
PID	$K_p = -58.37$, $K_i = -1.067$, $K_d = 0$ (for water level loop), $K_p = 0.1$, $K_i = 0.1$, $K_d = 0$ (for PDR loop)	Minimize the error for tracking the water level and PDR references, set to 0.15 m and 2, respectively, while avoiding actuator saturation.
H_∞	The controller order, calculated using the MATLAB *hinfsyn* function, and the selected weight functions can be found in [3].	Weights chosen for a trade off between performances for setpoint tracking and disturbance injection. Controller solved by D-K iteration method.
MPC	Prediction horizon = 600; control horizon = 30; sampling time $T_s = 0.2$ s; matrix calculation for Q and R are provided [9] .	Input constraints: underflow and overflow valve opening $0.1 \leq U_{Vu} \leq 1$, $0.03 \leq U_{Vo} \leq 1$, actual level limited to $0.1\ m \leq U_{Vu} \leq 0.2$ m
MRAC	The adaptation gain and reference model can be found in [4].	Minimize steady-state error; actuator limits $0 \leq U_{Vu}, U_{Vo} \leq 1$.

4. Experimental Results and Discussions

The effectiveness of each control solution is assessed according to the tracking performance of the corresponding closed-loop system. Meanwhile, by using the data from the deployed online OiW sensors, the system's de-oiling efficiency and the oil discharge ("leakage") rate are also evaluated. In the rest of this section, a comprehensive evaluation of the controllers across several key performance aspects is performed. First, Section 4.1 focuses on controller performance in terms of system outputs and control inputs. Then, Section 4.2 evaluates the controllers' behavior under intermittent measurement conditions. Finally, Section 4.3 examines performance with respect to OiW concentration measurements. Together, these analyses provide a clear comparison of each controller's strengths and limitations under varying operational scenarios.

4.1. Performances in Terms of Outputs and Inputs

The system's output responses and input signals subject to different control solutions are illustrated in Figure 6. During this period, the considered system sent a slugging feeding flow to the TPGS system as shown in Figure 3. Evaluated from the perspective of (output) tracking performance, the PID and MRAC solutions outperform the H_∞ control and MPC solutions in terms of water level management. But the PID solution could not make the controlled system track the PDR reference. There are high-frequency oscillations in both the PDR performance and the control input V_o. However, thanks to its flexibility, the MRAC solution still demonstrated a superior PDR performance, although its underflow control signal V_u could respond a little too rapidly for some short periods.

The H_∞ control solution exhibited its fundamental control principle, i.e., mainly prompted towards the worst scenario while "relaxing" for non-worst scenarios, and thereby, both the output responses and the control signals are the least intensified among all solutions, and this could be beneficial in terms of fatigue loadings to the entire system, thereby prolonging the facility's lifetime.

The most sophisticated MPC solution did not show superior performance in terms of tracking performance, and even led to the worst case in terms of water level tracking accuracy. However, the MPC solution led to the smoothest behaviors toward both output responses and both control signals. Unlike the H_∞ solution that exhibited intermittent control reactions, the MPC control took care of the system performance all the time and thereby led to the smoothest behaviors.

Compared to all advanced control solutions, the baseline PID control strategy utilized a two-SISO (decoupled) control structure to manage the water level and PDR performances. It can be observed that this solution is sometimes inadequate in dealing with the inherently coupled dynamics between the water level and PDR control loops.

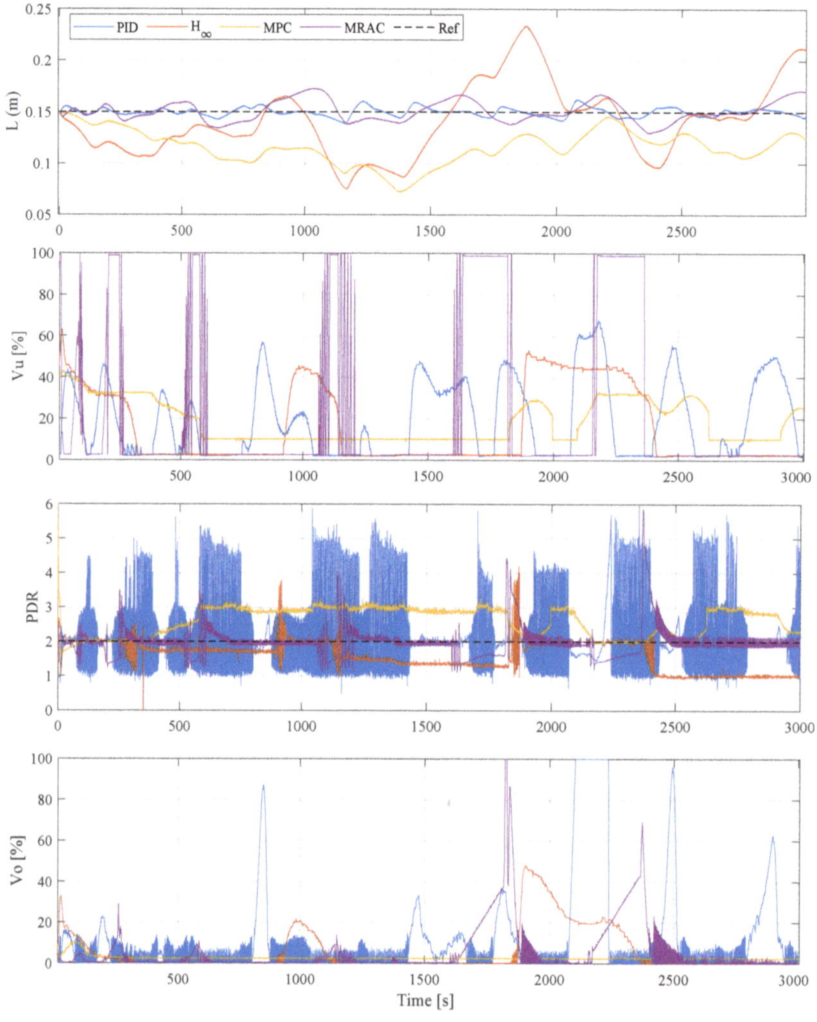

Figure 6. Dynamic performances of the water interface and the PDR along with the underflow/overflow valve opening degree $(V_u)/(V_o)$ subject to different control solutions.

4.2. *Performances in Terms of Intermittent Measurement*

All flow rates at the inlet, underflow, and overflow outlets of the hydrocyclone system are also measured using electromagnetic flow-meters/Coriolis flow-meters, though they are often not measured in the actual real-world installations due to the fact that they are just some intermittent variables, and also the high expenses of relevant flow-meters. Hereby, these variables are measured to gain a better understanding of the system's functionalities. These measured signals are plotted in Figure 7. It can be seen that the inlet flow rate Q_i and the underflow rate Q_u have quite similar flow patterns, regardless of the control solution applied; this is mainly due to the large difference in the sizes of the underflow and overflow control valves.

The dynamic flow split is also plotted in Figure 7. The PID solution led to large high-frequency oscillations in flow split due to its difficulty in maintaining a PDR tracking performance as shown in Figure 6. The MRAC and MPC solutions tried to maintain a constant flow split for most of the operating time, but clearly the MPC solution led to the smoothest performances in terms of both underflow and overflow rates, and thereby the smoothest flow split dynamics as well.

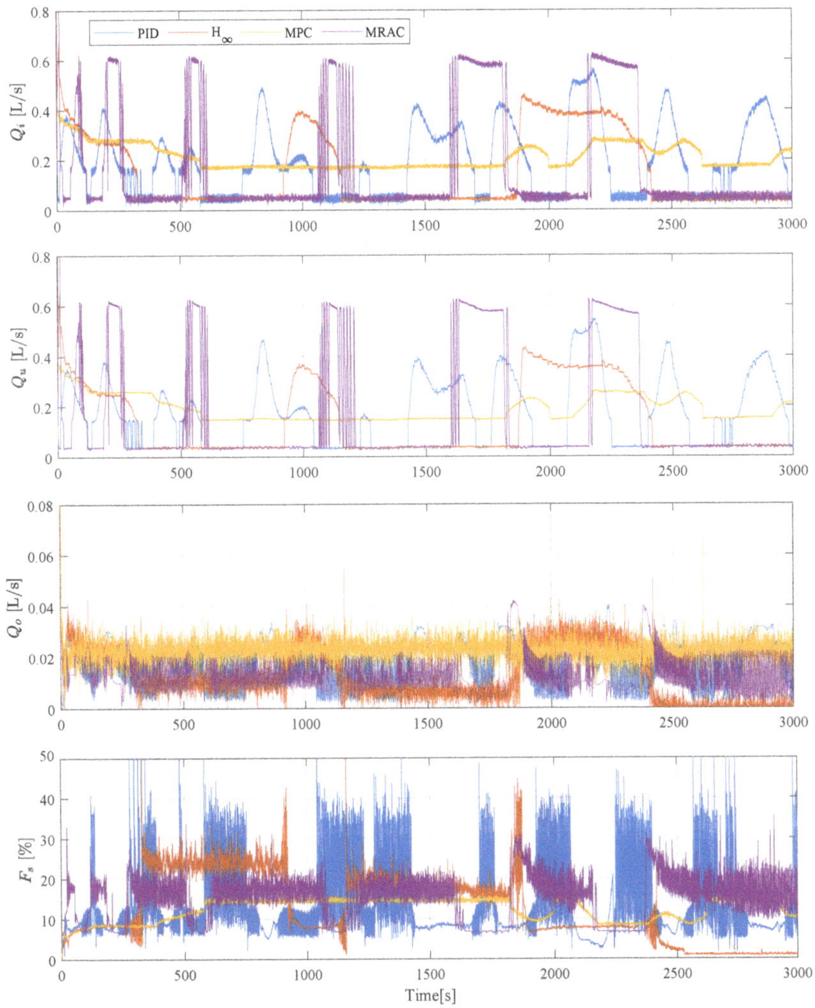

Figure 7. Hydrocyclone's inlet flow rate (Q_i), underflow rate (Q_u), overflow rate (Q_o), and calculated flow split (F_s).

4.3. Performances in Terms of OiW Measurement

The OiW concentrations measured by the OiW sensors installed online and thus the calculated de-oiling dynamic efficiency are plotted in Figure 8. As it has been noted in [6], the decrease in de-oiling efficiency can be attributed to either excessively high or critically low inlet flow rates, both of which disrupt optimal operating conditions.

In general, the MPC solution demonstrated the best steady efficiency among all solutions, which can be confirmed from the concentration measurement C_u at the underflow outlet. The de-oiling efficiencies of the PID and MRAC are more or less at the same level, and their dynamic oscillations have similar size and patterns. The dynamic oscillation of de-oiling efficiency led by the H_∞ solution is the slowest one, with a clear gradual decline until a low point is reached, followed by the de-oiling efficiency immediately increasing to an upper point (due to the intermittent control reaction). This pattern repeated throughout the testing period.

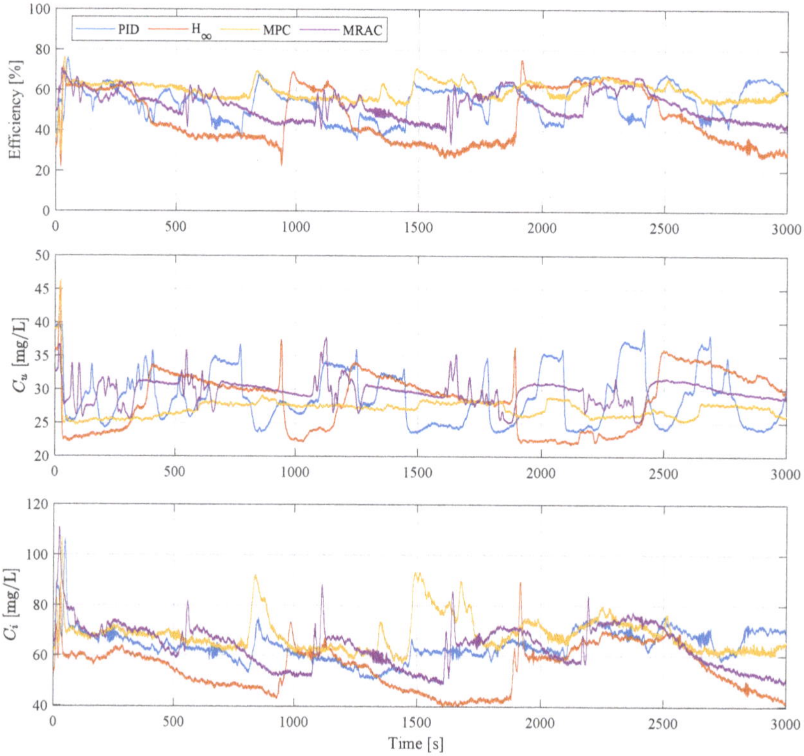

Figure 8. Calculated de-oiling efficiency, measured inlet OiW concentrations (C_i) and underflow OiW concentrations (C_u).

The oil leakage rate through the underflow stream is calculated based on the online OiW measurement (C_u) and the flow rate measurement installed in the underflow by Equation (14):

$$Q_{u,oil}(t) \hat{=} C_u(t) Q_u(t), \tag{14}$$

where $C_u(t)$ denotes the instantaneous underflow OiW concentration, and $Q_u(t)$ represents the volumetric flow rate. As illustrated in Figure 9, the MPC solution maintained a relatively constant discharge rate compared to other control solutions. In contrast, the MRAC solution was too proactive during certain periods, e.g., around $t = 600$ s. This is possibly due to the fact that this linear MRAC solution did not account for inherent control saturation. In contrast, the MPC control solution is designed by explicitly taking saturation into account, which naturally leads to an advantage in this context.

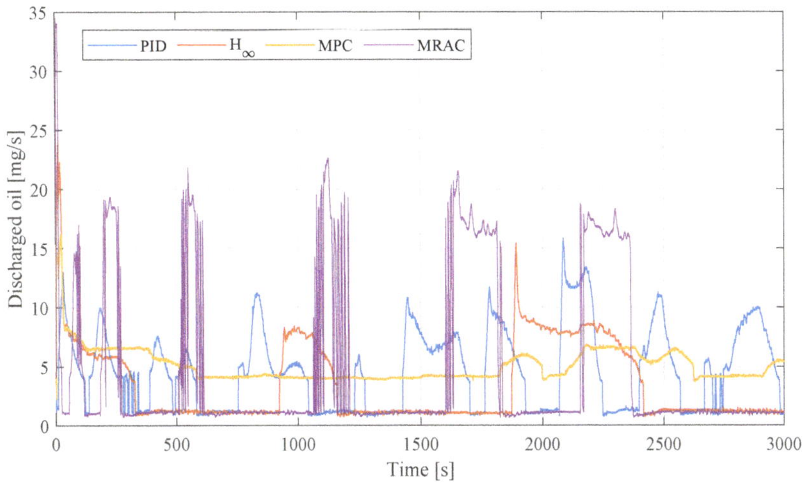

Figure 9. Oil leakage rates through the underflow stream of the hydrocyclone system.

4.4. Statistical Analysis

The mean values of the hydrocyclone inlet flow rate, underflow rate, overflow rate, flow split, system efficiency, water level tracking error, and PDR tracking error are listed in Table 2. In general, the MPC solution achieved the best mean efficiency of 59.96%. The mean efficiency induced by the MPC is 5.40 p.p. (percentage points) and 13.42 p.p., 7.42 p.p higher than that induced by the PID, H_∞, and MRAC solutions, respectively. However, in terms of reference tracking, the MPC approach demonstrated some difficulty in accurately following the water level and PDR, as indicated by its relatively high values.

Table 2. Means of inlet flow rate, underflow rate, overflow rate, flow split, mean efficiency, water level error, and PDR error.

Controller	$\overline{Q}_i$ [L/s]	$\overline{Q}_u$ [L/s]	$\overline{Q}_o$ [L/s]	$\overline{F}_s$ [%]	$\overline{\varepsilon}$ [%]	$\overline{Error_{WL}}$ [cm]	$\overline{Error_{PDR}}$ [-]
PID	0.1934	0.1774	0.0160	13.29	54.62	0.32	0.6663
H_∞	0.1497	0.1376	0.0122	12.31	46.54	2.90	0.4339
MPC	0.2112	0.1884	0.0228	12.26	59.96	3.38	0.6054
MRAC	0.1731	0.1582	0.0149	15.02	52.54	0.82	0.2636

Figure 10 shows the relative errors in terms of tracking errors and control efforts subject to the four different control solutions. As presented in Equation (15), the relative value for a variable $X(t)$ is defined as the ratio of a specific value to the largest value within a predefined set.

$$X^r(t) \triangleq \frac{X(t)}{\max(X(t))}, \text{ subject to } t \in [0, T] \tag{15}$$

It can be observed in Figure 10 that the PID and MPC solutions exhibited the minimum and maximum relative error for water level tracking, respectively. This is also observed in the results shown in Figure 6. In contrast, the MRAC solution showed the best performance for PDR tracking, while PID showed the poorest performance in this regard. In general, the MRAC solution has been proven to be highly effective in tracking references of both the water level and the PDR compared to the other methods. In terms of control effort, H_∞ and MPC outperform the other control methods, whereas the PID requires significantly high control efforts for manipulating both the underflow and the overflow valves.

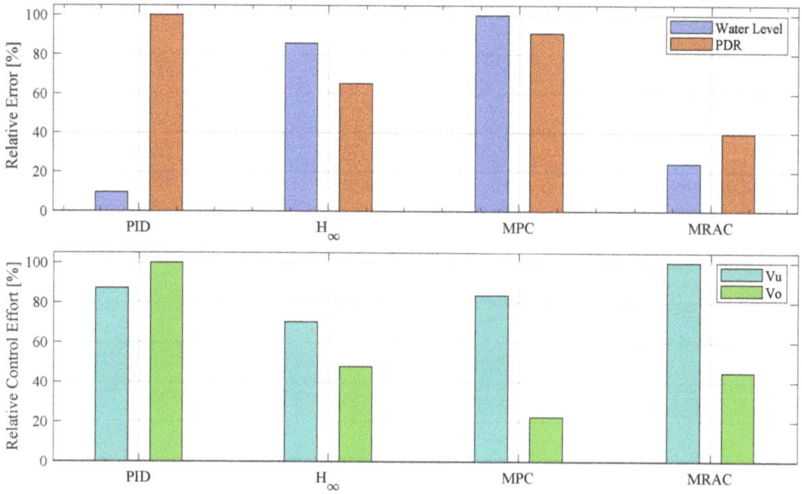

Figure 10. Comparison of relative errors and control efforts via different control solutions.

Tables 3–5 present the test results obtained for the separation process in a hydrocyclone under four control strategies. These results include the total volumes of liquid, oil, and water at different points in the process (inlet, underflow, and overflow), as well as the total mass of oil, the oil mass per liquid volume, and the water-to-liquid volume ratio.

Table 3. Total volumes of liquid $\sum Q_i$, oil $\sum Q_i^{oil}$, and water $\sum Q_i^{water}$ of the hydrocyclone inlet stream, as well as the total mass of oil $\sum M_i^{oil}$, total amount of oil mass pr. liquid volume $\sum M_i^{oil} / \sum Q_i$, and total amount of water pr. liquid volume $\sum Q_i^{water} / \sum Q_i$ in the hydrocyclone inlet stream.

Controller	$\sum Q_i$ [L/s]	$\sum Q_i^{oil}$ [mL]	$\sum Q_i^{water}$ [L]	$\sum M_i^{oil}$ [g]	$\sum M_i^{oil}/\sum Q_i$ [mg/L]	$\sum Q_i^{water}/\sum Q_i$ [L/L]
PID	580.1365	41.7063	580.0948	36.7433	63.3	0.9999
H_∞	449.2251	29.5740	449.1955	26.0547	58.0	0.9999
MPC	633.5406	47.6665	633.4929	41.9942	66.3	0.9999
MRAC	519.1526	38.6772	519.1139	34.0746	65.6	0.9999

Table 4. Total volumes of liquid $\sum Q_u$, oil $\sum Q_u^{oil}$, and water $\sum Q_u^{water}$ of the underflow valve, as well as the total mass of oil $\sum M_u^{oil}$, total amount of oil mass pr. liquid volume $\sum M_u^{oil} / \sum Q_i$, and total amount of water pr. liquid volume $\sum Q_u^{water} / \sum Q_i$ in the underflow valve.

Controller	$\sum Q_u$ [L]	$\sum Q_u^{oil}$ [mL]	$\sum Q_u^{water}$ [L]	$\sum M_u^{oil}$ [g]	$\sum M_u^{oil}/\sum Q_i$ [mg/L]	$\sum Q_u^{water}/\sum Q_i$ [L/L]
PID	532.2308	15.5696	532.2152	13.7168	23.6	0.9174
H_∞	412.7405	11.3749	412.7292	10.0212	22.3	0.9188
MPC	565.0892	16.7060	565.0725	14.7180	23.2	0.8919
MRAC	474.3282	14.6640	474.3135	12.9190	24.9	0.9136

Table 5. Total volumes of liquid $\sum Q_o$, oil $\sum Q_o^{oil}$, and water $\sum Q_o^{water}$ of the overflow valve, as well as the total mass of oil $\sum M_o^{oil}$, total amount of oil mass pr. liquid volume $\sum M_o^{oil} / \sum Q_i$, and total amount of water pr. liquid volume $\sum Q_o^{water} / \sum Q_i$ in the overflow valve.

Controller	$\sum Q_o$ [L]	$\sum Q_o^{oil}$ [mL]	$\sum Q_o^{water}$ [L]	$\sum M_o^{oil}$ [g]	$\sum M_o^{oil}/\sum Q_i$ [mg/L]	$\sum Q_o^{water}/\sum Q_i$ [L/L]
PID	47.9057	26.1367	47.8796	23.0264	39.7	0.0825
H_∞	36.4845	18.1992	36.4663	16.0335	35.7	0.0812
MPC	68.4514	30.9604	68.4205	27.2761	43.1	0.1080
MRAC	44.8244	24.0132	44.8004	21.1556	40.8	0.0863

The inlet stream data shows that the highest total inlet liquid volume ($\sum Q_i$) was achieved by the MPC controller (633.5406 L), followed by the PID (580.1365 L), MRAC

(519.1526 L), and H_∞ (449.2251 L) controllers. Correspondingly, MPC and PID also had a higher total oil mass entering the system, around 41.9 g and 36.7 g, respectively, while the H_∞ and MRAC had slightly lower oil loads. The oil concentration per unit volume in the inlet was highest for MPC (66.3 mg/L) and MRAC (65.6 mg/L), suggesting that they handled more oil-rich inflows. Despite these differences, all controllers maintained a very high water fraction in the inlet (>0.9999), as expected in de-oiling applications. As noted earlier, achieving optimal separation efficiency in a hydrocyclone relies on maintaining an adequate flow rate to generate a strong centrifugal field. This fact can be seen by comparing the results presented in Tables 2 and 3, where the MPC controller achieves higher efficiency and higher total liquid volume compared to the other control strategies.

Across controllers, the underflow stream handled most of the separated liquid, with MPC showing the highest total underflow volume (565.1 L), followed by PID (532.2 L), MRAC (474.3 L), and H_∞ again with the lowest volume (412.7 L). The oil mass removed through the underflow was highest for PID (13.7 g) and MPC (12.7 g), while MRAC and H_∞ had slightly lower values. The oil concentration in the underflow remained similar across controllers, around 22–24 mg/L, which aligns with the underflow's role in carrying concentrated oil away. The water fraction here decreased to around 0.89–0.92, reflecting higher oil content relative to the inlet.

As listed in Table 5, the overflow stream consistently carried a smaller liquid volume, with MPC again showing the largest overflow (68.8 L), followed by MRAC (48.4 L), PID (47.9 L), and H_∞ the lowest (36.4 L). Despite its lower volume, the overflow had a higher oil mass per unit volume, particularly for MPC (27.2 g) and MRAC (21.7 g), resulting in oil concentrations of approximately 40 to 43 mg / L. H_∞ had the lowest overflow oil mass and concentration (16.0 g and 35.7 mg/L). The water fraction in the overflow was notably lower (0.08–0.086), highlighting its role in rejecting more oil relative to water.

4.5. Comparative Analysis

Table 6 summarizes the comparative performance of the four control strategies tested based on experimental results from the pilot-scale offshore PWT system. Moreover, the industrial suitability of each method is also demonstrated by providing some indicators from practical considerations. As mentioned before, the results show that MRAC achieved the highest tracking accuracy, while PID lagged in tracking of PDR due to challenges in handling multivariable interactions. MPC and H_∞ required the lowest control effort, which could be advantageous for actuator longevity, whereas MRAC and PID demanded higher actuation. From an industrial perspective, PID remains the most practical choice for straightforward applications due to its low implementation cost, ease of tuning, and compatibility with existing hardware, despite offering moderate performance. H_∞ control, while highly robust and efficient in handling disturbances, is less commonly deployed industrially because of its high tuning complexity and cost. MPC provides a strong balance between high de-oiling efficiency, robustness, and operational feasibility, especially in well-instrumented facilities, though it demands a complex setup and significant resources. MRAC stands out for its excellent tracking accuracy and adaptability at a moderate cost, making it particularly suitable for dynamic offshore environments with varying system parameters. Ultimately, the selection among these methods depends on the specific industrial priorities regarding cost, complexity, and performance requirements.

Table 6. Comparative summary of control strategies for offshore PWT pilot plant.

Performance Indicator	PID	H_∞	MPC	MRAC
Tracking Accuracy	Medium-Low	Low	Low	High
Control Effort	High	Low	Low	High
De-oiling Efficiency	Medium	Medium-Low	High	Medium
Robustness	Medium	High	High	Medium-High
Tuning Complexity	Low	High	High	Medium–High
Adaptability	Low	Medium-Low	Medium	High
Implementation Cost	Low	High	High	Medium-High

Indicators such as cost constraints, desired control performance, ease of tuning, and adaptability to changing conditions vary across applications, guiding the choice between simpler methods like PID and more advanced solutions like MRAC or MPC. Operators prioritizing low cost and simplicity may favor PID, while those requiring high robustness or adaptability might opt for H_∞, MPC, or MRAC despite increased complexity or expense. Therefore, aligning control strategy selection with operational goals and resource availability is crucial for successful industrial implementation.

A multi-criteria decision-making (MCDM) approach can be used to identify the optimal method based on operator priorities. In this study, an MCDM procedure through the Weighted Sum Method (WSM) was employed to compare four control strategies based on qualitative performance ratings listed in Table 6. The proposed approach is as follows:

Step 1: Convert qualitative performance values into numerical scores, on a 1–5 scale. For benefit indicators (e.g., tracking accuracy, robustness), *Low* and *High* are scored as 1 and 5, respectively. For cost indicators (e.g., implementation cost, tuning complexity), the scoring is reversed, so that *Low* is 5 and *High* is 1.

Step 2: Apply scenario-specific weighting to each indicator to reflect different operational priorities. The weighted performance for each controller is calculated by multiplying the score of each indicator by its corresponding weight.

Step 3: Sum the weighted scores across all indicators to obtain a total score for each control strategy in each scenario. The strategy with the highest total score is considered the most suitable for that scenario.

$$S_i = \sum_{j=1}^{7} w_j \cdot r_{ij}, \tag{16}$$

where r_{ij} is the converted score for controller i on indicator j and w_j is the weight for that indicator in the given scenario.

Four weighting scenarios were defined to reflect different operational priorities. In Scenario 1, weights of $\{0.1, 0.1, 0.1, 0.2, 0.2, 0.1, 0.2\}$ were applied to the seven indicators, prioritizing robustness, low tuning complexity, and low cost. Scenario 2 assigned weights of $\{0.2, 0.1, 0.2, 0.2, 0.1, 0.1, 0.1\}$, giving more emphasis to tracking accuracy, de-oiling efficiency, and robustness. Scenario 3 used weights of $\{0.1, 0.2, 0.2, 0.1, 0.1, 0.1, 0.2\}$, focusing on low control effort, high efficiency, and low cost, while Scenario 4 applied weights of $\{0.1, 0.2, 0.1, 0.2, 0.1, 0.2, 0.1\}$, targeting robustness, low control effort, and adaptability. The total scores for the four methods were 3.3, 2.4, 2.8, and 3.2 in Scenario 1; 2.8, 2.5, 3.2, and 3.5 in Scenario 2; 2.9, 2.5, 3.2, and 3.0 in Scenario 3; and 2.5, 2.9, 3.4, and 3.3 in Scenario 4, for PID, H_∞, MPC, and MRAC, respectively.

Figure 11 illustrates the weighted performance comparison of different control strategies under four distinct weighting scenarios. Each radar chart corresponds to a unique prioritization of evaluation criteria, allowing visual identification of the most suitable control method for different operational objectives. The analysis showed that PID is best when simplicity and cost-effectiveness are most important (Scenario 1), MRAC excels when tracking and adaptability dominate (Scenario 2), and MPC leads when efficiency and low control effort are valued (Scenarios 3 and 4). This highlights that the "best" choice of control method is highly dependent on the weighting of performance criteria, and therefore on the specific operational priorities of the end user.

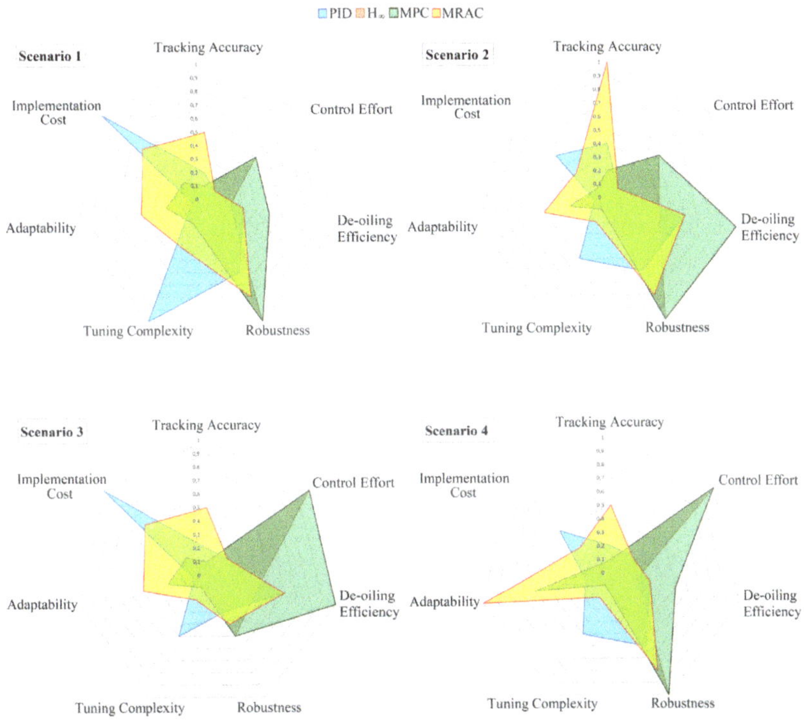

Figure 11. Scenario-based weighted performance comparison of control strategies for offshore PWT.

5. Conclusions

This study presented a comprehensive evaluation of various control strategies, including PID, H_∞, MPC, and MRAC, implemented in a pilot-plant setup at the AAU laboratory. The performance of these control solutions was assessed based on tracking accuracy, efficiency, compliance with operational constraints, and control effort. Among the tested strategies, MRAC demonstrated superior performance in reference tracking, outperforming the other controllers in terms of precision and dealing with the inherent system correlation. However, when considering overall efficiency, MPC emerged as the most effective solution. In terms of control effort, MRAC and PID required higher levels of actuation compared to H_∞ and MPC. This higher control effort is attributed to the aggressive tuning of the MRAC and PID controllers, which prioritize tracking performance over valve relaxation. In contrast, H_∞ and MPC exhibited lower control effort, as these controllers allow for the relaxation of control valves while maintaining system stability and efficiency.

Overall, the findings highlight the trade-offs between tracking performance, efficiency, and control effort among different control strategies. MPC, in particular, stands out as a balanced and practical solution for complex industrial applications, combining efficient operation with constraint adherence and minimal control effort. Future research could focus on further optimizing these control strategies and evaluating their applicability to larger-scale industrial systems. In this context, nonlinear system identification using machine learning techniques applied to recorded real-time laboratory data could be investigated to enable more effective control designs and enhance overall system performance. Moreover, the lab-to-industry gap can be addressed through a scalable validation framework combining larger pilot-plant tests, high-fidelity digital twins, and hardware-in-the-loop simulations. This approach enables realistic disturbance testing, safe evaluation of extreme conditions, and gradual progression to limited industrial trials. Such a framework ensures control strategies are robust and practical for full-scale offshore PWT applications.

Author Contributions: Conceptualization, M.K. and Z.Y.; methodology, M.K. and Z.Y.; software, M.K.; validation, M.K., Z.Y. and S.J.; formal analysis, M.K. and Z.Y.; investigation, M.K.; resources, Z.Y.; writing—original draft preparation, M.K.; writing—review and editing, M.K., Z.Y. and S.J.; visualization, M.K.; supervision, Z.Y.; project administration, Z.Y.; funding acquisition, Z.Y. All authors have read and agreed to the published version of the manuscript.

Funding: This research was partially funded by the Danish EUDP project named OiW Control by 3D spectroscopy, and partially supported by the Danish (DTU) Offshore Technology Center via the DIC project.

Data Availability Statement: The original contributions presented in this study are included in the article. Further inquiries can be directed to the corresponding author.

Acknowledgments: The authors would like to thank the partial support from the Danish Energy Technology Development and Demonstration Programme (EUDP) and the partial support from the Danish (DTU) Offshore Technology Center for this work. Thanks also go to Simon I. Andersen, Isabelle M.A. Viegas, and Benaiah Anabaraonye from Danish Offshore Technology Center, Kristian Nielsen and Kristian Rode from SHUTE Sensing Solutions A/S, for many valuable discussions. Special thanks also goes to Mohsen Sadeghi from AAU Energy for technical assistance during the experimental work.

Conflicts of Interest: The authors declare no conflicts of interest.

References

1. Husveg, T.; Rambeau, O.; Drengstig, T.; Bilstad, T. Performance of a deoiling hydrocyclone during variable flow rates. *Miner. Eng.* **2007**, *20*, 368–379. [CrossRef]
2. Jahanshahi, E.; Skogestad, S. Nonlinear control solutions to prevent slugging flow in offshore oil production. *J. Process Control* **2017**, *54*, 138–151. [CrossRef]
3. Durdevic, P.; Yang, Z. Application of H_∞ Robust Control on a Scaled Offshore Oil and Gas De-Oiling Facility. *Energies* **2018**, *11*, 287. [CrossRef]
4. Kashani, M.; Jespersen, S.; Yang, Z. Robust Adaptive Control of the Offshore Produced Water Treatment Process: An Improved Multivariable MRAC-Based Approach. *Water* **2024**, *16*. [CrossRef]
5. Colman, D.A.; Thew, M.T. Correlation of separation results from light dispersion hydrocyclones. *Chem. Eng. Res. Des.* **1983**, *61*, 233–240.
6. Thew, M. FLOTATION | Cyclones for Oil/Water Separations. In *Encyclopedia of Separation Science*; Wilson, I.D., Ed.; Academic Press: Oxford, UK, 2000; pp. 1480–1490. [CrossRef]
7. Skogestad, S.; Postlethwaite, I. *Multivariable Feedback Control: Analysis and Design*; John Wiley & Sons, Inc.: Hoboken, NJ, USA, 2005.
8. Li, S.; Zhang, Y.; Niu, J.; Luo, X.; Wang, J.; Tian, C. Finite-time fault-tolerant H_∞ control for offshore de-oiling hydrocyclone systems with input saturation via reinforcement learning. *Ocean Eng.* **2025**, *340*, 122188. [CrossRef]
9. Hansen, L.; Bram, M.V.; Pedersen, S.; Yang, Z. Performance Comparison of Control Strategies for Plant-Wide Produced Water Treatment. *Energies* **2022**, *15*, 418. [CrossRef]
10. Vallabhan, M.; Holden, C. Non-Linear Control Algorithms for De-Oiling Hydrocyclones. In Proceedings of the 2020 28th Mediterranean Conference on Control and Automation (MED), Virtual, 15–18 September 2020; pp. 85–90. [CrossRef]
11. Vallabhan, M.K.; Matias, J.; Holden, C. Feedforward, cascade and model predictive control algorithms for de-oiling hydrocyclones: Simulation study. *Model. Identif. Control* **2021**, *42*, 185–195. [CrossRef]

12. Vallabhan, K.G.M.; Holden, C.; Skogestad, S. Deoiling Hydrocyclones: An Experimental Study of Novel Control Schemes. *SPE Prod. Oper.* **2022**, *37*, 462–474. [CrossRef]
13. Jespersen, S.; Yang, Z.; Hansen, D.S.; Kashani, M.; Huang, B. Hammerstein–Wiener Model Identification for Oil-in-Water Separation Dynamics in a De-Oiling Hydrocyclone System. *Energies* **2023**, *16*, 7095. [CrossRef]
14. Tao, G. Multivariable adaptive control: A survey. *Automatica* **2014**, *50*, 2737–2764. [CrossRef]
15. Bagherpoor, H.; Salmasi, F.R. Robust model reference adaptive output feedback tracking for uncertain linear systems with actuator fault based on reinforced dead-zone modification. *ISA Trans.* **2015**, *57*, 51–56. [CrossRef] [PubMed]
16. Narendra, K.; Annaswamy, A. *Stable Adaptive Systems*; Dover Books on Electrical Engineering, Dover Publications: Garden City, NY, USA, 2012.
17. Guo, J.; Liu, Y.; Tao, G. A Multivariable MRAC Design Using State Feedback for Linearized Aircraft Models with Damage. In Proceedings of the 2010 American Control Conference, Baltimore, MD, USA, 30 June–2 July 2010; pp. 2095–2100. [CrossRef]
18. Hua, C.C.; Leng, J.; Guan, X.P. Decentralized MRAC for large-scale interconnected systems with time-varying delays and applications to chemical reactor systems. *J. Process Control* **2012**, *22*, 1985–1996. [CrossRef]
19. Song, G.; Tao, G. Partial-state feedback multivariable MRAC and reduced-order designs. *Automatica* **2021**, *129*, 109622. [CrossRef]
20. Wen, L.; Tao, G.; Liu, Y. Multivariable adaptive output rejection of unmatched input disturbances. *Int. J. Adapt. Control Signal Process.* **2016**, *30*, 1203–1227. [CrossRef]
21. Selfridge, J.M.; Tao, G. Multivariable Output Feedback MRAC for a Quadrotor UAV. In Proceedings of the 2016 American Control Conference (ACC), Boston, MA, USA, 6–8 July 2016; pp. 492–499. [CrossRef]
22. César, S.D.; De Jager, D.; Njoya, M. The role of hydrocyclone and induced gas flotation technologies in offshore produced water deoiling advancements. *Pet. Res.* **2024**, *10*, 342–351. [CrossRef]
23. Backi, C.J.; Skogestad, S. A Simple Dynamic Gravity Separator Model for Separation Efficiency Evaluation Incorporating Level and Pressure Control. In Proceedings of the 2017 American Control Conference (ACC), Seattle, WA, USA, 24–26 May 2017; pp. 2823–2828. [CrossRef]
24. Liu, Y.; Cheng, Q.; Zhang, B.; Tian, F. Three-phase hydrocyclone separator—A review. *Chem. Eng. Res. Des.* **2015**, *100*, 554–560. [CrossRef]
25. Meldrum, N. Hydrocyclones: A Solution to Produced-Water Treatment. *SPE Prod. Eng.* **1988**, *3*, 669–676. [CrossRef]
26. Jiang, M.; Zhao, L.; He, J.; Zhou, F. Pressure Drop Ratio–An Important Performance Parameter in Liquid-Liquid Hydrocyclone Separation, Volume All Days. In Proceedings of the International Ocean and Polar Engineering Conference, Montreal, Canada, 24–29 May 1998.
27. Tao, G. *Adaptive Control Design and Analysis*; John Wiley & Sons, Ltd.: Hoboken, NJ, USA, 2003. [CrossRef]

processes

MDPI

Article

A 31–300 Hz Frequency Variator Inverter Using Space Vector Pulse Width Modulation Implemented in an 8-Bit Microcontroller

Gustavo Cerda-Villafana [1,*], Adam Birchfield [2] and Francisco Javier Moreno-Vazquez [1]

[1] Department of Electronics Engineering, University of Guanajuato, Carretera Salamanca-Valle de Santiago km 3.5+1.8, Comunidad de Palo Blanco, Salamanca 36787, Mexico
[2] Department of Electrical and Computer Engineering, Texas A&M University, College Station, TX 77843, USA
* Correspondence: gcerdav@ugto.mx

Abstract: With the advancement in power electronics technology, variable-frequency drives have been widely adopted for motor operation due to their inherent benefits: control performance, extending equipment life, and energy savings. The most used technique is Sine Pulse Width Modulation, as it solely requires the modification of the reference signal (sine wave). However, Space Vector Pulse Width Modulation offers lower total harmonic distortion. Therefore, this study presents a technique for the control of induction motors operating in open-loop mode, utilizing a two-level voltage source inverter with a frequency range of 31 to 300 Hz. The proposed control system is modified to encompass between 930 and 1848 switching periods, varying the number of switching periods along with the frequency variation. This approach allows the use of a single LCL filter across the whole frequency spectrum. It is adapted for implementation in an 8-bit microcontroller, which has its inherent limitations, yet it achieves performance levels similar to those found in high-level processors like FPGAs and DSPs. The signals generated by the microcontroller are captured by a DAQ card and input into a MATLAB/Simulink model to observe and analyze the performance of the proposed control system.

Keywords: voltage source inverter; Space Vector Pulse Width Modulation; 8-bit microcontroller; frequency variation

1. Introduction

Variable-frequency drives are used in numerous industrial and commercial applications. They control electric motors in fans, air compressors, pumps, CNC machining operations, extruders, air conditioners, etc. [1]. They allow control performance, and as it is estimated that in the United States above 60% of all electrical energy is used on electrical motors, energy savings achieved by controlling their operational frequency have a significant impact. That is why their adoption keeps growing, and several proposals of variable-frequency drives (VFDs) can be found in the literature [2–13], most of them using FPGAs or DSPs to facilitate the implementation of efficient control systems. There are also some works developed to implement dc/ac inverters with SVPWM using low-cost microcontrollers [14–16], even microcontroller manufacturers have presented application notes with their own 8-bit microcontrollers [17–19]. These works are developed on a fixed operational frequency, 50 or 60 Hz. In [20], they present a variable-frequency controller, though only in the range 20–50 Hz with a maximum resolution of 8 bits for every 60°. Another variable-frequency controller in [21] has a range from 0.5 to 50 Hz, though they do not indicate how the times for T_0, T_1, and T_2 (described in Section 2) are implemented in the microcontroller, whether using delay functions or interrupts. There are papers using SVPWM for frequency variators working with frequencies above the nominal 50 or 60 Hz, but they use carrier-based SVPWM. This method is easier to implement but requires the generation of three sine waves, which makes it suitable for parallel computing devices like FPGAs. In sequential processing, SVPWM generated by timing the sections of a switching period is not merely preferred but necessary.

Industry applications of induction motors controlled by VFDs include conveyor systems, blowers [22], mixing and agitating equipment, extrusion machines [23], swimming pool filtration systems, pressure booster pumps, and HVAC systems [24]. In [25], they even describe the speed of the motors in the cement industry: centrifugal pumps with a range of 200–1800 RPM for vertical turbine pumps (VTPs) and 1800–1960 RPM for horizontal centrifugal pumps (CHPs). Most industrial VFDs have a range of 1–400 Hz, though there are some products reaching even 655 Hz [26–28].

In recent years, the main objective of research on VFDs is precise frequency control [8,10–12,29,30], frequency and torque [31], minimization of torque ripple, and THD [9,13]. All of these papers use SPWM or carrier-based SVPWM. Therefore, this work presents a VFD with a range from 31 to 300 Hz. Although the frequency range is less than the ranges found in industrial VFDs, the authors limited the range due to the constraints imposed by software, in this case, the language code xC8, version 2.5. The microcontroller could handle a higher range if it was programmed using assembly language. But once the system is evaluated, it can be implemented on industrial applications because, as will be shown in Section 3, its performance is comparable to those obtained using FPGAs or DSPs.

The main contributions of this paper are as follows:

(A) A detailed insight into the theoretical analysis of SVPWM and its implementation on an 8-bit microcontroller.

(B) A variable number of switching periods, decreasing with increasing frequency to have between 930 and 1838 per cycle. This is being conducted to reduce the switching frequency variation to have one LCL filter for the whole frequency range.

(C) Three changes to the switching sequences usually implemented in the literature. Although it increases the switching losses (switching more than a bit at a time), it reduces THD by 17%.

Following the introduction, Section 2 describes the theory of SVPWM and its implementation on an 8-bit microcontroller with the modifications necessary for both the limitations of an 8-bit microcontroller and to reach an industrial level performance. Section 3 presents the results from the system which, as indicated previously, are compared to implementations on higher processing devices. Section 4 gives the conclusion of this work.

2. Development of the Proposed Framework

2.1. Motivation

The motivation of this proposal is to provide an efficient SVPWM implementation for a variable-frequency driver based on an 8-bit microcontroller. The program is written in C code to allow its understanding and possible manipulation and improvement.

2.2. Theory

Space vector theory is well developed and implemented in many examples found in the literature [32,33]. Here, we present the basic principles necessary to understand how it works and, in the next subsection, how it is implemented on an 8-bit microcontroller.

Any three-phase electricity system consists of three AC voltages of identical frequency and similar amplitude but with a 120° phase shift from one another, as is represented by Equation (1).

$$\begin{aligned} V_a &= V_m \sin(\omega t) \\ V_b &= V_m \sin(\omega t + 2\pi/3) \\ V_c &= V_m \sin(\omega t - 2\pi/3) \end{aligned} \tag{1}$$

where V_m is the maximum value of the phase voltage, and ω is the angular velocity of the phase voltage. The basic implementation of the Clarke or α–β transformation allows the representation of this three-phase balanced system by one rotating vector generated by two components. The α–β transformation is given by

$$\begin{bmatrix} V_\alpha \\ V_\beta \end{bmatrix} = \frac{2}{3} \begin{bmatrix} 1 & -1/2 & -1/2 \\ 0 & \sqrt{3}/2 & -\sqrt{3}/2 \end{bmatrix} \begin{bmatrix} V_a \\ V_b \\ V_c \end{bmatrix} \tag{2}$$

where V_α and V_β are the horizontal and vertical components of the α and β axes. This transformation brings a 3D space into a 2D space without loss of information. The vector sum of V_α and V_β generates the rotating vector V_f, as shown in Figure 1.

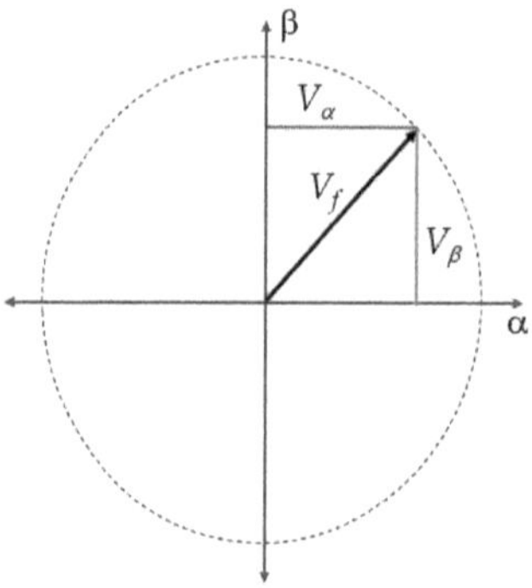

Figure 1. V_f and its components V_α and V_β.

Now, the Park transformation can be applied to vectors V_α and V_β using Equation (3):

$$\begin{bmatrix} V_d \\ V_q \end{bmatrix} = \begin{bmatrix} \cos(\theta) & \sin(\theta) \\ -\sin(\theta) & \cos(\theta) \end{bmatrix} \begin{bmatrix} V_\alpha \\ V_\beta \end{bmatrix} \tag{3}$$

where θ is the instantaneous angle of the $\omega/(2\pi)$ frequency. When a synchronous reference frame is selected, that is, $\theta = \omega t$, the dq frame rotates with the same angular speed as the system frequency, and it generates constant values for V_d and V_q. If the system is balanced, $V_q = V_m$. Representing a three-phase system with two dc quantities has several advantages and applications, from making mathematical computations much easier to facilitating inverter-motor control.

Now, to obtain a three-phase balanced system with a specific frequency and each phase at a specific amplitude, the inverse Park and Clarke transformations are applied, and it starts by defining the values of V_s and ω to define a rotating vector V_F by the following equation:

$$V_F = V_s e^{j\omega t} \tag{4}$$

where ω is the angular frequency of the system, and Vs is the maximum voltage of the signal, which will be taken from a DC source. For the implementation of the inverse transformation using a digital system, the simplest method to generate a three-phase voltage source from a DC source is through a two-level voltage source inverter like the one shown in Figure 2.

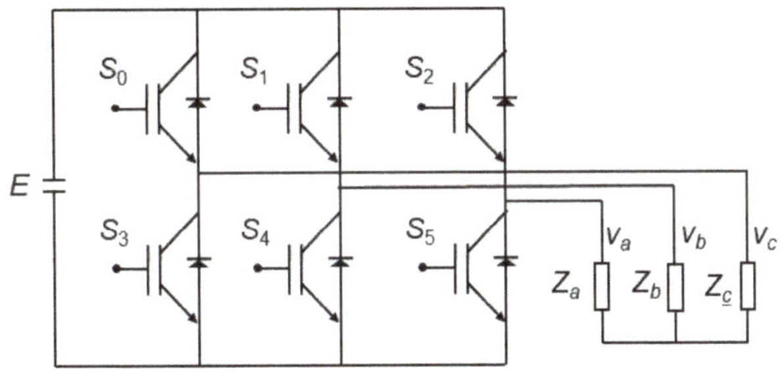

Figure 2. Two-level voltage source inverter architecture.

The operation of the two-level voltage source inverter is performed by activating the switching elements, which usually are IGBTs or MOSFETs. They are ON when receiving a digital signal '1' and OFF when the digital signal is '0'. Switches S_0 and S_3 are connected to phase a, S_1 and S_4 to phase b, and S_3 and S_5 to phase c. Because any pair of switches cannot be ON (short circuit) or OFF (no conduction) at the same time, they are operated in opposite states; therefore, the analysis that follows shows only the upper switches, S_0, S_1, and S_2. With two possible states for each switch, there are 8 combinations: 000, 001, 010, 011, 100, 101, 110, and 111. The first and the last combinations are considered non-conductive states. The other six states are commonly arranged in a sequence where only one switch changes state at a time (this is conducted to reduce losses and high-frequency harmonics); but, in this work, the sequence is kept according to the order they appear in Figures 3 and 4 because it brings the THD 17% lower (from 79.06% to 62.25%). This sequence generates the vectors represented by Equation (5).

$$\begin{aligned} V_{ak} &= V_s e^{\frac{j(k-1)\pi}{3}} \\ V_{bk} &= V_s e^{\frac{j[(k-1)\pi-2\pi]}{3}} \qquad for\ k = 1,\ 2,\ 3,\ 4,\ 5,\ 6 \\ V_{ck} &= V_s e^{\frac{j[(k-1)\pi+2\pi]}{3}} \\ V_{ak}, V_{bk}, V_{ck} &= 0 \qquad\qquad for\ k = 0,\ 7 \end{aligned} \tag{5}$$

Here, each one of the integer values (1 to 6) represents a 60° section. A graphical representation is shown in Figure 3. It can be observed that the maximum value of V_k is $2/3\ V_s$ where V_s corresponds to the voltage value of the DC source E.

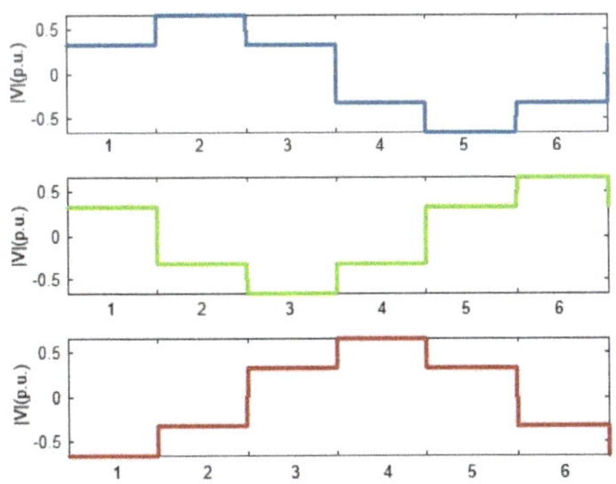

Figure 3. Graphical representation of Equation (5) for discrete value k, phase a-blue line, phase b-green line and phase c-red line.

The digital values, or states, necessary to activate the power switches and generate the output waveforms in Figure 3 with the normalized voltage values are presented in Table 1.

Table 1. Relation between switch states and voltages from Figure 3.

State $S_2S_1S_0$	**Van**	**Vbn**	**Vcn**
011	$Vdc/3$	$Vdc/3$	$-2Vdc/3$
001	$2Vdc/3$	$-Vdc/3$	$-Vdc/3$
101	$Vdc/3$	$-2Vdc/3$	$Vdc/3$
100	$-Vdc/3$	$-Vdc/3$	$2Vdc/3$
110	$-2Vdc/3$	$Vdc/3$	$Vdc/3$
010	$-Vdc/3$	$2Vdc/3$	$-Vdc/3$
000, 111	0	0	0

Table 1 and V_F from Equation (4) can now be represented on a hexagon where each state corresponds to one of the six vectors, and vectors 000 and 111 are at the origin of the coordinate system, as illustrated in Figure 4.

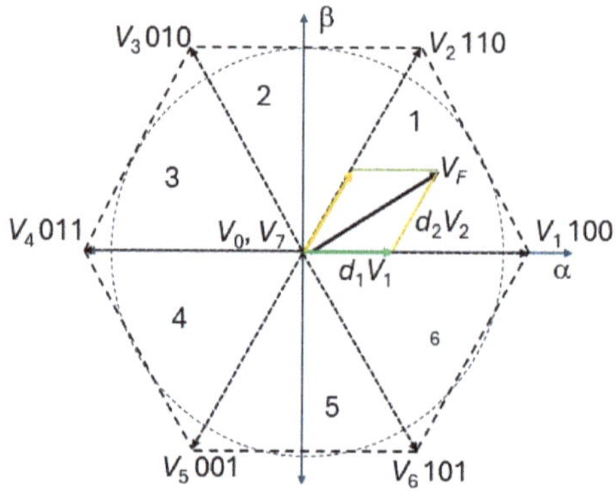

Figure 4. Voltage space and reference vectors.

Each vector is composed of its V_α and V_β parts, which are obtained by their switching state through the transformation described by Equation (6).

$$\begin{bmatrix} V_\alpha \\ V_\beta \end{bmatrix} = \frac{Vdc}{3}\begin{bmatrix} 2 & -1 & -1 \\ 0 & \sqrt{3} & -\sqrt{3} \end{bmatrix}\begin{bmatrix} S_2 \\ S_1 \\ S_0 \end{bmatrix} \tag{6}$$

Their angle and magnitude are derived from Equations (7) and (8):

$$\theta = \tan^{-1}\left[\frac{V_\alpha}{V_\beta}\right] \tag{7}$$

$$V = \sqrt{{V_\alpha}^2 + {V_\beta}^2} \tag{8}$$

The output voltage waveforms shown in Figure 3 are not desirable for many applications [34,35]; therefore, to generate a purer sine wave, Space Vector Pulse Width Modulation is implemented. The objective is to generate the reference vector V_F from the two vectors forming a sector. Using Sector 1 as an example, V_1 and V_2 are the vectors to be applied. It is performed by placing inside every sector one or more PWM periods, switching between states 100 and 110 and including states 000 and 111 to fill the whole period.

The time the system stays in each state varies as the reference vector V_F moves from V_1 to V_2 and the magnitude wanted for V_F. Figure 5 shows V_F and its components on V_1 and V_2.

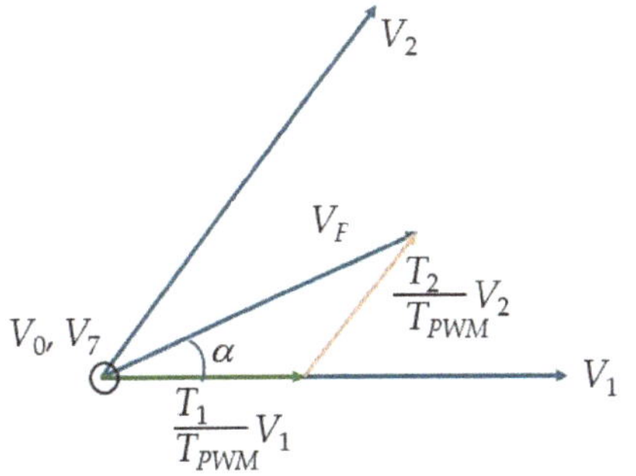

Figure 5. Reference vector and its component vectors.

From Figure 5, if T_1V_1/T_{PWM} and T_2V_2/T_{PWM} are the fractions from V_1 and V_2 used to generate V_F, Equation (9) is obtained:

$$\begin{aligned} T_1 &= \frac{\sqrt{3}T_{PWM}|V_F|}{V_{dc}} \cdot \left(\sin\left(\frac{\pi}{3} - \alpha\right)\right) \\ T_2 &= \frac{\sqrt{3}T_{PWM}|V_F|}{V_{dc}} \cdot (\sin(\alpha)) \\ T_0 &= T_{PWM} - (T_1 + T_2) \end{aligned} \tag{9}$$

where T_{PWM} is the period of the PWM, and α is the angle of V_F as it moves from V_1 to V_2. For the other sectors, if the reference vectors are rotated to the side vectors of each sector, the equations will be the same. The relation $|V_F|/V_{dc}$ is known as the modulation index. $|V_F|$ must be, as maximum, $2/3V_{dc}$, otherwise it is called overmodulation, which is a topic not covered in this work.

Equation (9) specifies the time of each state within a PWM period. Plotting T_1, T_2, and T_0, as α goes from 0 to 60°, and considering $|V_F| = 2/3(2/3V_{dc})$ and $T_{PWM} = 1$, is shown in Figure 6.

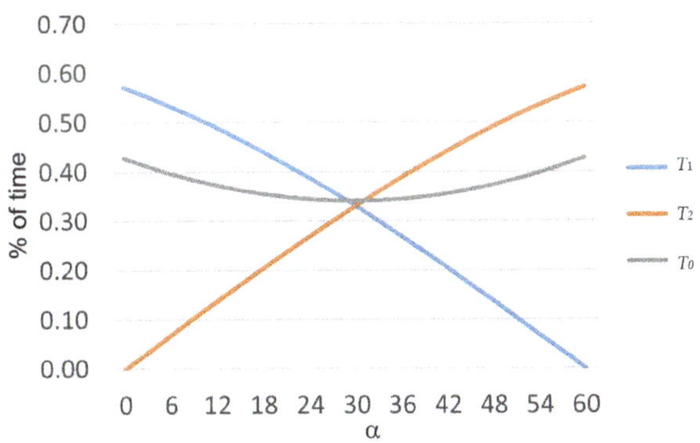

Figure 6. T_1, T_2, and T_0 from = 0° to 60° as percentage of a PWM period, according to Equation (9).

To reduce the computational demands for the microcontroller, T_0, T_1, and T_2 are linearized; therefore, T_1 will decrease and T_2 will increase by fixed values as α goes from $0°$ to $60°$. Therefore, instead of a look-up table requiring several jumps in the memory code, there will be only subtraction and addition operations. T_0 will be left at a fixed value.

Within each sector, there could be one or more switching periods, depending on the selected PWM frequency. There are several methods to generate a switching sequence [33]. The method used in this work is the symmetrically generated SVPVM. This method is not the simplest to implement or with the lowest switching losses but generates lower total harmonic distortion. The plot for one switching period is shown in Figure 7.

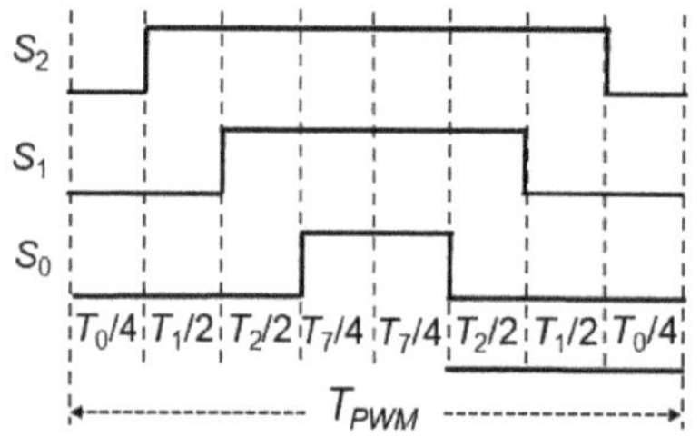

Figure 7. Symmetrically generated SVPWM for controlling the switches of a two-level voltage source inverter.

When applying a frequency variation to a motor, there are some factors to consider, like torque and voltage, when varying the operational frequency. In this work, a constant torque is applied for frequencies below 60 Hz [36,37] and constant voltage for frequencies above 60 Hz. The constant torque is performed by reducing the voltage while reducing frequency, as seen in Figure 8. It is conducted by increasing T_0 in Equation (9), inversely proportional to the frequency from 60 Hz, to the minimum of 31 Hz.

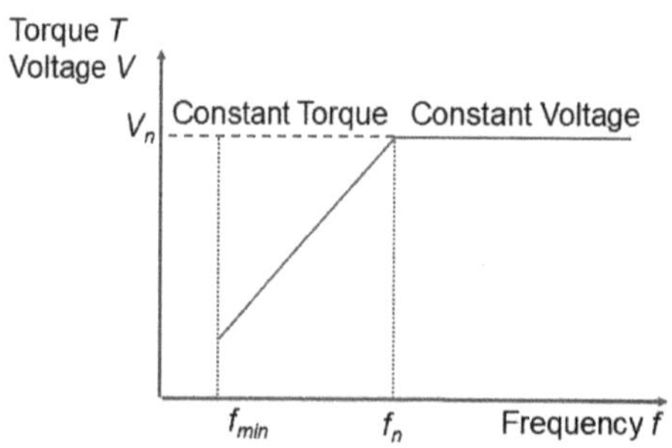

Figure 8. Torque and voltage as function of frequency, indicating the reduction in voltage below the nominal frequency of 50–60 Hz to keep constant torque.

Now that the description of the theory and outlines of a variable-frequency motor controller have been presented, the implementation is described in the following section.

2.3. Implementation

The algorithm is implemented in an 8-bit microcontroller, the PIC18F45K50. This microcontroller has two 16-bit comparator: one is for counting the time of each sector, and the other counts the time of each switching section, i.e., T_1, T_2 and T_0. The frequency of the microcontroller is 48 MHz, which is its maximum operational frequency. This frequency allows most of the instructions at assembler level to be executed in 0.0833 μS.

SVPWM is performed by calculating the times for T_0, T_1, T_2, and T_7. T_0 and T_7 are defined when selecting the output frequency. T_1 and T_2 vary within each sector as the reference vector moves along such sector. From Equation (8), the sine of α, from 0 to $\pi/3$, is needed. This is usually performed by a look-up table; however, it requires multiple instruction jumps so the curves in Figure 5 have been linearized as mentioned previously. With this approach, with the increment of α, T_1 decreases and T_2 increases, both by the same amount, making them just addition and subtraction operations.

When the user selects a new frequency, which is sent via serial communication, the program calculates the following parameters:

Time per sector in microseconds: T_s = 166666/frequency;

Time of a switching period: $T_{PWM} = T_s/N$;

10% of T_{PWM} for $T_0 = T_{PWM} \times 0.025$ (0.1 divided by 4) when frequency ≥ 60;

$T_0 = T_{PWM} \times [0.025 + (60\text{-frequency}) \times 0.00375]$ when frequency < 60;

T_1 starts at its maximum value = $N \times (T_{PWM} - 4 \times T_0)/(2(N+1))$;

T_2 starts at its minimum value: $T_2 = T_1/N$;

T_7 includes both sides of the PWM sequence: $T_7 = 2 \times T_0$;

Decrements to T_1 and increments to T_2: Increments = T_1/N.

N is the number of switching periods according to Table 2. This is implemented to have a frequency range of switching periods between 930 and 1838. The times for T_0, T_1, and T_2 are adjusted to finish the switching periods at the same time as each sector. T_1 and T_2 must have a minimum value of 6 µs to allow the program to go to the next interrupt. A modulation index (M.I.) of 0.9 is the maximum value that allows the program to finish the interrupt functions.

Table 2. Number of PWM cycles per sector vs. frequency.

Frequency (Hz)	Number of Switching Periods	Time Range for Every Switching Period (µs)
31–34	9	597–544
35–38	8	595–548
39–44	7	610–541
45–51	6	617–544
52–61	5	641–546
62–77	4	672–541
78–102	3	712–544
103–154	2	809–541
155–300	1	1075–555

The change from one switching state to another was conducted by interrupts generated by a 16-bit comparator module. A comparator was selected because, instead of generating an interrupt on overflow, like a timer, it is performed when a counter equals a preset value, resetting the counter, and when changing the preset value, the counter already has been counting, generating precise time values for every switching state.

The flow chart of the algorithm implemented to switch states is illustrated in Figure 9.

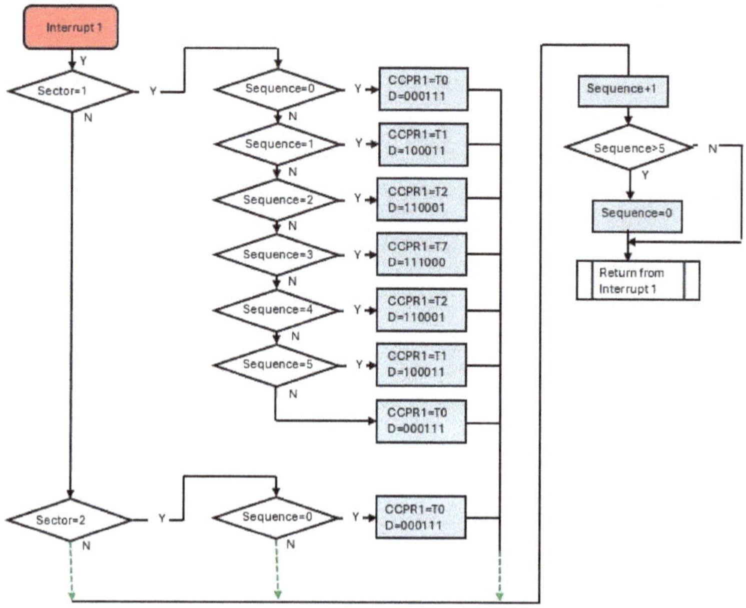

Figure 9. Flow chart for the changing states by an interrupt.

The flow chart of the algorithm for changing sectors is illustrated in Figure 10.

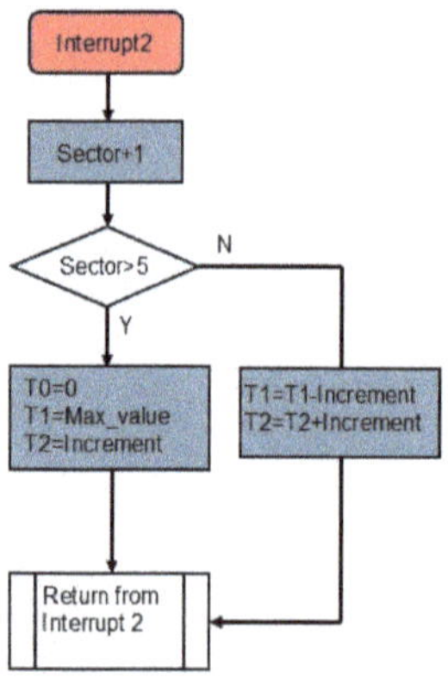

Figure 10. Flow chart for changing sectors by an interrupt.

3. Results

The focus of this section is to demonstrate that the system generates digital signals, as presented in Figure 7, with precise times, and the frequency variation is achieved for the proposed range from 31 to 300 Hz. The first tests were performed in a virtual environment, using Proteus software, version 8.3. This environment allows visualization of the digital signals generated by the microcontroller, therefore, proving the program works correctly, changing frequency as specified by the user, and that timing is right. In Figure 11, a sample of outputs S_2, S_1, and S_0 shows times for each period. Taking, for example, S_2 (yellow line), its period takes approximately 56 μs. According to Table 2, there are five cycles per sector, and six sectors generate a cycle of the output voltage, hence, the frequency is 59.52 $\approx$ 60 Hz.

When analyzing the signals in MATLAB/Simulink, version R2020b, it is shown that an exact value of 60 Hz is achieved.

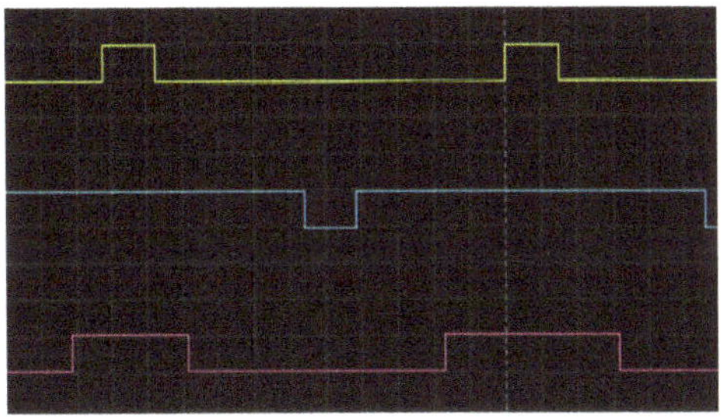

Figure 11. Signals S_2 (yellow line), S_1 (blue line), and S_0 (red line), for a 60 Hz voltage output (50 µs per division).

The changing values of T_1 (decreasing) and T_2 (increasing) are shown in Figure 12. The states included are 000, 100, 110, and 111. Because there is only a 1-bit change between vectors 100 and 110, the only signal showing variation in the plot is S_1 (blue line).

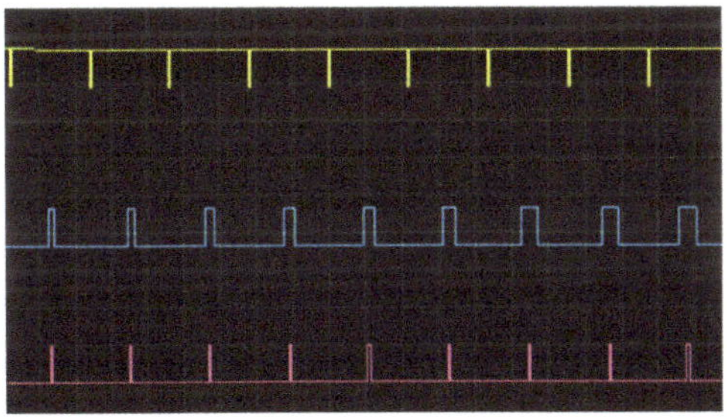

Figure 12. Signals S_2 (yellow line), S_1 (blue line), and S_0 (red line), for a 60 Hz voltage output (200 µs per division).

Programming and testing the microcontroller were performed on a board developed by our research team, as shown in Figure 13. This board allows for programming the microcontroller using a toolkit or through the USB port. It was designed to minimize electromagnetic interference (EMI) and to connect a Bluetooth module, LCD screen, and an SD card. A full description of the board and the Gerber files can be found in [38].

Figure 13. The development board employed to implement the control system.

The signals, generated by the microcontroller, were captured by a data acquisition board from National Instruments, the DAQ USB-6259, for a time frame of 0.1 s. They were then fed to a virtual system in MATLAB/Simulink. These signals activated the switching MOSFETs, forming a two-level voltage source inverter, like the one presented in Figure 2. The output voltage waveforms from the voltage source inverter are shown on Figure 14.

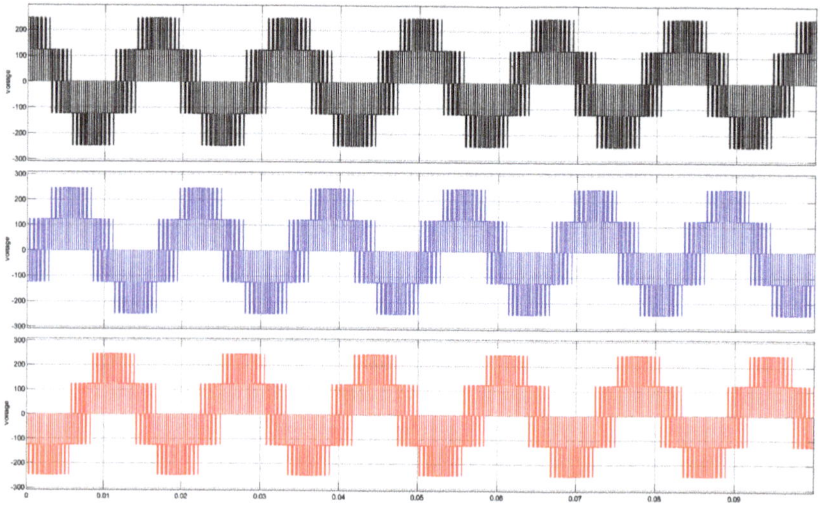

Figure 14. Voltage waveforms at the output of the two-level voltage source inverter, phase *a*-black, phase *b*-blue and phase *c*-red.

The performance of the proposed system is analyzed by measuring the percentage of THD and comparing it to other systems found in the literature. The selected operating frequency to measure THD is 50 Hz; although, at 60 Hz, the value is similar. Table 3 presents the values of THD for a modulation index (M.I.) of 0.9.

Table 3. Comparison of % THD for a modulation index of 0.9.

PWM Technique	This Work	[39]	[40]	[41]	[42]
Operating frequency	50 Hz	50 H		60 Hz	50 Hz
SPWM			77.28	80.33	77.74
SVPWM	61.47	57.28	63.85	64.61	52.05

In [39], their system works at 50 Hz, and in [40,41], they present values at 1 and 0.8 M.I., but it shows a value at 0.9 M.I. by using the rule of proportionality. In [42], their system also works at 50 Hz. In [40–42], they show the performance of Sine Pulse Width Modulation, which presents a THD 21–49% above SVPWM, indicating clearly the better performance of SVPWM for reducing THD.

In this work, as was explained at the end of Section 2.2, the M.I. at 31 Hz is 0.465 and it generates a THD of 128.2.%, which is consistent with the values shown in [39,40] when they decrease the M.I.

For application specifically on induction motors, Table 4 presents works found in the literature: in [39], they do not specify the operating frequency of their proposed system and only present a system using SVPWM, and in [43,44], they compare SPWM and SVPWM using a fixed frequency.

Table 4. Comparison of % THD for application on induction motors (M. I. of 0.9).

PWM Technique	This Work	[39]	[43]	[44]
Operating frequency	50 Hz	-	50 H	
SPWM			68.64	89.34
SVPWM	61.47	57.28	57.80	60.53

To demonstrate the performance of the system with an LCL filter, one was added to smooth the signals. The values for the inductors and capacitor were derived from [45], and the values calculated are $L_1 = 0.024$ H, $L_2 = 0.0144$ H, and $C = 15.00$ μF. The values of the parameters for the virtual system are presented in Table 5, and the voltage waveforms at the output of the LCL filter are shown in Figure 15.

Table 5. Values of parameters of the virtual system.

Parameter	Value	Parameter	Value
Output phase voltage	260 V	Source voltage	370 V
Active power	1.3 kW	Switching frequency	900 Hz

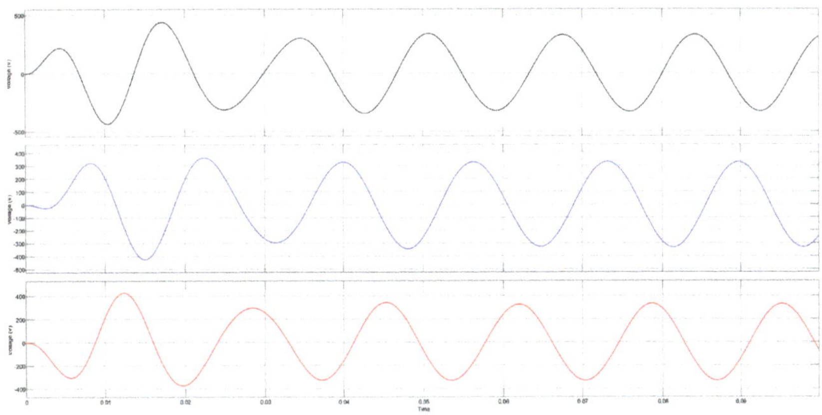

Figure 15. LCL filter output voltage waveforms for 60 Hz, phases *a*, *b* and *c* are represented by black, blue and red lines, respectively.

From Figure 15, two waveform peaks are measured, and it shows that, for 60 Hz, the system generates a precise frequency, as seen in Figure 16.

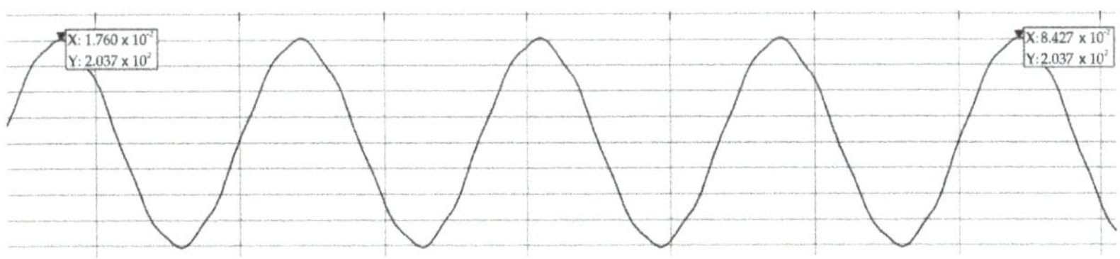

Figure 16. Voltage waveform measurement at 60 Hz (horizontal axis in seconds and vertical axis in volts).

Next, the system frequency is changed to operate at 31 Hz to observe its behavior as it is shown in Figure 17. The frequency operates at 30.97 Hz, a deviation of 0.097%. The output phase voltage is reduced to 119 V to keep a constant torque as was expected.

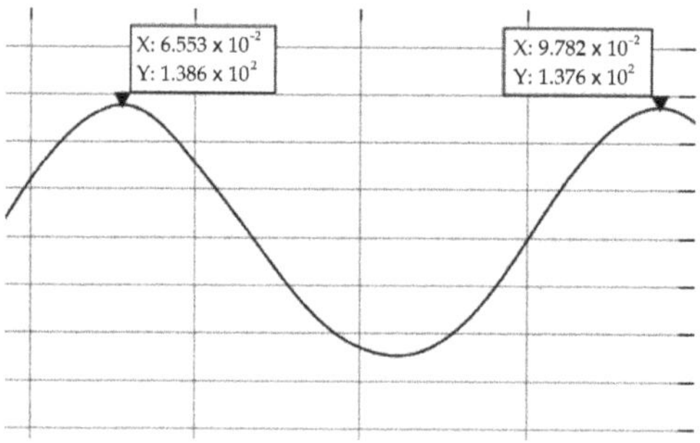

Figure 17. Voltage waveform measurement at 31 Hz (horizontal axis in seconds and vertical axis in volts).

Finally, the system frequency is changed to operate at 300 Hz, and its voltage waveforms are shown in Figure 18. The frequency operates at 299.4 Hz, with a deviation of 0.19%. This is the maximum deviation from the nominal frequency in the entire range. The output pick phase voltage is 216.7 V.

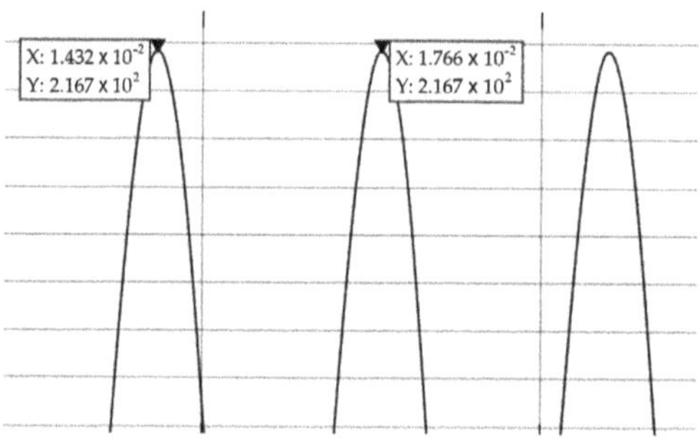

Figure 18. Voltage waveform measurement at 300 Hz (horizontal axis in seconds and vertical axis in volts).

The voltage waveforms for 60 Hz, shown in Figure 16, have a THD of 2.43%; for 31 Hz, observed in Figure 17, the THD was 4.02%; and for 300 Hz, presented in Figure 18, the THD was 0.43%. Table 6 presents THD values across various frequencies along the entire range the system can generate. These values have been obtained with a lineal load of 1.3 kW. For a non-lineal load with 500 var added to the active power, the THD for 60 Hz increases to 3.68%, showing the system performance. The IEEE 519–2014 Standard [46] recommends a maximum voltage THD below 8% for systems with a voltage of 1000 volts or less; therefore, the generated system voltage output waveforms adhere to this recommendation, showing the attenuation performance of the filter.

Table 6. THD after the LCL filter for different frequencies.

	31 Hz	**60 Hz**	**100 Hz**	**200 Hz**	**300 Hz**
THD	4.02%	2.43%	1.20%	1.47%	0.43%

According to [47], the resonance frequency of LCL filters is given by

$$f_{res} = \frac{1}{2\pi}\sqrt{\frac{1}{r(1-r)LC}}$$

where $L = L_1 + L_2$, and $r = L_1/L$. The resonance frequency for the LCL filter is 433.16 Hz, therefore, there is no damping requirement.

Finally, a virtual test bench was implemented in MATLAB/Simulink to evaluate the system using an induction motor with parameters identical to an industrial motor from GE, model number 5KAF145SAA202A, catalogue number M9301. The parameters of the motor are described in Table 7. The system worked at a frequency of 60 Hz. Figure 19 shows the rotor speed and current of phase *a* with 100% load. The behavior of the motor demonstrates that the system can operate an induction motor properly.

Table 7. Induction motor parameters.

Model	**5KAF145SAA202A**	**Unit**
Rated power	2	HP (1491.4 watts)
Rated voltage	230	V
Rated current	5.2	A
Maximum speed	1730	RPM at 60 Hz
Maximum speed	2595	RPM at 90 Hz
Speed range const torque	0–1725	RPM
Speed range const hp	1730–2595	RPM
SF 1.0	POLES 4	
LOAD%	100	75
% EFF	86.52	86.94
% PF	78.54	68.82
AMPS	2.74	2.35
Torque	8.2321	N-m

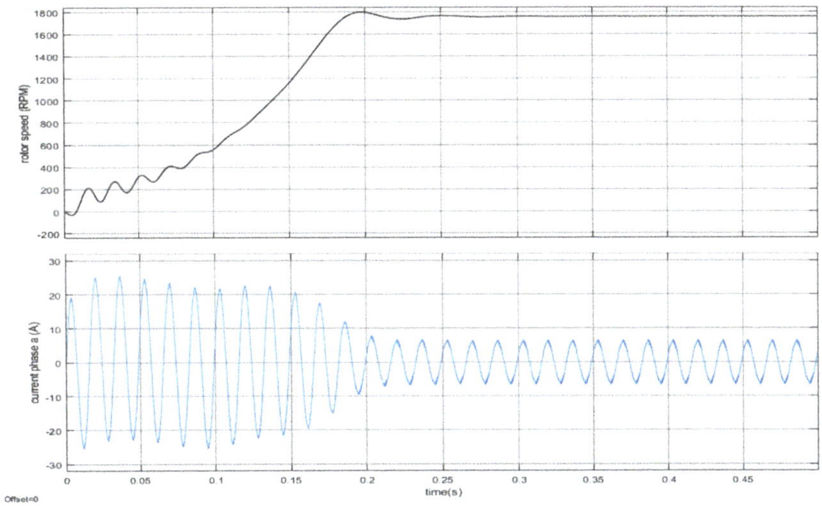

Figure 19. Rotor speed and current of phase from motor described in Table 6.

4. Conclusions

This paper presented a functional control system for a variable-frequency driver. The system was implemented in an 8-bit microcontroller programmed using language C. The aim was to offer a motor control system to generate the switching signals of a two-level inverter with variable frequency in the range of 31 to 300 Hz. The control includes a voltage reduction, inversely proportional to the frequency reduction for frequencies below 60 Hz to maintain constant torque. The number of switching periods decreased from nine to one with an increase in frequency, as is shown in Table 2, to enable the use of one LCL filter for the entire range of frequencies the system can generate. Two 16-bit comparators from the microcontroller were used for counting every sector and PWM cycles, obtaining a high-precision output frequency. A change in the switching sequence, following the order vectors presented in Figure 4, generated a THD reduction of 16.81%, and doing so allows this system to be used in industrial applications because it obtains a performance comparable to systems developed on high-level systems like FPGAs and DSPs. Besides, it offers power electronics students the opportunity to follow the theory for SVPWM, acquire the description of the implementation, and download the code to program a microcontroller from the PIC18 family with only having to adjust the configuration lines. Or they can adapt the C code to any microcontroller that can be programmed with language C.

Author Contributions: Conceptualization, G.C.-V.; methodology, G.C.-V.; software, G.C.-V. and F.J.M.-V.; validation, G.C.-V. and A.B.; formal analysis, G.C.-V. and A.B.; investigation, G.C.-V. and F.J.M.-V.; resources, A.B.; data curation, G.C.-V.; writing—original draft preparation, G.C.-V. and F.J.M.-V.; writing—review and editing, G.C.-V. and A.B.; visualization, G.C.-V. All authors have read and agreed to the published version of the manuscript.

Funding: This research received no external funding.

Data Availability Statement: The original contributions presented in this study are included in the article. Further inquiries can be directed to the corresponding author.

Conflicts of Interest: The authors declare no conflicts of interest.

References

1. Bose, B.K. The past, present, and future of power electronics [Guest Introduction]. *IEEE Ind. Electron. Mag.* **2009**, *3*, 7–11. [CrossRef]
2. Wu, H.; Hu, X. Reserch of SVPWM Variable Frequency Speed Regulation System Based on TMS320LF2407A. In Proceedings of the 2008 International Symposium on Intelligent Information Technology Application Workshops, Shanghai, China, 21–22 December 2008; pp. 1041–1044. [CrossRef]
3. Sandulescu, P.; Idkhajine, L.; Cense, S.; Colas, F.; Kestelyn, X.; Semail, E.; Bruyere, A. FPGA implementation of a general Space Vector approach on a 6-leg voltage source inverter. In Proceedings of the IECON 2011-37th Annual Conference of the IEEE Industrial Electronics Society, Melbourne, VIC, Australia, 7–10 November 2011; pp. 3482–3487. [CrossRef]
4. Madasamy, P.; Pongiannan, R.K.; Ravichandran, S.; Padmanaban, S.; Chokkalingam, B.; Hossain, E.; Adedayo, Y. A Simple Multilevel Space Vector Modulation Technique and MATLAB System Generator Built FPGA Implementation for Three-Level Neutral-Point Clamped Inverter. *Energies* **2019**, *12*, 4332. [CrossRef]
5. Sapkota, Y. Design of FPGA-Based PWM Techniques for Inverters. Master's Thesis, Youngstown State University, Youngstown, OH, USA, December 2023.
6. Waleed, U.; Waseem, M.; Shaukat, H.; Ijaz, A.; Almalaq, A.; Mohamed, M.A. An Efficient FPGA Based Scalar V/f Control Mechanism of Three Phase Induction Motor for Electric Vehicles. In Proceedings of the 2021 31st Australasian Universities Power Engineering Conference (AUPEC), Perth, Australia, 26–30 September 2021; pp. 1–6. [CrossRef]
7. Liyanage, A.; Nagrial, M.; Hellany, A.; Rizk, J. Speed Control of Induction Motors Using V/f Control Method. In Proceedings of the 2022 International Conference on Electrical and Computing Technologies and Applications (ICECTA), Ras Al Khaimah, United Arab Emirates, 23–25 November 2022; pp. 424–429. [CrossRef]
8. Wibowo, P.S.; Riyadi, S. SPWM Volt/Hz Based Speed Control of Induction Motor. In Proceedings of the 2019 International Seminar on Application for Technology of Information and Communication (iSemantic), Semarang, Indonesia, 21–22 September 2019; pp. 482–486. [CrossRef]
9. Sieklucki, G.; Sobieraj, S.; Gromba, J.; Necula, R.-E. Analysis and Approximation of THD and Torque Ripple of Induction Motor for SVPWM Control of VSI. *Energies* **2023**, *16*, 4628. [CrossRef]

10. Tuknyasrirut, S.; Srilad, S.; Suksri, T. Comparison Power Quality of the Voltage Source Inverter type SVPWM and SPWM Technique for Induction motor Drive. In Proceedings of the SICE Annual Conference 2008, Tokyo, Japan, 20–22 August 2008; pp. 241–246.
11. Ali, A.K.; Ercelebi, E. Simulation and Study of SVPWM Inverter for (VFD) Applications. *Int. J. Electron. Electr. Eng.* **2017**, *5*, 158–170. [CrossRef]
12. Patel, J.J.; Kubavat, A.M.; Jhala, M.B. Speed Control of a Three Phase Induction Motor Using PWM Inverter. *Int. J. Eng. Dev. Res.* **2014**, *2*, 503–507.
13. Dems, M.; Komeza, K.; Szulakowski, J.; Kubiak, W. Increase the Efficiency of an Induction Motor Feed from Inverter for Low Frequencies by Combining Design and Control Improvements. *Energies* **2022**, *15*, 530. [CrossRef]
14. Doan, N.S.; Tsvetkov, A.N.; Nguyen, T.H. Study and implementation of space vector pulse width modulation inverter on an Arduino. *E3S Web Conf.* **2021**, *288*, 01059. [CrossRef]
15. Badran, M.A.A.; Tahir, A.M.; Faris, W.F. Digital Implementation of Space Vector Pulse Width Modulation Technique Using 8-bit Microcontroller. *World Appl. Sci. J.* **2013**, *21*, 21–28.
16. Vivek, G.; Biswas, J.; Muthavarapu, A.K.; Nair, M.D.; Barai, M. Optimised Design and Implementation of SVPWM Switching sequences for Five level VSI. In Proceedings of the 2018 IEEE International Conference on Power Electronics, Drives and Energy Systems (PEDES), Chennai, India, 18–21 December 2018; pp. 1–6. [CrossRef]
17. AVR435: BLDC/BLAC Motor Control Using a Sinus Modulated PWM Algorithm. ATMEL Application Note. Available online: https://ww1.microchip.com/downloads/en/Appnotes/doc7671.pdf (accessed on 20 February 2025).
18. AN2154: Space Vector Modulation Using 8-bit ST7MC Microcontroller and ST7MC-KIT/BLDC Starter Kit. STMicroelectronics Application Note. Available online: https://www.st.com/resource/en/application_note/cd00055518-space-vector-modulation-using-8bit-st7mc-microcontroller-and-st7mckitbldc-starter-kit-stmicroelectronics.pdf (accessed on 23 February 2025).
19. AP0803620: Optimized Space Vector Modulation and Over-Modulation with the XC866. Available online: https://www.infineon.com/cms/en/product/gated-document/optimized-space-vectormodulation-and-over-modulationwith-the-xc866-description-db3a304412b407950112b4199b7b28b3/ (accessed on 23 February 2025).
20. Midhun, G.; Sandhya, P. Hardware Implementation of a Dual Two-Level Inverter with SVPWM Using Pic Microcontroller. *J. Res. Eng. Appl. Sci.* **2016**, *4*, 185–190.
21. Datta, S.; Chandra, A.; Chowdhuri, S. Design and development of an 8 bit microcontroller based space vector PWM inverter fed volt/Hz induction motor drive. In Proceedings of the 2016 2nd International Conference on Control, Instrumentation, Energy & Communication (CIEC), Kolkata, India, 28–30 January 2016; pp. 353–357. [CrossRef]
22. Hamim, M.Z.; Salimin, S.; Bakar, A.A. Analysis of Variable Frequency Drive for Induction Motor using Matlab Software. *J. Adv. Res. Appl. Mech.* **2024**, *116*, 117–129. [CrossRef]
23. Available online: https://www.bluebayautomation.com/blog/motion-control-gearing-mechanics-12/variable-frequency-drive-usage-in-industrial-applications-71 (accessed on 14 April 2025).
24. Available online: https://www.plctechnician.com/news-blog/variable-frequency-drives-basics-and-common-applications-vfds (accessed on 15 April 2025).
25. Sauer, B.J.; Brady, P.A. Application of AC Induction Motors with Variable Frequency Drives. In Proceedings of the 2009 IEEE Cement Industry Technical Conference Record, Palm Springs, CA, USA, 29 May–5 June 2009; pp. 1–10. [CrossRef]
26. PowerFlex Low Voltage AC Drives. Available online: https://literature.rockwellautomation.com/idc/groups/literature/documents/br/pflex-br008_-en-p.pdf (accessed on 18 April 2025).
27. FRENIC-Mini (C2) Series User's Manual. Fuji Electric Co., Ltd.. Available online: https://americas.fujielectric.com/wp-content/uploads/2016/07/FEA-ACDR-BR-105a_web.pdf (accessed on 25 May 2025).
28. Altivar Process ATV600. Available online: https://download.schneider-electric.com/files?p_Doc_Ref=EAV64318&p_enDocType=User+guide&p_File_Name=ATV600_Programming_Manual_EN_EAV64318_12.pdf (accessed on 16 April 2025).
29. Zellouma, D.; Bekakra, Y.; Benbouhenni, H. Field-oriented control based on parallel proportional–integral controllers of induction motor drive. *Energy Rep.* **2023**, *9*, 4846–4860. [CrossRef]
30. Sengamalai, U.; Anbazhagan, G.; Thamizh Thentral, T.M.; Vishnuram, P.; Khurshaid, T.; Kamel, S. Three Phase Induction Motor Drive: A Systematic Review on Dynamic Modeling, Parameter Estimation, and Control Schemes. *Energies* **2022**, *15*, 8260. [CrossRef]
31. Mencou, S.; Yakhlef, M.B.; Tazi, E.B. Advanced Torque and Speed Control Techniques for Induction Motor Drives: A Review. In Proceedings of the 2022 2nd International Conference on Innovative Research in Applied Science, Engineering and Technology (IRASET), Meknes, Morocco, 3–4 March 2022; pp. 1–9. [CrossRef]
32. Kubeitari, M.; Alhusayn, A.; Alnahar, M. Space Vector PWM Simulation for Three Phase DC/AC Inverter. *Int. J. Electr. Comput. Eng.* **2012**, *6*, 1402–1407. [CrossRef]
33. Neacsu, D.O. Space Vector Modulation-An Introduction. In Proceedings of the Tutorial at IECON2001, 27th Annual Conference of IEEE, Denver, CO, USA, 29 November–2 December 2001; pp. 1538–1592.

34. Patin, N. DC/AC Converters. In *Power Electronics Applied to Industrial Systems and Transports 2*, 1st ed.; ISTE Press–Elsevier: London, UK, 2015; pp. 36–100.
35. Jaen-Cuellar, A.Y.; Elvira-Ortiz, D.A.; Resendiz-Ochoa, E.; Saucedo-Dorantes, J.J. Analysis of the Effects Produced by Pure Sine and Modified Sine Inverters in an Induction Motor. In *New Trends in Electric Machines-Technology and Applications*; IntechOpen: London, UK, 2023. [CrossRef]
36. Evon, S.T.; Oakes, B. Variable frequency drive principles and practices (above NEMA) AC motors for variable frequency application. In Proceedings of the Conference Record of 1999 Annual Pulp and Paper Industry Technical Conference (Cat. No.99CH36338), Seattle, WA, USA, 21–25 June 1999; pp. 94–108. [CrossRef]
37. Song-Manguelle, J.; Ekemb, G.; Mon-Nzongo, D.L.; Jin, T.; Doumbia, M.L. A Theoretical Analysis of Pulsating Torque Components in AC Machines with Variable Frequency Drives and Dynamic Mechanical Loads. *IEEE Trans. Ind. Electron.* **2018**, *65*, 9311–9324. [CrossRef]
38. Available online: https://sites.google.com/view/beedev (accessed on 25 May 2025).
39. Balasubramanian, G.; Chandrasekar, P.; Alexandar, S.A. Variable Frequency Drive operated Air Blower in Air Handling Unit of Heating, Ventilation and Air Conditioning Systems. In Proceedings of the 2022 IEEE Delhi Section Conference (DELCON), New Delhi, India, 11–13 February 2022; pp. 1–7. [CrossRef]
40. Kumar, K.V.; Michael, P.A.; John, J.P.; Kumar, S.S. Simulation and Comparison of Spwm and Svpwm Control for Three Phase Inverter. *Arpn J. Eng. Appl. Sci.* **2010**, *5*, 61–74.
41. Islam, M.T.; Ayon, S.I. Performance Analysis of Three-Phase Inverter for Minimizing Total Harmonic Distortion Using Space Vector Pulse Width Modulation Technique. In Proceedings of the 2020 23rd International Conference on Computer and Information Technology (ICCIT), Dhaka, Bangladesh, 19–21 December 2020; pp. 1–4. [CrossRef]
42. Sharma, R.; Goyal, D.K.; Singh, H. Analysis of Space Vector PWM for Three Phase Inverter and Comparison with SPWM. *Int. J. Adv. Res. EEIE* **2016**, *5*, 533–540.
43. Hussein, T.A.; Mohamed, L.A. Detailed Simulink Implementation for Induction Motor Control Based on Space Vector Pulse Width Modulation SVPWM. *Indones. J. Electr. Eng. Comput. Sci.* **2021**, *22*, 1251–1262. [CrossRef]
44. Ramarao, G.; Karunakar, S.; Kumari, D.V.; Adithya, M.V.; Gnaneswar, D. Simulation and Comparative Analysis Between Svpwm and Spwm Techniques for Speed Control of 3 Phase Induction Motor. In Proceedings of the 2024 IEEE International Conference on Smart Power Control and Renewable Energy (ICSPCRE), Rourkela, India, 19–21 July 2024; pp. 1–6. [CrossRef]
45. Hasan, S.; Suherman; Saragi, D.P.; Siregar, Y.; Al-Akaidi, M. The LCL Filter Design for Three Phase DC-AC Voltage Converter. *J. Phys. Conf. Ser.* **2021**, *1783*, 012067. [CrossRef]
46. IEEE. IEEE Recommended Practice and Requirements for Harmonic Control in Electric Power Systems (IEEE 519-2014). 2014. Available online: https://ieeexplore.ieee.org/xpl/RecentIssue.jsp?punumber=6826457 (accessed on 23 March 2025).
47. Kim, Y.-J.; Kim, H. Optimal Design of LCL Filter in Grid-Connected Inverters. *IET Power Electron.* **2019**, *12*, 1774–1782. [CrossRef]

MDPI AG
Grosspeteranlage 5
4052 Basel
Switzerland
Tel.: +41 61 683 77 34

Processes Editorial Office
E-mail: processes@mdpi.com
www.mdpi.com/journal/processes

www.ingramcontent.com/pod-product-compliance
Lightning Source LLC
Chambersburg PA
CBHW061545220926
51850CB00005B/255